MEASURING IN THE METRIC SYSTEM

LENGTH

1 Kilometer = 1000 meters (also written 10^3 meters) = 0.62 mile

1 Meter (abbreviated m) = $\frac{1}{1000}$ kilometer (also written 10^{-3} kilometer) = 39.37 inches

1 Centimeter (abbreviated cm) = $\frac{1}{100}$ meter (also written 10^{-2} m)

1 Millimeter (abbreviated mm) = $\frac{1}{1000}$ meter (also written 10^{-3} m)

1 Micron (symbolized by μ) = $\frac{1}{1000}$ mm (also written 10^{-3} millimeter) = one millionth of a meter (written 10^{-6} m)

1 Nanometer (abbreviated as nm; also called 1 millimicron, which is symbolized as mμ) = $\frac{1}{1000}$ micron = one billionth of a meter (written 10^{-9} m)

1 Ångström (abbreviated Å) = $\frac{1}{10}$ nanometer = $\frac{1}{10,000}$ micron = one ten-billionth of a meter (written 10^{-10} m)

MASS

1 Kilogram = 1000 grams (also written 10^3 grams) = 2.20 lbs

1 Gram (abbreviated g) = $\frac{1}{1000}$ kilogram (also written 10^{-3} kilogram)

1 Milligram (abbreviated mg) = $\frac{1}{1000}$ gram (also written 10^{-3} gram)

1 Microgram (abbreviated μg) = $\frac{1}{1000}$ milligram (also written 10^{-3} milligram) = one millionth of a gram (written 10^{-6} g)

1 Nanogram (abbreviated ng) = $\frac{1}{1000}$ microgram (also written 10^{-3} microgram) = one billionth of a gram (written 10^{-9} g)

VOLUME

1 Liter (abbreviated l) = 1000 milliliters (also written as 10^3 milliliters) = 1.06 liquid quarts

1 Milliliter (abbreviated ml) = $\frac{1}{1000}$ liter (also written 10^{-3} l) = 1 cubic centimeter (abbreviated cc or cm^3). One ml (= 1 cc) of water at 4°C weighs 1 gram

1 Microliter (abbreviated μl) = $\frac{1}{1000}$ milliliter (also written 10^{-3} milliliter) = one millionth of a liter (also written 10^{-6} l). One microliter of water at 4°C weighs 1 milligram.

LIFE: CELLS, ORGANISMS, POPULATIONS

LIFE

CELLS, ORGANISMS, POPULATIONS

EDWARD O. WILSON
Harvard University

THOMAS EISNER
Cornell University

WINSLOW R. BRIGGS
Carnegie Institution of Washington

RICHARD E. DICKERSON
California Institute of Technology

ROBERT L. METZENBERG
University of Wisconsin

RICHARD D. O'BRIEN
Cornell University

MILLARD SUSMAN
University of Wisconsin

WILLIAM E. BOGGS

SINAUER ASSOCIATES, INC. • PUBLISHERS
Sunderland, Massachusetts

LIFE: CELLS, ORGANISMS, POPULATIONS

Library of Congress Catalog Card Number: 76-40649
ISBN: 0-87893-943-1

CONTENTS

PREFACE

Life, as its abbreviated title suggests, is a short version of *Life on Earth*. The parent book, more than 1000 pages long, was an ambitious attempt to present all of the fundamentals of biology that could be covered in a full-year course. *Life* provides a shorter and simpler coverage of the subject, suitable for one-semester or one-quarter courses and a broader audience. Supplemented by readings and lectures, it can also be used as the core book for a full-year course.

Life has been written to retain what we and many users regard as the best features of *Life on Earth*. Most important, each of the eight authors has retained responsibility for the subjects in which he is most competent. In one sense we found it harder to write a short book than a long one. We were concerned with giving the student a well-balanced book that would cover the key ideas of cellular, organismic and population biology, but this involved some hard decisions about what to omit. Such decisions are particularly difficult when a book is the work of several authors, each convinced that his area of research is both important and fascinating to beginning students.

We kept the overall organization of the parent book, starting in Part I (The Cell) with molecules, cells and genetics, and ascending the hierarchy of biological complexity. Part II (Multicellular Life) focuses on the physiology of multicellular organisms; Part III (The Origin and Diversity of Life) on the dawn of life and the diversity of microorganisms, plants and animals on our planet; and Part IV (The Strategy of Evolution) on the interactions of populations with one another and with the environment. Part V (Alternate Futures) summarizes our divergent views on the future of life.

We decided to abridge *Life on Earth* selectively rather than by across-the-board cuts. The chemistry in Part I was considerably shortened and simplified to make it accessible to readers unfamiliar with the physical sciences. We also eliminated many special topics in genetics, while keeping much of the original coverage of molecular and Mendelian genetics.

In Part II we condensed development from two chapters to one and rearranged the sequence of topics beginning with molecules and cells. The result is a chapter in which the cell is viewed as a self-assembling molecular machine, and the organism as a self-assembling aggregation of cells. Also, we added a new section on immunobiology to chapter 11 (The Internal Environment). In response to suggestions made by users of *Life on Earth,* we have assembled most of the parent book's material on hormones in one place (Chapter 12) and added new material. Since hormones are of such importance to topics in other chapters, such as to plant growth in the chapter on development, and to the menstrual cycle in the chapter on reproduction, we have included extensive cross-references to these chapters. Also, we have considerably expanded the listing of hormones in the index. Finally, in abridging Part II we were particularly careful to retain all of the original material on human biology. Although *Life* does not place undue emphasis on this atypical two-legged mammal, we have tilted the emphasis more toward human biology to make the book fit the needs and interests of students who may never take another course in biology.

In Part IV we dropped the chapter on molecular evolution. Although it was one of the most highly praised chapters in *Life on Earth,* we felt that it could not be clearly understood without more chemical background than we could offer in a short book. The other chapters have been shortened by almost half, mostly by deleting specialized topics.

The two chapters that comprise Part V have been updated to reflect the industrial nations' increasing concern with energy sources and costs in the last few years. Otherwise these chapters remain unchanged, and they continue to reflect the authors' disagreement about man's stewardship of life on our planet.

We have paid particular attention to clarity of language. We have assumed that most students studying from this textbook are intellectually curious but will have had little or no previous background in science. For these readers we have tried to present ideas simply, logically and interestingly.

There are many new illustrations in *Life,* and many of the illustrations carried over from *Life on Earth* have been modified. The photographs were chosen with the same care as those in *Life on Earth.* Instead of leafing through piles of stock photographs, including many of the old textbook standbys, we searched for new pictures that illustrated a structure in a fresh or particularly revealing way. Most of the new diagrams and drawings were created from scratch for this book. Each figure illustrates an important point, and each is cited at the appropriate place in the text. No picture is included solely because it is handsome, and there are no portfolios of simply picturesque plants and animals.

We have modified the list of readings at the end of each chapter, adding new titles, updating bibliographic information, and deleting the more specialized works. Also, because a first-year biology course introduces the student to almost as many new words as a course in a foreign language, we have included a hefty glossary and an extensive index.

Is *Life* then "an entirely new book" or just a condensation? We prefer to think of it, almost in a biological sense, as the offspring of its parent. The two books resemble each other in many ways, but there are important differences, most of which have been mentioned above. Some chapters have been completely rewritten while others have merely been revised and shortened; in the process, few paragraphs have survived unchanged.

As we noted in the preface to *Life on Earth,* biology is nearing its destiny as a unified science, and it can be presented that way to a beginning student. To an increasing degree biology has become the focus of man's hopes, the potential source of knowledge that might solve his most pressing problems: food, population, ecology, health and, in a deeper sense, the definition of man's place in nature. An important goal of a good liberal arts education is to know what science is, to understand how scientists think and how discoveries are made, and thus to be able to distinguish the sound and useful from the shoddy and the untrue. This is the spirit in which *Life* was written, and we hope that spirit will communicate itself to the reader.

Edward O. Wilson
Thomas Eisner

LIFE: CELLS, ORGANISMS, POPULATIONS

introduction

the third planet

A year and a half ago when we were coming home on Apollo 8, we looked at the Earth, and mentioned that the Earth was really the only place we had to go to. It was the only place that had color. It was the only place that we could see in the universe that had life, that had warmth, that was home to us. But I was safely tucked in a nice warm spacecraft with all systems functioning, and I really wasn't too worried. On Apollo 13 on the way home the situation was just a little different. I recalled the same words that I had said a year and a half ago, and I wondered just when and how we would get back.

JAMES A. LOVELL, JR.

Few men have ever left the planet Earth, and fewer still have had to wonder seriously if they could return safely. Most of us take our planet more or less for granted. Science fiction writers such as Jules Verne and H.G. Wells imagined the neighboring planets as populated with life, humanoids and even civilizations.

Today we are more knowledgeable about the possible existence of life elsewhere in the solar system. Mercury, the planet nearest the Sun, is a hot cinder turning slowly on its axis. Surface temperatures on the second planet, Venus, are close to that of melting lead. The fifth and sixth planets, Jupiter and Saturn, are surrounded by dense gaseous atmospheres that solar radiation may never penetrate. The three outer planets appear cold and inhospitable. Our own Moon turns out to be an airless, waterless, lifeless gray sphere, the product of four eons of meteoric bombardment and vulcanism (*Figure 1*). Only Mars and perhaps the moons of Jupiter and Saturn offer a slim hope for life of even a primitive kind. The puzzling data from the two Viking landers now parked on the Martian surface does not rule out the existence of some form of microbial life on the red planet, but further experiments will be needed to confirm or disprove this possibility. The dawn of space exploration has lent support to the medieval notion that Earth may be unique, at least in the solar system. Viewing our planet from the Apollo 8 spacecraft, astronaut James A. Lovell, Jr. uttered one of the most eloquent comments to come out of the space program: "The Earth from here is a grand oasis in the great vastness of space."

What makes Earth such an oasis in an otherwise inhospitable solar system? Of the nine planets, only the Earth has oceans, and they have been the cradle of life. Is life inevitably linked to seas? Or is life a natural stage in the evolution of matter? What features are unique to life wherever it may be found, and what features are accidents of the history of life on this particular planet? Looking beyond the solar system, man would be rash indeed to suppose that ours is the only inhabited planet in the Universe. But until we discover a form of life elsewhere in the Universe (or until it discovers us), the only answers to these questions must come from comparisons of the features common to living systems on our own planet.

THE EARLY EARTH

Our solar system, including the planet Earth, evolved about 4.5 billion years ago. Evidently the planets were built up by the gravitational attraction and aggregation of cold dust and particles into clumps of

EARTH FROM SPACE. Africa dominates photograph on opposite page. Spain, Italy, the Black Sea, the Caspian Sea and central Asia are visible to the north at the edge of the cloud cover. Photo was made during Apollo 11 flight in 1969.

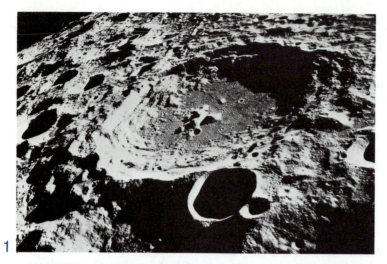

1

DARK SIDE OF MOON was photographed from Apollo 11 command module in lunar orbit. Large central crater is about 50 miles wide. On moon such craters may persist for billions of years.

solid matter. As the Earth grew, the weight of the outer layers compressed the center. This pressure, and energy liberated by the decay of radioactive elements, heated the interior until it finally melted. The settling of heavier elements produced the fluid iron and nickel core of the planet. Around this core, which has a radius of 2200 miles, lies an 1800-mile-thick mantle made of dense silicate minerals. Covering this is a lighter crust, up to 25 miles thick under the continent, but only 3 miles thick under the ocean floor. Only from the so-called Kimberlite diamond pipes that erupt to the surface in South Africa and a few other sites, has Man ever obtained a specimen of the minerals that comprise the mantle.

The scarcity of helium and other noble gases on Earth compared with their abundance on

2

BARRINGER METEOR CRATER in Arizona is about three quarters of a mile across and 600 feet deep. It was probably dug by a meteorite about 100 feet in diameter and weighing 100,000 tons, which struck the Earth at a velocity of about 36,000 miles per hour. Geological processes on Earth have erased all but most recent craters. This one is about 25,000 years old.

the Sun suggests that Earth lost its original atmosphere. The planet was once a sterile rocky ball with neither atmosphere nor oceans. A new atmosphere arose in time from the outgassing of the mantle and crust. Water vapor from the interior condensed into the seas that now cover most of the Earth.

The distribution of sea and land in those distant times was totally different from now. Since the beginning of the planet, the Earth's crust has been in constant motion. Sea beds have been uplifted and folded into mountain ranges, and gradually eroded down into plains which have been flooded again. Continents have drifted slowly apart. The history of the Earth is one of constant and ceaseless change. Evidence of external accidents, such as impact craters from meteors, have been quickly obscured; today only the most recent ones survive (Figure 2).

By contrast, the Moon and Mars are essentially static worlds. There are volcanoes on both, and Mars does have mountain ranges and deep rift valleys. Still, these two planets have few geophysical processes comparable with continental drift and the folding of the Earth's crust. The Moon, Mars and Earth increase in mass in a regular progression: Mars is 8.7 times the mass of the Moon, and Earth is 9.4 times the mass of Mars. Both the Moon and Mars were so small that the heat produced by collision as they were being formed, or subsequently generated in their interiors by gravitational pressure and radioactive decay, could rapidly escape. To use an analogy from nuclear physics, the Moon and Mars were below the critical mass necessary to build up enough heat to keep the core molten. If their interiors melted at all, they probably solidified again as heat was radiated into space from the planetary surface. They became solid balls of rock, and many of the processes that shape the surface of the Earth either never developed or never lasted for long. The small size and weak gravitational fields also meant an early loss of what initial atmospheres the Moon and Mars possessed, which meant there was little erosion from wind and rain. One sign of the lack of molten cores in these planets is their absence of magnetic fields.

The oldest rock and soil samples brought back from the Moon by the Apollo 11 and 12 expeditions have been dated as 4.5 billion years old. Both expeditions found basaltic lava 3.4 to 3.6 billion years old, indicating that there was a widespread outpouring of lava

The most ancient rocks found on Earth, more than 3.5 billion years old, already bear traces of living organisms.

over the lunar "seas" during a relatively narrow time span. For the first billion years of its history, at least, the Moon was not a static ball of rock. It has been suggested that Mars and perhaps even the Moon might have been covered early in their histories with shallow seas which were later lost as water molecules escaped the planets' weak gravitational pull and vanished into space. The scars of over four billion years of history are visible on the Moon as a heavily cratered surface, and parts of Mars also appear surprisingly Moon-like (*Figure* 3). Earth, by contrast, was large enough to develop differently.

This then, is the stage on which life was to play out its drama — a water-covered planet, with shifting land masses in constant geological turmoil. Some time around four billion years ago, the first actors appeared on stage. The oldest rocks found on Earth, over 3.5 billion years old, already bear traces of living organisms. How and why the first living systems appeared is the subject of Chapter 16. The oceans teemed with life, but the exposed land remained nearly as sterile as the Moon. The land was bathed in deadly ultraviolet radiation from the Sun. Life was confined to depths in the seas (below 5–10 meters) that would shield out this radiation. One effect of the "invention" of photosynthesis was the creation of an ozone layer in the upper atmosphere which blocked the ultraviolet rays. When this happened, life spread to shallow coastal waters and eventually to the land.

Photosynthesis evolved in response to a shortage of natural high energy compounds for food. When this new method of tapping solar energy to synthesize sugars developed, life divided into two classes of organisms: those that made their own food (plants) and those that ate the foodmakers (animals). Both plants and animals moved onto the land in their own ways: animals by migration of individuals, and plants by migration of spores and seeds.

The invasion of the land cleared the way for the major steps in the history of life (*Chapters 16–20*). In the animal kingdom this involved the rise of amphibians, reptiles, mammals, the special type of mammals known as primates, and finally that peculiar primate called Man. With the coming of Man, life crossed another threshold comparable to the synthesis of the first life, the evolution of photosynthesis, and the conquest of land. Man became the first thoroughly cultural animal, by virtue of his ability to communicate by speech and writing. A large part of his heritage became externalized in myths, traditions, custom, and law, rather than being predominately internal and genetic. To use an only slightly exaggerated image, human genes were supplemented by libraries.

The view of the Earth that faces page 1 shows the birthplaces of two later critical innovations in Earth's history: Man and civilization. Some of the earliest fossil remains of man, dating back three million or more years, come from East Africa, seen here as a patch of dark gray approximately two inches south of the Nile delta. Man probably learned to plant grasses and domesticate animals 10,000 years ago in the Armenian and Iranian highlands north and east of Mesopotamia, and moved down into the river valleys of Mesopotamia,

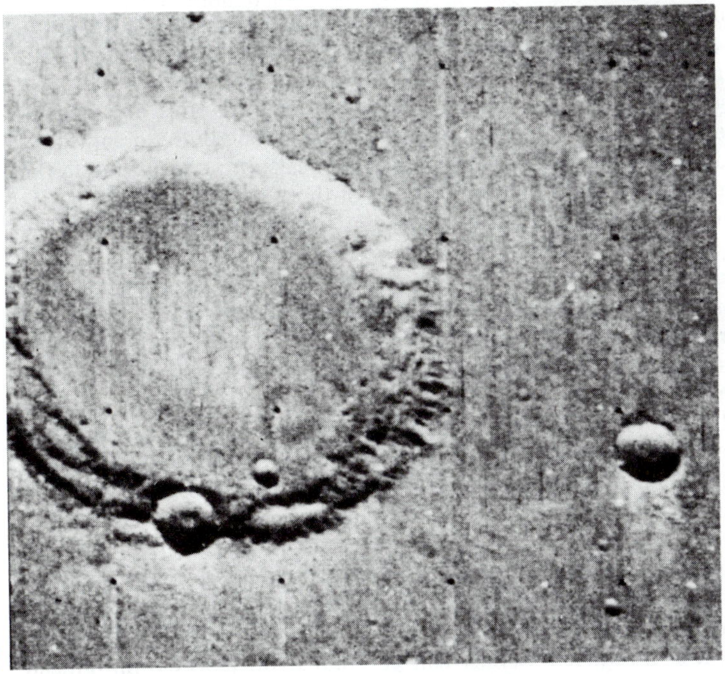

3

CRATERS ON MARS photographed by Mariner 6. Area shown is 63 by 45 miles. Much of the Martian surface appears moonlike, but some regions show Earthlike features such as dried-up river beds.

the Nile, and the Indus as he developed irrigation techniques. The effects of irrigation farming in Mesopotamia and the Nile valley are visible even from space as dark zones of vegetation. It puts things in their proper perspective to notice that the direct works of Man are invisible from space, and all that can be seen are his effects on the life of the planet.

CHANGE, EVOLUTION, AND INTELLIGENCE

The process of evolution is essentially conservative, not innovative. The "strategy" of evolution is not to promote change, but to counteract disadvantageous changes in the environment. To use an old example, fish did not first scrabble out of the water and across mud flats because they wanted to open up new frontiers on land, but because they wanted to find new ponds where they could live in their old way when their former homes dried out. Reptiles did not develop leathery shelled eggs to armor plate their embryonic offspring, but rather to enclose them in a portable aquatic environment — a marine home away from home, so to speak. To cite an earlier and more speculative example, tiny chemical systems in the primordial seas did not surround themselves with membranes to become more concentrated, but to avoid dilution. A biologist, ever wary of imputing a purpose to evolution, would probably say that chemical systems that spontaneously synthesized a membrane barrier that prevented them from being diluted and destroyed thereby increased their chances to remain intact, grow, and develop.

These are all examples of HOMEOSTASIS (from the Greek HOMEO, same and STASIS, state). Homeostasis is the tendency of a system to maintain its current state despite outside disturbances. A well-regulated thermostat is a mechanical homeostat. It turns the furnace on not to make the house hotter, but to keep it from cooling off when the outside temperature drops. Evolution is the history of one subterfuge after another adopted by living organisms to compensate for changes in their environment. Without such changes there could be no evolution. The changeable Earth teems with countless varieties of living things, while the constant Moon remains lifeless. Evolution is a mechanism for dealing with long term environmental changes. It can produce a strain of rabbits with unusually heavy fur as a new ice age commences over a period of many generations, but it cannot grow a heavy fur coat on an individual rabbit when unexpected cold weather comes. It can develop mammals such as whales that are adapted to the waters that their amphibian forebears left 450 million years ago, but it cannot teach a mountain goat to swim. The great advantage of Man over the other animals is that his intelligence — a product of his highly developed nervous system — permits him to make rapid responses to rapid changes in his environment. Man can put on a fur coat or learn to swim, and can build heated houses and submarines. Intelligence is a trait which allows its possessor to adapt to drastic changes within a single lifetime. In a sense, we are the product of the instability of our planet.

SCHRÖDINGER'S QUESTION

In 1944, the physicist Erwin Schrödinger wrote a small book entitled, *What is Life?* In it he asked whether or not simple molecules, as the chemist knew them, could possibly be enduring enough to act as the means of storing and transmitting genetic information. This was certainly not the first time that the question "What is life?" had been asked; almost every major religion and philosophy has had a try at answering it. Schrödinger's query helped to stimulate a surge of interest that led to the discovery of the structure of DNA and to the new field of molecular genetics.

Man is still asking Schrödinger's question, and will continue to do so for many years. In a sense this entire book asks that question. A biologist can list the properties and characteristics that appear in one living organism or another, but when asked to come up with the minimum list of those criteria which separate all living things from all nonliving things, he can only give a provisional reply. The reply is better than it would have been a generation ago, because more is known about what living organisms can and cannot do, and more is understood about what these activities accomplish. It has become clear that some activities that appear superficially different, such as running from a predator or producing millions of seeds, may have much the same effect in ensuring the survival of a species. Typical living things METABOLIZE, DEFEND themselves, GROW, REPRODUCE and EVOLVE. Often only one or two of these properties can be observed. An adult apple, for example, metabolizes; isolated vines can reproduce only when introduced into the right host. But at some stage in the life cycle of every living thing, at least two of the above processes occur.

All living organisms carry out reactions that release chemical energy; they use some of this energy immediately, and store the remainder. All organisms degrade (break down) materials taken in from their surroundings, and use the degraded pieces as building blocks to synthesize their own giant molecules. Four classes of macromolecule are found in terrestrial life: proteins, carbohydrates, lipids, and nucleic acids. Extraterrestrial organisms might be imagined using other kinds of macromolecules, but the processes by which living systems transform raw materials taken from the environment — degradation, energy extraction, and synthesis — appear to be fundamental to all life. These chemical processes are described as METABOLISM.

Living organisms need security to ensure the survival of the species. Not every individual need survive; only enough to produce the next generation. One DEFENSE mechanism of animals is flight. Another is a retreat to an isolated ecological niche; the sulfur bacteria that inhabit deep oil wells are relatively safe from predators. Still another security mechanism for a species, if not for an individual, is simple fecundity. Out of millions of salmon eggs, only a minute number are needed to perpetuate the species. The number of offspring that a pair of individuals produces is related to the probability that a given offspring will survive. Human beings, for example, have no need for millions of children per couple, because humans have evolved better ways of increasing the percentage of children who reach maturity. No forethought or master planning is implied here, only two different life patterns, both of which confer a high survival value on their species because each in its own way ensures that offspring endure to continue the line. Many of the behavior patterns associated with different species are really alternate mechanisms for guaranteeing the same thing: survival of that particular life form.

GROWTH AND REPRODUCTION are other properties of living things. No organism lives forever in a single state. Amoebas and other simple organisms grow for a while and then bud or divide into two or more individuals who then carry life independently. Such organisms never die of old age; each of the offspring is equally young. Higher organisms have evolved the more complex pattern of sexual reproduction; the instructions and machinery for making new individuals are passed on as the DNA-containing egg and sperm. New organisms grow, initially free of the accumulated defects and breakdowns that will eventually bring death to their parents. The group of organisms gains a new lease on life. In sexually reproducing organisms, these new individuals will age and deteriorate, but only after initiating yet another generation. The reproductive machinery for making accurate copies of an organism, whether by simple budding and division or by sexual reproduction, is an essential component of living organisms.

The entire advantage of sexual reproduction, however, does not lie in periodically restarting new individuals and thereby beating the aging process. Reproduction is not simple copying. Occasional mistakes can occur in the copying process, or can result from damage to the genetic material by radiation or certain chemicals. These genetic mistakes or mutations are copied in their offspring. A pure strain of identical individuals, if it existed, would soon develop variety as one generation followed another. Sexual reproduction speeds up this variation process, for it reshuffles the genetic material. Offspring are random mixtures of the characteristics of both parents. No child is ever precisely like either parent, and in a population of basically similar interbreeding organisms, sexual mating and shuffling of hereditary material means a constant supply of new variations on a common theme. This variation makes EVOLUTION possible. If all individuals were identical and perfectly adapted to one set of environmental conditions, then when conditions changed, all would be equally (and possibly lethally) handicapped. But with variations in a population, a shift in conditions may mean that some formerly less favored individuals become the best adapted, and increase in number relative to the former "favorites." With enough genetic variety in a population, the odds are that some members can survive all but a catastrophic change in

*With enough genetic variety in a population,
the odds are that some members can survive all
but a catastrophic change in the environment.*

*Life is a very special phenomenon, a behavior pattern
that matter exhibits when it reaches
a certain level of organization and complexity.*

environment. This counterpoint of variation and selection is the essence of evolution.

Many nonliving systems meet one or two of the five criteria. Crystals in solution use materials from the surrounding medium to grow, and can reproduce themselves from fragments if crushed. But their growth is a sterile accumulation of identical molecules. The molecules that are used in growth are the molecules that diffuse to the surface of the crystal. A crystal has no metabolism. Computers can be designed and programmed to imitate many of the purposeful actions of higher animals. They can be taught to make decisions and to learn from past experience. But they cannot yet reproduce, and a science-fiction movie in which a world master computer takes steps to defend itself against being shut off by humans is fortunately only fiction. Robots and other clockwork imitations of life are not worthy of discussion, but more subtle chemical systems developed by A. I. Oparin, Sidney Fox, and others, mimic some of the metabolic and growth properties of microorganisms. Nothing that man can now assemble or can conceive of assembling in the near future, however, is likely to show so many of the properties of metabolism, self-protection, growth, reproduction, and inherited variation that one would be compelled to describe it as living.

Life is a very special phenomenon, a behavior pattern that matter exhibits when it reaches a certain level of organization and complexity. Scientists cannot duplicate the phenomenon, but they can observe from a distance, and disassemble its parts for study, whether the part is a population, an individual, an organ or a molecule. They can propose ideas about how living systems work and why they exist as they do, and check the simpler of these ideas experimentally. The problem is complicated because we are part of the system under study — a dilemma that makes the study of life even more challenging.

READINGS

R.N. BRACEWELL, *The Galactic Club: Intelligent Life in Outer Space*, San Francisco, Freeman, 1974. A short, highly readable paperback by a Stanford astrophysicist who believes that man is not alone.

F. CRICK, *Of Molecules and Men*, University of Washington Press, 1966. Thoughts on the chemical basis of life, by the co-discoverer of the structure of DNA.

S.H. DOLE, *Habitable Planets for Man*, New York, Blaisdell, 1970.

F. HOYLE, *Of Men and Galaxies*, Seattle, University of Washington Press, 1964. Three public lectures by one of our most noted and articulate astronomers. The last two, "An Astronomer's View of Life," and "Extrapolations into the Future" are especially relevant to this Introduction.

C. SAGAN, *The Cosmic Connection*, New York, Dell, 1975. Fact and speculation about extraterrestrial life, by an astronomer who believes that it is an idea whose time has come.

E. SCHRÖDINGER, *What is Life?*, New York, Cambridge University Press, 1962 (1st edition 1944). The inspiration for much of post-war molecular biology. Not as revolutionary as it once was, but still interesting because of its author.

R. SIEVER, "The Earth," *Scientific American*, September 1975. The evolution of the early Earth. The entire September issue is devoted to a series of articles on the solar system, with special emphasis on information gained from recent space probes.

PART ONE THE CELL

One important property of life on Earth is that it is modular. The smallest unit of life is a complex but reasonably standard entity called a cell, and larger forms of life are built up from aggregations of cells. A small bit of chemically reactive "soup" is not alive — without the protection of a cell membrane it would rapidly be diluted out of existence. The closed boundaries of a cell permit concentrations and arrangements of chemical components that would be vanishingly improbable in an open solution. The limits of a cell keep the improbable from becoming impossible.

Large organisms do not consist of a single enormous cell but of many small ones. The upper limit on cell size is set by the slow rate at which substances diffuse through liquids. Organisms rely largely on diffusion to move molecules from place to place within the cell. Except for very short distances, the rate of diffusion is simply inadequate to keep pace with biochemical machinery that must carry out thousands of reactions per second. Moreover, a single giant cell would not have enough surface area for adequate exchange of gases, nutrients and waste products. The cytoplasm in the interior of the cell would suffocate, starve, or poison itself.

In multicellular organisms a basic cell pattern is laid down in the genetic material, and specialized variations of that pattern emerge as the organism develops from a one-celled fertilized egg to a many-celled adult. One cell of an embryo will eventually give rise to epidermis and another to internal organ tissue, although the exact reason remains a mystery.

A cell is a miniature machine for living — a tiny chemical factory that takes both small and large molecules from the environment and rearranges them into the stuff of life. Part I of this book focuses on the cell and the biochemical processes that occur within it, particularly upon the concept of genetic information, and how this information is encoded within a single molecule of DNA — the master control that regulates the chemistry of life.

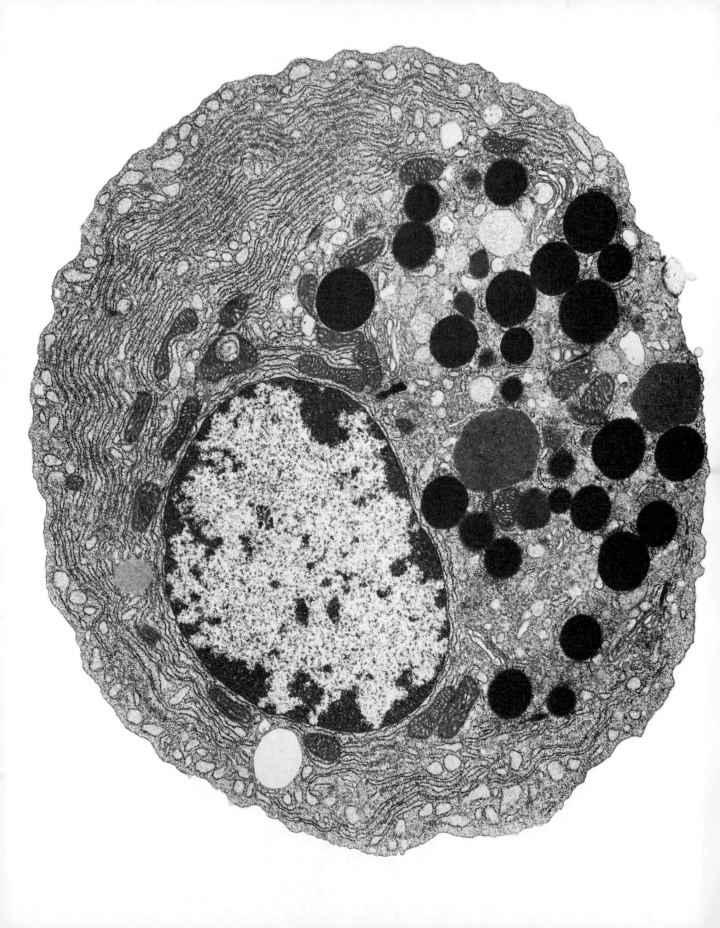

1

Any living cell carries with it the experiences of a billion years of experimentation by its ancestors. You cannot expect to explain so wise an old bird in a few simple words.

MAX DELBRÜCK

the architecture of the cell

Robert Hooke, the 17th century physicist, natural philosopher and inventor, introduced the word cell into the language of biology. Using the crude microscopes of his day, Hooke looked at a very thin slice of cork and saw a grid which reminded him of a monastery or a prison, with rows of cells to hold the inmates. Much later, in 1839, Matthias Schleiden, a botanist, and Theodore Schwann, a zoologist, promulgated the cell theory: the idea that all living things are composed of cells.

Cells vary greatly in size, from an ostrich egg, with a volume of about 10^{15} cubic microns, to parasitic bacteria called Chlamydozoa, with a volume of 10^{-2} cubic microns. (A micron, represented by the Greek letter μ, is a unit of length convenient for expressing the sizes of cells. One micron equals 10^{-4} cm. The period at the end of this sentence is about 200 μ in diameter. *See Box A.*) The great majority of cells have a volume of one to 1000 μ^3, and there are physical and chemical reasons for this. The volume and mass of a spherical cell increase with the cube of the radius, but the surface area available for intake of nutrients, exchange of gases, and excretion of wastes increases only as the square of the radius. A large cell has a much lower surface-to-volume ratio than a small one, and therefore a smaller absorptive surface per unit of mass. This simple fact sets an upper limit on the size of most cells. The ostrich egg is no exception — most of its bulk consists of stored food; the biologically active part of the egg, which will become the future embryo, is a tiny speck on the surface of the yolk. A single cell the size of a cow would have roughly the same surface area as all cells of a shrimp. In other words, this imaginary giant cell would have the mass of a cow and the absorptive capacity of a shrimp — biologically an untenable situation, to say the least.

Because all cells face certain common problems, they have many architectural features in common. All cells have a surface membrane that regulates molecular traffic into and out of the cell, maintaining an internal environment compatible with the chemistry of life. The control of life processes depends on a complicated set of chemical instructions which is passed from one generation of cells to the next.

All cells store their hereditary information in the form of threadlike molecules of DNA. In some kinds of organisms the hereditary apparatus is segregated from the rest of the cell, enclosed in a membrane-bounded nucleus. In other organisms the DNA is localized within a region called the nucleoid, which is not separated from the rest of the cell by a membrane. This distinction divides all cells into two great classes: EUCARYOTIC CELLS (from the Greek for true nucleus) and PROCARYOTIC CELLS (primitive nucleus). Bacteria and bluegreen algae are procaryotes. All other organisms are eucaryotes. Eucaryotic cells are generally much larger and more complicated than procaryotic cells, and they contain several structures collectively called ORGANELLES, that are not found in procaryotes. The evolutionary development of the eucaryotic cell was the breakthrough that made possible the evolution of multicellular organisms, including man. The first three

EUCARYOTIC CELL. Photograph on opposite page, made with an electron microscope, reveals many of the structures typical of animal cells. Large nucleus fills much of the left side of cell; around it can be seen several peanut-shaped mitochondria. Wavy lines in the cytoplasm at left are the endoplasmic reticulum. Black discs in the right half of the cell are secretion granules.

A
MEASURING IN THE METRIC SYSTEM

All the natural sciences, including biology, require frequent measurements using metric units. The reader would be wise to memorize the following divisions and special terms applied to them. Also, we recommend practice in the multiplying and dividing involved in order to get a better feel for the relationships of the units.

LENGTH

1 Kilometer = 1000 meters (also written 10^3 meters) = 0.62 mile

1 Meter (abbreviated m) = $\frac{1}{1000}$ kilometer (also written 10^{-3} kilometer) = 39.37 inches

1 Centimeter (abbreviated cm) = $\frac{1}{100}$ meter (also written 10^{-2} m)

1 Millimeter (abbreviated mm) = $\frac{1}{1000}$ meter (also written 10^{-3} m)

1 Micron (symbolized by μ) = $\frac{1}{1000}$ mm (also written 10^{-3} millimeter) = one millionth of a meter (written 10^{-6} m)

1 Nanometer (abbreviated as nm; also called 1 millimicron, which is symbolized as mμ) = $\frac{1}{1000}$ micron = one billionth of a meter (written 10^{-9} m)

1 Ångström (abbreviated Å) = $\frac{1}{10}$ nanometer = $\frac{1}{10,000}$ micron = one ten-billionth of a meter (written 10^{-10} m)

MASS

1 Kilogram = 1000 grams (also written 10^3 grams) = 2.20 lbs

1 Gram (abbreviated g) = $\frac{1}{1000}$ kilogram (also written 10^{-3} kilogram)

1 Milligram (abbreviated mg) = $\frac{1}{1000}$ gram (also written 10^{-3} gram)

1 Microgram (abbreviated μg) = $\frac{1}{1000}$ milligram (also written 10^{-3} milligram) = one millionth of a gram (written 10^{-6} g)

1 Nanogram (abbreviated ng) = $\frac{1}{1000}$ microgram (also written 10^{-3} microgram) = one billionth of a gram (written 10^{-9} g)

VOLUME

1 Liter (abbreviated l) = 1000 milliliters (also written as 10^3 milliliters) = 1.06 liquid quarts

1 Milliliter (abbreviated ml) = $\frac{1}{1000}$ liter (also written 10^{-3} l) = 1 cubic centimeter (abbreviated cc or cm³). One ml (= 1 cc) of water at 4℃ weighs 1 gram

1 Microliter (abbreviated μl) = $\frac{1}{1000}$ milliliter (also written 10^{-3} milliliter) = one millionth of a liter (also written 10^{-6} l). One microliter of water at 4°C weighs 1 milligram.

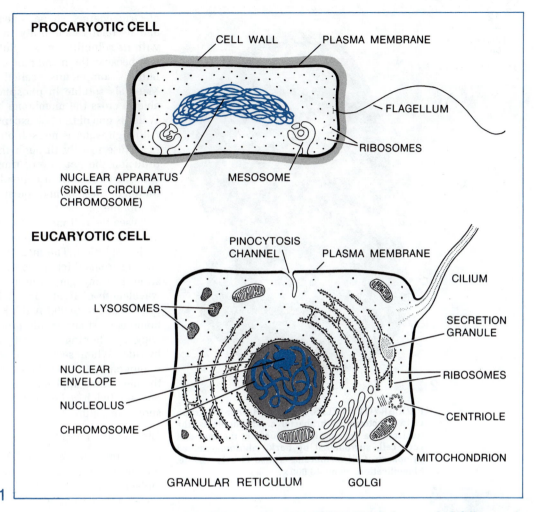

PROCARYOTIC CELL

CELL WALL PLASMA MEMBRANE

FLAGELLUM

RIBOSOMES

NUCLEAR APPARATUS
(SINGLE CIRCULAR
CHROMOSOME)

MESOSOME

EUCARYOTIC CELL

PINOCYTOSIS
CHANNEL PLASMA MEMBRANE

CILIUM

LYSOSOMES

SECRETION
GRANULE

NUCLEAR
ENVELOPE

RIBOSOMES

NUCLEOLUS

CENTRIOLE

CHROMOSOME

MITOCHONDRION

GRANULAR RETICULUM GOLGI

1

PROCARYOTIC AND EUCARYOTIC CELLS show marked differences in complexity and internal organization. Schematic diagram depicts a typical bacterium and a typical animal cell.

illustrations in this chapter (*opening page and Figure 1 and 2*) compare the most important features of "typical" procaryotic and eucaryotic cells. By the end of the book the reader will understand that the idea of a "typical" cell is an oversimplification, rather like the notion of a typical human being, but these illustrations at least indicate what to look for. Notice that both types of cell contain many minute particles called ribosomes. These are the sites of protein synthesis, and all true cells contain

them. All procaryote ribosomes are about the same size, and are considerably smaller than eucaryote ribosomes.

Viruses, the cause of many diseases in higher organisms, are much simpler than bacteria or bluegreen algae, but they cannot properly be called cells. Viruses do not have ribosomes, and must use the ribosomes of a host cell to synthesize viral proteins. Some biologists classify viruses as procaryotes, but the differences between viruses and bacteria or

*The cell membrane is not simply a sieve
that excludes large molecules and admits small ones.*

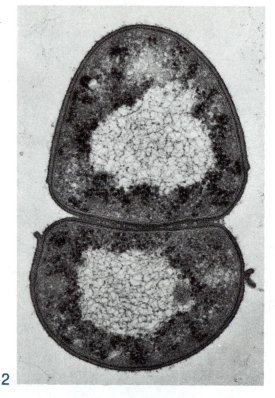

2

PROCARYOTIC CELLS. In this electron micrograph of two bacteria (Arthrobacter crystallopoietes) the nuclear area appears in light tones at the center of each cell, surrounded by dark, ribosome-packed cytoplasm. Cell membrane is also clearly visible. Magnification, about 46,000×.

electric charge, shape, chemical properties, and its relative solubility in water compared with its solubility in fats. Solubility is important because the membrane consists largely of fatlike compounds called phospholipids. Materials soluble in phospholipids are often able to cross the membrane by free diffusion. Water is one of the few exceptions to this rule. Although water is not soluble in fats, it seems to diffuse readily through the membrane. In diffusion the net rate of transport cannot be controlled, because it depends on the concentration of the substance outside and inside the membrane.

To see the cell membrane in any detail, one must examine cells at high magnification and high resolution. The membrane is only 70 to 80 Ångströms thick. (Abbreviated Å, the Ångström is a unit generally used to measure the wavelength of light. One Å is 10^{-8} cm. This page is about 70,000 Å thick.) The cell membrane is most commonly prepared for microscopy by staining the cell with osmium tetroxide. When sections of stained cells are photographed under the electron microscope, the membrane appears as three bands, a light one sandwiched between two dark ones, as shown in Figure 3.

TRANSPORT: PASSIVE AND ACTIVE

Compounds that are not fat-soluble must be transported into the cell by specific carrier molecules embedded in the membrane. The carrier and compound diffuse across the membrane together; then the compound dissociates from the carrier. This process is called CARRIER-FACILITATED DIFFUSION. The carriers are protein molecules, sometimes called permeases because they make the cell permeable to a particular compound. Carrier-facilitated diffusion can work in either direction. If a compound is at higher concentration outside a cell than inside, the net flow will be inward. If the compound is at higher concentration inside, the net flow will be outward. Obviously, carrier-facilitated diffusion cannot be used to create a high concentration of a nutrient inside a cell if the external concentration of that nutrient is low. Carrier-facilitated diffusion is sometimes called PASSIVE TRANSPORT.

One can distinguish free diffusion from carrier-facilitated diffusion by a careful study of the effect of concentration on the rate of entry of a compound into a cell — the number of molecules transported per minute. The most characteristic difference is that systems that

3

bluegreen algae are enormous. A virus is simply a packet of hereditary material wrapped in a coat of protein (and in some cases a few additional substances). The coat serves only to protect the hereditary material and to introduce it into a cell of a susceptible host. The coat is not comparable in function with the membrane of a true cell. Biologists have long debated whether viruses are "alive," but this remains largely a matter of definition (*Box B*).

THE CELL MEMBRANE

This structure, sometimes called the plasma membrane, regulates the passage of materials into and out of the cell. The cell membrane is not simply a sieve that excludes large molecules and admits small ones. Whether or not a molecule is admitted depends on its size,

CELL MEMBRANE looks like a railroad track under the electron microscope. This illustration depicts the membrane of a red blood cell. The space between the dark lines is about 25 Ångströms wide.

operate by carrier-facilitated diffusion become saturated when all of the available carrier molecules are in use all the time. Then further increases in the concentration of the "passenger" molecules will produce no increase in their rate of entry into the cell. Systems that operate by free diffusion are not saturable. Since cells can control the number of molecules of each kind of carrier that they build into their membranes, they can control the maximum rate of carrier-facilitated transport of each sort of molecule. But cells cannot

B

ARE VIRUSES ALIVE?

Viruses lie in a semantic twilight zone between living and nonliving organisms. They are essentially short segments of genetic material, DNA or RNA, protected by overcoats of protein. They can be crystallized and they will not grow or propagate in isolation, which suggests that they are nonliving (photograph at left). They do not

CRYSTALLIZED VIRUS. Tiny rods that constitute this crystal are viruses that cause mosaic disease in tobacco plants. Individual viruses are visible at bottom.

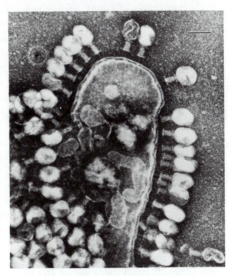

VIRUSES ATTACKING BACTERIUM. Dozens of bulb-shaped bacteriophage viruses have attached themselves to the cell wall of the sausage-shaped E. coli bacterium.

carry out any metabolic processes alone. On the other hand viruses are capable of infecting bacteria or higher plants and animals and of seizing control of their metabolic machinery. An intestinal bacterium, when infected by a bacteriophage (bacteria eater), shuts down some of its own chemistry and begins copying the phage DNA and making the component parts for new bacteriophage viruses (photograph at right). After half an hour or so, the bacterium bursts and several dozen new viruses emerge into the surrounding medium, ready to infect new bacteria. But from the time such a virus leaves its "incubator" until it finds another bacterium to infect, it undergoes no detectable chemical change. A lone virus can do nothing, but a virus plus a bacterium can do all the things that are required in a generalized living organism. Is the virus alive? Is the virus-plus-bacterium alive? The bacterium would be considered alive without the virus. It is probably best to regard the virus as a degenerate parasite which has lost many of its ancestors' independent functions. The virus now makes the bacterium perform these functions for it. We humans do the same thing when we eat fruit to obtain Vitamin C, which we can no longer synthesize but which our preprimate ancestors could. It should be clearly understood, however, that a virus is a degenerate organism which has lost functions. It is not a "half-way house" in the evolution of life from nonliving matter.

control the final internal concentration, because this depends only on the external concentration.

Certain compounds can also be brought into cells by a mechanism known as ACTIVE TRANSPORT, which can move molecules from a region of low concentration to one of high concentration — in the direction opposite from the one dictated by diffusion. This requires an expenditure of energy as if the transported molecules were being "pumped" into or out of the cell. Active transport enables a cell to accumulate certain nutrients in concentrations that are hundreds of times higher than those outside. Moreover, cells can also actively pump out certain substances, such as sodium ions, even though the concentration may be higher outside than inside the cell. Different cells may transport the same metabolite in different ways. For instance, glucose is transported actively during its absorption by cells of the intestine, but by carrier-facilitated diffusion in muscle cells. The mechanisms of the three types of transport are depicted in Figure 4.

The cell membrane creates and maintains a certain constancy of the cell contents in the face of fluctuations in the outside medium. It also creates a degree of biochemical privacy from other cells of the same organism.

CELL WALLS AND OSMOSIS

When Hooke examined his slices of cork, he saw that there were no inhabitants in his "cells." Hooke was actually viewing the CELL WALLS laid down by occupants who had long since died. Unlike cell membranes, cell walls are not found in all forms of life. They are characteristic of higher plants, fungi, and bacteria, but are not universally present even in these organisms. Cell walls help to give plants rigidity, a property that is advantageous to a stationary creature, but would be a handicap to motile animals.

It is possible to dissolve the cell walls with the help of certain enzymes, or to grow cells under conditions that inhibit the formation of walls. The resulting cells, enclosed only by their cell membranes, are called PROTOPLASTS.

Their properties reveal some interesting things about the role of cell walls. First, protoplasts are spherical, though the walled cells from which they originated may be rodlike, cubelike, or more complex in shape. Evidently the cell wall molds the shape of these cells. In the absence of the wall, the protoplast simply assumes the form which has minimum surface area per unit volume (animal cells, which do not normally possess a wall, are not called protoplasts, and their shape is determined by other factors). Second, protoplasts are easily damaged by OSMOSIS, an important process that requires a few words of explanation.

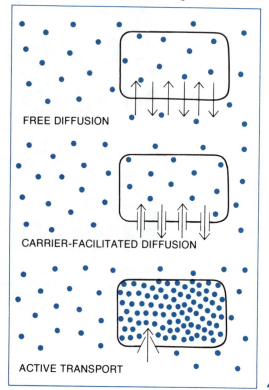

FREE DIFFUSION

CARRIER-FACILITATED DIFFUSION

ACTIVE TRANSPORT

4

THREE KINDS OF TRANSPORT. In free diffusion of substances (dots) through the cell membrane, the final concentration of substance inside the cell becomes the same as the concentration outside. The same is true of carrier-facilitated diffusion. In active transport, the final concentration inside the cell will be much higher than concentration outside, because substance is being "pumped" in.

*Cell walls help to give plants rigidity,
a property that is advantageous to a stationary creature,
but would be a handicap to motile animals.*

If one fills a membrane, such as a cellophane sausage casing, with the white of a raw egg and ties it closed at the top he will have a crude but useful model for studying osmosis — the movement of water across a cell membrane. Because the egg white contains proteins, salts, and small molecules, the concentration of water molecules in the egg white is a bit lower than in pure water. If the bag is immersed in a tub of pure water, the membrane swells, stretches, and may even break. Obviously, extra water has entered the bag. This osmotic flow can be explained on the grounds that each time a water molecule hits the bag, it has a certain probability of passing through it. Since the concentration of water inside the bag is less than the concentration outside, more water molecules will enter the bag than will leave it. The ions of salts and small uncharged molecules will also pass through the membrane, but the protein molecules are trapped inside because they are too large to penetrate the small pores in the cellophane (*Figure 5*).

Most normal cells of plants, fungi, and bacteria can survive and multiply under HYPOTONIC conditions; that is, when the surrounding medium has a lower concentration of dissolved molecules and ions than the inside of the cell. The wall limits the tendency of water to flow into the cell by confining the cell to a definite volume. Walled cells can even survive in distilled water. A protoplast, by contrast, swells as the outside medium is made progressively more hypotonic. Ultimately it bursts, killing the cell and spilling its contents into the medium. A protoplast is stable only under ISOTONIC conditions — an osmotic equilibrium in which there is no net flow of water either into or out of the cell. This is true as well of most animal cells. If any cell is put into a HYPERTONIC medium — one with a higher concentration of solutes (dissolved particles) than the inside of the cell — such as a strong salt or sugar solution, water will diffuse out of the cell. In animal cells, this shrinkage causes the surface of the cell to wrinkle. In plant cells, the mass of protoplasm surrounded by the cell membrane pulls away from the wall. This process, called plasmolysis, reveals that the cell wall is much more porous than the cell membrane and does not act as a serious barrier to small molecules or ions (*Figure 6*). In many cases, however, the wall does impede the diffusion of large molecules, notably proteins.

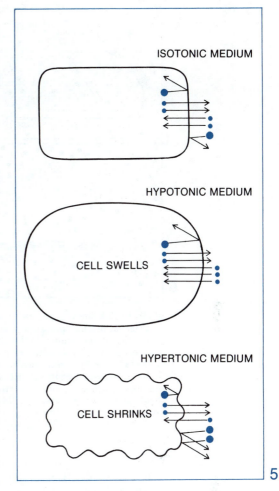

ISOTONIC MEDIUM

HYPOTONIC MEDIUM

CELL SWELLS

HYPERTONIC MEDIUM

CELL SHRINKS

5

OSMOSIS can cause animal cells to swell or shrink. Cell at top is in osmotic equilibrium with the surrounding medium; concentrations of non-diffusible molecules (large disks) inside and outside the cell are equal. Cell at center swells because it has a higher internal concentration of non-diffusable molecules, causing water molecules (small disks) to diffuse into the cell. Cell at bottom shrinks because higher external concentration of non-diffusible molecules causes water molecules within the cell to diffuse outward.

CONTRACTILE VACUOLES

Animal cells, lacking tough walls, cope differently with the problem of osmotic flooding. Many protozoans (one-celled animals) live in fresh water. Water flows in through their cell membranes because the concentration of solutes in the cytoplasm is higher than the concentration of solutes in the water in which they live. This hypotonic environment would cause a fatal bloat if it were not for the contrac-

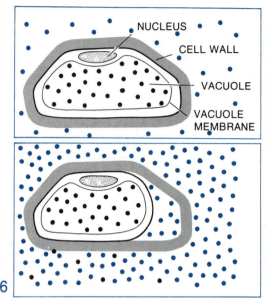

6

PLANT CELL shows a different response to changes in external osmotic pressure. For simplicity, water molecules are not shown in this drawing. Plant cell consists of a layer of cytoplasm around a central vacuole. In a hypotonic medium (top) cell membrane is pressed firmly against cell wall, which is rigid but permeable. In a hypertonic medium (bottom), cytoplasm shrinks, pulling cell membrane away from the wall. Shrinkage of cell stops when it is in osmotic equilibrium with its environment.

tile vacuole, an organelle that serves as the cell's excretory organ (*Figure 7*). The vacuole swells and shrinks in a steady cycle, slowly ballooning as water collects in it, then rapidly contracting as it expels its contents through the cell membrane, then ballooning again. It is not yet known how the cell pumps water into the contractile vacuole, but it is obvious that energy must be expended in a process that moves water from a region where it is less concentrated (inside the cell) into a region where it is more concentrated (outside).

OTHER VACUOLES

In plant cells, as much as 90 per cent of the volume of the cell may be occupied by a large central vacuole. Between the vacuole membrane and the plasma membrane lies only a thin layer of cytoplasm pressed against the cell wall. The vacuole seems to serve in part as a cellular sewer into which the cytoplasm flushes toxic waste products. The vacuole also

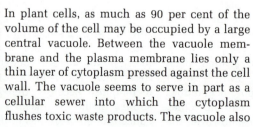

7

CONTRACTILE VACUOLE serves as excretory organ of one-celled animals, and opens to the outside of the cell (bottom). A three dimensional view of a contractile vacuole appears on page 205.

serves as a reservoir for some sugars, amino acids, and even proteins that can be taken up by the cytoplasm as needed. Some of the pigments that give plants their color are dissolved in the fluid contents of the vacuole. Many of the most interesting plant colors — the reds of Indian war paint, borscht, and chianti, for example — are located in the vacuole.

The high concentration of particles in the central vacuole of plant cells produces a tendency for water to flow into the cell, creating a high internal pressure. This turgor pressure is important both in the growth of plants and in maintaining their erect posture. An increase in the number of cells in the plant accounts for only part of the plant's growth; the enlargement of nondividing cells by the intake of water is responsible for much of the increase in size.

Some of the vacuoles in eucaryotic cells originate from a remarkable transport process called PHAGOCYTOSIS (cell eating). Phagocytic cells such as our own white blood cells can ingest solid material into pockets in their cell membranes, then pinch off the pockets, sealing them inside the cell. Our white blood cells devour bacteria in this way, thus helping to protect us from bacterial infection (*Figure 8*). Some free-living cells, such as the amoeba, obtain all their food in this way.

Cells also engulf liquid material from their surroundings, a process called PINOCYTOSIS (cell drinking). In higher animals pinocytosis plays a part in the transport of materials from the circulating blood through the thin cellular walls of capillaries into the surrounding tissues. Phagocytosis and pinocytosis are such similar processes that they are often lumped together and called ENDOCYTOSIS.

THE ENDOPLASMIC RETICULUM

Eucaryotic cells contain a collection of internal membranes called the endoplasmic reticulum. These membranes are folded in complicated ways to form plump fingerlike projections or flat stacked pockets. The endoplasmic reticulum varies from time to time within a cell and shows vastly different appearances from one type of cell to another. It is capable of continual reorganization, breaking old connections and forming new ones. The membranes can break up into vesicles and rejoin one another. Under the electron microscope, some of the reticular membranes are heavily peppered with ribosomes (*Figure 9*). Not all ribosomes in eucaryotic cells are bound to the en-

ENDOCYTOSIS. This highly schematic diagram depicts a cell engulfing a food particle. Once engulfed, the particle is enclosed within a former segment of the cell membrane.

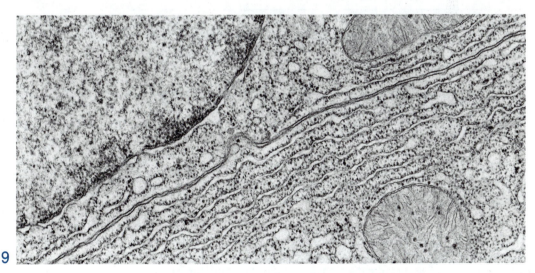

8

1
2
3

doplasmic reticulum; unattached ribosomes float freely in cellular fluids. Ribosomes attached to the endoplasmic reticulum are involved in the synthesis of proteins for export from the cell, while the free ribosomes participate in the synthesis of proteins that remain inside the cell.

Part of the endoplasmic reticulum is attached to a structure near the nucleus called the GOLGI COMPLEX, named for an Italian pathologist who first described it in 1898. Studies with radioactive tracers (*Box C*) demonstrated that the Golgi complex is involved mainly in packaging chemical products for export from the cell.

MITOCHONDRIA

All eucaryotic cells contain small, double-membrane-bounded organelles called MITO-CHONDRIA (singular: mitochondrion). They carry out a variety of functions in different cells, but they always have one function in common. They are the power plants of eucaryotic cells. Mitochondria tend to be most

numerous in regions of the cell that consume large amounts of energy. For example, in muscle cells, mitochondria lie close to the contractile elements; in cells that secrete proteins they are closely associated with the endoplasmic reticulum; in brown fat cells, which provide heat for a hibernating animal, they are abundantly distributed throughout the cytoplasm.

The outer of the two membranes of a mitochondrion forms a smooth sack-like covering that does not connect with any of the other membranes of the eucaryotic cell (*Figure 10*). The inner membrane, quite different in structure and chemical composition, is elaborately infolded to form shelf-like pockets called cristae (singular: crista). Because of the folding, the inner membrane has a much larger area than the outer one. The inner membrane seems to contain the enzymes responsible for many of the most important metabolic activities of the mitochondrion, and it encloses a region called the matrix, which contains a small amount of DNA and the biochemical machinery for duplicating it. It also contains ribosomes very similar to those of procaryotes, which are distinctly smaller than the ribosomes found elsewhere in the eucaryotic cell. Biologists have speculated that mitochondria

9

ENDOPLASMIC RETICULUM is clearly visible in this electron micrograph of parts of two cells from the pancreas of a bat. Membrane separating the two cells extends from bottom left to upper right. Parallel and immediately to the right of it lie membranes of the endoplasmic reticulum, richly peppered with ribosomes. Large structure at top left photo is a cell nucleus.

and chloroplasts were once free-living organisms. In the course of evolution, according to this line of argument, some primitive eucaryotic cells engulfed bacteria and blue-green algae and eventually formed a symbiotic relationship with them, a relationship that eventually became hereditary. There is a growing body of evidence to support this hypothesis (*Box D*).

CHLOROPLASTS

The chloroplasts of green plants are the site of photosynthesis — the conversion of carbon dioxide and water to sugar. Chloroplasts also carry out a process called photophosphorylation, which converts the energy of light to a source of chemical energy called ATP. Chloroplasts are larger than mitochondria, but they are basically similar in structure (*Figure 11*). A chloroplast is bounded by a double membrane, and the smooth outer membrane does not connect with other membranes of the cell. The inner membrane system is even more elaborately folded than the inner membrane of the mitochondrion, and the structure of the chloroplast is therefore more complicated. Like mitochondria, chloroplasts contain DNA and the apparatus for replicating it. Chloroplasts also contain ribosomes that resemble those of mitochondria and of procaryotes.

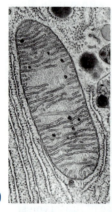

10

MITOCHONDRION has an outer and inner membrane. Inner one is folded to form partitions called cristae. Between the cristae lies the matrix, which is in a semifluid state.

LYSOSOMES

The cells of plants and animals also contain vesicles called lysosomes that appear to be the functional equivalent of the cyanide capsule that every good spy is supposed to carry. Lysosomes contain enzymes known collectively as acid hydrolases, which under slightly acid conditions can quickly dissolve all of the major molecules that comprise the cell. These enzyme packages are produced in the endoplasmic reticulum. One obvious function of these granules is to accomplish the self-destruction of injured cells or cells that have outlived their usefulness, such as the cells of the tadpole's tail when the tadpole undergoes its metamorphosis into a tailless adult. If the cell is injured or exposed to certain chemicals, including a variety of hormones and dyes, the lysosomal membrane will rupture and empty its lethal contents into the cell.

The lysosome is also involved in the digestion of materials taken into the cell in food vacuoles. When a phagocytic cell engulfs a bacterium, for example, the food vacuole then fuses with a lysosome and the enzymes of the lysosome digest the bacterium into small molecules that can be used as nutrients. Undigested fragments of the bacterium may be kept inside the vacuole for a while or dumped outside the cell.

C

RADIOACTIVE
TRACERS

Chemical pathways in the cell can be traced with the aid of isotopes — atypical atoms that differ from ordinary atoms of the same element only in the weight of their nuclei. Many of the atoms involved in the chemistry of the cell exist in nature in two or more forms, one of which is a radioactive isotope. Hydrogen, for example, exists in three forms. The most abundant isotope of hydrogen has a nucleus consisting of a single proton. About one of every 7000 hydrogen atoms has a nucleus containing both a proton and a neutron. This isotope is deuterium, commonly called heavy hydrogen. One of every 10^{17} hydrogen nuclei contains a proton and two neutrons. This isotope, known as tritium, is radioactive.

The chemical behavior of radioisotopes is almost identical to that of the nonradioactive forms. Cells metabolize radioisotopes identically with the non-radioactive compounds. Amino acids can be labeled by incorporating tritium into them. When cells are provided with labeled amino acids, they will use them to synthesize radioactive proteins. One way to find out where new proteins are being made in the cells is to let the newly-synthesized proteins take their own picture (auto-radiography). If a thin slice of tissue containing radioactive proteins is coated with a photographic emulsion, the radiation will expose the emulsion in those areas where the radioisotope is concentrated. The emulsion is then developed just like an ordinary photograph and viewed under a microscope, revealing the location of the radioactive protein within the cell.

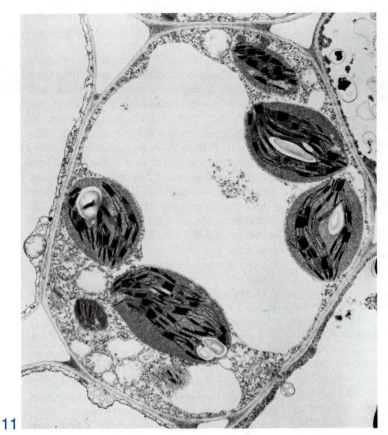

CHLOROPLASTS are large dark ovals in this electron micrograph of a plant cell. The light objects within chloroplasts are starch grains. Large vacuole at center is typical of plant cells.

FIBERS AND TUBULES

Eucaryotes possess a variety of fibrous or cablelike materials that contribute to the shape and rigidity of cells and tissues, or that move parts of the cell in relation to one another. Some of these materials lie outside of the cells and others inside. Collagen, a tough, fibrous material found between the cells of skin and tendon, is a structural material that strengthens the skin and ties muscle to bone. Elastin is a stretchy fibrous protein that gives bounce to our connective tissues and, in the ligaments, connects the bones to one another without locking them tightly together. Keratin toughens the outermost layer of the skin, and also occurs in hair and horn.

A number of other fibrous proteins, especially those involved in movement, are found inside cells. The contractile elements of muscle cells are fibers composed of bundles of myofibrils, which in turn consist of fibrous filaments.

Many other kinds of cell exhibit contractile behavior. Such cells commonly contain bundles or networks of threadlike elements called microfilaments, which in some ways resemble the fibers in skeletal muscle. Microfilaments facilitate a variety of movements in cells. Another type of movement, called CELL STREAMING, is facilitated by networks of microfilaments attached to the inside of the cell membrane. By alternately contracting and re-

---

D

MITOCHONDRIA, CHLOROPLASTS AND EVOLUTION

The striking resemblance of mitochondria and chloroplasts to procaryotic cells has encouraged biologists to speculate that these organelles originated with the infection of some primitive eucaryotic cell by procaryotes. In the case of the chloroplast, the inner membrane resembles the cell membrane of a blue-green alga in that it is the site of photosynthesis and photophosphorylation. The inner membrane of the mitochondrion resembles the cell membrane of many bacteria, in that the cell membrane of bacteria is also the site of oxidative phosphorylation. There are many other interesting parallels. For example, many antibiotics inhibit the growth of procaryotes by interfering with their protein synthesis. These same antibiotics also interfere with protein synthesis by mitochondria and chloroplasts, but do not inhibit protein synthesis in other parts of the eucaryotic cell, such as in the ribosomes of the rough endoplasmic reticulum.

What about the outer membrane of the mitochondria and chloroplasts? The simplest hypothesis to explain how procaryotes got into a primitive eucaryotic cell in the first place is that they were devoured by the process of endocytosis. Since endocytosis involves surrounding the particle to be ingested with a piece of the cell membrane, it is a reasonable guess that the outer membrane of mitochondria and chloroplasts originated from the cell membrane of the primitive eucaryote. This has given rise to further speculation that the essential innovation that made eucaryotic life possible was the invention of endocytosis.

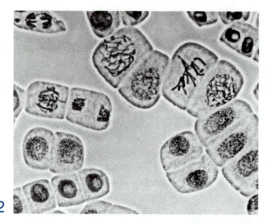

12

VARIOUS STAGES OF MITOSIS are visible in this photomicrograph of rapidly dividing cells from the root tip of a bean plant. Individual chromosomes can be clearly seen in the nuclei of cells at top.

laxing, they stir the contents of the cell. Some cylindrical cells have a ring of microfilaments at one end. When the microfilaments contract the cell becomes conical. Sheets of tissues are often constructed of cylindrical cells packed closely together. When the "purse-string" microfilaments at one end of all of the cells contract simultaneously, the flat sheet assumes the shape of a hollow bulb. Such cell movements are a common feature of embryological development.

In addition to microfilaments, eucaryotic cells contain many microtubules. These are not some sort of conduit, as their name implies, because they do not seem to be truly hollow. They are made up of small protein subunits, and apparently help to maintain cell shape. They are also involved somehow in moving chromosomes during cell division. Microtubules are also components of cilia and flagella, hairlike projections that enable some eucaryotic cells to move through water, or to create currents in fluids (*Chapter 14*).

THE NUCLEUS

Near the center of the cell lies a fibrous tangle of chromosomes, the repository of most of the cell's genetic information. In procaryotic cells, this region is called the NUCLEOID, and in eucaryotic cells, the NUCLEUS (*Figure 12*). A similar verbal distinction is made in referring to the genetic apparatus of the eucaryote and procaryote. Cytologists speak of CHROMOSOMES

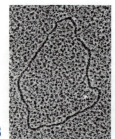

13

RING-SHAPED DNA appears in many viruses, bluegreen algae and bacteria as well as in mitochondria and chloroplasts. This electron micrograph depicts DNA from polyoma virus.

in eucaryotes, but often use the term GENOPHORE when referring to the corresponding structures in procaryotes.

The genophores of procaryotes are made simply of deoxyribonucleic acid (DNA), a threadlike molecule about 20 Å thick and enormously long. In the bacterium *Escherichia coli*, an inhabitant of the human intestine, the genetic apparatus is a single molecule of DNA about 10 million Å long — the molecule is 500,000 times longer than it is thick. The bacterium itself is about one micron thick and about four microns long. Thus, the space into which the long thread of DNA is packed in the bacterial nucleoid is very small relative to the length of the DNA molecule. It is not surprising that the molecule appears in electron photomicrographs as a hopeless tangle of fibers.

When bacterial cells are gently lysed (broken open) to release their contents, the bacterial chromosome sometimes comes untangled and spreads out to its full length. Many DNA molecules in viruses, procaryotes, mitochondria, and chloroplasts show up as closed rings (*Figure 13*). It is reasonable to suppose that the ring structure serves some functional purpose, but so far that purpose is a mystery.

The nucleus of the eucaryotic cell is surrounded by an envelope consisting of two unit membranes. The nuclear membrane is perforated by pores (*Figure 14*). At the edges of these pores, the inner and outer membrane join to form a continuous structure. In some electron photomicrographs the pore is covered by a thin plate or by a plug. The nuclear envelope is not simply a sieve that lets things pass in and out freely. All that is definitely known about the pores is that they exist; their function is obscure.

GENOPHORES AND CHROMOSOMES

When cells divide, each of the daughter cells needs genetic information to govern its biochemical processes. This requires a mechanism for copying the genetic information and distributing replicas to each daughter cell during cell division. In procaryotes, the separation is relatively simple (*Figure 15*). The two separating genophores are somehow attached to a new wall that begins to form between the dividing halves of the cell. The cell wall grows outward in both directions from the point of attachment, moving the two genophores apart. By the time the cell divides the two genophores are completely separated.

In eucaryotes cell division is much more complicated. Eucaryotes carry a great deal more genetic information than procaryotes, and they do not carry it in a single piece. The amount of DNA found in the cells of living things is roughly proportional to the complexity of the organisms. One simple virus, designated ϕX174, contains 2.6×10^{-18} grams of DNA; the bacterium *Escherichia coli*, a much larger and more complex organism, contains 4.1×10^{-15} grams of DNA; one human cell contains 6.4×10^{-12} grams of DNA. The DNA content of the three organisms represents a ratio of about 1:1580 : 2,460,000. In this sense Man is more than two million times more complex than a virus and over a thousand times more complex than a bacterium. The total length of DNA in a cell of E. coli is one mm; the total length in a human cell is 1.5 meters — about five feet. To organize that string of information within the tiny dimensions of the nucleus, a filing system more complex than that of procaryotes had to evolve.

Part of nature's solution was to organize the DNA in the eucaryote nucleus into several separate CHROMOSOMES. The word means "colored body", and refers to their tendency to absorb certain dyes. In addition to DNA, chromosomes also contain proteins and some ribonucleic acid (RNA). Some proteins of chromosomes are of a characteristic type, called HISTONES, which at cellular pH carry a positive charge. These are attracted to the negatively charged nucleic acids and form a coating of protein over the long strands of DNA. There are also some nonhistone proteins that seem to be involved in turning genes on and off.

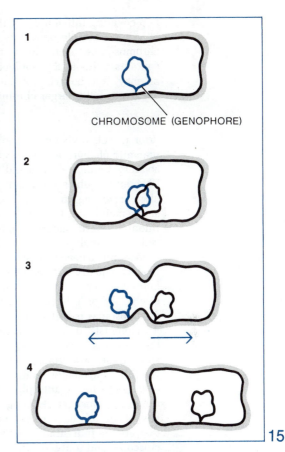

1

CHROMOSOME (GENOPHORE)

2

3

4

15

CELL DIVISION IN PROCARYOTES involves the formation of a new wall between the dividing cells. Genophores are attached to the new wall, and as it grows outward in both directions, it pulls the genophores apart. When cells divide, genophores are completely separated.

When chromosomes are spread out and examined under the electron microscope, they represent a distressingly complicated picture. Depending on how they have been prepared for microscopy, the chromosomes appear as a twisted, knotted webwork of fibers 250 Å thick. The core of each 250 Å fiber is a thread of DNA 20 Å in diameter, covered by a histone coat.

The DNA is apparently coiled like a spring and then further twisted and folded back upon itself to form the 250 Å fiber. This efficient packaging produces a bundle of DNA compact enough to fit into the nucleus, but the complicated folding has confounded all efforts to understand its structure. Does the chromosome contain one long crumpled and folded molecule of DNA? Or does it consist of several short DNA molecules glued end to end or laid side by side and twisted along with their pro-

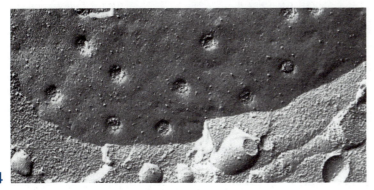

14

PORES appear as pits in the dark nuclear membrane at the top of this electron micrograph. Lighter-toned section of cytoplasm lies at bottom of picture. Magnification, about 70,000×.

tein covering into a long, multistranded rope? In yeast, a simple eucaryote with very small chromosomes, each chromosome apparently contains a single DNA molecule; but this may or may not be true of more complicated eucaryotes with larger chromosomes (*Box E*).

MITOSIS

Prior to cell division, chromosomes undergo dramatic changes and separate. A cell that is about to divide makes an additional copy of its DNA and doubles the quantity of everything else, becoming in effect two cells within a single membrane. Cell division consists of partitioning these materials into two daughter cells. Organelles such as ribosomes and mitochondria need be distributed only approximately equally between the daughter cells, but the chromosomes must be distributed precisely. The orderly process of distributing the chromosomes is called MITOSIS.

Cytologists find it convenient to classify the process into four stages: PROPHASE, METAPHASE, ANAPHASE and TELOPHASE. The "resting" stage between divisions is called INTERPHASE. During interphase, when chromosomes are most actively engaged in the business of controlling the biochemistry of the cell, they are not individually visible under the microscope and only a suggestion of diffuse

fibers can be seen. During prophase, the appearance of the chromosomes begins to change (*Figure 16*). They become visible first as thin fibers. Then the fibers become thicker, and the tangle of chromatin begins to resolve itself into a number of compact, sausage-shaped objects. As the chromosomes thicken, it becomes evident that each is composed of two parts lying side by side. These two CHROMATIDS, each a complete copy of the original chromosome, are attached to each other at a spot called the CENTROMERE. After jostling about in the nucleus, the chromosomes eventually begin to align themselves so that all the centromeres lie in a plane through the center of the cell: the metaphase plane (or metaphase plate). Late in prophase, the nuclear membrane begins to break up and disappear and the so-called spindle apparatus begins to form. Some of the microtubules extend from the centromeres, which lie in the metaphase plane, toward the poles of the mitotic apparatus, where the tubules more or less converge. Other tubules run straight across the cell from one pole to the other. It is thought that the spindle microtubules pull the chromosomes toward the metaphase plane, and later toward the poles. The poles mark the future locations of the nuclei of the two daughter cells.

In animal cells, the centriole seems to play some part in the formation of the spindle. Centrioles normally occur in pairs, oriented at right angles to each other. At interphase, a pair of centrioles lies near the nuclear envelope. During prophase they divide and a pair migrates toward each pole of the cell, as the newly-forming spindle becomes visible behind them. Except for fungi, algae, and some

MITOSIS in an animal cell, as shown here, is typical of the process in the cells of virtually all eucaryotes. Interphase is the term used to designate growth stages between mitotic divisions. During interphase chromosomes are not readily visible. Mitosis is divided into four stages: prophase, metaphase, anaphase and telophase. Labels on diagram indicate events characteristic of each stage.

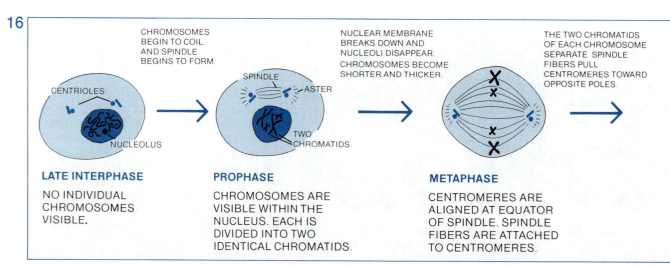

16

CHROMOSOMES BEGIN TO COIL, AND SPINDLE BEGINS TO FORM.

NUCLEAR MEMBRANE BREAKS DOWN AND NUCLEOLI DISAPPEAR. CHROMOSOMES BECOME SHORTER AND THICKER.

THE TWO CHROMATIDS OF EACH CHROMOSOME SEPARATE. SPINDLE FIBERS PULL CENTROMERES TOWARD OPPOSITE POLES.

CENTRIOLES

SPINDLE

ASTER

NUCLEOLUS

TWO CHROMATIDS

LATE INTERPHASE

NO INDIVIDUAL CHROMOSOMES VISIBLE.

PROPHASE

CHROMOSOMES ARE VISIBLE WITHIN THE NUCLEUS. EACH IS DIVIDED INTO TWO IDENTICAL CHROMATIDS.

METAPHASE

CENTROMERES ARE ALIGNED AT EQUATOR OF SPINDLE. SPINDLE FIBERS ARE ATTACHED TO CENTROMERES.

ferns, plant cells do not seem to contain centrioles, although they do form spindles.

At the start of metaphase, the nuclear envelope has dispersed, the nucleoli have disappeared, the spindle is complete, and the chromosomes lie in the metaphase plane with several spindle fibers connecting the centromere of each chromosome to the poles of the spindle. This stage is the easiest time to count the chromosomes and to determine their shapes. Each organism contains a characteristic number of chromosomes, ranging from two to almost 400. In the fruit fly *Drosophila melanogaster* the number is 8; in humans it is 46. The number, however, is not a measure of the complexity of an organism. Sugar cane has 80 chromosomes, for example, and dogs have 78. It is frequently possible to identify certain metaphase chromosomes by their sizes and the positions of their centromeres, making it possible to associate certain genetic anomalies with visible abnormalities in specific chromosomes.

At anaphase, the centromeres seem to split into two, and the chromatids in each pair separate from one another and move quickly apart, one toward each pole, as if drawn by their associated spindle fibers. At telophase, the newly separated chromosomes reach their poles and begin to revert to their previous fibrous appearance. The spindle disappears, except for some microtubules which remain between the separating chromosomes, and the nuclear envelope reforms around the chromosome bundle at each pole. At the center of the cell, along the metaphase plane, the process of cell separation begins. In animal cells, the cell surface begins to constrict, as if a belt were being tightened around it, pinching the old cell into two new ones. Microfilaments play a major role in this process. In plant cells, where a stiff cell wall interferes with this sort of pinching, new cell membranes form between the two daughter cells along the plane previously defined by the metaphase plate, and a new cell wall is deposited between the two new cell membranes. Cytologists think of cell division as two distinct processes: the separation of identical copies of the chromosomal content of the parent cell — MITOSIS or CARYOKINESIS; and the physical separation of the cells — CYTOKINESIS.

The chromosomes duplicate themselves during interphase. The time when the cells are actively engaged in DNA synthesis should correspond to the time when they are making copies of their genetic complement of chromosomes. Using mitosis and DNA synthesis as milestones, one can divide the division cycle of cells into four stages (*Figure 17*). Mitosis (M) is followed by a gap (G1) during which the new daughter cells make do with the DNA that they inherited from the parent cell. It should not be supposed that the cells are doing nothing during this time — they are simply not making DNA. Next, there is a period of DNA synthesis (S), during which the amount of DNA in the cell must double in preparation for the next cellular division. Between S and mitosis, there is a second gap (G2), during which the cell is preparing itself for mitosis.

Experiments have shown that the length of the cell cycle and the relative lengths of the four stages vary greatly from tissue to tissue and from organism to organism. In many kinds of cells, the M phase comprises about 5–10

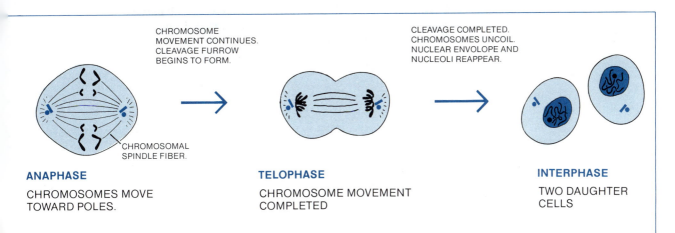

CHROMOSOME
MOVEMENT CONTINUES.
CLEAVAGE FURROW
BEGINS TO FORM.

CLEAVAGE COMPLETED.
CHROMOSOMES UNCOIL.
NUCLEAR ENVOLOPE AND
NUCLEOLI REAPPEAR.

CHROMOSOMAL
SPINDLE FIBER.

ANAPHASE

CHROMOSOMES MOVE
TOWARD POLES.

TELOPHASE

CHROMOSOME MOVEMENT
COMPLETED

INTERPHASE

TWO DAUGHTER
CELLS

percent of division cycle; the G1 phase, 30–40 percent; the S phase, 30–50 percent; and the G2 phase, 10–20 percent. Interphase comprises about 90 percent of the entire cycle, and about half of interphase is spent on chromosome duplication.

THE NUCLEOLUS

Eucaryotic nuclei contain one or more spherical bodies called NUCLEOLI, which are not surrounded by a membrane. During prophase the nucleolus disappears. When daughter nuclei

---

E
THE LIMITATIONS OF MICROSCOPES

The layman tends to assume that anything in the cell can be seen through the microscope. He might suppose that, if the image in the microscope were too small, he could magnify it further by viewing it through a second microscope, and so on. This doesn't work because one needs not only magnification, but also resolution: the power to distinguish detail. Resolving power is limited by the wavelength of the light illuminating the subject. The limit of resolution of a microscope — the smallest detail that it can form into a discrete image — is a dot about half the wavelength of the illuminating light. The shortest wavelength of light that the eye can detect is about 4000 Å and lies toward the violet end of the spectrum. Thus, the smallest object that one can resolve using an ordinary light microscope is about 2000 Å in diameter. This means that plasma membranes (about 75 Å thick), ribosomes (about 200 Å in diameter), microtubules (about 250 Å in diameter), DNA molecules (about 20 Å in diameter), and many of the other cellular structures discussed in this chapter are too small to see or photograph with a light microscope.

The electron microscope illuminates a specimen with a beam of electrons whose wavelength permits a resolution of about four to five Å. Theoretically the instrument can resolve even a single atom, but the magnetic lenses that focus the electron beam are not perfect, and the "color" of the electron beam (the energy of the electrons in the beam) is not uniform, so the finest details visible in the best electron photomicrographs are about two to four times the size of individual atoms.

A microscope must also distinguish the subject from the background — the image must have contrast. This is a particularly troublesome requirement because most cells and their components are relatively transparent. The observer faces the problem of the cartoon cat trying to catch a mouse who rubs himself with "vanishing cream." The problem for the cat is not resolving the image — the mouse is quite large enough for that; the problem is contrast. The cat throws ink or flour at the mouse in an effort to improve the contrast. For the same reason, the cytologist stains cells with dyes. This almost always kills the cell. The phase contrast microscope uses a clever optical trick to produce contrasty images of unstained living cells, but its resolving power is no greater than that of other light microscopes.

Examination of cells under the electron microscope requires more drastic chemical treatment. First the cells are usually fixed by an embalming process intended to keep them looking in death as they looked in life. Then they are stained, usually with a heavy metal such as osmium, to produce contrast. Finally they must be sliced into thin sections because out-of-focus layers of thick slices complicate the picture. Such harsh techniques for preparing specimens frequently provoke questions about whether a particular structure in a picture is characteristic of a living cell, or whether it is merely an artifact — an accidental byproduct of the preparation process. The problem of artifacts has haunted microscopists since Leeuwenhoek's time, and as instruments become more sophisticated and preparation techniques more rigorous the problem shows no sign of going away.

begin to form at telophase, new nucleoli emerge near specific sites on the unfolding chromosomes.

The nucleolus is rich in RNA, and electron microscopists have found that its periphery contains granular material similar in size and staining properties to ribosomes. It has been recently demonstrated that the nucleolus contains RNA identical to that extracted from ribosomes. In addition, it contains DNA that seems to represent multiple copies of the chromosomal gene that directs the synthesis of ribosomal RNA. It seems reasonable to suppose that the nucleolus is the site of ribosome synthesis.

There is a limit to what one can learn about cells by looking at them. Many of the most important activities of cells occur at the chemical level — a level that cannot be observed even with the electron microscope. The next few chapters discuss the structure and function of the essential molecules of life.

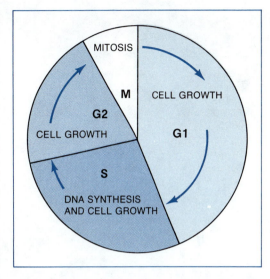

17

CELL CYCLE is divided into four stages. The circle graph indicates relative length of each stage. During G1 the cell grows but does not synthesize DNA. During S the amount of DNA in the cell is doubled. G2 is a period of further growth that leads to mitosis, M, and formation of daughter cells.

READINGS

C.J. AVERS, *Cell Biology*, New York, D. Van Nostrand Company, 1976. An excellent and up-to-date text on cell structure, biochemistry, and genetics. Plenty of experimental de-

D.W. FAWCETT, *The Cell, Its Organelles and Inclusions: An Atlas of Fine Structure*, Philadelphia, W.B. Saunders Company, 1966. A collection of electron micrographs of many types of cells, with commentary on the activites of the structures shown. A strikingly beautiful book by one of the masters of the art of electron microscopy.

T.L. LENTZ, *Cell Fine Structure: An Atlas of Drawings of Whole-Cell Structure*, Philadelphia, W.B. Saunders Company, 1971. A marvelous collection of ink drawings of cells tail shows not only what cytologists know, but how they learned it.
as they appear in the electron microscope. It is possible in a drawing to do something that is impossible in a photograph; namely, to include in one picture all of the structures found in a given type of cell, even those structures that are hardest to find or that only appear when special fixation or staining techniques are used. One of Lentz's pic-

tures is the equivalent of a whole portfolio of electron micrographs.

A.G. LOEWY AND P. SIEKEVITZ, *Cell Structure and Function*, 2nd Edition, New York, Holt, Rinehart, and Winston, 1970. The emphasis in this fine book is on function. The level is fairly advanced, but the writing is clear and concise.

A.B. NOVIKOFF AND E. HOLTZMAN, *Cells and Organelles*, New York, Holt, Rinehart, and Winston, 1970. The emphasis in this book is on the structure of cells, how the structure is studied, and, finally, how the functions of various structures are determined. Much information, economically presented.

L. ORCI AND A. PERRELET, *Freeze-Etch Histology: A Comparison Between Thin Sections and Freeze-Etch Replicas*, New York, Springer-Verlag, 1975. A beautiful introduction to a relatively new technique for showing the detailed structure of membrane surfaces.

C.P. SWANSON, *The Cell*, 3rd Edition, Englewood Cliffs, N.J., Prentice-Hall, 1969. A short, well-written introduction to cell structure and function.

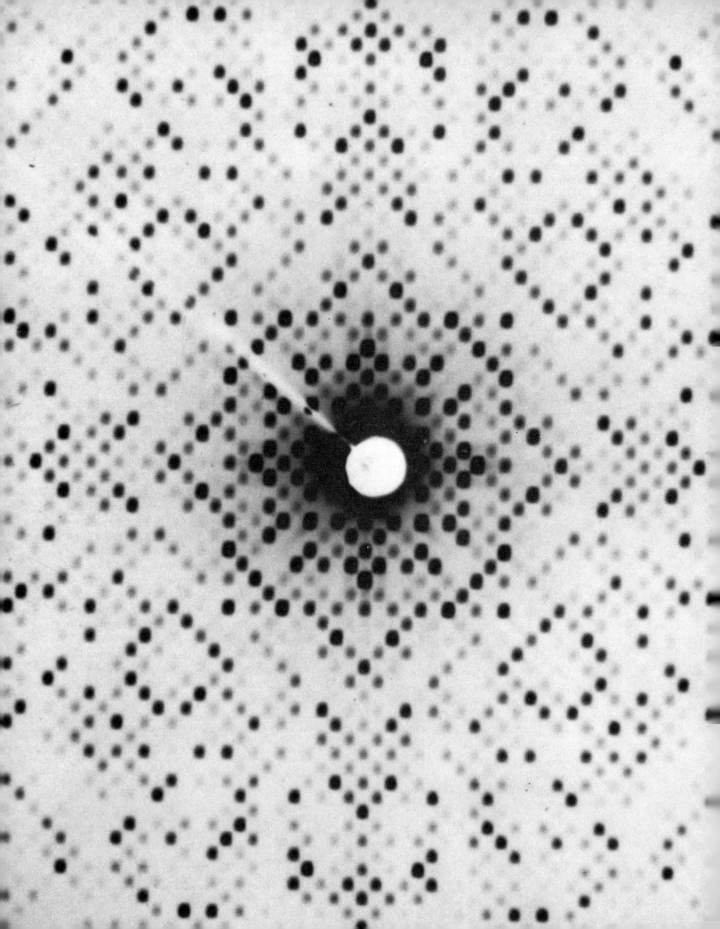

2

molecules

In my hunt for the secret of life, I started my research in histology. Unsatisfied by the information that cellular morphology could give me about life, I turned to physiology. Finding physiology too complex I took up pharmacology. Still finding the situation too complicated I turned to bacteriology. But bacteria were even too complex, so, I descended to the molecular level, studying chemistry and physical chemistry. After twenty years' work, I was led to conclude that to understand life we have to descend to the electronic level, and to the world of wave mechanics. But electrons are just electrons, and have no life at all. Evidently on the way I lost life; it had run out between my fingers.

ALBERT SZENT-GYÖRGYI,
PERSONAL REMINISCENCES

The notion that life might be explainable solely in terms of chemistry and physics troubles many biologists. One cannot study life at the level of atoms and molecules, they maintain, because in the act of dissecting out molecules and reactions for examination, one destroys the phenomenon he seeks to study. They and Szent-Györgyi are essentially correct. According to the definition of life in the first chapter, the smallest unit of life is the intact cell. When one dissects the cell he is no longer looking at a living system. The secret of life lies not only in the structure of biological molecules, but also in the way they are organized within the cell. British astrophysicist Fred Hoyle expressed this idea with insight and clarity in his science-fiction novel, *The Black Cloud:* "... the distinction between animate and inanimate is more a matter of verbal convenience than anything else. By and large, inanimate matter has a simple structure and comparatively simple properties. Animate or living matter on the other hand has a highly complicated structure and is capable of very involved behavior."

Biochemists view living organisms as chemical systems. Molecular biologists go one step further, emphasizing organization and spatial arrangement. Their two great triumphs have been the successful explanation of the machinery of heredity (at the molecular level) and of the action of enzymes. Both of these discoveries have depended on knowing how the chemical groups in a giant molecule are arranged in space. X-ray studies of crystalline samples have revealed the structures of DNA and proteins, and electron microscopy has disclosed the fine structures of many organelles of the cell. There is still a tantalizing no-man's land between the upper limit of resolution possible with x-ray crystallography and the lower limit of electron microscopy; this prevents investigators from seeing the molecular structure of a membrane or the arrangement of enzymes within a mitochondrion. But molecular biology and biochemistry have become thoroughly structural. The phrase, "the molecular architecture of life" is no longer a flight of rhetoric; the modern biologist means it literally.

THE MOLECULES OF LIVING ORGANISMS

The chemistry of life, at least on this planet, is the chemistry of water and carbon: water, because it is the most abundant molecule in the cell, comprising on the average about 70 percent of its weight (*Figure 1*); carbon, because about 95 percent of the dry weight of a cell consists

DIFFRACTION PATTERN on opposite page was produced by x-rays beamed through a crystal of lysozyme, an enzyme that causes the breakdown of the cell walls of bacteria. Computer analysis of diffraction patterns has made it possible to work out the detailed three dimensional structure of complicated enzyme molecules, such as the one depicted in Figure 18.

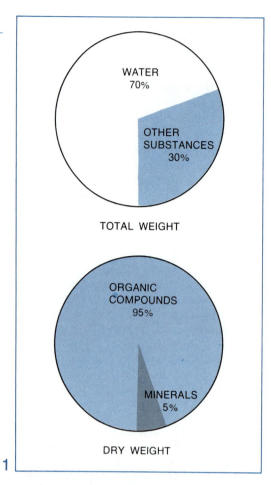

WATER
70%

OTHER
SUBSTANCES
30%

TOTAL WEIGHT

ORGANIC
COMPOUNDS
95%

MINERALS
5%

DRY WEIGHT

1

COMPOSITION OF CELLS. Living cells consist mainly of water (top graph). The dry weight of the cell (bottom) consists mainly of organic compounds.

romolecules, include PROTEINS, CARBOHYDRATES, LIPIDS, and NUCLEIC ACIDS. Some of the macromolecules are polymers — compounds formed by the linkup of a number of small molecules. Proteins, for example, are long-chain polymers of AMINO ACIDS. Proteins serve as structural material (hair, muscle, skin); as carriers of small molecules (the protein hemoglobin, for example, transports oxygen); and most important, as compounds that speed up chemical reactions in the cell (ENZYMES). Carbohydrates include simple sugars and their polymers. The simple sugar glucose is a small organic molecule, but its polymer cellulose is a long-chain compound that reinforces the rigid walls of plant cells. Starch, the principal energy storage compound in plants, is also a polymer of glucose subunits, but they are linked together in a slightly different way. Lipids are a mixed bag of compounds. Anything that can be extracted from plant or animal tissues by organic solvents (such as alcohol) is called a lipid. The phospholipids, an important type derived from fats, are widely found along with proteins in membranes. Other lipids, such as steroids and carotenoids, act as hormones — chemical signals that transmit information from one part of the organism to another — or as antenna pigments that detect outside stimuli such as light. Nucleic acids are enormous polymers that store the hereditary blueprints for the synthesis of proteins.

Many of the small organic molecules in the cell are monomers, the subunits of giant polymers. Some small molecules, particularly adenosine triphosphate (ATP), serve as storage units for chemical energy. Others carry electrons, small molecules, or chemical groups from place to place within the cell. Some organisms (notably plants and bacteria) can synthesize all of the organic compounds that they need. Most animals cannot, and require an outside source either for the final compound or for the raw material from which it can be synthesized. Vitamins, for example, are small organic molecules that must be obtained in the animal diet, and the list varies from one species to another. Ascorbic acid (vitamin C) is required by humans and other primates, but most other mammals can make their own.

of organic (carbon-based) molecules. Life began in the sea, and the properties of water shape the chemistry of all living organisms. Reactions in solution are much more rapid than reactions between solids, and complex molecules can behave in solution in ways that they cannot behave in a solid or a gas. Most reactions of the cell occur in watery surroundings.

If all the water is removed from a cell, the residue contains four main types of large organic molecule, about 100 different small organic molecules, and a few mineral salts. The large organic molecules, often called mac-

Carbon, nitrogen, oxygen, hydrogen, phosphorus and sulfur are the most abundant elements in large biological molecules.

At the bottom of the scale of chemical complexity are the ions (charged particles) and atoms that all living organisms need from outside sources. These include sodium, potassium, magnesium, calcium, manganese, iron, copper, zinc, molybdenum, phosphorus, chlorine, and iodine. Some elements, such as calcium, are required in quantity for bones and teeth in vertebrates, but most are required only in minute amounts. Metals such as iron, copper and zinc are essential to the function of enzymes, and are bound to the surface of the enzyme molecule. Other metals, notably sodium and potassium, are important in the process of nerve conduction.

Given sources of carbon, nitrogen, oxygen, hydrogen and other elements, plants can synthesize all the molecules they need. An organism that can synthesize its own proteins and carbohydrates is called an AUTOTROPH, from the Greek for self-feeder. Animals have lost much of the capability to manufacture their own food and must obtain the semi-finished raw materials in the food they eat. They are classed as HETEROTROPHS, mixed-feeders. It is more efficient and economical for an organism to drop the genetic and biochemical machinery for making a substance that can easily be obtained from the environment. The fats that a man eats are used partly for energy and partly to make lipids. Proteins in the diet are digested to amino acids, and some of these are relinked in a different order to make the proteins of the body. Carbohydrates are degraded to simple sugars like glucose, and stored again in the form of polymers. None of these large molecules is used by animals just as is — all are chopped up into their subunits and the subunits are then reassembled into complex molecules of the kinds that make up the animal.

SMALL MOLECULES

The most important elements in living systems are shown in Table I. Carbon, nitrogen, oxygen, hydrogen, phosphorus and sulfur are the commonest elements in large biological molecules. Sodium and potassium occur mainly as ions in solution. The other elements are needed in small amounts in special compounds.

If cells are the brick in the architecture of living organisms, then atoms and molecules are the clay from which the brick is fashioned, and they impose fundamental limits on the mechanisms of life. Atoms are built from two components: a positively charged nucleus and enough negatively charged electrons to make the entire atom electronically neutral. The nucleus itself is comprised of two kinds of particles: protons and neutrons, which have approximately the same mass. Protons have a charge of +1, electrons a charge of −1, and neutrons a charge of 0.

The number of protons in the nucleus of an atom is called its ATOMIC NUMBER, and the total number of protons and neutrons is, to a first approximation, its ATOMIC WEIGHT. The atomic number is important in determining the chemical properties of an atom, and all atoms with the same atomic number are classified as the same chemical element. All atoms of gold, for example, have exactly 79 protons in their nuclei. All carbon atoms have six protons, all nitrogen atoms have seven, and all oxygen atoms have eight.

Some atoms of a chemical element are

ELEMENTS FOUND IN LIVING SYSTEMS are listed below, together with their properties. Carbon, nitrogen, oxygen, hydrogen phosphorus and sulfur are main components of large biological molecules.

I

ATOMIC NUMBER	SYMBOL	NAME	ATOMIC WEIGHT	PROPERTIES OF PURE ELEMENT
1	H	HYDROGEN	1.01	LIGHT, COLORLESS GAS
6	C	CARBON	12.01	HARD SOLID (DIAMOND, GRAPHITE)
7	N	NITROGEN	14.01	COLORLESS GAS
8	O	OXYGEN	16.00	COLORLESS GAS
9	F	FLUORINE	19.00	PALE GREENISH GAS
11	Na	SODIUM	23.00	REACTIVE SILVER METAL
12	Mg	MAGNESIUM	24.31	LIGHT, SILVERY METAL
15	P	PHOSPHORUS	30.97	WHITE, RED, OR YELLOW NONMETAL
16	S	SULFUR	32.06	YELLOW SOLID
17	CL	CHLORINE	35.45	YELLOW-GREEN GAS
19	K	POTASSIUM	39.10	LIGHT, SILVER-WHITE METAL
20	Ca	CALCIUM	40.08	SOFT, SILVERY METAL
25	Mn	MANGANESE	54.94	HARD, BRITTLE METAL
26	Fe	IRON	55.85	SILVERY GREY METAL
29	Cu	COPPER	63.54	MALLEABLE REDDISH METAL
30	Zn	ZINC	65.37	BLUEISH-WHITE METAL
34	Se	SELENIUM	78.96	RED OR GREY SEMIMETAL
42	Mo	MOLYBDENUM	95.94	TOUGH, SILVERY METAL
53	I	IODINE	126.90	VIOLET CRYSTALLINE NONMETAL

heavier than others because of extra neutrons in the nucleus. These variant atoms are called ISOTOPES (*Figure 2*). The number of neutrons has little effect on the chemical properties of an element, but the effect on the physical properties is more pronounced. The rate of diffusion of a gas, for example, depends to some extent on its mass. More important, the presence or absence of a neutron or two makes some nuclei unstable, and they tend to break down by emitting energy in the form of radiation. Biologists use radioactive isotopes to label a chemical compound and radiation detectors to trace its pathway through an organism or through a complicated series of reactions.

One of the biochemist's favorite tracers is carbon 14. Of all the atoms of carbon found in nature, 98.9 percent have six protons and six neutrons in their nuclei. This isotope is called carbon 12 (written C^{12}). A much smaller number of carbon atoms, 1.1 percent, have seven neutrons instead of six, and a very few have eight neutrons. These isotopes are written C^{13} and C^{14}, respectively. Carbon 14 is unstable and breaks down spontaneously into another element, nitrogen 14, with the release of radioactivity. Its half life — the time required for half of any given quantity of the isotope to disappear — is 5,570 years. A small amount of C^{14} is present in all samples of natural carbon. In effect the slowly decaying C^{14} is a built-in clock that makes it possible to determine the age of ancient organic materials. It has proved to be a valuable tool in establishing the time scale of relatively recent evolutionary events.

CHEMICAL BONDS

Atoms of some elements, such as sodium, tend to lose an electron and become positively charged IONS. Other atoms, such as chlorine, tend to gain an extra electron and become negative ions. Positive and negative ions can join to form IONIC BONDS. Common table salt, sodium chloride, is written Na^+Cl^- because each sodium atom has lost an electron to become a positive ion and each chlorine atom has gained an electron to become a negative ion (*Figure 3*). When sodium chloride dissolves in water the ions are still present and the salt solution will conduct electricity well because of the presence of charged ions. If the water evaporates and the salt crystallizes, the Na^+ and Cl^- ions settle down into a regular crystalline lattice (*Figure 4*). No one sodium ion is associated with a particular chloride ion, but there are equal numbers of both, so the crystal as a whole is electrically neutral.

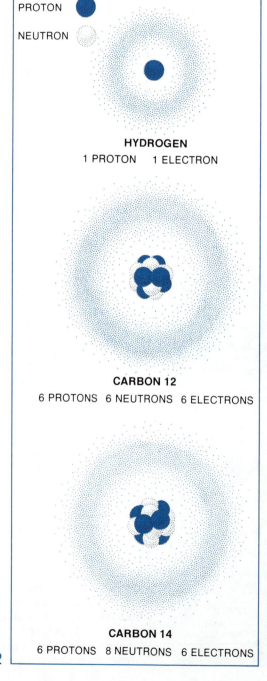

PROTON

NEUTRON

HYDROGEN

1 PROTON   1 ELECTRON

CARBON 12

6 PROTONS   6 NEUTRONS   6 ELECTRONS

CARBON 14

6 PROTONS   8 NEUTRONS   6 ELECTRONS

STRUCTURE OF ATOMS. Electrons are represented as clouds around the central nuclei. The size of the nucleus is exaggerated in this diagram. The electron cloud in a typical atom is 100,000 times the diameter of the nucleus.

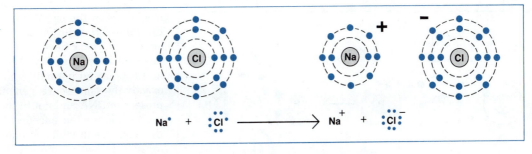

3

IONIC BONDS in sodium chloride. Sodium atom loses an electron to become a cation (positive ion) and chlorine atom gains an electron to become an anion (negative ion). Chemical shorthand notation at bottom shows electrons around chloride ion (right) grouped into four pairs.

In an ionic bond, one atom loses an electron to another, and electrostatic forces between the positive ion (the electron donor) and the negative ion (the acceptor) makes the ions attract one another. This attraction is ordinarily expressed in terms of the number of calories of energy required to break the bond.

Elements such as carbon, nitrogen, oxygen and phosphorus tend to share electrons rather than donating or accepting them. These shared electrons form COVALENT BONDS, which differ from ionic bonds in several important ways. They connect specific individual atoms, unlike the forces between ions in a crystal of table salt or between Na^+ and Cl^- ions in a salt solution. This assemblage of connected atoms is called a MOLECULE (Figure 5), and the sum of the weights of the atoms that comprise the molecules is called its MOLECULAR WEIGHT. The atoms in a molecule are held in definite positions, which can be predicted from quantum theory. The orientation of the covalent bonds determines the shape of the molecule, and the shape in turn determines many of its chemical properties — a fact that is highly relevant to living systems.

A chemist cannot count or weigh individual atoms and molecules, but he needs some way of being sure that he can measure out equal numbers of atoms (or molecules) of two different substances. The task is greatly simplified by the use of a unit called the MOLE. A mole of any substance is an amount whose weight in grams is numerically equal to its atomic or molecular weight. Now this is the handy part:

no matter what the substance — water, carbon, or even uranium — a mole of it contains the same number of molecules (or atoms) as a mole of anything else. For example the atomic weight of C^{12} is 12, and the molecular weight of water, H_2O, is the sum of the weights of its component atoms: $1 + 1 + 16 = 18$. One mole of carbon (diamond or graphite) is 12 grams; it contains the same number of molecules as one mole of water, which is 18 grams. The mole is one of the most useful concepts in practical chemistry.

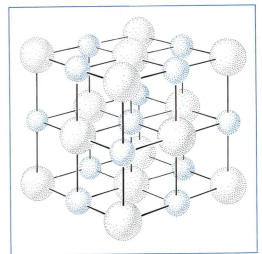

4

CRYSTAL STRUCTURE of sodium chloride. Sodium ions (color) and chloride ions (grey) are packed into a cubic lattice in crystals of rock salt. Here atoms have been reduced in size for clarity; actually they touch one another.

The DNA molecule is the master control of the cell, but the only processes that it controls directly are its own replication and the synthesis of proteins.

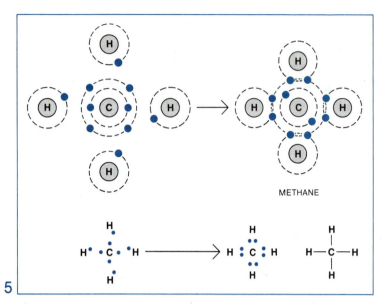

METHANE

5

METHANE MOLECULE is held together by covalent bonds. The carbon atom needs four electrons to complete its outer shell, and the hydrogen atom needs one more electron. In forming a molecule of methane the atoms gain the missing electrons by forming four electron pairs (top). Shorthand notation below shows the same reaction. Covalent bond formed by electron pair is usually represented as a straight line between atoms (lower right).

Methane, CH_4, is a molecular compound in a sense that sodium chloride, NaCl, is not. The carbon and four hydrogen atoms in methane are tightly bound into a molecule that behaves in most cases as a unit. The covalent bonds that can form between carbon, nitrogen, oxygen and hydrogen are about as strong as ionic bonds. A third type of bond, of particular importance in living systems, is the hydrogen bond (Figure 6). Although such linkages are weaker than ionic and covalent bonds, they are important in the cell because they are so common. Hydrogen bonds are largely responsible for maintaining the structure of most giant molecules, particularly proteins and nucleic acids.

GIANT MOLECULES

By far the largest of the four types of giant molecules found in living systems is DNA (deoxyribonucleic acid). Instructions for making a new organism lie in the DNA donated by the parents to the fertilized egg. The chemical

6

HYDROGEN BOND is a weak ionic bond. If hydrogen is bonded to another atom, such as oxygen, with a great affinity for electrons, the electron pair of the bond will shift closer to the oxygen, leaving the hydrogen with a slight positive charge and the oxygen with a negative charge (top). The hydrogen atom will be attracted to another atom having a slight negative charge, such as the oxygen of another water molecule or a carbonyl oxygen in a protein chain (bottom).

machinery of the egg is necessary to express this information, just as the electronic machinery of a playback deck is necessary to express music recorded on a tape. Clearly the music is on the tape and not built into the recorder. In exactly the same sense, genetic information is present in DNA and not in the cell around it.

The information encoded in DNA governs the sequence in which amino acids are strung together to make a protein chain. The way in which this chain folds after it has been put together, and the exact positions of important chemical groups at the active site of an enzyme, are determined by the sequence in which amino acid side groups occur along the chain. There are no three-dimensional templates that guide the folding of a protein molecule or shape the geometry of an enzyme surface. Once synthesized, the protein molecule assumes its intricate three-dimensional shape automatically, as a result of hydrogen bonding and the interactions among neighboring amino acids.

The DNA molecule is the master control of the cell, but the only processes that it controls directly are its own replication and the synthesis of proteins, including enzymes. Enzymes then direct the thousands of reactions that occur every second in the cell. Enzymes are biological CATALYSTS: compounds that speed up chemical reactions without being permanently changed in the process. Their power to control cell chemistry is a result of their astonishing efficiency. The proper enzyme can speed up a reaction by as much as 10^8 times — the difference between a reaction time of one second versus 38 months. Each enzyme catalyzes only one particular type of reaction, and if the enzyme is absent, the reaction proceeds so slowly that for all practical purposes it does not occur at all.

DNA does not contain the blueprint for making other giant molecules such as lipids and carbohydrates. Instead, enzymes made from instructions in the DNA catalyze a series of reactions that fashion these macromolecules from an assortment of smaller ones. Unlike DNA and proteins, the lipids and carbohydrates from various organisms are more or less the same; these giant molecules are merely passive raw materials that do not carry information.

LIPIDS AND CARBOHYDRATES

The word lipid is really a functional term that denotes any greasy substance that can be

The only current way in which man can make nutritional use of cellulose, although admittedly not an unpleasant one, is to feed grass to cows or other ruminants, and then to eat the beefsteak. Direct degradation with the enzyme cellulase, which is made by bacteria in the cow's stomach, is successful in the laboratory but not commercially. Cows are still cheaper than chemists.

One biochemist, while looking for a better tire cord material at the American Viscose Corporation, did discover how to make a form of cellulose which was at least edible although nutritionally worthless. O. A. Battista sheared cellulose molecules mechanically in a Waring blender and produced a low-molecular-weight cellulose product that could be used in diet foods in place of flour. It was not absorbed in the digestive tract and had no taste, no calories, no vitamins, no nutritional or food value whatever. Cookies and cakes made with the material would stave off hunger pangs for a few hours, fooling the stomach into equating "full" with "fed." It was hailed by Life magazine in 1961 as the answer to the average American's tendency to overeat, but sank into rapid oblivion. At the time there seemed to be a consensus that our food-processing industries produce enough nutritionally worthless foods without even trying.

More recently, however, "edible" cellulose has found a use in diet foods and as a substitute for milk and ice cream in milk shakes sold in cheap drive-in restaurants. The inventor received a promotion and a pay raise, but as is often the case in industrial laboratories, had signed his patent rights away in advance for the token sum of one dollar, making him ineligible to share in any financial bonanza from the use of his product in diet foods. As Life concluded, it was somehow fitting that the inventor of nonfood be paid in nonmoney.

extracted from animal or vegetable tissues by organic solvents such as alcohol, ether, chloroform and benzene. Lipids are a mixed bag of compounds with a wide variety of uses: energy storage, digestion, chemical signaling, membrane structure, bone formation, and photosensitive pigments for growth and vision. The cell membrane appears to be a double layer of phospholipids sandwiched between two layers of protein molecules. β-carotene, a pigment that serves as a light-gatherer in photosynthesis, belongs to a family of lipids that also includes vitamin A, which is crucial to the chemistry of vision in humans. Cholesterol, a lipid synthesized in the liver, is the raw material for hormones such as testosterone, estrogens, and cortisone. Fats and fatty acids are the chief molecules for long-term energy storage in animals. Containing twice as much chemical energy per gram as carbohydrates or proteins, they are literally the gasolines of the body. This efficient energy storage system is especially appropriate in mobile animals, where weight is at a premium. Stationary plants store energy primarily in carbohydrates, particularly starches. Most lipids can be synthesized in the animal body; the few that cannot be synthesized, such as the fat-soluble vitamins A, D, E and K, must be obtained in the diet.

Carbohydrates are a class of organic compounds whose name is derived from the old and erroneous idea that they were compounds of only carbon and water (actually hydrogen is also an important component). The most important kinds of carbohydrates are sugars, starches, glycogens and celluloses. Some carbohydrates store energy, while others are important structural materials, particularly in plants. SUGARS such as glucose are efficient energy storage molecules, although only half as good as fats on a weight basis. STARCHES and GLYCOGENS are long-chain polymers of glucose, and are the media for carbohydrate

Cellulose is the standard building material for woody stalks, fibers, and all types of cell walls in plants. It is by far the most common organic compound on Earth.

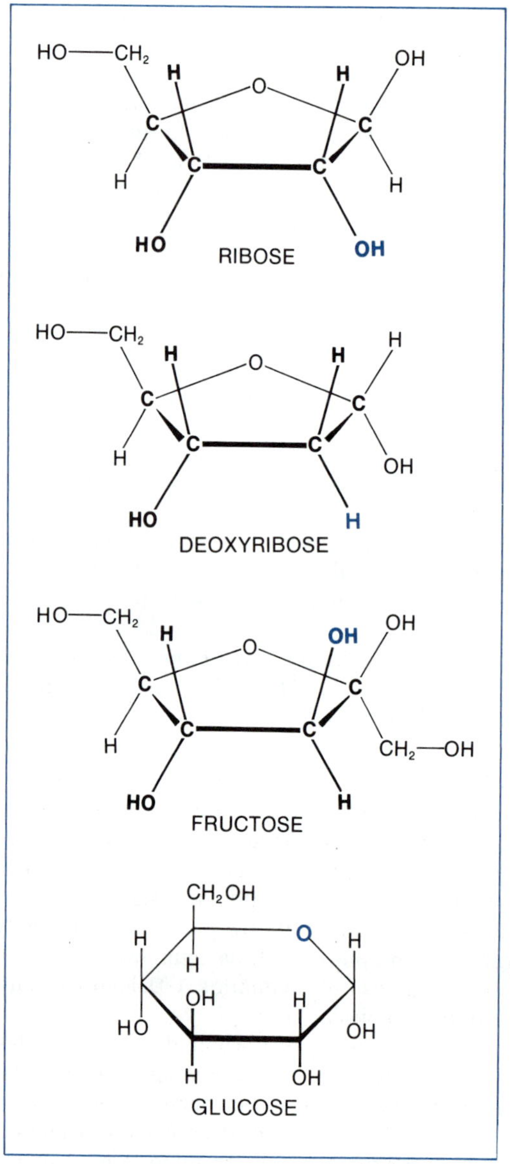

The structures shown in Figure 7 are labeled: RIBOSE, DEOXYRIBOSE, FRUCTOSE, GLUCOSE.

7

SIMPLE SUGARS include the pentoses ribose, deoxyribose and fructose, which have five-sided rings, and glucose, which has a six-sided ring.

The linkage between sugar units in cellulose is a simple evolutionary trick to defeat the enzymes that digest starch. Very few organisms can digest cellulose. But for cows and termites, whose guts harbor microorganisms that can synthesize the enzyme cellulase, even cellulose is a good meal (*Box A*).

Sugars are classified according to the number of carbon atoms they contain. Five-carbon sugars are called pentoses. Sugars with more carbons are hexoses and heptoses. To the biologist, perhaps the most important are two pentoses, RIBOSE and DEOXYRIBOSE, and two different hexoses, FRUCTOSE and GLUCOSE. In biological systems these sugars exist as closed rings of atoms. Ribose has the five-membered ring structure shown in Figure 7, which is closed by an oxygen atom. Ribose is particularly interesting to biologists because it is incorporated into RIBONUCLEIC ACID or RNA. Deoxyribose (*Figure 7*) lacks the oxygen of the hydroxyl group at C_2. The similar polymer of deoxyribose is DEOXYRIBONUCLEIC ACID (DNA).

Fructose is a six-carbon hexose, although it has the five-membered ring of the pentoses (*Figure 7*). Glucose is a hexose with the same chemical composition as fructose, $C_6 H_{12} O_6$, but with a six-membered ring.

The sugars depicted so far are monosaccharides: sugars with one ring. Disaccharides (*Figure 8*) are common in nature, especially sucrose, which is found in cane,

energy storage in plants and animals, respectively. CELLULOSE is a slightly different polymer of glucose, which performs the same kind of supporting and structural role in plants that the calcium skeleton does in vertebrates. Chitin, a carbohydrate derivative, accomplishes the same purpose in insect exoskeletons, and even the cell walls of bacteria have carbohydrate components. Organisms on this planet seem to have fashioned their bodies from edibles, so that one creature's architecture is another's dinner.

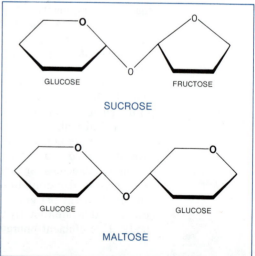

SUCROSE

MALTOSE

8

DISACCHARIDES consist of two simple sugars linked together. Sucrose is made up of glucose and fructose units; maltose, of two glucose units. Maltose linkage also occurs in starch. This diagram shows only the basic ring structure of the sugars.

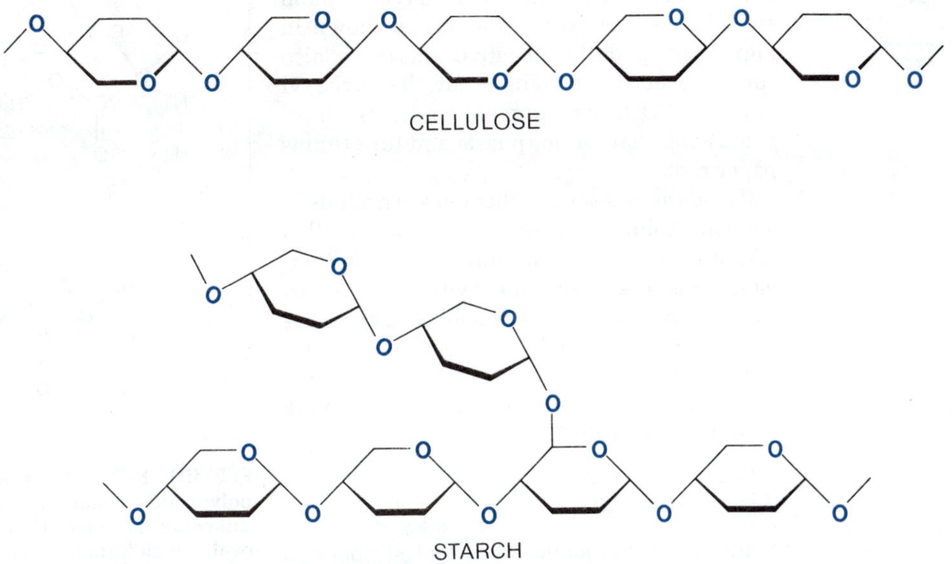

CELLULOSE AND STARCH are long-chain polysaccharides. Cellulose is a straight-chain polymer of glucose units bonded together as shown at top. Starch is a highly branched polymer of glucose. Very few organisms have enzymes that can break the links in cellulose.

beets and a great many other plants. Maltose is the disaccharide breakdown product of starch, and is encountered in the brewing process.

Maltose consists of two joined glucose rings. Starch is a very long chain of maltose units (*Figure 9*). Sugars can also be linked to form very long polymers called polysaccharides. Starch and cellulose are both polymers of glucose, but they are very different compounds. Amylases, the enzymes that make and break the bonds of starch, cannot touch those of cellulose. Plants use the same basic pieces to build both their food and its container, but there is no chance for confusion.

Cellulose chains are long, straight, and unbranched, with several thousand glucose molecules per chain. The purest source is cotton, which is 90 percent pure cellulose. Cellulose is the standard building material for woody stalks, fibers, and all types of cell walls in plants. It is by far the most common organic compound on the planet, containing over half the carbon involved in plant life.

Starch chains are branched like a thicket, with tens of thousands of glucose monomer units per molecule. A similar branched-chain glucose polymer, GLYCOGEN, is sometimes called animal starch. Glycogen is used for energy storage in liver and muscle. Animals store a much smaller amount of energy in glycogen than in fats, but the glycogen reserves are used to stabilize the level of sugar in the blood, possibly because the energy in glycogen can be extracted more rapidly than the energy in fats.

AMINO ACIDS AND PROTEINS

Proteins are long-chain polymers of amino acids. At the simplest level, the nature of these giant molecules can be explained in terms of acid-base chemistry. An acid is a compound that ionizes in water, releasing one or more protons (hydrogen ions). Figure 10 depicts the ionization of acetic acid, a typical organic acid with a carboxyl group (—COOH) at one end. Because the proton has a positive charge, the resulting acetate ion has a negative charge. The double bond that formerly existed be-

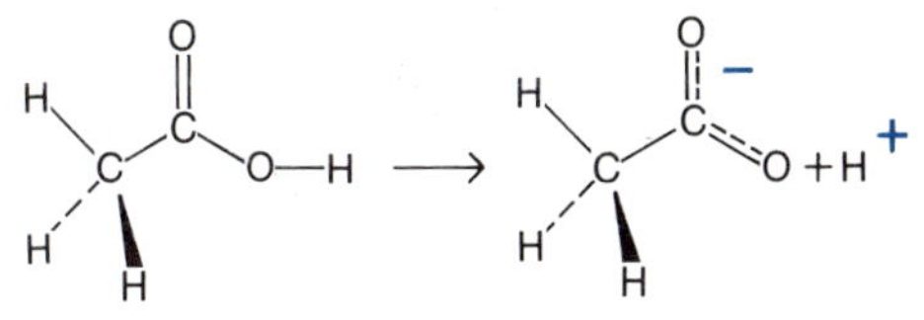

IONIZATION OF ACETIC ACID. When a carboxylic acid ionizes and loses a hydrogen ion, the resulting negative charge is spread over both oxygen atoms of the carboxylate ion.

tween the carbon and one oxygen atom spreads over both oxygen atoms as shown in Figure 10, and the negative charge is also spread over the remainder of the carboxyl group, —COO. Most organic acids, like inorganic acids, have a sharp taste and turn litmus paper red.

If acids are substances that release protons in aqueous solution, BASES are substances that take up protons from solution. Common inorganic bases are sodium hydroxide, NaOH (commonly known as lye), and potassium hydroxide, KOH. These bases dissociate in solution and the resulting hydroxyl ions combine with hydrogen ions already present, removing them from the solution.

$$NaOH \rightarrow Na^+ + OH^-$$
$$OH^- + H^+ \rightarrow H_2O$$

Solutions of inorganic bases feel slippery to the touch because the hydroxide ion makes soap from the fats in the skin of the fingertips. Methylamine, CH_3NH_2, is shown in Figure 11. Amines act as bases by combining with protons (hydrogen ions). The positively charged proton is attracted to the pair of electrons on the nitrogen atom, and the resulting covalent bond is identical to the two original N—H bonds. The entire $—NH_3^+$ group as a whole possesses a positive charge.

The AMINO ACIDS are an important class of compounds that have both a carboxylic acid group and an amine group attached to the same carbon atom, called the α-carbon. Also attached to the α-carbon are a hydrogen atom

PEPTIDE BOND links amino acids to form a polypeptide chain. At top, two amino acids are linked by the removal of the atoms in the colored oval, which form a molecule of water. R represents any side group. Below, the resulting peptide or amide bond.

and a side chain that gives the amino acid its distinctive chemical properties. Twenty amino acids are the building blocks of the giant protein molecules of living organisms. In aqueous solutions of amino acids both the amine and carboxyl groups are ionized, as shown in Figure 11. The carboxyl group has lost a proton and the amine group has gained one, forming a double-ended ion. In the synthesis of proteins the carboxyl group from one amino acid is linked to the amine of the next, splitting out water and forming a peptide bond (*Figure 12*).

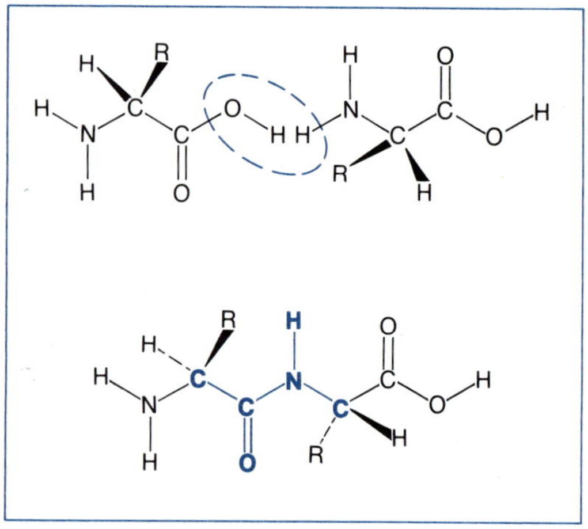

THREE-DIMENSIONAL STRUCTURE

As noted earlier, the information coded in DNA specifies only the order of amino acids in the protein chain. The side chains of the 20 different amino acids (*the R's in Figure 12*) show a wide variety of chemical properties. The order of amino acids in a protein determines how the protein molecule will fold. Perhaps the most straightforward classification of amino acids is based on whether they are usually found on the inside of a folded protein molecule, on the surface, or both. This depends mainly on the affinity of the various amino acids for water.

The structures of the 20 amino acids are shown in Figure 13. Valine, leucine, isoleucine, phenylalanine and methionine are hydrophobic; that is, they tend to move away from water. Because most enzymes and similar proteins operate in aqueous solution, the parts

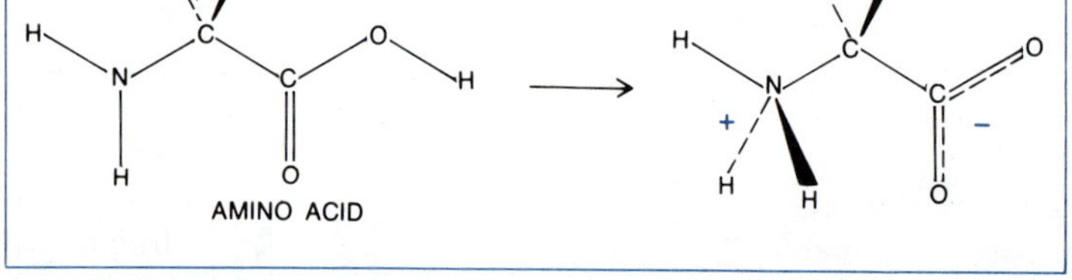

AMINES AND AMINO ACIDS. The nitrogen atom in methylamine has a lone pair of electrons that can attract a proton from solution to form a positive ion. Amino acids have both carboxyl and amine groups attached to same carbon atom. Side groups (R) give each amino acid its individual properties. In aqueous solution, both ends of amino acid are ionized.

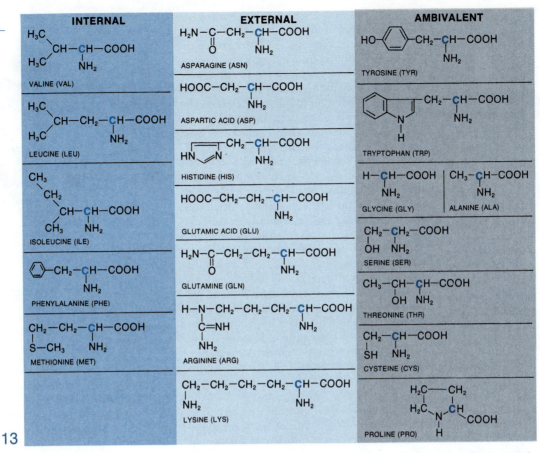

INTERNAL

VALINE (VAL)

LEUCINE (LEU)

ISOLEUCINE (ILE)

PHENYLALANINE (PHE)

METHIONINE (MET)

EXTERNAL

ASPARAGINE (ASN)

ASPARTIC ACID (ASP)

HISTIDINE (HIS)

GLUTAMIC ACID (GLU)

GLUTAMINE (GLN)

ARGININE (ARG)

LYSINE (LYS)

AMBIVALENT

TYROSINE (TYR)

TRYPTOPHAN (TRP)

GLYCINE (GLY)   ALANINE (ALA)

SERINE (SER)

THREONINE (THR)

CYSTEINE (CYS)

PROLINE (PRO)

13

20 AMINO ACIDS are grouped according to their usual location in protein molecules. Internal amino acids usually lie in the interior of proteins because their large and hydrophobic side groups are most stable when distant from the aqueous environment. External amino acids, which have charged or polar side groups, almost always lie on the outside of a protein, in contact with the aqueous environment. Ambivalent amino acids are found in both locations. Their side chains are either hydrophobic but small, or polar but uncharged (glycine has no side chain at all). The carbon atom to which the amino group is attached is printed in color.

of the chain containing the hydrophobic amino acids tend to be folded into the interior of the molecule.

Six amino acids are strongly hydrophilic; they tend to move toward water. Asparagine, aspartic acid, glutamine, glutamic acid, lysine and arginine tend to become oriented toward the aqueous environment at the surface of a protein molecule. As their names indicate, two of these compounds are acidic. Lysine and arginine are basic, while asparagine and glutamine are neutral.

The other amino acids apparently can occur either on the surface or in the interior of the protein molecule. Their side chains are uncharged. Some of them — serine, threonine, tyrosine and tryptophan — form hydrogen

bonds whenever they lie in the interior of a protein. Two cysteine side chains can be oxidized so that their sulfur atoms are joined by a covalent bond in a disulfide bridge. Hydrogen bonds and disulfide bridges are important in determining how a protein chain folds. The glycine side chain is important because of its absence (in this case, R is just a hydrogen atom). Glycines often appear in tight corners in the interior. Alanine is technically a hydrocarbon, but its small size tends to minimize its water-repellent properties, and it is frequently found on the surface of protein molecules.

Most of the chemical and catalytic properties of the protein molecule arise from its elaborate structure. One of the standard forms that proteins assume is the helix shown in

Figure 14. The helix is shaped by hydrogen bonds that extend from one turn of the chain to the turns above and below. The bonds serve to make a rigid cylinder from a floppy chain. The α-helix occurs in a class of fibrous structural proteins called keratins, which includes most of the protective tissues found in animals, such as nails and claws, skin, hair, and wool. Contractile muscle protein is also made of two protein components, actin and myosin, of which at least the myosin is arranged in an α-helix. Hair can be stretched because this involves only the breaking of hydrogen bonds in an α-helix, and the bonds are remade when the tension is released and the helix reforms.

Silk fibers are based on a second basic structure, the pleated sheet. Here the protein chains are almost completely extended and are bound into sheets by hydrogen bonds from one chain to a neighbor. Weaker forces tend to hold the sheets in three-dimensional stacks. Silk is supple for the same reason that a ream of typing paper held at both ends is flexible — the sheets can slip over one another in bending. But silk is not as stretchable as wool, because appreciable stretching requires that protein chains be snapped.

The collagen that is found in cartilage, tendons, the underlayers of the skin, and the cornea of the eye, is a protein based on yet

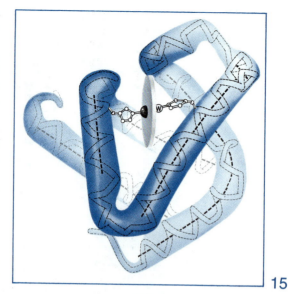

15

MYOGLOBIN is a globular protein. Tubular elements are α-helices. Disk at center is a heme group seen edge-on.

another structural pattern — a twisted three-stranded polypeptide chain. The three chains are twisted around one another like the strands of a cable. Hydrogen bonds run from each chain to the other two, producing a structure that is strong, rigid and unstretchable.

Some of the most important proteins are globular. Their polypeptide chains are folded back and forth on themselves like a ball of spaghetti to build globular molecules with typical diameters of 25 to 100 Å or more. Most of the chains in globular proteins are 80 to 400 amino acids long. Some globular proteins are much larger, but they are merely aggregates of smaller subunits. Apparently it is inefficient for a biological system to code its DNA for enormous protein chains when aggregates of shorter chains will serve as well.

Myoglobin is a typical example of a globular protein (*Figure 15*). Its function is to store O₂ in the tissue until needed. It has 153 amino acids in one chain, with no disulfide bridges, and is somewhat unusual in consisting almost entirely of α-helices. Its eight helices make a pocket that encloses a heme group — an iron-containing organic ring structure. In myoglobin hydrophobic side chains on the inner sides of the helices help to ensure that the helices fold against one another correctly as the molecule is formed. Myoglobin illustrates the way that the amino acid sequence specified by

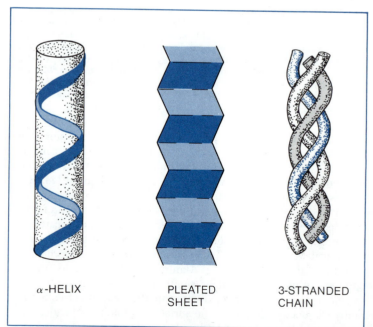

α-HELIX   PLEATED   3-STRANDED
            SHEET   CHAIN

14

POLYPEPTIDE CHAINS of many proteins fold themselves into one of the three forms shown here. Shape of chain is a result of sequence of amino acids along the chain, and hydrogen bonding between amino acids.

DNA determines the three-dimensional folding of a protein.

Hemoglobin, a protein that brings oxygen from the lungs to the tissues and delivers it to myoglobin for storage, illustrates another aspect of the relation between protein structure and function (*Figure 16*). As hemoglobin takes up or releases oxygen its four chains slightly shift their relative positions. Ionic bonds are broken, exposing buried side chains that enhance the binding of molecular oxygen. Each subunit of hemoglobin is folded exactly like a myoglobin molecule because they are both evolutionary decendants of a common oxygen-binding ancestor protein. But on the surfaces where subunits come in contact — regions which on the myoglobin molecule are exposed to the aqueous surroundings — hemoglobin has hydrophobic side chains where myoglobin has hydrophilic ones. Again, the chemical nature of the side chains, as coded by DNA, helps in deciding how the molecule will fold and pack in three dimensions.

Globular proteins are generally either carriers or enzymes. Myoglobin and hemoglobin are carriers (or storehouses) of oxygen. The cytochromes carry energy. Other proteins transfer small molecules or chemical groups from one reaction to another or from one part of the cell to another. The gamma globulins are antibodies — two-headed globular proteins whose function is to bind foreign proteins (such as the coat of a virus) into an insoluble clump, out of harm's way. The enzymes control the rates and pathways of all the reactions of an organism (*Chapter 3*).

THE ENZYME MOLECULE

Until about a decade ago biochemists knew very little about the behavior of enzymes at the molecular level. The concept of surface catalysis was familiar from industrial chemistry. Substrates (the molecules acted upon by the catalyst) bind to a solid surface before reacting, and dissociate from it afterward. Chemists generally agreed that the substrates of enzymes, too, were momentarily bound to an active site on the surface of the enzyme molecule.

Enzymes are far more specific in their action than inorganic catalysts like platinum. The remarkable ability of the enzyme to select exactly the right substrate was explained by the assumption that the binding of the substrate to the site depended on a precise interlocking of molecular shapes. In 1894 the great German biochemist Emil Fischer compared the fit between enzyme and substrate to that of a lock and key. Fischer's model persisted for more than half a century with only indirect evidence to support it (*Box B*).

The first direct evidence came in 1965, when David Phillips and his colleagues at The Royal Institution in London succeeded in crystallizing the enzyme lysozyme and, using the technique of x-ray crystallography, determining its structure. Since then the structures of more than two dozen other enzymes have been solved by x-ray diffraction studies.

This work has revealed a great deal about how enzyme molecules are designed, how they work, and to some extent how they are controlled. Enzymes turn out to be globular proteins with molecular weights ranging from 10,000 to several million. Those toward the smaller end of the scale consist of a single folded polypeptide chain; the largest contain several identical chains. Many of the active sites contain metal ions that enhance the reaction, particularly by helping to bind the substrate or to withdraw electrons.

Although it is probably an oversimplification to say that to understand one enzyme molecule is to understand them all, the x-ray studies suggest that most enzymes behave in very similar ways. Each enzyme attacks a bond having very specific characteristics. The explanation for this fastidious specificity, which underlies the entire chemical strategy of living organisms, is that the structure of the active site is molded to fit the substrate molecule. The binding of the substrate depends on the same

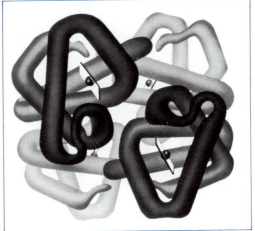

16

HEMOGLOBIN MOLECULE consists of four polypeptide chains, each similar in length and folded like the myoglobin molecule in previous figure.

Long after the discovery of enzymes biochemists refused to accept the idea that such remarkably efficient catalysts could be made of unglamorous materials akin to egg albumen and hair. In 1900 two chemists, Pekelharing and Ringer, crystallized a protein from digestive juices which they claimed was the enzyme pepsin. Seen in retrospect, it probably was. But they and others were unable to repeat the experiment, and their conclusions were rejected. It was then said that enzymes were "neither fats, carbohydrates, nor proteins, but an entirely new and unknown class of compounds." The great biochemist Richard Willstätter of the University of Munich purified many enzymes, including peroxidase, saccharase, lipases and amylase, and shaped the field of protein chemistry in the first three decades of this century. He maintained that enzymes were unknown catalytic substances associated with colloidal protein. (If one substitutes "active sites" for "enzymes," and deletes the word "colloidal," it turns out that he was not too far wrong.) Evidence for this was the observation that catalytic activity persisted even when purification had progressed so far that none of the usual tests for proteins would give a positive reaction. It never occurred to biochemists of the time that these protein enzymes could be so active that the catalytic tests were far more sensitive than the tests for protein.

James B. Sumner, a young chemist at Cornell University, isolated the enzyme urease from jack beans in 1926, and crystallized what he claimed was the pure protein enzyme. The response from Willstätter and his school was immediate and derisory. In a confrontation at a seminar at Cornell, Willstätter sarcastically denied the validity of all of Sumner's work. He maintained that Sumner had only crystallized the "carrier protein," and had let the enzyme slip through his fingers. But Sumner did not wilt under fire. He and others could repeat his work, and over the next few years he demonstrated that the repeatedly recrystallized protein lost none of its enzymatic activity.

When John Northrop, then at the Rockefeller University, crystallized the enzyme pepsin as a protein in 1930, resistance to the new idea began to crumble, although even four years later, diehards were refusing to equate enzyme and catalytic protein. The late 1930's saw the crystallization of many other enzymes — all proteins. This led to the first hesitant steps in what was to be an enormously powerful tool, x-ray analysis of protein crystal structure. Belated recognition of Sumner's work came with the award to him and Northrop in 1946 of the Nobel Prize for Chemistry. Sumner died in 1955, having had the satisfaction of seeing the heresy for which he was vilified in 1927 become the cornerstone of enzymology.

forces that maintain the folded structure of the enzyme protein — hydrogen bonds, the electrostatic attraction and repulsion of charged chemical groups, and the interaction of hydrophobic (water-repelling) groups.

PORTRAIT OF AN ENZYME

One enzyme whose mechanism has been studied extensively is carboxypeptidase. It belongs to a family of digestive enzymes that are synthesized in the pancreas for secretion into the intestine, where they attack the polypeptide chains of proteins in food. The chains are broken into amino acids that can be absorbed through the intestinal walls. Digestive enzymes HYDROLYZE a polypeptide bond; that is, they break the chain by adding a water molecule across it, as shown in Figure 17. Although the main function of these enzymes is to digest proteins, they will also attack other substrates having the same kind of bonds (*Box C*). Carboxypeptidase, as its name implies, severs one residue at a time from the carboxyl (COOH) end of the chain.

Carboxypeptidase has a molecular weight of 34,600, a fairly typical size for single-chain enzymes. Its chain is a polymer of 307 amino acids including approximately 2500 atoms of C, N, O, and S; nearly 2000 hydrogen atoms; and one metal atom: zinc. The folding of the

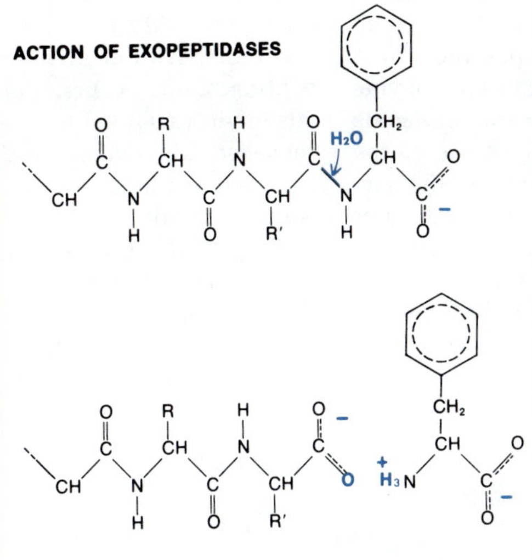

ACTION OF EXOPEPTIDASES

PEPTIDE CHAIN IS BROKEN by digestive enzymes. The process is called hydrolysis (splitting with water) because a water molecule is added across the severed ends of a peptide bond (color). Carboxypeptidase, the enzyme depicted in Figure 18, is especially effective in removing amino acids with bulky side groups from the end of the chain, as shown in the above illustration.

chain is represented schematically in Figure 18, with each of the 307 amino acids represented only by its α-carbon atom, numbered in sequence from the amino end of the chain. The peptide group (—CO—NH—) that connects α-carbons is shown as a straight line. Only the most important side chains are drawn. Another drawing of the molecule is shown in Figure 19, with only the catalytically important features represented.

The functional heart of the enzyme is its active site, a bowl-shaped depression visible at the upper right of Figure 18, and emphasized in the schematic Figure 19. At the bottom of this depression sits the essential zinc atom. Zinc, like carbon, commonly has four bonds. One edge of the active site is folded over to make a pocket (residues 243–251); the interior of the pocket is lined with hydrophobic (water-repelling) side chains. The opposite rim of the active site opens into a groove running down the side of the molecule (between residues 279–282 and 12–16 on the left, and 125–117 on the right in Figure 16). The polypeptide chain of the substrate aligns itself in this groove, with its carboxyl end in the active site.

The arrangement of electrons in the atoms comprising the active site clamps the substrate

---

C

ENZYMOLOGY IN THE SERVICE OF MANKIND

Subtilisin is a protein-digesting enzyme from Bacillus subtilis. It is remarkably heat-stable, remaining active in hot water when most other enzymes would be destroyed. This is why subtilisin achieved such popularity in recent years as a presoak laundry agent or as a component in washday detergents. Subtilisin will digest protein stains such as chocolate and blood that have traditionally been hard to remove with soap or detergents. It will also digest silk, wool, or human epidermis if given a chance.

This is not the first time that enzymes have been used in a product, as opposed to being used to prepare it, as in brewing or baking. The carbohydrate chemist H. S. Paine achieved immortality of a sort in 1924 for a patent that made him the Father of Liquid Center Chocolates. Have you ever wondered, as you ate a liquid-center cherry candy, how the manufacturers managed to wrap the chocolate coating around the center while it is fluid? The answer, thanks to H. S. Paine, is that they don't have to — the center is solid when the candy is made. Paine discovered the trick of mixing a minute quantity of the enzyme invertase with the sugar of the fondant around the cherry. After the candy is chocolate-coated and packaged, the enzyme slowly hydrolyzes the disaccharide sucrose and converts it to the monosaccharides glucose and fructose. The monosaccharides are hygroscopic — they slowly absorb moisture through the chocolate coating until finally they dissolve and the center liquefies. Invertase is so efficient an enzyme that the amount needed to "invert" the sucrose is too small to be tasted or otherwise detected by the customer. This is nearly as clever a trick as finding an enzyme that will grow ships in glass bottles. Paine, like many geniuses, lies unsung and unappreciated, but his work endures.

molecule to the enzyme and distorts the position of the electrons forming the peptide bond. Thus weakened, the bond is readily snapped and the cut end of the polypeptide chain momentarily becomes attached to the enzyme molecule. The addition of a water molecule across this temporary connection restores the enzyme to its original condition, and the severed polypeptide chain falls away. The water molecule, one might say, picks the polypeptide chain off the surface of the enzyme. The entire process is over in a hundredth of a second, and the enzyme is free to attack another substrate molecule.

The carboxypeptidase mechanism is typical of many enzymatic reactions. Nothing was done that could not have been accomplished eventually without the enzyme. Hydrolysis of peptide bonds occurs spontaneously in aqueous environments, but at a very slow rate because no convenient mechanism exists for a water molecule in neutral solution to attack a peptide bond; otherwise no protein chain could persist for long in the aqueous environment within living cells. Carboxypeptidase sharply accelerates the rate of peptide bond hydrolysis. Because of the specificity of the enzyme, not every peptide bond is hydrolyzed

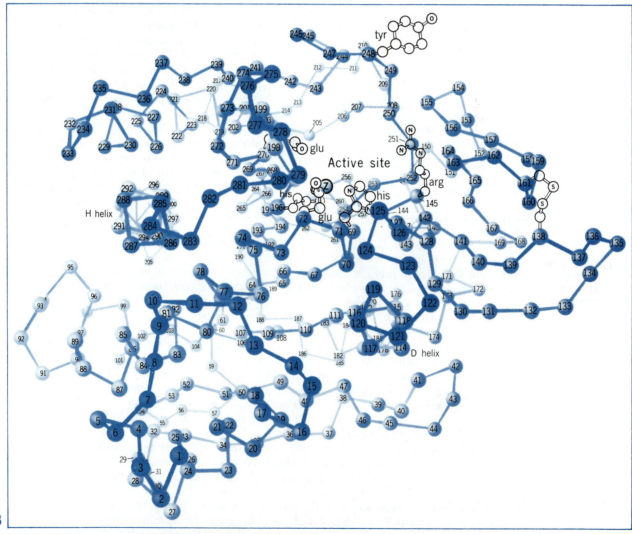

CARBON SKELETON OF CARBOXYPEPTIDASE. Framework of the molecule is a pleated sheet surrounded by eight α-helices. One disulfide bridge holds more extended regions of chain in place at right. Active site (depicted in detail in the two following illustrations) is a bowl-shaped depression at top.

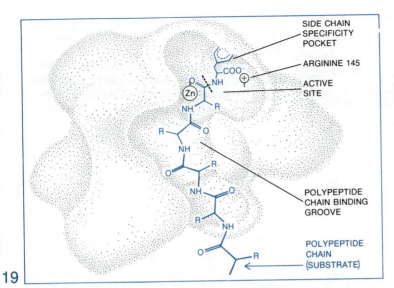

19

MAIN CATALYTIC FEATURES of carboxypeptidase. This schematic rendering of the molecule depicted in Figure 18 emphasizes the active site depression, specificity pocket, chain binding groove, zinc atom and chain-holding arginine 145. Polypeptide chain of substrate is in color.

at this new rapid rate; only those carboxyl-terminal bonds for which the final amino acid side chain is bulky and hydrophobic.

Certainly there is a great deal still to be learned about enzyme molecules, but there is no longer a reason to suspect that future research will disclose any radical surprises. At last biochemists understand enzymes for what they are: molecular-scale machines.

NUCLEOTIDES AND NUCLEIC ACIDS

Every living thing organizes matter from its environment into its own molecules. It is possible, although not particularly flattering to the human ego, to describe a man as the best possible local environment for the survival of human DNA. This is a gross oversimplification, but contains some element of truth. Not everything in a new organism comes solely from the DNA of its parents, and a DNA molecule without the protein-synthesizing machinery of a cell is as useless as a prerecorded tape without a playback system. Nevertheless, the information required to construct a new organism comes from the DNA.

The closely related ribonucleic acid, RNA, is a polymer whose building blocks are ribose molecules with organic bases attached to carbon atoms. DNA differs from RNA only in having one of its hydroxyl groups replaced by a hydrogen; in other words, the sugar is deoxyribose rather than ribose.

Four different organic bases are found in DNA: the purine rings adenine and guanine, and the pyrimidine rings thymine and cytosine (*Figure 20*). RNA uses uracil (with one less methyl group) rather than thymine. The combination of a purine or pyrimidine with ribose forms a NUCLEOSIDE; if the ring is adenine, for example, the nucleoside is adenosine (*Figure 21*). DNA and RNA are polymers of nucleotides in which the ribose (or deoxyribose) of each monomer is connected to the next by phosphate groups. The sugar-phosphate "backbone," with purine and pyrimidine side groups on the sugars, is diagrammed in Figure 22.

A nucleoside such as adenosine can add a phosphate group to form the molecule known as adenosine monophosphate, AMP. But most importantly for energy storage purposes, the nucleoside can add a second and a third phosphate to build adenosine di- and triphosphate, ADP and ATP. ATP is the main short-term energy storage molecule in cells. It is the source of instant energy for muscle contraction, active transport across membranes, activation of molecules for subsequent reaction, and almost every other uphill (energy-consuming) chemical process in the cell. The synthesis of ATP from ADP and inorganic

20

ORGANIC BASES of DNA and RNA. Adenine and guanine are purines. Uracil, thymine and cytosine are pyrimidines. In DNA and RNA, guanine on one chain is bonded to cytosine on another chain. Adenine is paired with thymine in DNA and with uracil (minus the CH_3 group) in RNA. Atoms involved in hydrogen bonding are printed in color.

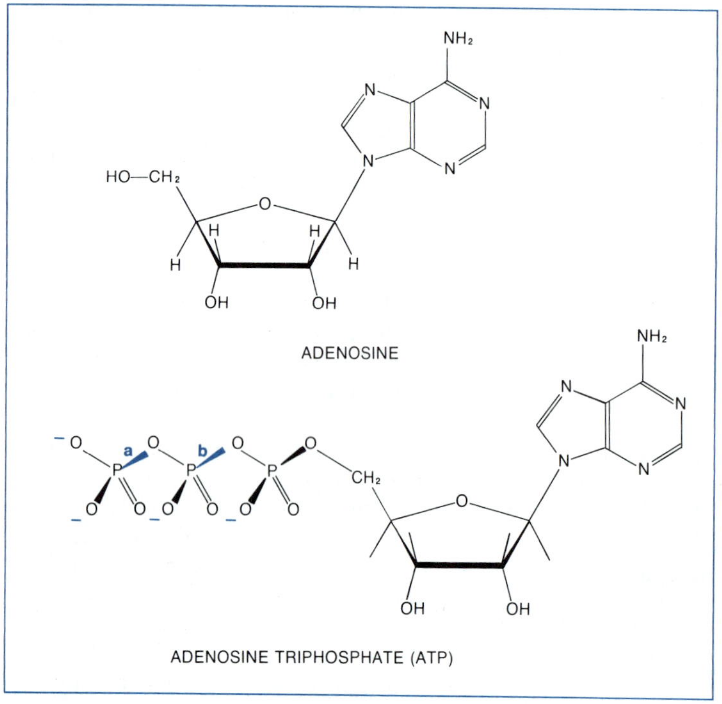

NUCLEOSIDES AND NUCLEOTIDES. The combination of a purine or a pyrimidine with ribose is called a nucleoside. Adenosine (top) is a nucleoside, and adenosine triphosphate (bottom) is a nucleotide. Letters *a* and *b* designate high-energy bonds in ATP.

tions. Some code words are merely punctuation marks that tell the protein-assembly machinery to stop work.

The two strands of the DNA ladder run in opposite directions, and the ladder is twisted about its long axis into a helix (*Figure 23*). The hydrogen-bonded base pairs are stacked parallel to one another like steps in a spiral staircase. The sugar-phosphate backbone runs around them. When DNA replicates, the two strands unwind and each makes a complementary copy of itself. The result is two daughter DNA molecules, each exactly like the parent, each with one new strand and one derived from the parent.

The structure of DNA was worked out in 1953 by Francis Crick, a British physical chemist, and James Watson, an American biochemistry postdoctoral fellow, as a result of

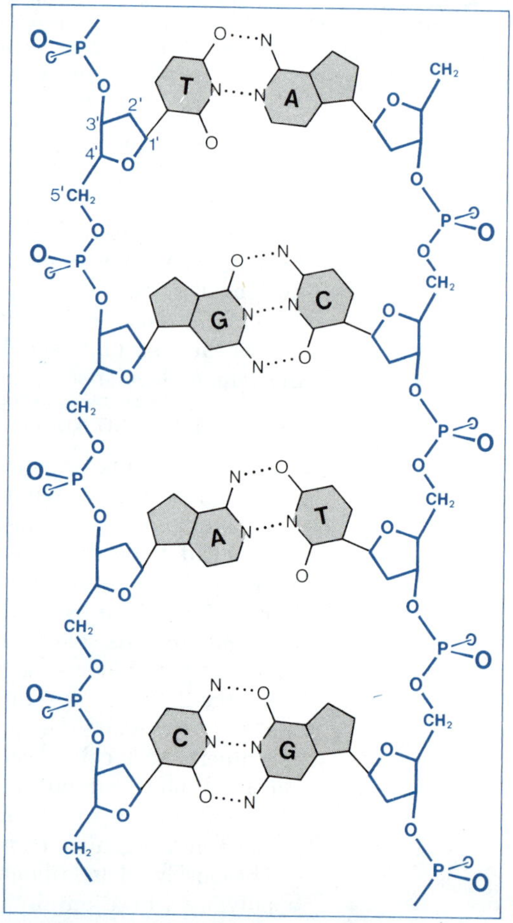

phosphate is the first step in packaging and storing energy obtained from the breakdown of foods. We shall look more closely at the role of ATP in Chapter 3.

One of the most essential properties of any set of instructions is that they can be read and copied. A book must be readable by those who intend to use its information. There should also be some way of making additional copies of the book. Reading and copying in DNA and RNA is possible because the organic bases "recognize" one another by means of hydrogen bonds between the rings (*Figure 22*). Adenine can only form a hydrogen bond with thymine (or with uracil in RNA), and guanine can only bond to cytosine. DNA is a double-stranded structure in which every base on one strand is hydrogen bonded and paired with its corresponding partner on the other strand. The secret of information storage is that every three successive bases along the chain comprise a code word for one amino acid. If there are four different letters in this alphabet, and every word has three letters in it, then obviously there are $4^3 = 64$ different code words for only 20 amino acids. The code is redundant; that is, most amino acids have several code combina-

BACKBONE OF DNA molecule consists of two chains oriented in opposite directions. The chains are built from phosphate groups and deoxyribose molecules. Base pairs connect the backbone chains like rungs of a ladder. Dots are hydrogen bonds.

23

MODEL OF DNA MOLECULE reveals the way that the ladderlike backbone is twisted into a double helix. Large white balls represent oxygen atoms, and small dark balls, hydrogen atoms. Carbon atoms appear as small black triangles, and phosphorus atoms as larger grey triangles. Nitrogen atoms appear as large dark balls. Only a small segment of the enormous molecule is shown.

their experiments in the Cavendish Laboratory at Cambridge University. Watson and Crick shared the Nobel Prize for medicine for 1961 with Maurice Wilkins, the London crystallographer whose x-ray photographs provided the evidence for the validity of the double helix. Because the two strands in DNA are complementary but opposite, molecular geneticists often differentiate between them by calling one "Watson" and the other one "Crick."

The reader has come a long way in this chapter, from protons and electrons to DNA. The progression is symbolic, for over the first one or two billion years of its history, our solar system came along precisely this path. The biologist George Wald has on occasions defined life as a property of molecules in a sufficiently intricate state of organization. This

chapter has described some of these molecules. The rest of this book emphasizes the consequences of their organization.

READINGS

R.E. DICKERSON AND I. GEIS, *Chemistry, Matter, and the Universe*, Menlo Park, Calif., W.A. Benjamin, Inc., 1976. An elementary and abundantly illustrated introduction to chemistry for the non-chemist.

R.E. DICKERSON AND I. GEIS, *The Structure and Action of Proteins*, New York, Harper & Row, 1969. A more extended treatment of protein structure, folding and physical and chemical properties. Profusely illustrated, including stereo pair drawings of protein molecules.

T.R. DICKSON, *Introduction to Chemistry*, 2nd Edition, New York, John Wiley & Sons, 1975. A good introductory textbook at a very simple level.

J.C. KENDREW, "The Three-Dimensional Structure of a Protein Molecule," *Scientific American*, December 1961. The structure of myoglobin, the first protein to be solved.

H. NEURATH, "Protein Digesting Enzymes," *Scientific American*, December 1964. Discussion of chymotrypsin and trypsin based on their amino acid sequences. Predates the x-ray structure analysis.

M.F. PERUTZ, "The Hemoglobin Molecule," *Scientific American*, November 1964.

D.C. PHILLIPS, "The Three-Dimensional Structure of an Enzyme Molecule," *Scientific American*, November 1966. The structure of hen egg white lysozyme.

C.U.M. SMITH, *Molecular Biology: A Structural Approach*, Cambridge, Massachusetts, M.I.T. Press, 1969. An extremely well-written introduction to the field, with constant emphasis on what is important to the biologist. Highly recommended.

R.M. STROUD, "A Family of Protein-Cutting Proteins," *Scientific American*, July 1974. The structural story of the digestive enzymes trypsin and chymotrypsin. A complement to the article by Neurath.

J.D. WATSON, *The Double Helix*, New York, Atheneum, 1968. The view by a young American post-doctoral fellow of the process by which he and his British coworkers solved the structure of DNA and ultimately won the Nobel Prize. Biased and opinionated, but honestly so, and revealing as to the motives that sometimes impel scientists to achieve. Excellent reading.

3

energy

It takes a membrane to make sense out of disorder in biology. You have to be able to catch energy and hold it, storing precisely the needed amount and releasing it in measured shares. A cell does this, and so do the organelles inside. Each assemblage is poised in the flow of solar energy, tapping off energy. . . . To stay alive, you have to be able to hold out against equilibrium, maintain imbalance, bank against entropy, and you can only transact this business with membranes in our kind of world.

LEWIS THOMAS,
THE LIVES OF A CELL

To sustain the processes of life the cell carries out thousands of chemical reactions per second. In the nonliving world many of these reactions would proceed sluggishly, if at all. Outside a living organism, in fact, many biological reactions would run spontaneously in the opposite direction. On a planet with an oxygen atmosphere, for example, large organic molecules are chemically unstable and tend to decompose gradually into carbon dioxide and water.

Living systems have circumvented the limitations of their planetary environment by devising a fourfold chemical strategy. The chemistry of life is elaborate in its working details but elegantly simple in principle. First, to drive reactions uphill — to run them opposite to their spontaneous direction — living systems plug themselves in to external sources of energy. Plants tap the energy of sunlight, and other organisms rely on the chemical energy of various foods. Second, to accelerate the tempo of slow reactions living systems invented enzymes, which are superbly efficient catalysts. Third, living systems have evolved feedback mechanisms to regulate their network of reactions. Here life displays an impressive economy of effort by using the catalysts as the instruments of control. Fourth, living systems exploit spontaneous reactions. Without enzymes, some of these reactions would proceed very slowly — so slowly as to be biologically useless. If a reaction is not spontaneous, enzymes cannot make it happen. But the right enzyme can greatly accelerate the rate of a spontaneous reaction, sometimes by a factor of a million or more. Cells often couple a spontaneous reaction to one that is not, diverting the energy of the spontaneous reaction to drive the nonspontaneous one.

Reaction rates also explain how the unstable organic molecules of living systems persist for so long a time. Although organic compounds tend to decompose spontaneously, the decomposition rates are usually very slow; to start them requires an input of energy, just as firewood requires a match. This input is called ACTIVATION ENERGY. The rate of biological reactions depends primarily on the concentration of reactants and on the height of the activation-energy "barrier" to reaction. Enzymes speed up reactions by providing alternate reaction pathways with lower activation-energy barriers.

MATTER AND ENERGY

Living systems require a continual supply of matter and energy: matter to build new structures and to replace those that have worn out; energy to assemble these structures and to perform biological work. The matter is obtained from the environment. The energy is drawn

THE SUN is ultimate source of all biological energy. In the telescope photograph of the solar disk on opposite page, white areas are solar flares and black strands are filaments of cooler gas.

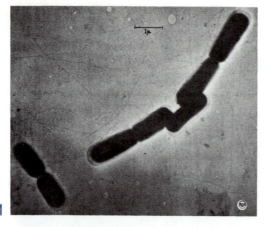

1

CLOSTRIDIUM BOTULINUM is an anaerobic soil bacterium that causes the deadly form of food poisoning known as botulism. This photomicrograph shows seven of these rod-shaped bacteria, each about two microns long.

tridia oxygen is not only useless but lethal, and they can flourish only in microenvironments devoid of air — deep in soil, in wounds, or in sealed cans of food.

Yeasts and many bacteria can survive either with or without oxygen. In their cells fermentation comprises the first few links in the energy chain. If oxygen is available, these organisms use the process of respiration to break down further the waste products of fermentation (*Figure 2*). Respiration not only enables them to burn (oxidize) their biochemical garbage, but also to harness the energy of the fire. By oxidizing a sugar molecule all the way to carbon dioxide and water, instead of stopping short at ethyl alcohol, a yeast cell can derive 19 times more energy per gram of food. If oxygen is not available such cells can live quite well

from three chemical processes: FERMENTATION, RESPIRATION and PHOTOSYNTHESIS. They are the biochemical engines that power the machinery of life.

Each process involves a chain of reactions that has been developed during countless millenia of trial-and-error evolution. Some links of the energy chain are much older than others. The first organisms on Earth probably were single-celled scavengers that inhabited tidal ponds, ingesting energy-rich molecules from the water and digesting them into smaller fragments by fermentation. One of the oldest and commonest fermentation pathways is the breakdown of sugars in the absence of oxygen into two- or three-carbon fragments such as ethanol and lactic acid. Because no oxygen is involved, the process is most often called ANAEROBIC GLYCOLYSIS (*glyco*-sugar, and *lysis*-splitting).

The chemical machinery for the fermentation of sugars survives in the cells of all living organisms. It remains the sole source of energy for some microorganisms, such as the soil bacterium *Clostridium botulinum*, which causes the deadliest form of food poisoning (*Figure 1*). The energy machinery of the Clostridia is a biochemical fossil that has survived for billions of years, dating back to a primitive Earth with an atmosphere devoid of oxygen. To Clos-

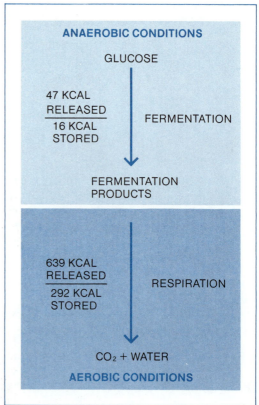

2

COMPARISON OF EFFICIENCY of anaerobic vs. aerobic metabolism. Aerobic systems can capture 19 times more energy from a molecule of glucose.

An active cell requires more than two million ATP molecules per second to drive its biochemical machinery.

Fats and fatty acids can store more than twice as much energy per gram as proteins and carbohydrates. A weight-saving strategy is important to animals, which must move about, carrying their energy supply with them. It is interesting to compare these compounds with other potential fuels such as hydrogen gas and octane (a gasoline). The burning of hydrogen yields 34 kcal of free energy per gram; octane, 11.4 kcal; stearic acid (a fatty acid), 9.5 kcal; alanine (a typical amino acid from proteins), 4.4 kcal; and glucose (a carbohydrate), 3.7 kcal. Gasoline would be an even more portable form of energy than fats or fatty acids, but gasoline has no functional group like a carboxylic acid, and thus could not readily be metabolized on demand. Hydrogen gas would be more than three times as efficient an energy carrier as fats on a weight basis, but an animal that stored energy in great sacs of hydrogen gas would be strange indeed.

by fermentation alone, although at much lower levels of biochemical efficiency (*Figure 3*).

Yeasts such as *Saccharomyces* (brewer's yeast) will grow and reproduce rapidly under aerobic conditions. If the oxygen supply is cut off, the yeast simply shuts down its aerobic reactions and maintains itself by fermentation. Winegrowers exploit this metabolic versatility by first aerating crushed grapes to encourage the yeasts to grow, then letting the must (the crushed skins and juice) stand in vats for several days while the yeasts convert the remain-ing grape sugar into alcohol by anaerobic fermentation. If the must were aerated continuously the yeasts would convert the sugar to carbon dioxide and water; the end product would be soda water instead of wine.

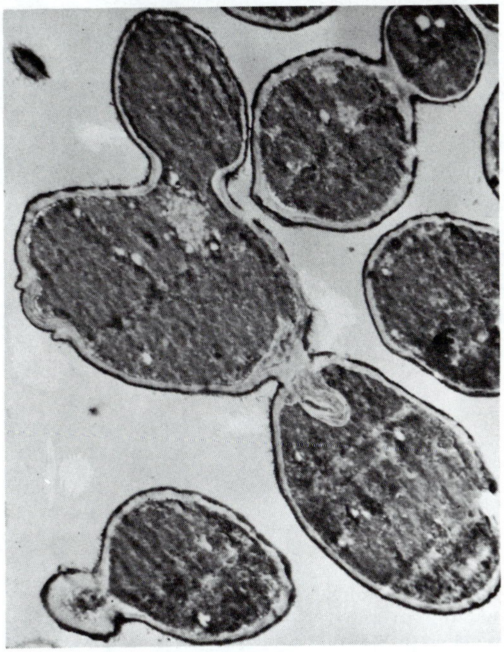

Human cells, and those of all other higher animals and plants, have lost the ability to switch the respiratory machinery on and off. Our cells have no simple way to dump the end product of this particular fermentation: lactic acid. When an athlete exercises vigorously, his quick energy comes from the rapid fermentation of glucose in muscle cells. But as lactic acid accumulates it produces fatigue and eventually muscle cramps. When the muscles have a chance to recover, the slower aerobic process eliminates some of the lactic acid by oxidizing it to carbon dioxide and water, and converts the rest back to glucose.

More than 20 enzymes control the various reactions in the energy chain. The enzymes for fermentation are dissolved in the cytoplasm of the cell, and those for aerobic respiration are incorporated into the structure of the mitochondrion, the powerhouse of eucaryotic cells (*Chapter 1*). The reactions take place in a series of steps. At several steps there is a small yield of "free" energy (energy that is not consumed by the reaction itself, and hence free to do biochemical work). The cell uses much of this energy to synthesize adenosine triphosphate (ATP).

All living cells rely on ATP molecules for the short-term storage of energy. An active cell requires more than two million molecules of ATP per second to drive its biochemical machinery. A fraction of the ATP is diverted to the synthesis of long-term energy storage compounds. Chapter 2 mentioned that living organisms store energy in fats and in starch, a

3

BREWER'S YEAST, Saccharomyces cerevisiae, is a fungus that converts sugar into alcohol by the process of fermentation. In this electron micrograph new cells are budding from larger parent cells.

long-chain polymer of glucose (*Box A*). Of course, as every nutritionist knows, any large molecule synthesized by the cell — a protein, for example — is a storehouse of energy, but energy storage is not its primary function. ATP can be considered as the coinage of energy exchange in living organisms, a useful "packet" that contains eight kilocalories of free energy. Starches and fats are a savings account at the energy bank. When plants need energy they usually draw on their deposits of starch, which they convert to glucose and oxidize (burn); the energy released during these reactions is used to make ATP. Similarly animals can draw on their deposits of fats, which they oxidize to carbon dioxide and water, using the energy released to make ATP.

Unlike animals, green plants make their own food and ATP directly. The invention of photosynthesis enabled them to tap a virtually unlimited source of free energy (sunlight) to synthesize new molecules of glucose and to form ATP. Sunlight is the ultimate source of all biological energy. Plants rewind the mainspring of life by creating food not only for themselves but also for all other organisms on Earth. Animals obtain their energy either second hand (by eating plants) or third hand (by eating other animals). Scavengers such as fungi and bacteria complete the cycle by feeding on the energy stored in the large organic molecules of dead plants and animals.

FERMENTATION

Although anaerobic fermentation reactions can begin with many different compounds, the commonest fuels are six-carbon sugars, particularly glucose. The cells of many microorganisms and most higher animals break down the glucose molecule into two three-carbon molecules of lactic acid. The overall reaction can be summarized as:

$$C_6H_{12}O_6 \rightarrow 2CH_3-\underset{\underset{OH}{|}}{C}H-\overset{\overset{O}{||}}{C}-OH$$
(GLUCOSE)   (LACTIC ACID)

$$2ADP + 2P \rightarrow 2ATP$$

$$C_6H_{12}O_6 + 2ADP + 2P \rightarrow$$
$$2\ CH_2-CH(OH)-COOH + 2ATP$$

GLYCOLYSIS begins with a series of pump-priming reactions and molecular rearrangements (top), splitting a glucose molecule into two molecules of phosphoglyceraldehyde. After being primed with a second phosphate group, the 3-carbon molecules transfer both phosphates to ADP to form ATP. Each step involves an enzyme. NAD is a carrier molecule.

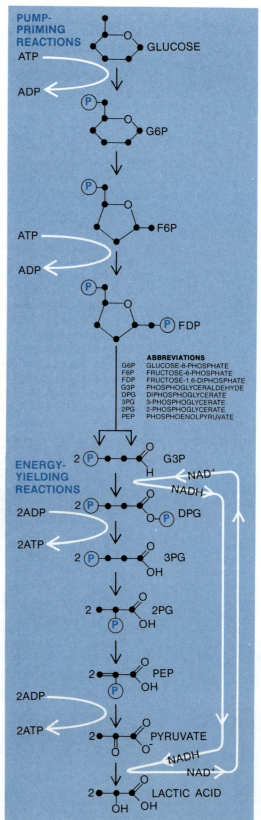

PUMP-PRIMING REACTIONS
ATP
ADP
GLUCOSE
G6P
F6P
FDP

ENERGY-YIELDING REACTIONS
G3P
2ADP
2ATP
DPG
3PG
2PG
PEP
2ADP
2ATP
PYRUVATE
LACTIC ACID
NAD⁺
NADH
NADH
NAD⁺

ABBREVIATIONS
G6P   GLUCOSE-6-PHOSPHATE
F6P   FRUCTOSE-6-PHOSPHATE
FDP   FRUCTOSE-1,6-DIPHOSPHATE
G3P   PHOSPHOGLYCERALDEHYDE
DPG   DIPHOSPHOGLYCERATE
3PG   3-PHOSPHOGLYCERATE
2PG   2-PHOSPHOGLYCERATE
PEP   PHOSPHOENOLPYRUVATE

Anaerobic cells use up to 19 times more glucose than aerobic cells to obtain the same amount of energy because they throw away their food when only about seven percent of its energy has been extracted.

Fermentation of glucose by yeasts follows the same reaction pathway almost all the way to lactic acid, but veers off in the final two steps to yield ethanol (a two-carbon alcohol) and carbon dioxide.

The conversion of glucose to lactic acid or to ethanol releases a great deal of free energy — too much energy in fact for the cell to capture in one step. If the energy were released all at once, most of it would be lost as useless heat. Cells have evolved a way to ferment glucose in 11 steps, each catalyzed by a different enzyme (*Figure 4*). It is not necessary to memorize the details, but the overall energy-extracting strategy is important.

The first few steps are essentially pump-priming reactions, and two of them require an input of energy. The glucose molecule is rearranged into fructose (*Chapter 2*) and two molecules of ATP are invested to attach two phosphate groups to the sugar, making it more reactive. The next few reactions split the six-carbon sugar into two three-carbon molecules, each primed with a phosphate group. The three-carbon molecule is phosphoglyceraldehyde, which will turn up again in the discussion of photosynthesis.

The last six reactions in Figure 4 are the real source of energy. They are commonly described as the oxidation of phosphoglyceraldehyde to lactic acid. Apparently there is a contradiction here: fermentation was previously defined as a process that involves no oxygen. In the strict chemical sense, the term "oxidation" does not necessarily imply the presence of oxygen; in fact the entire sequence of fermentation reactions discussed here can be thought of as oxidation without oxygen. Chemists define oxidation as a loss of electrons. Because the electrons have to go somewhere, another molecule must gain electrons — a process known as REDUCTION. In other words, every time something is oxidized, something else is reduced. Sometimes oxidation does involve adding oxygen; but often it involves subtracting hydrogen, or a loss of electrons. Oxidation-reduction reactions always occur in pairs, with electrons being transferred from reduced (hydrogen-rich) molecules to oxidized (hydrogen-poor) ones.

The oxidation of phosphoglyceraldehyde to diphosphoglycerate in the sixth reaction is accompanied by the reduction of an important "carrier" molecule known as NAD (*Box B*):

$$NAD^+ + 2H \rightarrow NADH + H^+$$

The sole function of NAD is to accept hydrogen atoms and free energy from compounds

B

CARRIERS AND VITAMINS

One half of NAD looks very much like a molecule of ATP. It is easy to imagine that several energy carriers involving adenine, ribose, phosphates, and other groups, might have evolved from a common (and less efficient) precursor. The structure of several other carrier molecules can be described as: adenine — ribose — phosphate — phosphate — Groups X. Part of the Groups X is often a compound that humans can no longer synthesize for themselves, and so is classed as a vitamin. If Groups X are a ribose and a nicotinamide ring, the compound is NAD. Nicotinic acid, or NIACIN, *is one of the vitamin B_2 complex. Another member of this same vitamin complex is* RIBOFLAVIN, *which is involved in another carrier.* FLAVIN ADENINE DINUCLEOTIDE, *or FAD. Pantothenic acid, another member of the vitamin B complex, is used in a carrier of acetate groups called* COENZYME *A. Other vitamins have similar uses. Humans need only small amounts of these vitamins because these carrier molecules are recycled through the metabolic machinery, and need to be replaced only when they are accidentally destroyed.*

being oxidized, or to donate hydrogen atoms and energy to compounds being reduced. The diphosphoglycerate is eventually converted to pyruvate, and in the process four ADP molecules are converted to ATP. The final reaction, the conversion of pyruvate to lactic acid, oxidizes the NADH to NAD⁺ and recycles it.

A balance sheet on the entire process would show that two molecules of ATP are used per molecule of glucose, but four are returned — a net gain of two ATP. The process yields 47 kcal of free energy; 16 kcal are recovered and stored in the bonds of ATP, and the other 31 kcal are wasted as heat. This is not very efficient on two counts. First, only 34 per cent of the energy liberated in glycolysis is recovered; and second, lactic acid is too rich in energy to throw out as a waste product. If the cell could oxidize lactic acid or ethanol to carbon dioxide and water, it could obtain another

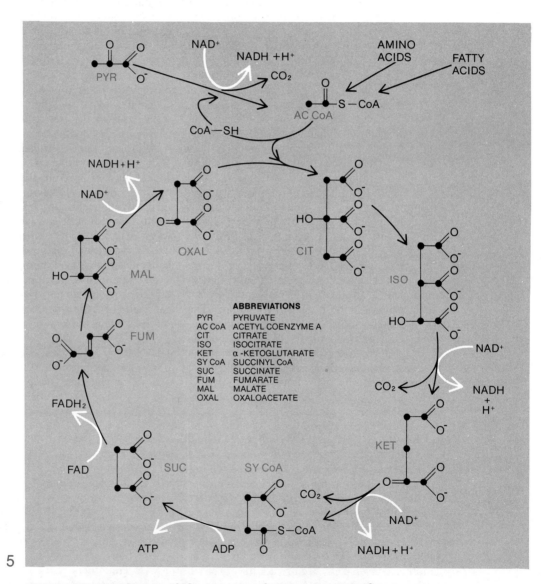

ABBREVIATIONS
PYR   PYRUVATE
AC CoA   ACETYL COENZYME A
CIT   CITRATE
ISO   ISOCITRATE
KET   α-KETOGLUTARATE
SY CoA   SUCCINYL CoA
SUC   SUCCINATE
FUM   FUMARATE
MAL   MALATE
OXAL   OXALOACETATE

5

CITRID ACID CYCLE begins with the conversion of pyruvate to acetate (the acetyl attached to coenzyme A, top). The acetate is degraded to CO_2 in a series of steps. Free energy released in controlled amounts reduces two types of carrier molecule: NAD and FAD. Reduced forms of these molecules enter the respiratory chain, where their free energy is stored in ATP. Acetate from digestion of fats and proteins can also enter the cycle via CoA (upper right).

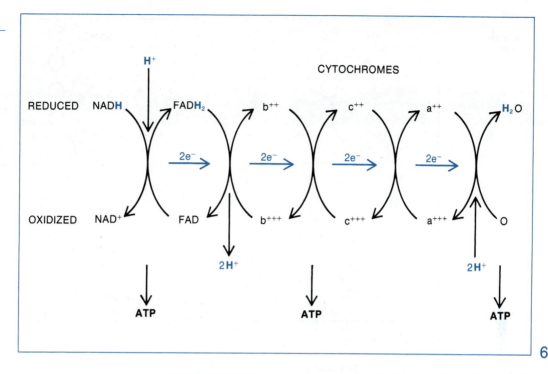

6

RESPIRATORY CHAIN. NADH reduced during citric acid cycle is reoxidized and recycled. It transfers two electrons to oxygen in a series of steps. The free energy released is captured by the cell and stored in ATP.

640 kcal from glucose. Anaerobic cells use up to 19 times more glucose than aerobic cells to obtain the same amount of energy because anaerobic cells throw away their food when only about seven percent of its energy has been extracted.

Anaerobic glycolysis is thought to be the oldest of the three energy pathways discussed in this chapter — the first to evolve. It was efficient enough to serve the needs of simple one-celled organisms on a primitive planet whose environment provided an abundant supply of reduced molecules such as glucose. Indeed it serves the needs of many microorganisms alive today. But the emergence of plants drastically changed the conditions of life on the early Earth, as the process of photosynthesis began to dump ever-increasing tonnages of oxygen into the atmosphere (*Chapter 16*). The availability of oxygen favored the evolution of a new breed of organism, one that could use its food much more efficiently. The new organisms achieved this efficiency by inventing the process of respiration.

RESPIRATION

Respiration burns the biochemical garbage of glycolysis. The starting point is not lactic acid but pyruvate. A series of reactions called the CITRIC ACID CYCLE oxidizes pyruvate until nothing is left but carbon dioxide and water. The cycle produces more NADH and ATP, and involves a new carrier, FAD (flavin adenine dinucleotide). Eventually the energy carried by NADH and $FADH_2$ is converted to ATP by a special reaction pathway called the RESPIRATORY CHAIN.

The steps of the citric acid cycle are shown in Figure 5. The biochemical details are complex and will be omitted here, but as with glycolysis, the overall energy strategy is important. To enter the cycle, pyruvate is oxidized to a derivative of acetic acid called acetyl coenzyme A (Ac-CoA), and in the process a molecule of NAD^+ is reduced to NADH. The cycle proper begins with the transfer of a two-carbon acetate ion from Ac-CoA to a four-carbon oxaloacetate, converting it to citrate. During a full turn around the cycle two molecules of carbon are stripped away and excreted from the cell as carbon dioxide. The accompanying hydrogens are used to reduce NAD^+ and FAD. For each pyruvate ion that enters the cycle one molecule of ATP is made directly, three molecules of NAD^+ are reduced, and an FAD molecule is reduced to $FADH_2$.

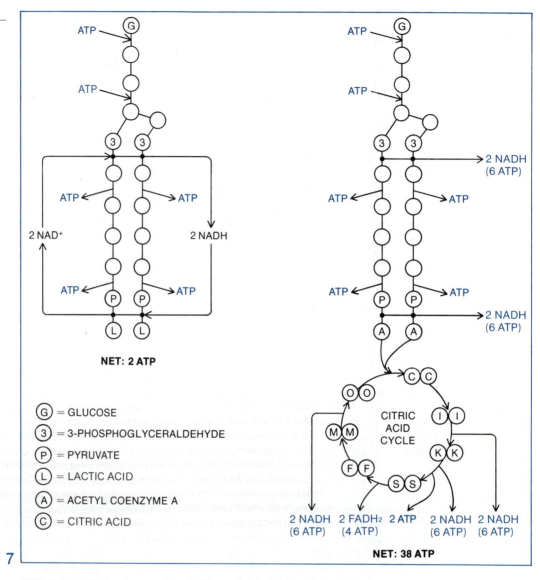

7

YIELD OF ATP from anaerobic glycolysis (left) and glycolysis plus respiration (right). Circles represent intermediate compounds, and energy carriers are indicated in color. When the closed NAD loop of anaerobic glycolysis is broken in respiration, the NADH produced after phosphoglyceraldehyde (upper right) becomes available for ATP production.

The NADH and $FADH_2$ molecules are then funneled into the respiratory chain, where they are oxidized and recycled. The free energy that they carry is captured to make ATP.

The components of the respiratory chain are shown schematically in Figure 6. The oxidation of each NADH releases 52 kcal of free energy — too much to be trapped in a single step. As NADH is reoxidized FAD is reduced to $FADH_2$, releasing enough free energy to make one ATP from ADP and phosphate. The next three steps involve the oxidation and reduction of pigmented proteins: cytochromes a, b and c. The function of these highly specialized molecules is to release energy step by step, enabling the cell to make two more molecules of ATP. Of the total 156 kcal of energy in the three NADH molecules, 72 kcal are recovered and stored as nine molecules of ATP. One ATP molecule is synthesized directly by the citric acid cycle. The energy of the FAD molecule that was reduced to $FADH_2$ during the cycle is trapped to make two more

8

MITOCHONDRION contains the respiratory machinery of the cell. This electron micrograph shows the pitted inner membrane of a mitochondrion. Particles on the surface of the membrane contain the enzymes of the respiratory chain. Magnification, about 80,000 ×.

increase in efficiency: the harnessing of the energy of the two NADH molecules produced previously during glycolysis. The 52 kcal of free energy in each of these molecules can now be harnessed to make six more ATP, raising the total to 36. These plus the net of two ATP produced by anaerobic glycolysis bring the grand total of ATP obtained from one mole of glucose to 38. The processes of glycolysis and respiration are summarized in Figure 7, and their relative efficiencies are compared in Table I.

All of these reactions occur within the cell. The enzymes for anaerobic glycolysis float freely in the cytoplasm. The enzymes of the citric acid cycle and the components of the terminal respiratory chain are confined in mitochondria (*Box C*). Mitochondria are barely large enough to be seen under the light microscope. An electron micrograph of a typical mitochondrion is shown in Figure 8. A single cell may contain as few as a dozen or as many as 100,000 mitochondria. They are self-contained "power packs" for the cell, taking in pyruvate and oxygen and returning carbon dioxide, water, and ATP. They are particularly abundant in cells that need large amounts of energy, such as those of the heart and other active muscles.

Mitochondria have a double membrane wall. The outer membrane is usually smooth, but the inner membrane is folded back and forth into parallel layers of CRISTAE, seen in cross section in Figure 8. These infoldings give the inside membrane an enormous surface area. The enzymes of the citric acid cycle float freely in the matrix (the interior soup) of the mitochon-

molecules of ATP. In other words, each turn of the cycle produces 12 molecules of ATP (1+9+2). But this is only part of the story.

The NADH molecule produced during the conversion of pyruvate to Ac-CoA — the transition step from glycolysis to respiration — eventually liberates enough free energy to make three more molecules of ATP, for a total of 15. Because one molecule of glucose makes two of pyruvate, this total must be doubled to 30. One of the major innovations of respiration is the use of a different acceptor for hydrogen when NADH is reoxidized. Some bacteria use sulfate ions, nitrate ions, or even iron atoms to reoxidize and recycle NADH, but aerobic cells use oxygen. The availability of oxygen as an acceptor for hydrogen atoms allows a further

COMPARISON OF EFFICIENCY of anaerobic and aerobic metabolism shows the overwhelming superiority of aerobic systems in terms of free energy available for storage.

I

	ANAEROBIC GLYCOLYSIS	GLYCOLYSIS PLUS RESPIRATION
FREE ENERGY OF REACTION	47.3 kcal	686 kcal
ATP SYNTHESIZED	2	38
FREE ENERGY STORED	16.2 kcal	308 kcal
EFFICIENCY OF CONVERSION	34%	45%
FRACTION OF TOTAL AVAILABLE FREE ENERGY STORED	24%	45%

One of the more intriguing unproven hypotheses of biology is the suggestion that mitochondria and chloroplasts are the descendants of bacteria that once lived in symbiotic partnership with their host cell, and gradually lost their independence. As the author of a recent review said about these ideas: "Some theories are born respectable; others have respectability thrust upon them." The proposal that mitochondria are the degenerate remains of bacteria is at least 70 years old, but lay in disrepute for many years for lack of supporting evidence. In the last decade, supported by new evidence, the old theory has been hauled out for a fresh look.

Mitochondria and chloroplasts are roughly the size of bacteria, and they have similar membrane structure. They each contribute a specific function to the host cell: mitochondria the ability to follow glycolysis by aerobic respiration, and chloroplasts the ability to synthesize glucose with solar energy. It is easy to imagine that some early anaerobic cell solved the problem of gaining more free energy from glucose, not by developing its own respiratory pathways, but by entering into partnership with aerobic bacteria, supplying them with glucose and taking some of their ATP for the cell's own needs. Examples of symbiotic cooperation are not rare among organisms alive today. The lichens that grow on rocks and tree trunks are not a single organism but a meshwork of two very different organisms: an alga and a fungus. Several kinds of animal, ranging from one-celled paramecia to the reef-building corals of tropical seas, derive at least part of their food from photosynthetic algae that live within their cells.

The bacterial-origin theory was resurrected by the discovery that both mito-chondria and chloroplasts contain their own DNA, separate from that in the nucleus of the cell, and that mitochondrial DNA apparently contains the genetic information for some of its inner membrane proteins. Moreover, mitochondrial DNA is of the same circular form as the DNA of bacteria. Mitochondria also have their own ribosomes to translate messenger RNA into protein. Even more striking is the observation that the ribosomes of mitochondria, chloroplasts, and bacteria are all of similar size — slightly smaller than the ribosomes in the cytoplasm of the cell. Cytoplasmic ribosomes are rendered inoperative by the compound cycloheximide. Mitochondrial, chloroplast, and bacterial ribosomes are unaffected by cycloheximide but are all poisoned by chloramphenicol. The structures of bacterial and mitochondrial membranes and their permeability to ions and small molecules are quite similar, and much less like the corresponding properties of cell membranes.

Mitochondria are apparently not made anew from nuclear DNA when sperm and egg unite to create a new organism. Instead, they are carried along in the cytoplasm of the egg, and grow and divide autonomously. However, the enzymes of the citric acid cycle and the respiratory chain heme proteins are made at cellular ribosomes under the control of the DNA in the cell nucleus, and diffuse into the mitochondria after synthesis. The mitochondrial DNA apparently codes for the structural proteins that serve as the rack into which the various respiratory enzymes fit. If the mitochondria are truly former bacteria which have lost one biological function after another as they assumed a greater dependence upon their host, then one would expect that the mitochondria of primitive organisms might retain more of their original DNA and the functions it governs. And true enough, the mitochondria of the bread mold Neurospora crassa and similar microorgan-isms contain as much as six to seven times the amount of DNA as do the mitochondria of higher plants and animals.

It now appears as if animals and plants themselves might have arisen from symbiotic relationships between anaerobic, nonphotosynthetic cells, and aerobic or photosynthetic bacteria whose remains can still be seen as mitochondria and chloroplasts. The case is not yet proven, but the evidence is becoming quite persuasive.

drion. The respiratory chain flavoproteins and cytochromes are bound to the inner membrane. They are apparently present in equal amounts, and may occur in orderly assemblies with cytochrome b next to flavoprotein, cytochrome c next to b, and a next to c, just as they are used in the respiratory chain. This orderly arrangement may be partly responsible for the increased efficiency of energy conversion in respiration.

FEEDBACK CONTROL

Glycolysis and the citric acid cycle involve more than 20 reactions, each catalyzed by a specific enzyme. All of these enzymes must act in unison. What mechanism coordinates their action?

Glycolysis, the citric acid cycle and the respiratory chain are regulated by FEEDBACK CONTROL of enzymes. Certain products of the later reactions in a particular pathway, if present in excess, can combine with one of the enzymes that catalyzes an earlier reaction, changing the shape of the enzyme molecule and thus suppressing its activity, or in some cases, stimulating the enzyme to even greater activity. An excess of the products of one branch of a synthetic pathway can slow down the steps in that pathway, or can speed up reactions in a parallel branch and divert raw materials away from its own synthesis (*Figure 9*). These positive and negative feedback control mechanisms are used at many points in the energy-extracting processes.

The main control point for the citric acid cycle is the conversion of isocitrate to α-ketoglutarate by means of the enzyme isocitrate dehydrogenase (*step 4, Figure 5*). ATP and NADH are feedback inhibitors of this reaction, and ADP and NAD⁺ are activators. If too much ATP is accumulating, or if NADH is being produced faster than it can be used by the respiratory chain, the isocitrate reaction is blocked and the citric acid cycle is shut down. This would lead to a pileup of large amounts of isocitrate and citrate, except that the conversion of acetyl coenzyme A to citrate is also

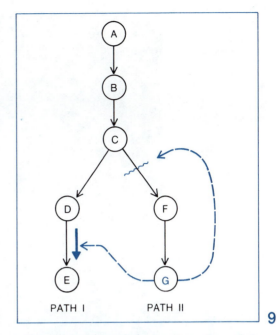

PATH I   PATH II

9

FEEDBACK CONTROL. Products of reaction path II, if present in excess, can shunt most of the intermediates into path I, either by inhibiting the enzyme that converts compound C into F, or by activating the enzyme that converts D into E. Black arrows represent chemical reactions. Colored arrows and wavy blocking line represent feedback controls.

blocked (*step 2*). The adverse effects of halting the isocitrate reaction are thus spread backward up the chain of reactions. A certain excess of citrate does accumulate, however, and this excess inhibits the fructose-6-phosphate reaction early in glycolysis. Thus if the citric acid cycle has been slowed down BECAUSE OF AN EXCESS OF ATP (not because of a lack of oxygen), glycolysis is shut down as well. Both processes resume when the ATP level falls.

It is wasteful to produce more acetyl coenzyme A than there is oxaloacetate to handle it in the citric acid cycle. It is also wasteful to shunt too much pyruvate into making oxaloacetate — the gear of the citric acid cycle — and to neglect to produce the fuel, acetyl coenzyme A. Enzyme feedback control maintains the proper balance among the uses of pyruvate by

One effect of photosynthesis is the liberation of a corrosive and highly reactive gas, molecular oxygen, which was probably lethal to many of the anaerobic organisms on the primitive Earth.

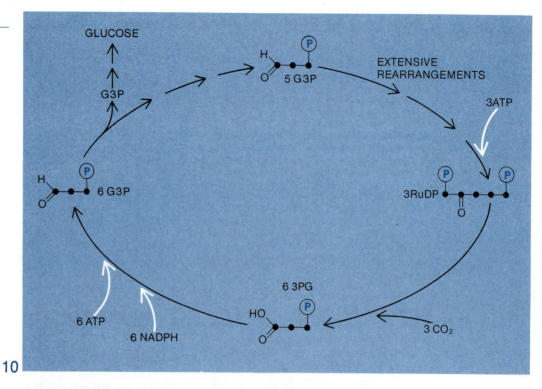

CALVIN-BENSON CYCLE, summarized in diagram above, is a major part
of the pathway by which plants synthesize glucose from carbon dioxide.
Three molecules of ribulose diphosphate (RuDP) combine with three
molecules of carbon dioxide (right) to yield six molecules of
2-phosphoglycerate (2PG). These are converted to six molecules of phosphoglyceraldehyde (G3P), five of which are rearranged (top right) and
recycled as RuDP. The net gain from the cycle is one molecule of G3P (top
left), which is used in the synthesis of glucose. The cycle comprises the
so-called dark reactions of photosynthesis. The energy to drive the cycle, in
the form of six molecules of ATP and NADPH, is provided by the light
reactions.

being sensitive to a shortage or an oversupply
of acetyl coenzyme A. All of the other feedback
controls have evolved because they make the
system more efficient, and hence contribute to
the survival of the species that carries them.

Feedback control so elegant as this creates
the illusion that it has been designed by a
systems analyst. In fact it is one of the most
intellectually satisfying examples of the tight
logic that can arise by the process of natural
selection, when selection pressure favors
efficient operation in the fierce competition
among organisms for limited resources.

PHOTOSYNTHESIS

If anaerobic glycolysis is considered as the first
great universal step in the evolution of the
energy metabolism of living organisms, respi-

ration is probably the third step. The aerobic
respiratory machinery could not have evolved
until a supply of oxygen was present. The primary source of O_2 as an essential component of
the biosphere is photosynthesis.

One effect of photosynthesis is the liberation
of a corrosive and highly reactive gas,
molecular oxygen, which was probably lethal
to many of the anaerobic organsims on the
primitive Earth. It is obvious that any organism
that could learn to tolerate oxygen, or better
still, to use it in some beneficial way, would
enjoy a tremendous advantage as oxygen from
photosynthesis began to accumulate in the
primitive atmosphere. Life's answer was the
invention of respiration.

Photosynthesis is the process by which
plants use energy from the Sun to make glu-

cose as a means of storing the energy for later use. The source of carbon atoms for the sugar is carbon dioxide, and the overall reaction is

$$6 \ CO_2 + 12(H) \rightarrow C_6H_{12}O_6 + 3 \ O_2.$$

The earliest and most primitive photosynthetic bacteria used hydrogen sources such as H_2S or even molecular hydrogen, H_2, to reduce CO_2 to sugars. At an early stage of evolution, however, organisms invented the machinery to obtain hydrogen from water, the most abundant source. This is the mechanism that is used in all higher green plants.

Energy is required to synthesize glucose in photosynthesis — at least 686 kilocalories per mole. Hydrogen atoms, no matter where they come from, are brought to the synthesis as the reduced form of nicotinamide adenine dinucleotide phosphate, NADP. This carrier molecule is similar to NAD and it carries virtually the same amount of energy when reduced, 52 kcal per mole.

The overall reaction for the synthesis of glucose can be written

$$6 \ CO_2 + 6 \ NADPH + H^+ \rightarrow$$
$$C_6H_{12}O_6 + 3 \ O_2 + 6 \ NADP^+.$$

Molecules of NADPH supply both hydrogen and energy, and ATP is a second energy source. As with oxidation of glucose, this simple equation summarizes an elaborate mechanism. The carbon dioxide reduction cycle, often called the Calvin-Benson cycle, is depicted schematically in Figure 10. This cycle begins when a molecule of CO_2 combines with a molecule of the five-carbon sugar ribulose diphosphate. The result of the cycle is the formation of glucose from CO_2. Several reactions in the Calvin-Benson cycle are just the reverse of glycolytic reactions. Although the enzymes are different, the duplication of chemistry is understandable. Photosynthesis is not a magical system grafted onto nonphotosynthetic metabolism. It evolved slowly, and it is certainly natural that its compounds and processes have been adapted largely from preexisting biochemical machinery. Any chemical reaction will go equally well in reverse if enough energy is applied to drive it uphill. The driving energy in photosynthesis is supplied by NADPH and ATP.

So far, nothing has been said about light or solar energy. The Calvin reactions can take place in the total absence of light so long as CO_2, ATP, and NADPH are present, and they are therefore called the DARK REACTIONS of

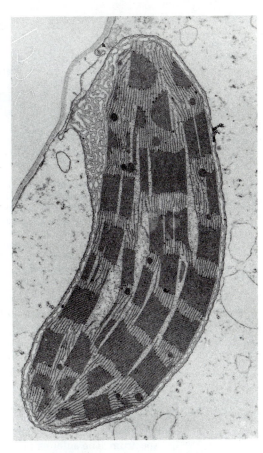

11

CHLOROPLAST is the site of photosynthesis, the process that converts the energy of sunlight into chemical energy. Parallel bands are membranes that contain photosynthetic pigments.

photosynthesis. The sole function of light is to maintain the supply of ATP and NADPH.

Just as cellular respiration takes place within mitochondria, so photosynthesis occurs in chloroplasts (*Figure 11*). The dark reactions occur in solution within the chloroplasts. The light-trapping reactions occur within stacked platelets called grana, visible in Figure 11. All of the chlorophylls and other light-absorbing pigments, and the enzymes and cytochromes that use this light energy to make ATP and NADPH, are intricately organized in these grana in a way that resembles the probable organization of the terminal respiratory chain in the walls of mitochondria.

LIGHT REACTIONS

Light is trapped in photosynthesis by several pigments: chlorophylls, carotenoids, and phycobilins. Some of the molecules of chlorophyll are located at REACTION CENTERS in the grana,

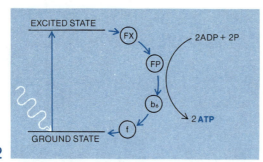

EXCITED STATE

FX

FP

2ADP + 2P

b₆

2 ATP

f

GROUND STATE

12

CYCLIC PHOSPHORYLATION uses the energy of sunlight to synthesize two molecules of ATP. Chlorophyll absorbs a photon of light (wavy arrow), which excites one of its electrons to a higher energy level. Plants harness the energy of the excited electron to drive a chain of reactions that leads to the synthesis of two molecules of ATP (right). FX is ferredoxin, FP flavoprotein, P inorganic phosphate, and b_6 and f are cytochromes.

where the actual separation of a negative and a positive charge occurs. The negative charge, an excited electron, can then be used to synthesize ATP and NADPH. Other chlorophyll and various pigment molecules spread around these reaction centers act as a light-gathering antenna system. Together they absorb light over a range of wavelengths that extends from the visible to the near-infrared region of the spectrum. The antenna pigments transfer their excitation energy to the chlorophylls at the reaction center.

The chlorophylls are particularly good absorbers of visible light. The electronically excited chlorophyll molecule becomes a very good reducing agent. It can pass its excited electron on to another molecule, thus reducing it. The second molecule can then reduce a third and so on, as the electron cascades down the energy scale. In CYCLIC PHOSPHORYLATION (Figure 12) an excited chlorophyll molecule reduces a molecule of ferredoxin, which is reoxidized as it reduces a flavoprotein. The flavoprotein is reoxidized as cytochrome b_6 is

13

PURPLE SULFUR BACTERIA of the genus Chromatium use hydrogen sulfide as their hydrogen source. Granules of sulfur are visible within cells.

reduced, b_6 reduces cytochrome f, and f passes the electron back to chlorophyll in the ground (unexcited) state. Along the way, the energy of excitation that had come from light is used to synthesize two molecules of ATP. Cyclic phosphorylation resembles the terminal respiratory chain (Figure 5), but with excited chlorophyll as the reducing agent instead of NADH.

For many photosynthetic bacteria, the ATP produced by cyclic phosphorylation suffices as an energy source. The bacterium Chromatium can use hydrogen gas as the source of hydrogen atoms for the synthesis of glucose. This scheme may have been an efficient one when the Earth's atmosphere had large amounts of free H_2, but is not a generally usable mechanism now. Organisms need a more dependable supply of hydrogen atoms for producing NADPH. Purple and green sulfur bacteria that live in hot sulfur springs use H_2S as their hydrogen source, either excreting solid sulfur or storing it as a byproduct (Figure 13). This system has the disadvantage of somewhat restricting the habitable range of the organisms that use it. An obvious and far more widespread source of hydrogen atoms is water, H_2O.

The LIGHT REACTIONS of photosynthesis in higher plants are diagrammed in Figure 14. Two kinds of reaction centers are involved: pigment systems I and II. Both systems contain chlorophyll a. Other pigments, such as chlorophyll b and carotenoids, are usually associated with system II, but may be involved in system I as well.

Excited electrons in system I can tumble back to the ground state, using the energy of excitation to synthesize ATP from ADP and inorganic phosphate. This is simply cyclic phosphorylation. But they can do something else: they can reduce $NADP^+$ to NADPH. This provides a supply of NADPH for the dark photosynthetic reactions, but leaves pigment system I electron-deficient. After one excitation of electrons, the reaction center must either find an alternate supply of electrons or shut down.

The machinery of photosynthesis consists of a set of dark reactions to synthesize glucose from CO_2, and a set of light reactions that fuel the synthesis by supplying ATP and NADPH.

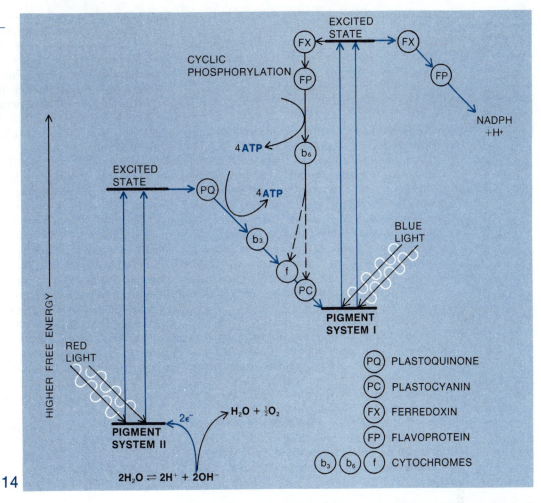

LIGHT REACTIONS of photosynthesis. Four photons of light are absorbed, two by pigment system II in the red part of the spectrum, and two by pigment system I in the blue region. Colored arrows show pathways of excited electrons. One molecule of water is oxidized to oxygen gas and its two hydrogens are used to reduce a molecule of $NADP^+$ to NADPH and H^+. The process also generates four molecules of ATP. An alternate pathway for excited electrons in pigment system I is cyclic phosphorylation (Figure 12), shown by dashed arrow at left center.

The alternate source is electrons from the other kind of reaction center (*Figure 14, left*). These excited electrons are passed to plastoquinone, cytochromes b_3 and f, plastocyanin, and finally to the pigments of system I and its chlorophyll a. During this fall from the excited state of system II to the ground (unexcited) state of system I, part of the energy released is harnessed to synthesize four ATP molecules.

But where did the electrons come from that were excited in system II? They came ultimately from water molecules, by the reaction

$$2 \, H_2O \rightarrow 4 \, H^+ + 4 \, \epsilon^- + O_2.$$

The overall result is that water is split into oxygen gas, hydrogen ions, and electrons. The electrons receive light energy in two different reaction centers, and release this energy to produce ATP and NADPH. The protons and electrons from water finally rejoin one another in the reaction

$$2 \, H^+ + 2 \, \epsilon^- + NADP^+ \rightarrow NADPH + H^+.$$

Oxygen gas is ejected as a waste product, just as sulfur was ejected by green sulfur bacteria.

In brief, the structure of the photosynthetic machinery consists of a set of dark reactions to synthesize glucose from CO_2, and a set of light reactions that fuel the synthesis by supplying ATP and NADPH. As a result photosynthetic organisms do not have to depend on other organisms for nourishment. All they need is sunlight, carbon dioxide, and a suitable source of hydrogen atoms, such as water.

LIGHT AND LIFE

The planet Earth can support life only so long as its supply of sunlight persists. The evolution of life on Earth has led from unorganized chemicals to highly organized creatures. Individually and collectively, living organisms seem to contradict the Second Law of Thermodynamics, which states that spontaneous processes tend to run downhill — that is, toward states of lower energy and greater disorder. How do living systems carry out thousands of spontaneous chemical reactions per second and still maintain their remarkable degree of organization?

Living systems do not violate the Second Law. In the language of the physicist, a living creature is an "open" thermodynamic system. It can maintain a high level of organization only by extracting free energy from the environment, either in the form of sunlight or of energy-rich molecules, and dumping back into the environment an assortment of low-energy breakdown products such as carbon dioxide.

Shutting off the Sun would have the same effect as enclosing the planet in a steel coffin.

All life presently on Earth would eventually dwindle to extinction. If life has evolved and become more complex, it is only because a virtually infinite source of free energy was at hand. But the Sun is slowly burning out.

The Earth-Sun system obeys the Second Law and is running down. The energy resources of the Sun are so vast that the comparatively small free-energy changes on Earth are negligible in comparison. With a radius of 4000 miles and a distance of 93 million miles from the Sun, the Earth intercepts less than one part in 10^{17} of the Sun's total radiation. This tiny fraction of the solar output has been enough to transform the planet from a lifeless water-covered sphere to the habitat of millions of species of living creatures.

According to current theories of stellar evolution, the Sun should continue to shine at about its present brilliance for tens to hundreds of millions of years. As it depletes the fuel in its core, its thermonuclear fires will begin to consume its outer shell. The Sun will change from a benign golden disk to a swollen red giant, enormously larger and brighter than it is today. Its heat will singe the Earth, boiling away the oceans and destroying all forms of life. Its fuel almost exhausted by this outburst, the dying Sun will cool and collapse, first to a white dwarf star, and finally to a dark cinder, still circled by the dead planets to which it once gave light and life.

READINGS

R.E. DICKERSON, *Molecular Thermodynamics,* New York, W.A. Benjamin, 1969. Chapter 7, "Thermodynamics and Living Systems," is a restatement of many of the ideas of this chapter, with more emphasis on the thermodynamics and more attention to the citric acid cycle.

H.A. KREBS AND H.L. KORNBERG, *Energy Transformations in Living Matter: A Survey,* Berlin, Springer-Verlag, 1957. The source of all of the thermodynamic data in this chapter. Individual free energy values have been questioned since 1957, but this is still the most complete and self-consistent set of data available.

A.L. LEHNINGER, *Bioenergetics,* 2nd Edition, Menlo Park, Calif., W.A. Benjamin, 1971. Thorough, and beautifully clear, but considerably more detailed than this chapter. Recommended as a follow-up.

L. MARGULIS, *The Origin of Eukaryotic Cells,* New Haven, Conn., Yale Univ. Press, 1970.

L. MARGULIS, "Symbiosis and Evolution," Scientific American, August 1971. The best current statements of the theory that mitochondria and chloroplasts were once separate but symbiotic organisms.

C.U.M. SMITH, *Molecular Biology: A Structural Approach,* Cambridge, Massachusetts, M.I.T. Press, 1969. Chapter 10, "Bioenergetics," has a particularly clear elementary introduction to photosynthesis.

L. STRYER, *Biochemistry,* San Francisco, W.H. Freeman, 1975. A clear and elementary biochemistry textbook.

W.B. WOOD, J.H. WILSON, R.M. BENBOW, AND L.E. HOOD, *Biochemistry: A Problems Approach,* Menlo Park, W.A. Benjamin, Inc., 1974. A serious book that teaches biochemistry through interesting problems. Hard work but lots of fun.

NOTES

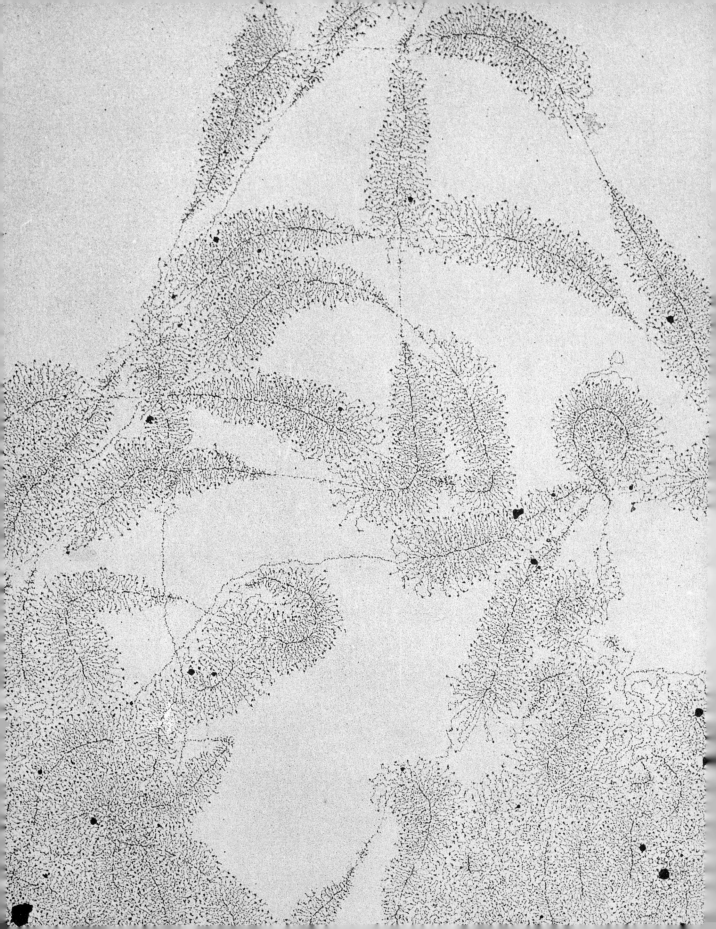

4

information, codes & genes

If, chemically, the components are the same and are synthesized by the same processes in all living beings, what is the source of their prodigious morphological and physiological diversity? And even more puzzling, how does each species, using the same materials and the same chemical transformations as all the others, maintain unchanged from generation to generation the structural standard that characterizes it and differentiates it from every other? We now have the solution to this problem. The universal components — the nucleotides on the one side, the amino acids on the other – are the logical equivalents of an alphabet in which the structure and consequently the specific functions of proteins are spelled out. In this alphabet can therefore be written all the diversity of structures and performances the biosphere contains.

JACQUES MONOD,
CHANCE AND NECESSITY

The human mind has long been intrigued by the idea of assembling a human being from a collection of parts. Anyone who set out to create an organism, molecule by molecule, from atomic raw materials would face two problems. He would need a tool that could manipulate individual molecules, and a blueprint that contains a great deal of information. In nature enzymes are the tool for manipulating molecules, and nucleic acids, principally DNA, provide the blueprint.

When this blueprint is passed from one generation to the next it must be reproduced accurately and distributed to the offspring in an orderly way. If it were not, offspring would not resemble their parents. Every living organism is the product of billions of years of natural selection. Any organism that differed sharply from its parents would almost certainly fail to survive. But if the blueprint were always copied with absolute fidelity, evolutionary progress would be impossible. Errors in copying do occur, from time to time, and once they have occurred, they are copied with the same high fidelity as the original instructions. Known as MUTATIONS, these errors are the only source of new genes. Together with sexual reproduction, which creates new combinations of existing genes, they are the mainspring of evolution (*Chapter 21*). If the mutation leads to a better-adapted organism, it will be preserved by natural selection and will become part of the new blueprint.

Every cell contains mechanisms for extracting the information in DNA and using it to direct cellular activities. In the main, the instructions that must emerge from the DNA will specify the order of amino acids in each protein chain, how much of each protein is to be made, and when to make it. DNA does not participate directly in the synthesis of protein; it makes disposable copies of appropriate parts of its information in the form of molecules called MESSENGER RNA. The messenger molecules directly control the sequence of amino acids in proteins. In short, there are three kinds of informational macromolecules in cells: DNA, RNA, and proteins. Information flows only from DNA to RNA to protein, never in the reverse direction (*Box A*).

DNA AND RNA MOLECULES form featherlike structures in this electron micrograph of genetic material from the egg of the spotted newt. The backbone of each "feather" is a long strand of DNA coated with protein. Many RNA molecules extend in clusters from the DNA strand. Transcription of genetic information begins at one end of the gene, with the RNA molecules growing progressively longer. Up to 100 RNA molecules are transcribed simultaneously from each gene.

MOLECULAR INFORMATION

DNA contains linear sequences of four different nucleotides, each of which can be considered as a symbol. Any sequence of symbols can carry information. Proteins consist of linear sequences of 20 different amino acids. The four nucleotides in DNA must specify the 20 amino acids in proteins, a problem rather similar to specifying the 26 letters of our alphabet with the dots and dashes of Morse code. The telegrapher's solution is to use various combinations of dots and dashes to construct a code for the letters. In converting information from nucleotide sequences in DNA to amino acid sequences in protein, nature's solution is to use combinations of nucleotides to specify amino acids. If each nucleotide were assigned to code a particular amino acid, only four different amino acids could be coded. If PAIRS of nucleotides were used to specify single amino acids, such that AA was one amino acid, AT was another, TA was yet another, and so on, 16 amino acids could be coded. If the DNA code were perfectly analogous to the Morse code, one could specify four of the amino acids by a single nucleotide apiece, and the remaining 16 by nucleotide doublets (two nucleotides). Nature does not employ a mixed code of this sort, but uses a code in which every amino acid is coded by a triplet of nucleotides. Sequences of three nucleotides can code for 64 amino acids. For a short time after it was first understood that the code must be based on triplets, biologists thought that the extra 44 triplets might be untranslatable by the cellular decoding machinery. It is now known that all 64 possible triplets mean something to the cell. Three of them carry the message: "Don't insert any amino acid; terminate the protein chain." The other 61 all specify one of the 20 amino acids, so that on the average, there are about three synonymous triplets to represent each acid.

A linear sequence of nucleotides in DNA specifies a linear sequence of amino acids in proteins. Because there is no information in proteins that is not also present in DNA, it might seem that DNA, like proteins, could catalyze the chemical reactions of living organisms. It cannot. By its chemical nature, DNA is unable to fold itself into the complicated functional shapes that characterize molecules like chymotrypsin and carboxypeptidase. When biologists first began to speculate on the nature of the hereditary material, it was not at all obvious that DNA was a likely candidate, since it seemed to have no biological activity.

THE SEARCH FOR THE HEREDITARY MATERIAL

In 1866, Gregor Mendel, the father of modern genetics, laid down the rules of hereditary transmission of traits. Meanwhile, chemists were starting to take cells apart and analyze their composition. DNA was discovered in pus cells and in fish sperm just three years after Mendel's discovery. Yet it was not until about 1950 that the idea of DNA as hereditary material was generally accepted. Since it was possible to see a relationship between DNA and hereditary material much earlier than 1950, why was this insight so long delayed? The chief reason was that biologists believed that the gene must be a complex structure if it was to perform the complicated job of bearing information. Chemical analysis suggested that

---

A

THE CENTRAL DOGMA

What is semihumorously referred to as The Central Dogma of information transfer is: "DNA makes RNA makes protein." In 1964, Howard Temin of the University of Wisconsin studied an RNA virus that causes a cancer of chickens known as the Rous sarcoma. The virus enters the chicken cell and subsequently causes the cell to make a DNA copy of the viral RNA. The afflicted cell does not burst, but changes permanently in shape, metabolism, and growth habits. The new DNA becomes part of the hereditary apparatus of the chicken cell. More recently, Temin and others have shown that the virus carries an enzyme for the manufacture of DNA, using viral RNA as an informational template. This discovery does not invalidate the Dogma. In its original and proper form, the Dogma merely stated that information in proteins could not flow back to nucleic acids. The flow of information from RNA to DNA does not contradict this.

DNA was rather an uncomplicated molecule. It was therefore shunned by the theoreticians who speculated on the chemical nature of the gene. The decades between 1900 and 1940 brought a great flowering of formal genetics (the rules for the transmission of hereditary information) but the chemical nature of the gene remained a mystery.

BACTERIAL TRANSFORMATION

In the 1940s Oswald T. Avery and his colleagues at Rockefeller Institute demonstrated that DNA could carry genetic information. They were investigating a phenomenon discovered by Frederick Griffith in 1928. When injected into mice, one strain of the *Pneumococcus* bacterium causes septicemia (blood poisoning) and usually death. Another strain of the bacterium is relatively harmless. The infective form always has a capsule: a complicated polysaccharide coating; the noninfective form does not. Griffith called the encapsulated strain the "S strain" because the colonies looked smooth on a culture plate, and the harmless, unencapsulated one the "R strain" because of its rough colonies (Figure 1). Griffith also found that when large numbers of S bacteria were grown outside of animals, he could obtain a few R bacteria. These proved to be mutants that had lost the ability to make a capsule. Mice injected with R bacteria and simultaneously with a large number of heat-killed S bacteria developed septicemia and died; live S bacteria were found in their blood. Later work was to show that this process, called TRANSFORMATION, was not a matter of breathing life back into the dead bacteria, but rather that something from the dead bacteria was converting R bacteria into S bacteria.

Avery and his group isolated the material responsible for the transformation. Instead of adding heat-killed S bacteria, they made an extract of heated S cells and started purifying the material by removing, one after another, those substances that did not cause transformation of R bacteria into S bacteria. Finally they were left with virtually pure DNA. Next

they did something even more impressive. Different kinds of S Pneumococci form chemically different capsules. If the geneticists extracted DNA from a particular kind of S strain and added it to R bacteria, the type of capsule made by the transformed bacteria was always identical with that of the bacteria that donated the DNA, indicating the hereditary information must be carried by DNA. Geneticists now know that virtually any genetic trait

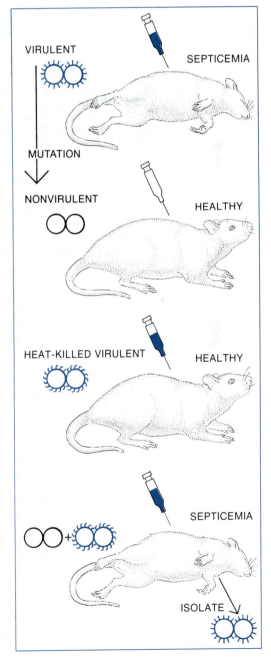

1

GENETIC TRANSFORMATION. Noncapsulated pneumococcus was transformed into a virulent capsulated strain (color) when mixed with dead virulent strain. Genetic material from dead bacteria entered the live ones, transforming some of them into capsulated strain. Mice injected with noncapsulated strain remained healthy. Mice injected with capsulated strain died of blood poisoning.

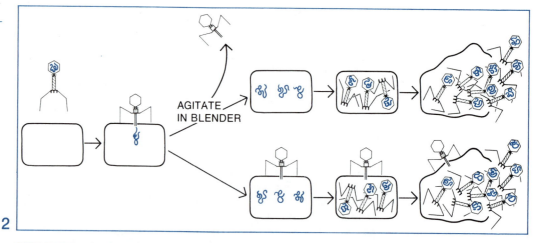

2

LIFE CYCLE OF BACTERIOPHAGE T2 and T4 growing on E. coli. Phage attaches itself by its tail fibers to the surface of the bacterial cell (left) and injects virus DNA (color) into the cell. The DNA programs the cell to produce more phage viruses. Eventually the cell bursts, releasing the phages to infect other cells. Empty coat of phage can be shaken loose from cell in a blender.

in *Pneumococcus* or in several other genera (singular, *genus*) of bacteria can be passed, via DNA, from one bacterium to another.

There was still room for skepticism about whether DNA could carry all of the hereditary information of any organism. In 1952, A.D. Hershey and Martha Chase of the Carnegie Institution of Washington proved it could. They studied the life cycle of the bacteriophage T2. A bacteriophage, or phage, is a virus that grows in a bacterial host cell (*Figure 2*). Bacteriophage T2 is a tadpole-shaped virus that consists of about 50 per cent DNA and 50 per cent protein. It can infect the bacterium *E. coli* (and a few very close relatives) by attaching by its tail to the bacterial cell wall. About 20 minutes later the bacterium bursts, releasing about 200 new T2 viruses. Hershey and Chase found that the protein coat remains on the outer surface of the bacterium: if the bacteria are put into a blender a few minutes after infection, the virus coats can be sheared off, leaving the DNA core of the viruses inside the bacteria. As before, the bacteria eventually burst, producing a new generation of phage particles. Evidently the protein husk of the phage acts only

as a syringe to inject the hereditary material into the bacterium. Hershey and Chase prepared bacteriophages in which the phosphorus of the DNA was labeled with radioactive phosphorus, P^{32}, and others in which the protein portion was labeled with radioisotope of sulfur, S^{35}. Since DNA contains no sulfur, and proteins generally contain little or no phosphorus, these two isotopes could be used to follow the fate of DNA and protein through the life cycle. The investigators found that if bacteria were infected with these phages, much of the P^{32} turned up in the new phages, but practically none of the S^{35}, demonstrating that the DNA alone was the link between generations.

DNA DUPLICATION

There are two possible ways to duplicate an object, such as a line of type. One is to set an identical line of type; the other is to make a mold of the type, and use the mold to cast a second line just like the first. The first process requires a complicated intermediary device, a human typesetter or an engraver. The second process — nature's way of duplicating DNA — can be executed directly by a relatively simple

When the DNA molecule duplicates itself the two strands separate, and each serves as a template for the manufacture of a complementary new strand.

machine. In effect, the cell makes a chemical mold of the sequence of nucleotides in a strand of DNA. The mold itself is made of nucleotides, but differs from the original. When a cast is made from the mold, the resulting nucleotide sequence is identical with the original. In nature, the original and the mold are wrapped around each other in the form of a double-stranded helix. One of the two complementary strands is whimsically called "Watson," the other "Crick." The nitrogenous bases in the two strands are always paired in such a way that an adenine (A) in one strand always lies opposite a thymine (T) in the other, and a guanine (G) always lies opposite a cytosine (C). Knowing the pairing rules that apply to the bases in the two strands, one can easily write the base sequence of one strand given the base sequence of the other. If the order of bases in Watson is A-A-G-T-C-T-C, for example, the order in the Crick strand must be T-T-C-A-G-A-G, with the sugar-phosphate backbone of the Crick strand running in the antiparallel (opposite) direction from that of the Watson strand (*Figure 3*). Thus each strand contains the information needed to specify the structure of the other. When the DNA molecule duplicates itself the two strands separate, and each serves as a template for the manufacture of a complementary new strand. In a solution containing the four nucleotide components of DNA, the template strand attracts the base of the next nucleotide to be added. An enzyme then moves one notch along the molecule like the slide of a zipper, forming a phosphodiester bond. This fastens the newly added nucleotide to the growing strand.

When investigators first started thinking about the duplication of the double helix, it was apparent that there were three general ways in which it might possibly happen. First, a particular Watson and a particular Crick might be destined to stay paired forever. At the time of DNA synthesis, the Watson would synthesize a new Crick, and the Crick would synthesize a new Watson. The new Crick and new Watson would likewise stay paired forever. This possibility was called conservative replication. Second, a particular Watson and a particular Crick might each pair with the new complementary strand whose synthesis it has just guided, so that every double helix of DNA would consist of an old strand and a new strand. This was called semiconservative replication. Third, the Watson and the Crick strands could be somehow broken into small

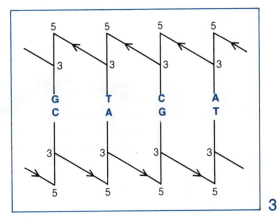

3

ANTIPARALLEL ORIENTATION of the two strands in a double helix of DNA. The direction of a strand depends on the direction of the phosphodiester bonds from the 3′ position of a deoxyribose sugar to the 5′ position of the next one. Phosphodiester bonds are shown as arrows and the sugars are shown as straight vertical lines.

fragments at the time of DNA synthesis and new materials introduced, so that both strands of every DNA molecule would always consist of a mixture of old and new materials. This was called dispersive replication (*Figure 4*).

THE MESELSON-STAHL EXPERIMENT

DNA is actually duplicated by a semiconservative mechanism. The experiment that demonstrated this fact was done in 1958 by Matthew S. Meselson and Franklin W. Stahl, then at the California Institute of Technology. Their experiment is a model of close reasoning and imaginative design. Their approach depended on a new technique, called density-gradient centrifugation, which they developed in collaboration with Jerome Vinograd. The technique enabled them to separate DNA molecules that differed slightly in density (weight per unit volume).

Meselson and Stahl realized that the density of DNA could be modified by incorporating into DNA heavy isotopes of its normal atoms. For example, nitrogen accounts for about 17 per cent of the total molecular weight of the DNA in *E. coli*. Replacing all of the N^{14} normally found in *E. coli* DNA with N^{15}, would increase the density of the molecule about 1.2 per cent. Meselson and Stahl found that they could grow *E. coli* on a medium in which the only source of nitrogen was ammonium chloride containing N^{15}, of high isotopic purity. Cells grown on this medium made "heavy"

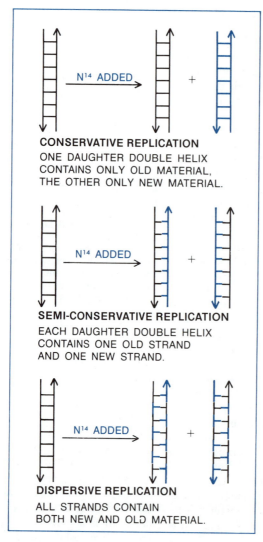

CONSERVATIVE REPLICATION
ONE DAUGHTER DOUBLE HELIX
CONTAINS ONLY OLD MATERIAL,
THE OTHER ONLY NEW MATERIAL.

SEMI-CONSERVATIVE REPLICATION
EACH DAUGHTER DOUBLE HELIX
CONTAINS ONE OLD STRAND
AND ONE NEW STRAND.

DISPERSIVE REPLICATION
ALL STRANDS CONTAIN
BOTH NEW AND OLD MATERIAL.

4

THREE TYPES OF REPLICATION are possible during the synthesis of new DNA molecules. The original DNA molecule is shown in black and the new material in color. The Meselson-Stahl experiment described in text indicated that the replication of DNA is semi-conservative.

DNA that could easily be separated from normal DNA by spinning the DNA solution in an ultracentrifuge (Figure 5).

After they had grown the bacteria for many generations in the N^{15} medium so that virtually all the DNA was N^{15} DNA, Meselson and Stahl abruptly transferred the culture of cells to a medium containing N^{14}. All DNA made after the transfer would contain the normal isotope; the average density of the DNA extracted from the bacteria would decrease as the cells resumed the synthesis of normal DNA. Samples were taken from the culture at various times after the change of medium, and DNA extracted from them was spun down in the centrifuge to trace the fate of the heavy N^{15}-labeled DNA. During the first generation of growth following the change of medium, the DNA progressively became lighter until it reached a density midway between that of the heavy DNA and normal DNA. After the next generation, half of the DNA was light and the other half showed an intermediate density corresponding to a hybrid of heavy and normal DNA. In subsequent generations, more and more DNA showed the normal density, but the amount of hybrid DNA remained unchanged. Evidently the heavy parental strands of DNA became associated during synthesis with an exactly equal amount of light DNA.

This conclusion was readily demonstrable by heating DNA to temperatures high enough to "melt" the hydrogen bonds that hold the two strands together. Meselson and Stahl found that when they heated hybrid DNA, they released equal amounts of light and heavy single-stranded DNA. The first duplication of DNA separates the N^{15}-labeled strands so that both enter into DNA of hybrid density. During subsequent cycles of DNA synthesis, the total amount of hybrid DNA remains constant, though it becomes a smaller and smaller fraction of the total DNA. This experiment has been repeated with a variety of organisms, including algae and human cells, and the result has always been the same: the synthesis of DNA is semiconservative.

GENOTYPE AND PHENOTYPE

Two organisms are said to have the same GENOTYPE if they carry identical hereditary information. They are said to have different genotypes if they have any differences in their hereditary information, even if an observer cannot readily tell them apart. The visible expression of genetic characteristics is the organism's PHENOTYPE. Phenotype is the product of the genotype interacting with the environment. Some expressions of the genotype are very resistant to the effects of the environment, such as blood type, or the color of the eyes. Others, such as stature at maturity, skin color, or athletic ability, are more readily influenced by environmental differences.

When a mutation occurs in the genotype, the change in the phenotype may be complex, such as an alteration in shape or behavior pattern, so that the phenotype gives little indication of the nature of the change in cellular instructions. However, as long as the altered phenotype is heritable and easily characterized, the superficial abnormalities of the affected individual serve as evidence of a change in his genes, even if the molecular anomaly remains unknown.

MUTATIONS

Mutations, heritable changes in the genetic information, are so slight that DNA chemists rarely can detect them by chemical analysis. But minute changes in the genetic material often lead to easily observable changes in the outward form and function of the individual. The detection of a mutation depends on the ability to observe its effects. Some mutations are obvious in humans — dwarfism, for example, or the presence of more than five digits on each hand. A mutation may be almost equally obvious in a microorganism — for example, one that results in a change in color, or in nutritional requirements. Other mutations may be virtually unobservable.

Mutations are the raw material of evolution, and the working materials of genetics. The only method for "labeling" a gene to discover its function is to find a mutation affecting that gene. In fact, geneticists frequently refer to mutations that alter the function of genes as "markers" because they provide a means for specifically studying the behavior of individual genes.

Some mutations involve extensive chemical changes in the structure of DNA. Others, called POINT MUTATIONS, change only a single nucleotide. Point mutations can generally REVERT — they can mutate back to the original form. Extensive mutations, often called CHROMOSOME ABERRATIONS, may be REARRANGEMENTS, which change the position or direction of a DNA segment without actually removing any genetic information, or DELETIONS, in which a segment of DNA is irretrievably lost.

All mutations are rare events. The observed frequencies of mutations are different for different organisms and for different genes within

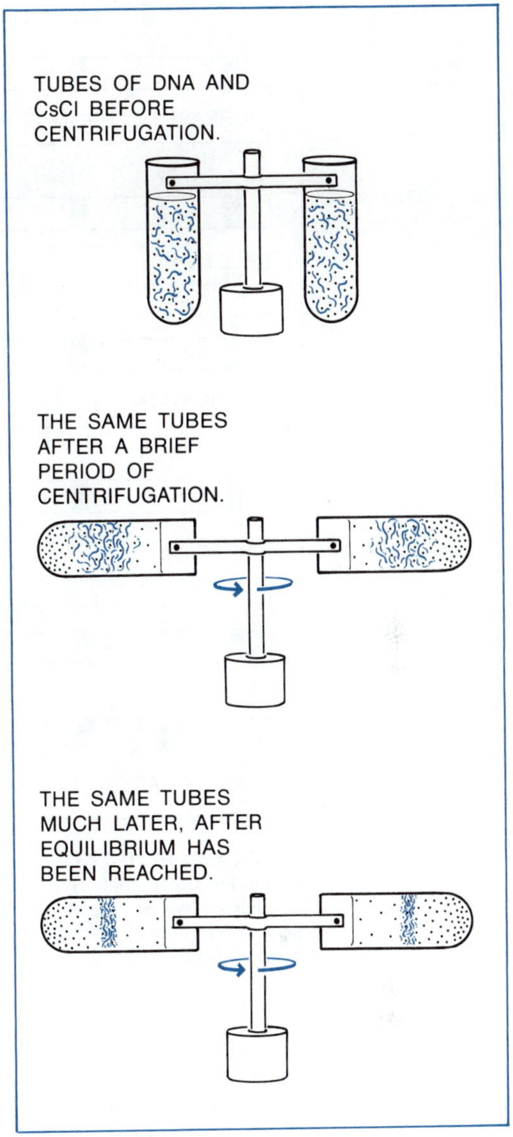

TUBES OF DNA AND CsCl BEFORE CENTRIFUGATION.

THE SAME TUBES AFTER A BRIEF PERIOD OF CENTRIFUGATION.

THE SAME TUBES MUCH LATER, AFTER EQUILIBRIUM HAS BEEN REACHED.

5

ULTRACENTRIFUGE can separate normal DNA from "heavy" DNA containing N^{15}. This schematic diagram depicts the technique known as density-gradient ultracentrifugation, in which the DNA molecules (color) are suspended in a solution of cesium ions (black dots) and spun until they reach an equilibrium position. In effect, this technique provides a very sensitive method of separating DNA molecules of slightly different weights.

Mutations are the raw material of evolution, and the working material of genetics. The only method for labeling a gene to discover its function is to find a mutation affecting that gene.

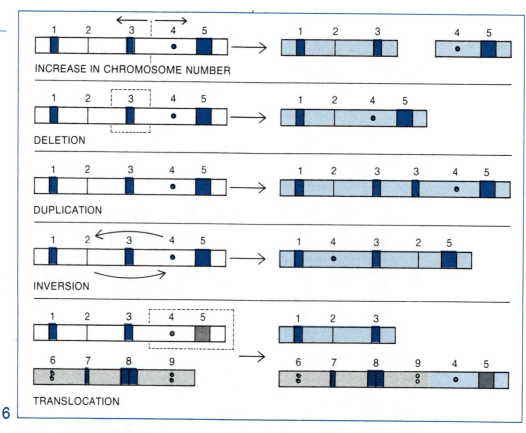

INCREASE IN CHROMOSOME NUMBER

DELETION

DUPLICATION

INVERSION

TRANSLOCATION

6

MODEL OF CHROMOSOME depicts five types of aberration. Chromosome resembles putty stick that can be broken and reassembled in a limited number of ways. Bands and numbers are included as visual reference points. When an altered chromosome (color) pairs with a normal one, they must undergo contortions to match up during synapsis and meiosis.

a given organism. Usually the frequency of mutation is lower than one mutation per 10^4 genes per DNA duplication, and sometimes the frequency is as low as one mutation per 10^9 genes. Geneticists speak of mutation frequencies per DNA duplication because it is likely that most mutations in nature occur during the copying of the genetic message. The majority are point mutations resulting from the substitution of one nucleotide for another or from the addition or deletion of a single nucleotide during the synthesis of a new DNA strand.

A number of chemicals can induce mutations. These MUTAGENS include purines and pyrimidines not found in natural DNA but enough like the natural bases that the cell can build them into its DNA. These "foreign" bases are mutagenic presumably because they are even more likely than the natural DNA bases to undergo the rare molecular alterations that cause mispairing. Certain other mutagenic

agents such as hydroxylamine and nitrous acid directly alter the structure of the natural bases in DNA, changing them to other bases that have a tendency to mispair.

The mechanism by which single base pairs are added to or subtracted from DNA is a mystery. One family of molecules is notorious for inducing such mutations: the acridine dyes, some of which are used in the treatment of malaria. Mutations resulting from the addition or deletion of single base pairs are known as FRAME-SHIFT mutations, because they interfere with the decoding of the genetic message by throwing the decoding apparatus out of register. Recall that the information in a gene is decoded by a mechanism that starts at one end of the genetic message and reads it as if it were a continuous sequence of three-letter words, each word corresponding to a particular amino acid in the desired protein product. If a base is added to the message or subtracted from it, the decoding process will work perfectly until it comes to a mutational change. From that point on the three-letter words in the message will be one letter out of register. In other words, such mutations shift the "reading frame" of the genetic message. Frame-shift mutations almost always lead to the production of completely nonfunctional proteins.

A simple analogy can illustrate the concept of frame-shift mutations. If a message contains only three-letter words there is no need to leave spaces between the words, so long as they are read in the right register:

THECOPSAWTHEMANAIMHISGUN.

Now consider the effect of deleting a single letter, analogous to the deletion of a single nucleotide:

THECOPSWTHEMANAIMHISGUN.

It is small wonder that a frame-shift mutation is so disruptive. An organism carrying such a mutant gene can survive only if the gene product affected is not an essential part of the cellular machinery or if the organism also carries another copy of the gene in its normal form. The effects of point mutations are usually much less drastic. They often change the genetic message so that one amino acid is substituted for another in the protein. Such a substitution may sometimes cause the protein to be completely nonfunctional, but usually it only reduces the protein's functional efficiency. Individuals who carry point muta-

tions may survive even though the affected protein is absolutely essential to life.

Chromosome aberrations all involve the breakage and rejoining of the genetic strands, with gross disruption of the sequence of genetic information. These major alterations, involving not one but hundreds or thousands of nucleotide pairs, are visible under the microscope. Chromosome aberrations are often regarded by students as rather difficult to master, but there is an easy way to comprehend them. With the aid of Figure 6, think of a chromosome as a stick of putty and imagine all of the ways to modify it without twisting or pulling it out of shape. The putty stick can be broken up to create a larger number of smaller sticks (increase in chromosome number). It can be joined to another putty stick (fusion, leading to reduction in chromosome number). One can take a piece out of the stick (deletion); double an existing piece (duplication); remove a piece, flip it over, and reinsert it (inversion); or transfer a piece to another stick (translocation). It is also possible simply to duplicate the entire stick. If the full set of chromosomes is duplicated — if the chromosome number per cell is doubled, tripled, or more — the result is called polyploidy. The way that mutation creates new genes is discussed briefly in Box B, and the crucial role of mutant genes in evolution is described in Chapter 21.

DECODING GENETIC INFORMATION

In double-stranded DNA, all of the genetic information is really present on each strand, but

---

The great majority of mutations are harmful to the organism that carries them, but once in a while a mutation improves the organism's adaptation to its ecological niche or enables it to invade a new niche. Duplication mutations may be the source of "extra" genes. The more complex creatures living on this planet seem to have more genes than the simplest creatures. Man, for example, has 1,000 times more genetic material than a bacterium. How do new genes arise in evolution? If whole genes are sometimes duplicated by the mechanism described in the text, the bearer of the duplication would have a surplus of genetic information that might be turned to good use. Subsequent mutations in one of the two copies of the gene would have no adverse effect on survival because the other copy of the gene would continue to turn out functional protein. Mutation after mutation could occur in the extra gene without ill effect. If the random accumulation of mutations in the extra gene should produce some useful message, natural selection would take full advantage of it. The result might be a new organism, an organism of greater complexity and versatility than its ancestors.

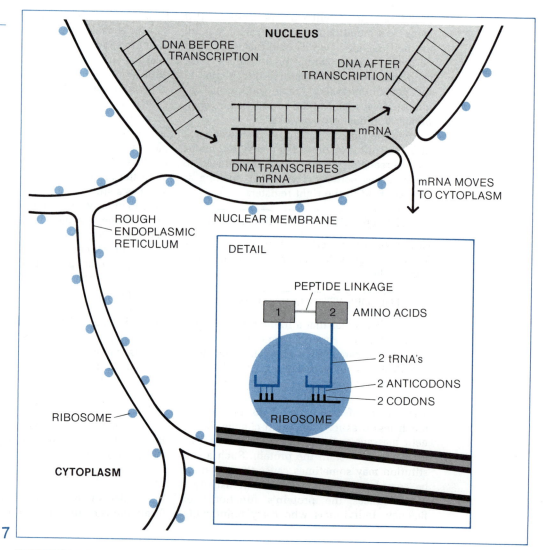

NUCLEUS

DNA BEFORE
TRANSCRIPTION

DNA AFTER
TRANSCRIPTION

mRNA

DNA TRANSCRIBES
mRNA

mRNA MOVES
TO CYTOPLASM

ROUGH
ENDOPLASMIC
RETICULUM

NUCLEAR MEMBRANE

DETAIL

PEPTIDE LINKAGE

1   2   AMINO ACIDS

2 tRNA's

2 ANTICODONS

2 CODONS

RIBOSOME

RIBOSOME

CYTOPLASM

7

FROM DNA TO PROTEIN. The steps of protein synthesis are summarized
in this schematic diagram. After transcription, mRNA moves into the
cytoplasm through a pore in the nuclear membrane. Only small segments
of the various nucleic acids are shown. The detailed view at center shows
the events that occur at a single ribosome. Two tRNA's, each carrying a
single activated amino acid, align themselves along the mRNA long
enough for a peptide bond (grey line) to form between the adjacent
amino acids. The details of transcription, translation and activation are
depicted in the next three diagrams.

in complementary form. One may ask whether
both the information from one strand and the
complementary information on the other
strand are used for the synthesis of protein.
Actually this would be comparable to trying to
devise a sentence that made good sense in English, and also made good sense in Russian
when read backward in the mirror. Therefore it
is not surprising that only one of the strands
provides information for protein synthesis.

Even this one strand is not used directly in
the process of protein synthesis. Its information is transferred by synthesizing a complementary molecule of RNA, using the DNA
strand as a template. This RNA is called MES-
SENGER RNA, abbreviated mRNA. The process
of synthesizing it is called TRANSCRIPTION.
This word was first applied to mRNA synthesis
by Sol Spiegelman, then at the University of
Illinois, who emphasized that the first step in

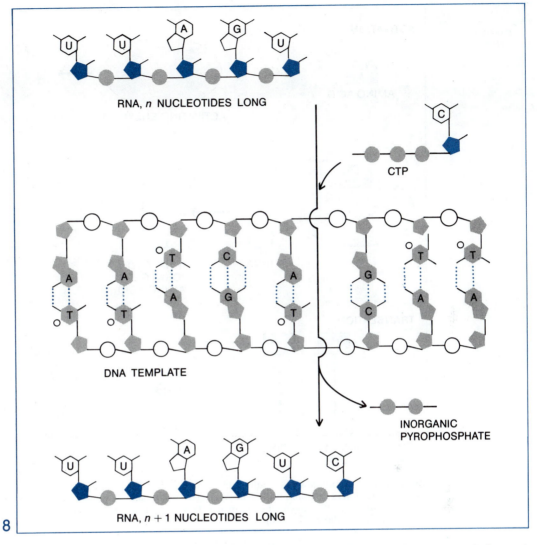

RNA, *n* NUCLEOTIDES LONG

CTP

DNA TEMPLATE

INORGANIC
PYROPHOSPHATE

RNA, *n* + 1 NUCLEOTIDES LONG

8

TRANSCRIPTION BY RNA POLYMERASE. Polymerase (vertical arrow) moves along DNA template. Here the upper strand of DNA is being transcribed. The resulting RNA is an exact copy of nontranscribed strand of DNA. Structures at top and bottom show addition of one nucleotide (cytosine) to a growing strand of RNA. Cytosine is originally present as cytosine triphosphate (CTP).

guiding the synthesis of proteins was merely a change of information from DNA to RNA. Since both are nucleic acids, the change is comparable to a change of script. Spiegelman contrasted this with the next step, in which the language itself is changed from the nucleotides in RNA to the amino acids in proteins, a process he called TRANSLATION. The terms are so vivid and appealing that they have stuck, no doubt permanently. A schematic overview of the entire process of protein synthesis appears in Figure 7.

Superficially the process of transcription is very much like the process of DNA synthesis. An enzyme called RNA polymerase attaches to the DNA molecule it is to copy, somehow choosing the correct strand of the double helix. It then moves along the strand, matching a particular base in the DNA strand (say, G) with the complementary base of a molecule of (ribo)nucleoside triphosphate, namely, cytosine triphosphate (CTP). The two terminal phosphate groups of CTP are split off, while the innermost phosphate becomes joined in an

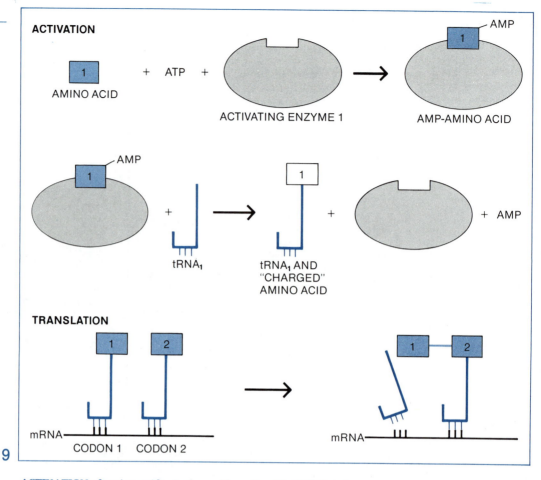

ACTIVATION

AMINO ACID + ATP + ACTIVATING ENZYME 1 → AMP-AMINO ACID

AMP

tRNA₁ → tRNA₁ AND "CHARGED" AMINO ACID + AMP

TRANSLATION

mRNA — CODON 1 CODON 2

9

ACTIVATION of amino acids. Amino acid reacts with ATP, forming an enzyme-amino acid complex (top). The 'charged' amino acid is plucked from the surface of the enzyme by the proper tRNA and transported to a ribosome for translation (bottom).

internucleotide (phosphodiester) bond with the nucleotide added to the growing chain just ahead of it. This adds a new (ribo)nucleotide to the chain. The enzyme moves one more (deoxyribo)nucleotide down the DNA chain, say to a T. It then pairs the complementary base, A, to it in the form of ATP, and so on (*Figure 8*). There is still much to be learned about transcription: how RNA polymerase picks the right strand of DNA to copy, how it knows where to start and where to stop, and even how the tightly-coiled double helix of DNA unravels enough to permit one of the strands to be copied.

The translation of mRNA into protein is much more complicated. The mRNA carries information in the form of nucleotide triplets called CODONS. Each codon stands for a particular amino acid. Translation of the message

results in the synthesis of a protein molecule with a sequence of amino acids corresponding exactly to the sequence of codons in the mRNA. The formation of peptide bonds is thermodynamically an uphill reaction. To synthesize proteins the cell must invest energy in the form of ATP. Protein synthesis begins with the activation of amino acids, a process that occurs in two steps. The first step, which occurs in free solution within the cell, is the formation of a high-energy intermediate compound, an AMP-amino acid (*Figure 9*). This reaction requires the presence of an activating enzyme. The cell contains some 20 different activating enzymes, each specific for one particular type of amino acid. Each catalyzes the formation of a different kind of AMP-amino acid, which remains attached to the enzyme until the amino acid is attached to a special sort of RNA called TRANSFER RNA (tRNA). The role of these tRNA molecules is to line up the amino acids in the order specified by the codons of the mRNA.

To simplify the picture temporarily, imagine that there are only 20 codons, one for each amino acid. (Actually there are 64 codons.) There is a separate kind of tRNA for each codon, and each of these tRNA's can recognize one specific activating enzyme bound to an AMP-amino acid. As shown in Figure 9, one end of the tRNA molecule then attaches itself to the appropriate amino acid, pulling it away from both the AMP group and the activating enzyme.

The next step, the decoding of the messenger, occurs on an elaborate biochemical workbench: the RIBOSOME. Any ribosome can act as the workbench for the synthesis of any protein. The reason is that the blueprint for the protein is not built into the ribosome but into the nucleotide sequence of the mRNA. The mRNA can be thought of as a series of codons with no punctuation marks between them. According to the rules of Watson-Crick pairing, each codon in mRNA will have an attraction

for a particular triplet (called an ANTICODON) in the tRNA helix. This will cause that tRNA, "charged" with its amino acid, to bind to the messenger. A series of codons in mRNA thus arranges amino acids in a definite order.

The process just described is the key to translation. The crucial molecule is the tRNA. This molecule must be recognized by the specific activating enzyme on one hand, and by the messenger on the other. The situation is analogous to the way a human driver and a car ignition "recognize" each other, with tRNA being the key. The driver recognizes the key by its octagonal plate at one end, and the car recognizes the key by the indentations in its other end. The result is decoding.

All that remains is to attach the amino acids to a growing protein chain. Protein synthesis starts at the end that will ultimately bear the free amino group. Once the chain has been started it grows, one amino acid at a time, toward the end that will ultimately bear the free carboxyl group. At any moment during growth, the most recently added amino acid (n) will still be attached to its tRNA (Figure 10). The tRNA, in turn, will still be paired by

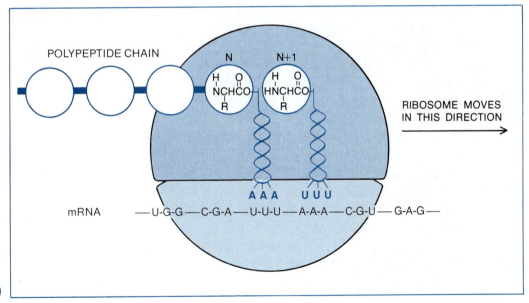

POLYPEPTIDE CHAIN

N   N+1

RIBOSOME MOVES
IN THIS DIRECTION →

A A A   U U U

mRNA   —U-G-G—C-G-A—U-U-U—A-A-A—C-G-U—G-A-G—

10

TRANSLATION OF mRNA. Ribosome (colored disk) slides along a molecule of mRNA, reading the codons that specify the next amino acid to be added to the end of a growing polypeptide chain. Ribosome is shown consisting of two subunits, a small one that binds to the mRNA and a large one on which peptide bonds are formed. Two spirals at center are molecules of tRNA shown with their specific anticodon triplets at one end. The polypeptide chain appears at top, with amino acids represented as circles. N is last amino acid on chain, N + 1 next to be added, and so on.

its anticodon to the n codon of mRNA. The n + 1 amino acid, also attached to a tRNA, then approaches, and the anticodon of this tRNA pairs with the n + 1 codon. Following this, the n amino acid lets go of its tRNA and attaches to the free amino group of the n + 1 amino acid, forming a peptide bond. The tRNA which has lost its amino acid drifts out of the way. The ribosome then moves along the messenger a distance of three nucleotides, putting the n + 1 amino acid (still attached to its tRNA) in the position on the ribosome formerly occupied by the n amino acid. The machinery is then ready for the n + 2 amino acid, and the whole cycle repeats.

Most messenger molecules are much longer than a ribosome, and therefore a single molecule of mRNA can be read by several ribosomes at once. The first ribosome threaded onto one end of the message will be the first to finish translating the message and to drop off the other end. The assemblage of a thread of mRNA with its beadlike ribosomes and their growing polypeptide chains is called a POLYRIBOSOME, or simply a POLYSOME. Large numbers of polysomes, and relatively fewer free ribosomes or ribosomal subunits, are found in cells that are synthesizing proteins very actively.

PUNCTUATION IN PROTEIN SYNTHESIS

To make a protein of a definite size, the cell not only must put the amino acids in the right order; it must also initiate and terminate the chain in the right places. Since a molecule of mRNA usually contains the information to make several proteins, start and stop signals are necessary to avoid synthesizing just a single gigantic protein. The nature of these signals has been described in part. The mRNA region between a start and a stop signal codes for an unbroken polypeptide or protein chain. The DNA corresponding to this stretch of mRNA is called a GENE, or a CISTRON. The word gene is far the older of these two words, and through the years has had a number of less precise meanings. It was for this reason that the word cistron was coined. However, the old word has reasserted itself, and is now used as a synonym for cistron. It is important that these words be understood as the unit which codes for a single polypeptide chain.

UNIVERSALITY OF THE CODE

Figure 11 shows the genetic code of E. coli expressed in mRNA language (codons). It was worked out painstakingly by a large number of enzymologists and organic chemists, especially Nirenberg, Khorana, and Ochoa. Work with organisms other than E. coli soon showed that the code was not just a local dialect used by that particular organism. The code proved to be identical in a eucaryotic bread mold, Neurospora crassa, and in yeast. mRNA has been isolated from the immature red blood cells of rabbits; such cells make almost nothing but hemoglobin, and much of their messenger is messenger for hemoglobin. When the rabbit messenger was combined with ribosomes, activating enzymes, and tRNA's from E. coli, the E. coli machinery proceeded to translate the rabbit mRNA into regular rabbit hemoglobin. This experiment, and analogous experiments in algae and a wide assortment of organisms suggest that all earthly creatures use a similar or probably identical code. Of course, this is not certain; the code has not yet been studied in snapdragons, or armadillos, or in most creatures. But no surprises are expected any more, and the matter is usually considered closed.

FIRST LETTER	SECOND LETTER				THIRD LETTER
	U	C	A	G	
U	PHENYLALANINE	SERINE	TYROSINE	CYSTEINE	U
	PHENYLALANINE	SERINE	TYROSINE	CYSTEINE	C
	LEUCINE	SERINE	(END CHAIN)	(END CHAIN)	A
	LEUCINE	SERINE	(END CHAIN)	TRYPTOPHAN	G
C	LEUCINE	PROLINE	HISTIDINE	ARGININE	U
	LEUCINE	PROLINE	HISTIDINE	ARGININE	C
	LEUCINE	PROLINE	GLUTAMINE	ARGININE	A
	LEUCINE	PROLINE	GLUTAMINE	ARGININE	G
A	ISOLEUCINE	THREONINE	ASPARAGINE	SERINE	U
	ISOLEUCINE	THREONINE	ASPARAGINE	SERINE	C
	ISOLEUCINE	THREONINE	LYSINE	ARGININE	A
	METHIONINE	THREONINE	LYSINE	ARGININE	G
G	VALINE	ALANINE	ASPARTIC ACID	GLYCINE	U
	VALINE	ALANINE	ASPARTIC ACID	GLYCINE	C
	VALINE	ALANINE	GLUTAMIC ACID	GLYCINE	A
	VALINE	ALANINE	GLUTAMIC ACID	GLYCINE	G

11

GENETIC CODE consists of three-letter words called triplets. Letters represent nucleotide bases uracil, cytosine, adenine and guanine. Each triplet specifies a single amino acid or a single instruction such as "end chain." To decode a triplet, read this grid across (left to right), then down, then across (right to left). Thus the triplet AUG designates the amino acid methionine (dark colored block). Expressed here in the language of mRNA, code is the same in all organisms, from bacteria to man.

How did the genetic code originate? There is nothing chemically obvious about why the codon UUU should mean phenylalanine, for instance, since there are no clear-cut affinities between the codons (or the anticodons) and the amino acids themselves. But such affinities may exist under conditions of pH, temperature, or concentration that have not yet been tried. If so, the code in Figure 10 may not be just one of millions of equally likely codes, and, if life is found in other planetary systems, it may employ a similar or identical code. The fact that, on this planet, the code seems to have remained the same in various organisms while all other cellular components have diverged through evolution should not, however, be taken as evidence that our code is the only one that can work.

A MAP OF THE CHROMOSOME

In the quest for a complete understanding of the anatomy and physiology of a living organism, the logical first step would be to determine the sequence of nucleotides in its DNA. So far biochemists have not succeeded in determining the complete structure of any natural DNA molecule. Because of the enormous size and complexity of even the simplest DNAs, this biochemical feat seems to be a long way off. The best technique for mapping the positions of the genes in a DNA molecule is GENETIC ANALYSIS.

The geneticist starts with a "normal" WILD TYPE organism. He can then introduce mutant genes and identify them by looking for offspring whose phenotype differs from that of wild type individuals. The goal is to map these mutations: to find out the order in which they occur along the DNA molecule and the distances between them.

The way he does this is in principle very simple. He performs a CROSS, or mating, between two individuals that carry two different mutations, and examines the progeny. There are four possible types of offspring. Some will be just like one mutant parent. These are called parental-type offspring. Some progeny will carry both mutations, and are called double mutants. Still others will carry neither mutation, and will be identical with the original wild type from which the mutants were derived. The double-mutant and wild-type progeny are called RECOMBINANTS.

The frequency of recombinant progeny compared with that of parental-type progeny is a measure of the distance on the DNA molecule between the two mutations. Why is this so? The process of recombination can be compared to the splicing of tape recordings by a deaf editor. This editor can line up the two tapes side by side so that they correspond to one another, note by note. Here and there, he cuts across both tapes with a scissors and rejoins them with a splice, producing a recombinant tape without ever listening to it. Now suppose that the violin soloist has a borrowed bow on the day that his orchestra records a concerto, and his instrument squawks in the middle of his first solo when the first tape is made. The conductor gamely carries on to the end of the movement, and then the orchestra starts over for a second try. As might be expected, the violinist does it again the next time through. If he muffs the same note on the second try the two tapes cannot possibly be put together to make a perfect performance. But if he muffs different notes on the two tapes, the deaf editor, using his scissors and splicer, will occasionally produce a flawless recording. And, of course, sometimes he will produce a recording that contains both mistakes. The likelihood of each depends on how far apart the two mistakes lie (*Figure 12*).

The genophores of viruses and bacteria (and the chromosomes of eucaryotes) seem to recombine in just this way, breaking at perfectly corresponding positions, exchanging pieces of genetic information, and then putting them back together again. The only way to detect such exchanges is to mark the genes with mutations (*Box C*).

PHAGE EXPERIMENTS

The modern concept of the gene grew out of experiments with the bacteriophage T4. Seymour Benzer, working at Purdue Univer-

Control of enzyme synthesis is the key to controlling the chemical pathways of the cell. By switching particular genes on and off, the cell controls not only the kinds of enzyme that it produces, but also the amounts.

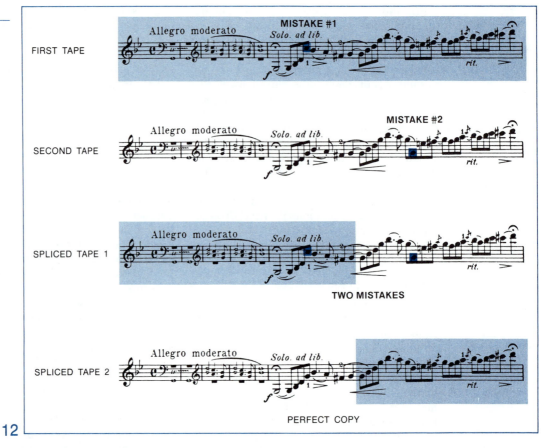

FIRST TAPE

SECOND TAPE

SPLICED TAPE 1

SPLICED TAPE 2

12

SPLICED TAPE RECORDINGS illustrate the process of genetic recombination. Illustration reproduces the violin solo from Max Bruch's G Minor
Violin Concerto. First and second tapes contain wrong notes (solid color)
at different points in the score. By splicing the two tapes together, editor
can produce either a perfect tape or one with both errors.

sity in the 1950's, studied the properties of the
rII mutants of T4. These mutants will grow on
E. coli of strain B but cannot grow on E. coli of
strain K. In this sense, all rII mutants are alike.
But genetic crosses, in which mixtures of two
genetically different phages are added to E.
coli B, thus infecting each bacterium with several phages, showed the rII mutations recombined with one another. That is, crosses between two different rII mutations generally
produced offspring that contained some WILD
TYPE (non-mutant) recombinant phages that
could grow on E. coli K. Benzer showed by
such crosses that rII mutations occupied hundreds of different sites on the T4 DNA
molecule. Because all of these sites were close
to one another, the proportion of wild-type
recombinants from these crosses was low.

Benzer then showed that his hundreds of
different rII mutants could be subdivided into
just two functional classes by a procedure
called a COMPLEMENTATION TEST. Benzer infected E. coli K cells with mixtures containing
two different rII mutants. Sometimes he found
that the two mutants could complement one
another so that phages were produced by cells
infected with the mixture of the two mutants.
In other tests, he found that a mixture of two rII
mutants was just as unsuccessful at producing
phage on E. coli K as either of the mutants
alone. By these tests, Benzer could lump all of
his rII mutants into two classes: A and B. Class
A mutants always complemented class B mutants but never complemented each other.
Class B mutants always complemented class A
mutants, but never each other.

The molecular mechanism of recombination is poorly understood at this time. How cells manage to cut and splice chromosomes so precisely is a mystery. It seems likely that the Watson-Crick pairing rules that govern the precise duplication of genetic information during DNA synthesis and the matching of tRNA's to codons in mRNA are also responsible for the perfect matching of chromosomal fragments when they rejoin during recombination. A highly speculative diagram of this process appears below.

One molecule in the diagram contains a mutation (M₁): a G···C has replaced an A···T. The other molecule contains a different mutation (M₂): a C···G has replaced a T···A. The two parental molecules are shown at top. After some unknown intermediate steps, the two molecules are cut, with their Watson-Crick strands broken in different places (middle). The fragments then pair in a new arrangement by forming hydrogen bonds between complementary single strands.

After a few more unknown events, the recombined strands have restored their continuity (bottom). Extra nucleotides have been discarded, missing nucleotides have been inserted, and interruptions in the Watson-Crick strands have been mended. The nucleotides shown in boldface type were not present in either of the parental nucleotides.

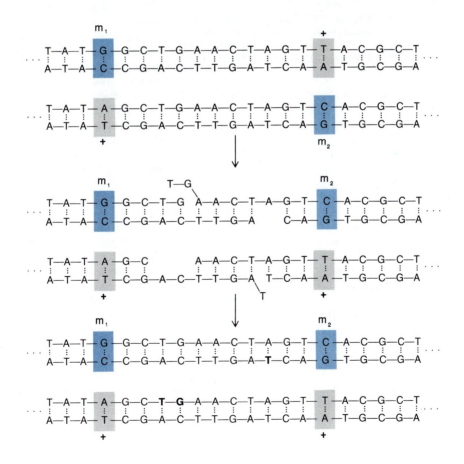

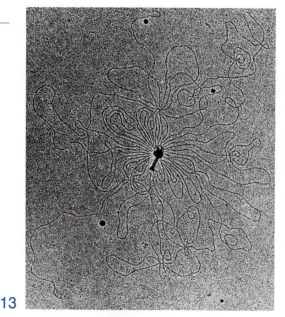

13

BACTERIOPHAGE VIRUS at the center of this electron micrograph lies tangled in a string of its own DNA. This classic picture gives some idea of the enormous length of a single DNA molecule.

Benzer concluded that two genetically controlled functions were needed for growth on *E. coli* K. All of the A-group *rII* mutations produced a defect in the A function, and the B-group mutations produced deficiencies in the B function. Because the crosses revealed that different A group mutations are located at different positions on the DNA molecule, Benzer concluded that the gene that governs the A function must be a SEGMENT of DNA — not just a point. Mutations that destroy the gene function can occur at many different positions along that segment. A similar conclusion can be drawn about the gene that controls the B segment. In other words, a gene is not a point or a blob; it is a line.

Phages and bacteria are favorite subjects for genetic experiments because they contain only a single genophore and their DNA is comparatively simple. A typical bacterium contains only about one thousandth as much DNA as a single human cell, and a typical bacteriophage virus contains perhaps one hundredth as much DNA as a bacterium. This simplifies the task of mapping their genes. But a genetic map can be drawn for any sexual organism in which heritable traits can be recognized. Although the principles remain the same in the genetic analysis of eucaryotes, their application can be

complex. Eucaryotic cells contain more than one chromosome, and frequently (as in humans) there is more than one copy of all the genetic information.

CONTROL OF CELL CHEMISTRY

The cell nucleus can be compared to a central computer that controls a network of sophisticated and highly automated biochemical machinery. The genes represent a library of programs stored in the computer's memory banks, programs that specify the precise nature of each protein synthesized by the cell. The computer monitors changing conditions both within the cell and in the external environment, and responds to this input of information by selecting and activating the appropriate genetic programs.

Control of enzyme synthesis is the key to controlling the chemical pathways of the cell. At any given moment most of the genes in a cell are switched off; in the language of the geneticist, they are not being EXPRESSED. By switching particular genes on and off, the cell controls not only the kinds of enzymes that it produces, but also the amounts. All the cells in a multicellular organism (except the gametes) contain the same set of genes (*Chapter* 6). The difference between a kidney cell and a liver cell is that each is expressing a different part of the overall genetic program.

The mechanisms that control gene expression are not nearly as well understood as those that underlie the coding, replication, and transmission of genetic information, but at least the overall strategies of the control mechanisms have become clear. One way that the cell can control the flow of a metabolite through a biochemical pathway is to decrease the activity of the enzyme molecules that catalyze that pathway. This control strategy, called FEEDBACK INHIBITION, is common in both bacteria and higher organisms (*Chapter* 3). Another way to reduce the flow is to decrease the synthesis of the unneeded enzymes by switching off the corresponding genes. This strategy is more efficient than making the enzymes and then deactivating them, and the mechanism for it is built into the design of all living cells.

Conversely, to increase the output of a metabolic pathway within a cell, nature has a choice of several control strategies, including: 1) to decrease the rate at which enzyme molecules are broken down within the cell, thus increasing the supply of enzyme molecules; 2) to increase the number of genes that code for a particular set of enzymes; 3) to speed up the rate of transcription, thus increasing the number of mRNA molecules involved in the synthesis of the needed enzymes; and 4) to speed up the rate of translation of the message, enabling the ribosomes to synthesize more of the enzyme proteins (*Figure 14*). Nature uses each of these mechanisms in one situation or another. The study of these control mechanisms is one of the frontiers of biological research, and some mechanisms are understood better than others.

CONTROL BY DEGRADATION

The amount of an enzyme within the cell depends not only on its rate of synthesis, but also on its rate of degradation (*Box D*). The cell can synthesize two enzymes at the same rate, but break them down at different rates. The one broken down more slowly will pile up to a higher level. Different proteins in the liver, for example, turn over at different rates. On the average, half of the liver protein of a rat turns over every six days, but some liver enzymes

have a half-life as short as 90 minutes (meaning that half of the molecules synthesized at a given moment will have been destroyed 90 minutes later). In some cases an enzyme is stabilized by one of its substrates. A high substrate concentration tends to decrease the degradation rate without affecting the rate of synthesis, so that the amount of the enzyme rises. The elevated level of enzyme, in turn, causes the substrate to be used up more rapidly, so that its concentration drops back toward normal levels. This self-adjusting system tends to smooth out fluctuations in the concentration of a substrate. Both the rate of synthesis and the rate of degradation of each enzyme seem to be under genetic control, enabling higher organisms to control their enzyme levels in many different ways.

MULTIPLE COPIES OF GENES

An obvious way for a cell to make more of one enzyme than of another would be to have more genes of one kind than of another, and to transcribe them all. This would require having more DNA per cell than would be necessary if

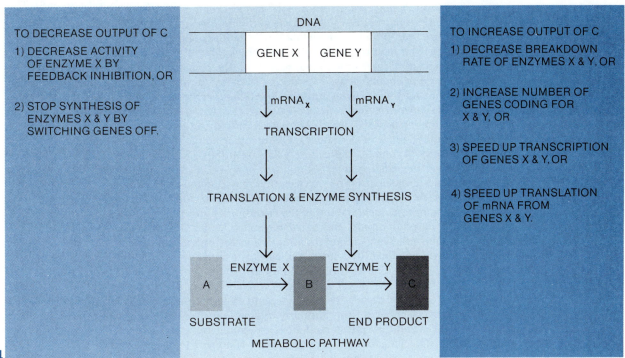

TO DECREASE OUTPUT OF C

1) DECREASE ACTIVITY
   OF ENZYME X BY
   FEEDBACK INHIBITION, OR

2) STOP SYNTHESIS OF
   ENZYMES X & Y BY
   SWITCHING GENES OFF.

DNA

GENE X | GENE Y

mRNA$_X$ mRNA$_Y$

TRANSCRIPTION

TRANSLATION & ENZYME SYNTHESIS

ENZYME X   ENZYME Y

A ——→ B ——→ C

SUBSTRATE   END PRODUCT

METABOLIC PATHWAY

TO INCREASE OUTPUT OF C

1) DECREASE BREAKDOWN
   RATE OF ENZYMES X & Y, OR

2) INCREASE NUMBER OF
   GENES CODING FOR
   X & Y, OR

3) SPEED UP TRANSCRIPTION
   OF GENES X & Y, OR

4) SPEED UP TRANSLATION
   OF mRNA FROM
   GENES X & Y.

14

ALTERNATE CONTROL STRATEGIES enable the cell to regulate the synthesis of a particular product. Diagram at center summarizes the steps of the synthetic pathway. Columns at left and right indicate how the cell can decrease or increase its output of end product C.

D

TURNOVER OF PROTEINS

Biologists of the late 19th century were influenced as much as were their lay contemporaries by the spirit of the Industrial Revolution, and tended to view the living animal as a sort of steam engine. Early in the 20th century, it became widely accepted that proteins were the permanent machinery of the human engine, and that carbohydrates and fats were the fuel shoveled into the firebox. In this view, an adult, nongrowing animal needed a little dietary protein to repair wear and tear on the machine; if more protein was ingested, the excess amino acids were thrown into the firebox along with the carbohydrates and fats.

Evidence that flesh-and-blood creatures differed from industrial machinery mounted rapidly in the late 1930's when isotopically labeled compounds became available. Rudolf Schoenheimer and David Rittenberg, working at Columbia University, found that if adult rats were given a dose of N^{15}-labeled amino acid, about ½ of the isotopic label showed up within a few days in the body proteins, about ¼ remained in the body in compounds other than proteins, and about ¼ was promptly excreted. The nitrogen atoms that were entering the body, by and large, were not the same ones that were leaving. It was soon found that proteins in virtually all organs of the body were in a constant state of flux, and that under the placid surface, the proteins of mammals, at least, are in a dynamic steady state. During the 1950's, this view was in turn challenged by several groups, principally that of Jacques Monod in Paris. Working with β-galactosidase in E. coli, Monod's group found that this enzyme does not turn over. If β-galactosidase is induced and then the inducer is removed (but growth is allowed to continue), the enzyme that was formed during the induction period is not scrapped; it is simply passed on to the daughter cells at each division. The total amount of enzyme remains the same, but there is progressively less and less per bacterium as growth proceeds. Who was right? Or were Schoenheimer and Rittenberg right about animals and Monod right about E. coli? It now seems likely that both groups were partly right. In 1958, Joel Mandelstam reported that although rapidly-growing E. coli proteins are metabolically stable, nondividing cells which are being starved of a required amino acid will break down (and resynthesize) as much as five percent of their protein per hour. With the benefit of hindsight, we can understand why it is advantageous for starving bacteria, but not for growing bacteria, to turn over their existing proteins. Breaking down proteins only to then resynthesize the same proteins is metabolically wasteful, and it is not likely to be necessary for growing bacteria, which are successfully using externally available nutrients to synthesize all their enzymes. On the other hand, starving cells are, by definition, already in trouble, and can only survive if they liquidate some of their capital by converting existing enzymes into free amino acids. The amino acids can then be used to make kinds of enzymes that may previously have been repressed. These new enzymes may be thought of as a last-ditch attempt to escape from the metabolic impasse that caused the cessation of growth.

Most cells in an adult animal are, in some respects, quite analogous to a starving bacterial cell. They are not dividing or increasing in mass. Therefore, they have no way to adjust their content of various enzymes except by breaking down existing proteins. This is true even in a well-fed animal, provided it is not growing. The situation is even more dramatic when an organism is in a state of starvation. For example, during the cocoon-bound metamorphosis of a caterpillar into a moth, the insect cannot eat. Similarly a tadpole fasts as it turns into a frog. Certain bacteria, when they are starved, develop spores that can survive for years in the dormant state. In each of these cases, there is no way to change over to a new set of proteins except to break down the existing ones.

there were only one gene of each kind. Chapter 1 pointed out that humans have about 1000 times as much DNA per cell as a typical bacterium. It pleases us to feel that we contain so much genetic information because we represent the zenith of evolutionary complexity. But the cells of several amphibians and fishes contain from 10 to 100 times more DNA than ours. The highest known DNA content per cell is found in the Congo eel *Amphiuma*, an ugly and stupid creature. Although there is a rough correlation between the amount of DNA per cell and the complexity of an organism, the idea cannot be pushed very far. Some amphibians that are closely related to one another, and are about equally complex, differ many-fold in the amount of DNA per cell. It seems unlikely that one of them makes vastly more kinds of enzymes than another. Instead it suggests that certain organisms have many copies of some or most of the genes, or that there may be a type of "non-informational" DNA that is much more abundant in some organisms than in their close cousins.

CONTROL OF TRANSCRIPTION

One of the most important and best understood mechanisms is the one that controls transcription. In the 1960's François Jacob and Jacques Monod, working at the Pasteur Institute in Paris, discovered that the digestion of lactose (milk sugar) in the bacterium E. *coli* involves a cluster of genes that they named an OPERON. The lactose operon includes three STRUCTURAL GENES (each coding for one of the three enzymes necessary for lactose metabolism), plus an OPERATOR and a PROMOTER gene.

These five genes lie side by side along the strand of bacterial DNA. Under the influence of the operator and the promoter the three structural genes are switched on and off as a group — an efficient design strategy that enables the cell to regulate the enzymes for an entire pathway all at once. When E. *coli* is grown on a medium that does not contain lactose the structural genes are not expressed, and the levels of the three enzymes within the cell are very low. But when lactose is added to the medium the structural genes are switched on, and the enzyme level rises abruptly.

Jacob and Monod suggested that the operator gene acts as a switch. In the lactose operon the operator is normally locked in the 'off' position by a repressor molecule synthesized by a REGULATOR gene that lies outside the operon. What switches the operator on? According to the Jacob-Monod model, a lactose molecule combines with the repressor and inactivates it, unlocking the operator. A molecule of RNA polymerase attaches itself to the promoter (a gene that marks the starting point for transcription) and slides along the DNA, transcribing a single molecule of messenger RNA that carries the instructions for all three enzymes (Figure 15). Ribosomes attach to this mRNA, and translation of the three structural genes into protein begins.

The lactose operon is called an INDUCIBLE system because the addition of a substrate molecule to the medium induces (switches on) the synthesis of enzymes by turning on genes that are usually off. Inducible systems are of adaptive value to organisms that must switch on a metabolic pathway in response to the presence of an outside agent such as lactose. It is equally valuable to an organism to be able to switch off the synthesis of certain enzymes. For example, if the amino acid tryptophan is present in the medium it is advantageous to be able to stop making all of the enzymes that are involved in trypotophan biosynthesis. When the formation of an enzyme is turned off in response to a biochemical cue, the enzyme is said to be REPRESSIBLE. Repressible systems, such as that for tryptophan synthesis, work by mechanisms similar to those of inducible systems, but with one important difference: the repressor molecule cannot shut off its operon unless it unites with a COREPRESSOR, which may be either the nutrient itself (in this case,

Many cells, especially those of higher organisms, control enzyme synthesis by shifting not the rate at which genetic information is transcribed into mRNA, but the rate at which RNA is translated by the ribosomes.

tryptophan) or a compound derived from it. If the nutrient is absent the operon is transcribed at a maximum rate. If the nutrient is present, the operon is turned off (*Figure 16*). As in the case of the lactose operon, the tryptophan system has a regulator gene and an operator. Mutations in the regulator gene can change the repressor so that it never binds to the operator, even when tryptophan is present. In such mutants the enzymes are made all the time — their synthesis cannot be repressed.

Later research has turned up many different types of operons in procaryotes, but no convincing examples in eucaryotes. At least 70 percent of the genes in *E. coli* are thought to occur as clusters of two or more functionally related cistrons, but in eucaryotes clusters seem to be relatively uncommon, and those that have been found seem to differ from operons in some important ways. Perhaps higher organisms have some quite different ways to control transcription and the steps that follow

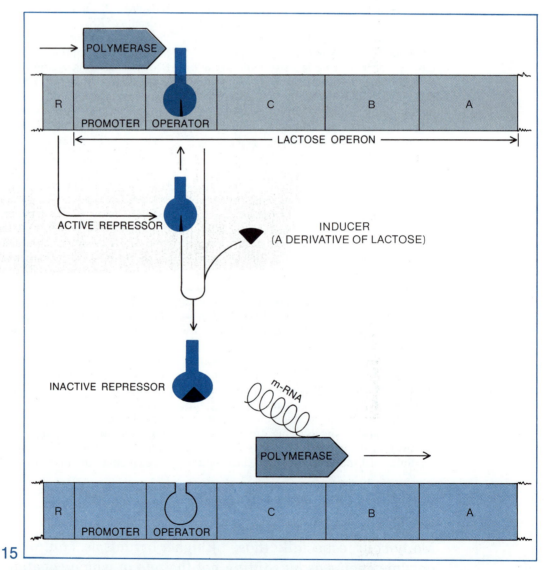

15

MODEL OF LACTOSE OPERON depicts the induction of enzyme synthesis. Repressor molecule, shown here as a flask-shaped object, binds to the operator (top) and prevents transcription by RNA polymerase. Inducer molecule can change shape of repressor so that it can no longer bind to operator. Removal of repressor allows transcription to proceed (bottom).

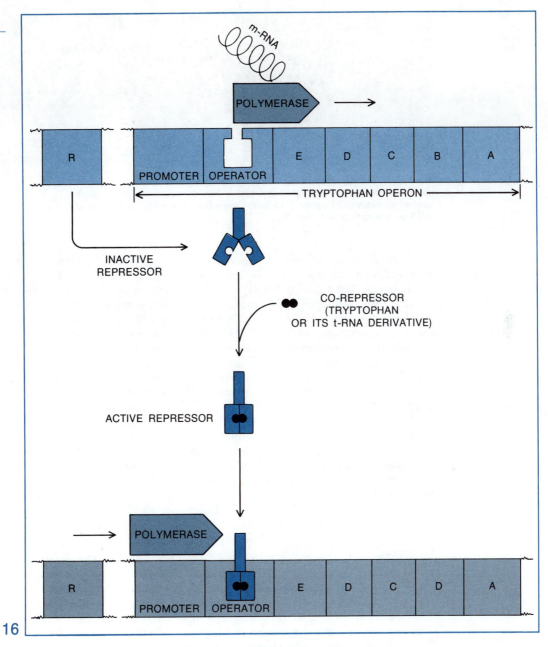

16

TRYPTOPHAN OPERON is a repressible system. Letters on operon denote structural genes controlled by promoter and operator. Repressor is activated by binding to a small co-repressor molecule, either tryptophan or a derivative of it. Active repressor binds to operator and stops transcription.

it. The definitive difference between procaryotes and eucaryotes is the presence in eucaryotes of a nuclear membrane. The function of the membrane is not well understood, but it may be important in the control of metabolism. For example, it might act as a selective barrier to the passage of messenger RNA from the chromosomes into the cytoplasm, the site of protein synthesis.

CONTROL OF TRANSLATION

Many cells, especially those of higher organisms, control enzyme synthesis by shifting not the rate at which genetic information is

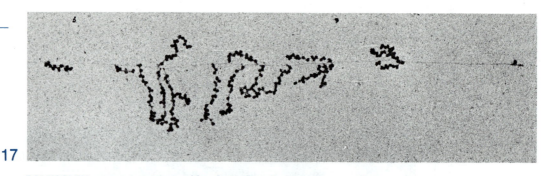

17

POLYSOMES appear as strings of beads in this electron micrograph. A strand of DNA from E. coli stretches like a thin wire across picture. Dark spot at extreme right appears to be the promoter site with a molecule of RNA polymerase attached to it.

transcribed into mRNA, but the rate at which RNA is translated by the ribosomes. Eggs of the sea urchin furnish a good example. In the unfertilized egg very little protein synthesis occurs once the egg is "ripe," even though it contains ribosomes, messenger RNA, and compounds necessary for activation of amino acids. After fertilization, polysomes are quickly formed and protein synthesis increases manyfold without any appreciable synthesis of RNA. Apparently mRNA is stored in the unfertilized egg, but in a masked or inactive form that the ribosomes cannot translate. Fertilization unmasks it, allowing the ribosomes to begin translation. Similar results have been obtained in experiments on many other creatures.

In bacteria it can be very difficult to distinguish transcriptional control from transla-tional control because the two processes are not completely separated in time and space. Translation of the messenger by ribosomes begins at one end of the molecule while the other end of the molecule is still being transcribed. Translation is necessary to pull the mRNA away from the DNA, and most inhibitors of translation also bring transcription to a halt. Since transcription and translation occur together in bacteria, one might hope to be able to see polysomes attached to the genophore. The remarkable electron micrograph in Figure 17 shows clearly a region of transcription (perhaps an operon) of the E. coli genophore, with polysomes protruding from a strand of DNA.

The next chapter applies the principles of genetics to "real" organisms — the multicellular eucaryotes that the reader is more likely to encounter in everyday life.

READINGS

W. HAYES, The Genetics of Bacteria and Their Viruses, 2nd Edition, New York, John Wiley & Sons, 1969. A magnificent text. This book covers all of microbial genetics. It is so clearly written that it can be read by a beginning student and so comprehensive that it is an indispensable reference for the expert.

F. JACOB AND E.L. WOLLMAN, Sexuality and the Genetics of Bacteria, New York, Academic Press, 1961. A translation of a French classic in microbial genetics. This is a technical book that deals in detail with bacterial sexuality, but the flow of ideas is smooth and orderly so that the book can be read rapidly with pleasure.

A. KORNBERG, DNA Synthesis, San Francisco, W.H. Freeman and Company, 1974. An up-to-date, readable review of the whole field.

F.W. STAHL, The Mechanics of Inheritance, 2nd Edition, Englewood Cliffs, N.J., Prentice-Hall, 1969. This is a general genetics text written by a microbial geneticist. The sections on microbial genetics are predictably strong. Challenging problems appear at the ends of the chapters.

G.S. STENT, Molecular Genetics: An Introductory Narrative, San Francisco, W.H. Freeman and Company, 1971. One of the most articulate and witty students of molecular biology has chosen, happily, to write an introduction to the subject.

J.D. WATSON, The Molecular Biology of the Gene, 3rd Edition, Menlo Park, W.A. Benjamin, 1976. A brilliant profile of the gene, written in deceptively simple language by one of the discoverers of the structure of DNA.

NOTES

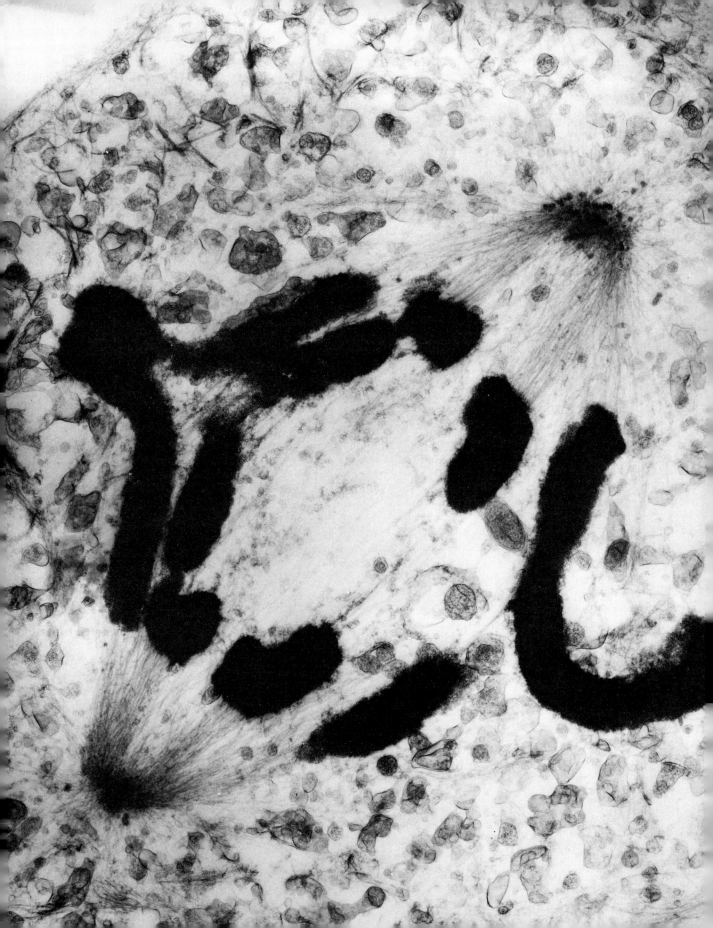

5

classical genetics

It seems to me that among the things we commonly see there are wonders so incomprehensible that they surpass all the perplexity of miracles. What a wonderful thing it is that this drop of seed from which we are produced bears in itself the impressions not only of the bodily form, but also the thoughts and inclinations of our fathers! Where can that drop of fluid contain that infinite variety of forms?

MICHEL DE MONTAIGNE

In 1578 the French essayist Michel de Montaigne was 45 years old and had begun to suffer from a bladder stone. His father too had had such a stone and Montaigne wondered what mechanism could explain the transmission of form and foible from one generation to the next. It would be almost 300 years before Gregor Mendel, a monk at the monastery at Brünn (then Austria, now Czechoslovakia), would discover the laws that govern the transmission of hereditary traits from generation to generation. If Montaigne's discomfort had goaded him to study of the problem of inheritance he could have performed Mendelian experiments with doves, cats, rabbits, or any other domestic animals that were at hand. It was not so much Mendel's knowledge that gave him the scientific edge over Montaigne as it was his outlook. That outlook was the product of the 17th, 18th and 19th centuries.

Until the 17th century science was dominated by the Aristotelian passion for classification. Then came the urge to measure and to experiment. Calculus was invented in the 17th century by Newton and Leibniz, giving birth to modern physics and astronomy. The work of the 18th century elucidated the physical and chemical properties of matter and the nature of electricity. The 19th century, in which most science was called engineering, saw the birth of the steam railway, the friction match, the gas light, the camera, the incandescent bulb, the internal combustion engine, and the phonograph.

Meanwhile biology had discovered experimentation, but the best efforts of biologists were devoted to description and classification. The cell theory had been advanced, and Darwin's work made it clear that an understanding of the laws of heredity was essential not only to explain the recurrence of bladder stones and baldness within families but also to explain the process of evolution. But biologists, slow to adopt the lessons of their bright colleagues in the other sciences, had scarcely learned to count.

Arithmetic was the basis for Mendel's success in discovering the laws of heredity and for the failure of his colleagues to recognize his discovery. When Mendel presented his results in 1865 to a meeting of the Brünn Natural History Society, his audience perhaps assumed that he was simply trying to overwhelm them with numerical mumbo jumbo. It was not until 1900, when biologists had finally learned the value of numbers, that Mendel's breakthrough was rediscovered by the biological world and appreciated for what it was.

Mendel's success rested upon his invention of a simple model that lent itself to a quantitative test. His favorite organism was the pea, which is a eucaryote. Eucaryotes in general can be considered to have a mother and a father, and the genetic contributions of the two parents to the makeup of the offspring are essentially equal. This means that an individual must have at least two copies of each gene, one from the

CHROMOSOMES of an animal cell during anaphase appear as bold dark strands at the center of the electron micrograph on opposite page. Thin fibers connected to the chromosomes are the microtubules of the mitotic spindle.

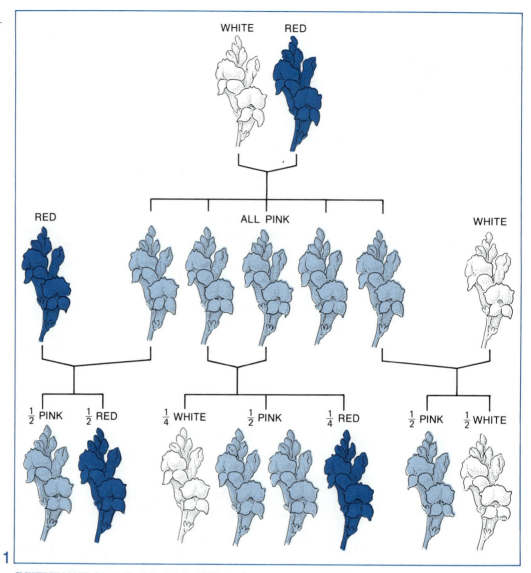

INHERITANCE OF FLOWER COLOR follows the classic Mendelian pattern. Cross between red and white snapdragons produces only pink offspring. This is due to mixing of pigments, not mixing of genes. When pink hybrids are crossed with one another, about half the progeny are pink, one fourth are pure red and one fourth pure white. There are no intermediate gradations of color.

mother and one from the father. Mendel based his model on this hypothesis. Because fertilization involves the fusion of two GAMETES (for example, one ovum and one sperm), at some stage in the production of gametes the number of genes must be reduced to half of the number in the rest of the parent's cell. Otherwise the number of genes per cell would double at each generation — a process that would obviously lead to chaos in a few generations. The produc-

tion of gametes with half the normal genetic content is accomplished by a special type of cell division called MEIOSIS, which is discussed later in this chapter.

MENDEL'S FIRST LAW

Mendel imagined that during the formation of gametes the parent organism randomly puts one copy of a particular gene into each gamete. When egg and sperm united, the new individual would get one copy from each parent. These two gene copies in the new individual might or might not be identical. If they are identical, the individual is said to be HOMOZYGOUS for this gene; if they are different, the individual is said to be HETEROZYGOUS. Different forms of a given gene (or CISTRON) are called ALLELES. If the individual is heterozygous, the phenotype will be determined by the dominant allele. If neither allele is dominant, the phenotype of the heterozygote will show the presence of both alleles. When the heterozygote offspring reproduces it will produce gametes carrying one allele or the other in equal numbers. Mendel called the separation of the two alleles during the formation of gametes SEGREGATION. Mendel's First Law is that ALLELES SEGREGATE FROM ONE ANOTHER DURING THE FORMATION OF GAMETES.

The consequences of this law can be examined with real organisms, beginning with one that makes an even better textbook case than Mendel's peas: the common snapdragon. Two common kinds of true-breeding snapdragons differ in flower color; one is red, the other is white. True-breeding means that red snapdragons crossed to red give only red offspring generation after generation, and white crossed to white give only white. Red and white are alleles of a single gene. When red snapdragons are crossed to white, all of the progeny are pink (*Figure 1*). The pink progeny are heterozygotes that contain an allele for red and an allele for white. Because the offspring resemble neither parent but have an intermediate phenotype, geneticists say that neither allele is DOMINANT. Alleles can be designated by letters. If *C* stands for color, the red allele can be designated C^r and the white allele C^w. The red homozygous parents have the genotype C^r/C^r. The white homozygous parents have the genotype C^w/C^w. The pink progeny have the genotype C^r/C^w.

Three new sorts of crosses are possible. The test of Mendel's model is whether it correctly predicts the results of these crosses. If white snapdragons are crossed with pink ones, pre-Mendelian biologists would predict that all of the progeny would be a pale pink because they believed that hereditary traits could

be blended like cans of paint. But on the basis of Mendel's First Law, one would predict that the white snapdragons would produce gametes bearing only C^w, and the pink snapdragons would produce equal numbers of two types of gametes: C^w and C^r. These two types would unite with C^w gametes from the white parent to yield equal numbers of two kinds of offspring: C^w/C^w, or white, and C^r/C^w, or pink. The pink offspring would be just as pink as the pink parent, and the white offspring would be pure white. The Mendelian prediction is correct. Another kind of cross, red × pink, works in the same general way. The progeny are not dark pink; rather, half the progeny are regular pink and the other half pure red.

What about a cross of pink × pink? Here each parent produces two kinds of gametes. Thus four kinds of fertilization are possible:

1. C^w pollen + C^w ovum → C^w/C^w (white)
2. C^w pollen + C^r ovum → C^w/C^r (pink)
3. C^r pollen + C^w ovum → C^r/C^w (pink)
4. C^r pollen + C^r ovum → C^r/C^r (red).

Since all of these four types of fertilization are equally probable, one would expect that in a large number of progeny, the ratio would be about 1 white: 2 pink: 1 red. This is the result depicted in Figure 1.

Peas are not quite so straightforward. They too have red and white flowers, and Mendel used these two allelic forms in some of his experiments. When he crossed red peas to white peas, the progeny were not pink; they were as red as the red parent. In peas the red allele is dominant. When one allele is clearly dominant over the other, geneticists use a capital letter to symbolize the dominant allele and a lower case letter to symbolize the recessive allele. Mendel's white peas had a genotype designated C^r/C^r or (omitting the *C*) *r/r* for short. The homozygous red parents had a genotype *R/R*; and the heterozygous red progeny had the genotype *R/r*. Mendel crossed the heterozygous red progeny to homozygous white plants of the original parental type. Half the progeny were red, and the other half were white. This confirms Mendel's expectation that half the offspring should be *R/r*, and the other half *r/r* (*Figure 2*).

Mendel's most revolutionary finding was that genes do not blend like cans of red and white paint.

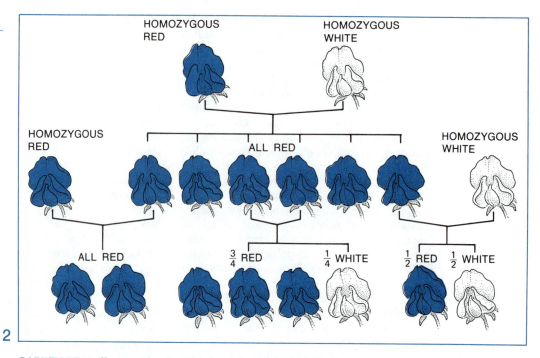

2

GARDEN PEAS illustrate dominant and recessive inheritance. Cross between a true-breeding white-flowered plant and a true-breeding red-flowered one yields only red-flowered heterozygous offspring (center). Crosses between red homo- and heterozygotes yields all red offspring (lower left). Cross between red heterozygote and white homozygote yields half red, half white offspring (lower right).

When Mendel crossed heterozygous red peas to themselves, he expected that there would be four different types of fertilization, all of them equally probable:

1. r pollen + r ovum →
   r/r (white homozygote)
2. r pollen + R ovum →
   r/R (red heterozygote)
3. R pollen + r ovum →
   R/r (red heterozygote)
4. R pollen + R ovum →
   R/R (red homozygote).

The ratio then is 3 red:1 white. Mendel went even further, and let the red offspring from this last cross pollinate themselves. He found that about ⅔ of them gave at least some white offspring, and therefore had to be heterozygous. The rest of the red plants gave only red offspring. Once again, this was in accord with Mendel's hypothesis. He had predicted that ⅔ of the red plants from the $R/r \times R/r$ cross would be heterozygous.

Mendel's most revolutionary finding was that genes do NOT blend like cans of red and white paint. The results may be red, as in peas,

or pink, as in snapdragons, but the genes sort themselves out in the next generation to give the original colors. In all, Mendel did essentially the same experiment with several other traits: tallness of the plant, shape of seed, shape of the ripe pod, color of the unripe pod, and position of the flower. None of these traits showed any blending. He concluded that the genes were discrete "particles" that could emerge unchanged when the heterozygote produced gametes. This discovery was of great importance to the understanding of the processes of evolution, because it plugged a gaping hole in the theory of evolution as it had been proposed by Darwin. Darwin realized of course that for natural selection to work at all it had to work on stable heritable traits. Unfortunately the genetic models current at Darwin's time implied that any adaptive heritable change in an individual would be rapidly diluted in the genetic material of later generations. Genetic variability, according to this reasoning, would tend to disappear before natural selection had a chance to alter the characteristics of the population. Mendel's demonstration of the particulate nature of the gene solved this problem and put Darwin's theory on a reasonable basis.

ASSORTMENT OF GENES

What happens in a cross between two parents that differ with respect to two or more genes? When the double heterozygote makes gametes, are the alleles from one of the parents packaged in one gamete and those from the other parent in the other gamete? Mendel's plants differed in two seed properties: one made spherical yellow peas and the other made misshapen green ones. The first type can be designated *S/S Y/Y*, indicating that it is homozygous both for the *S* allele of the spherical, or "shape" gene and for the *Y* allele of the "yellowness" gene. The second doubly homozygous variety is *s/s y/y*. All of the doubly heterozygous offspring from a cross between these two varieties would carry the genotype *S/s Y/y*.

There are two ways in which these doubly heterozygous plants might produce gametes. If the alleles maintained their original associations during gametogenesis, the double heterozygotes would produce only two kinds of gametes — *R Y* and *r y* — and the offspring would be of three types: ¼ *R/R Y/Y*, ½ *R/r Y/y*, and ¼ *r/r y/y*. In other words, the results of crossing the doubly heterozygous plants to themselves, or to one another, would be exactly like the result of the cross involving flower color. There would be no reason to suppose that seed shape and seed color were really regulated by two different genes, because spherical seeds would always be yellow, and misshapen seeds green.

Alternatively, the segregation of *S* from *s* could be independent of the segregation of *Y* from *y* during formation of gametes. This would mean that the doubly heterozygous plant would produce four kinds of gametes, all in equal numbers: *S Y*, *S y*, *s Y*, and *s y*. The fusion of two of these gametes at fertilization can give rise to nine different genotypes among the progeny:

1. S/S Y/Y	4. S/S Y/y	7. s/s Y/Y
2. S/s Y/Y	5. S/S y/y	8. s/s Y/y
3. S/s Y/y	6. S/s y/y	9. s/s y/y.

The progeny can thus have any of three possible genotypes for shape (*S/S*, *S/s*, or *s/s*) and any one of three possible genotypes for color (*Y/Y*, *Y/y*, or *y/y*).

The number of different phenotypes cannot be greater than nine (the number of genotypes). Actually it is less, because the *S* (spherical) allele is dominant to *s* (misshapen), and *Y* (yellow) is dominant to *y* (green). Then genotypes 1, 2, 3, and 4 above will all have the same phenotype: spherical, yellow; and genotypes 5 and 6 will both be spherical, green. Genotypes 7 and 8 will both be misshapen, yellow; and genotype 9 will be misshapen, green. Actually there are only four possible phenotypes because one allele for each of the two seed characteristics is dominant. There would be six possible phenotypes if *S* were dominant but *Y* were not, so that *Y/Y*, *Y/y*, and *y/y* seeds would form three distinct phenotypic classes.

The frequencies of the nine genotypes listed above are not all the same. Some genotypes can only be produced by one kind of fertilization event, but other genotypes can be produced by several different kinds of fertilization. For example, *S/S Y/Y* individuals can arise only from fertilization of *S Y* ova by *S Y* pollen, whereas *S/s Y/y* individuals can arise four ways: from fertilization of *S Y* ova by *s y* pollen, of *s Y* ova by *S y* pollen, of *s y* ova by *S Y* pollen, or of *S y* ova by *s Y* pollen. A convenient illustration of the possible kinds of fertilization events is given by the diagram in Figure 3. This type of square, fashioned in 1905 by R. C. Punnett, puts Mendel's hypothesis into a form that anyone can understand. Perhaps if Mendel himself had thought of it in

OVA GENOTYPES \ POLLEN GENOTYPES	SY	Sy	sY	sy
SY	SSYY 1	SSYy 4	SsYY 2	SsYy 3
Sy	SSYy 4	SSyy 5	SsYy 3	Ssyy 6
sY	SsYY 2	SsYy 3	ssYY 7	ssYy 8
sy	SsYy 3	Ssyy 6	ssYy 8	ssyy 9

3

PUNNETT SQUARE shows independent assortment of alleles for seed shape and seed color. Dominant S denotes smooth seed, recessive s misshapen seed. Dominant Y stands for yellow seed, recessive y for green seed. Color of squares indicates four phenotypes of offspring.

Mendel never saw chromosomes, and probably never imagined that cytologists would soon witness under the microscope the events that lead to the segregation of alleles: the process of meiosis.

1866, his work would have been readily accepted by his contemporaries.

In the Punnett square the four ovum genotypes are written vertically at the left of the square and the four pollen genotypes are written across the top. The offspring resulting from fusion of two gametes have the genotypes shown in the boxes where the appropriate ovum rows and pollen columns intersect. The number shown in each box is the genotypic class to which the offspring belongs. Note that Classes 1, 5, 7, and 9, which are homozygous for both genes, are all equally hard to produce. Each of these genotypes appears in only one of the 16 boxes of the Punnett square. Classes 2, 4, 6, and 8 (single heterozygotes) each appear in two boxes. And Class 3 (the double heterozygote) appears in four of the 16 boxes. Thus, the nine genotypic classes appear, respectively, in the relative proportions 1:2:4:2:1:2:1:2:1. Since the capitalized alleles

are dominant the various genotypic classes will give rise to the following phenotypes:

Classes 1, 2, 3, 4: spherical, yellow
Classes 5, 6: spherical, green
Classes 7, 8: misshapen, yellow
Class 9: misshapen, green

The proportion of the four possible phenotypic classes can, therefore, be determined from the Punnett square to be 9:3:3:1. The boxes of the Punnett square are tinted to indicate this distribution of phenotypes.

Mendel's experimental results agreed closely with the predicted ratio. They were the basis of his Second Law: that alleles of different genes are assorted independently of one another during gamete formation. This law is not as universal as the first; it applies only to genes that lie on separate chromosomes. It becomes a true law if one says that CHROMOSOMES ARE ASSORTED INDEPENDENTLY DURING FORMATION OF GAMETES.

MEIOSIS, depicted schematically below, begins like mitosis. Very early in prophase chromosomes begin to condense into easily visible form. Homologous chromosomes line up next to each other, then join to form X-shaped chiasmata. Red chromatids are maternal; black, paternal. By metaphase the nuclear membrane and nucleoli have disappeared. Anaphase is followed by telophase, then cells divide, halving the number of chromosomes. After another metaphase and anaphase (metaphase II and anaphase II, not shown) the two haploid cells again divide. Second meiotic division resembles mitosis. The overall result of the process is the production of four daughter cells from one parent cell. Daughter cells each have haploid number of chromosomes.

4

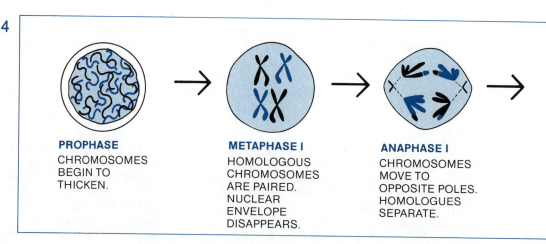

PROPHASE
CHROMOSOMES BEGIN TO THICKEN.

METAPHASE I
HOMOLOGOUS CHROMOSOMES ARE PAIRED. NUCLEAR ENVELOPE DISAPPEARS.

ANAPHASE I
CHROMOSOMES MOVE TO OPPOSITE POLES. HOMOLOGUES SEPARATE.

Mendel never saw chromosomes, and probably never imagined that cytologists would soon witness under the microscope the actual process that leads to the segregation of alleles. The process of segregation and assortment in eucaryotes is called MEIOSIS (from the Greek for "diminution"). Meiosis consists of two cell divisions with only one duplication of chromosomes. Four daughter cells result, each with only one copy of its chromosomes, instead of two, as in the parent cell.

The sequence of events in meiosis is depicted in Figure 4. The process begins as does mitosis. Chromosomes that have already duplicated in the interphase stage begin to shorten in the MEIOTIC PROPHASE to form condensed, easily visible chromosomes, each consisting of two identical chromatids joined together at the centromere. It is during this condensation of the chromosomes that the first unique event of meiosis occurs: the SYNAPSIS of homologous chromosomes. Recall that the cell contains two copies — homologues — of each chromosome, one from each parent. At the first meiotic division the two chromosomes in each homologous pair, each made up of two chromatids, line up next to each other and migrate together to the metaphase plane so that the centromere of one homologue lies just above the plane and the centromere of the other lies just below it. As the chromatids become more and more condensed, it becomes possible to see that there appear to be points at which the homologous chromatids connect to one another, forming cross-shaped patterns called CHIASMATA (singular, chiasma: a "crosspiece"). These appear to be the visible

manifestations of recombination between the homologous chromatids. It appears that at any one position on the chromosome, only one pair of homologous chromatids is involved in the formation of the chiasma.

At the meiotic METAPHASE the chromosomes have reached their maximum state of condensation, and the nuclear envelope and nucleoli have disappeared. This is followed by the FIRST MEIOTIC ANAPHASE, at which the homologous centromeres separate from one another, each carrying its attached chromosome. As the two chromosomes, each retaining its double structure, move apart, connections at the chiasmata move toward the end of the chromosomes until they run off the end, like a twist in two pieces of string as the strings are pulled apart. The chromosomes move to the two poles of the meiotic spindle. At TELOPHASE, when they have reached the two poles, they begin to grow less condensed and to enter into a brief MEIOTIC INTERPHASE, in which the nuclear membrane forms again. The chromosomes grow diffuse but do not replicate. The existence of this interphase is not a universal feature of meiosis. In some organisms, the chromosomes do not uncoil and the nuclear membrane does not reform at this point. In such organisms, the second meiotic metaphase and anaphase take place without further ado.

A second prophase follows rapidly. The chromosomes again condense. In this division there is no synapsis, because the homologous chromosomes were separated at the first meiotic anaphase, one going to each pole. The chromosomes migrate to the metaphase plate once again and the nuclear membrane disappears. The chromatids then separate at the centromere so that, at the SECOND MEIOTIC ANA-

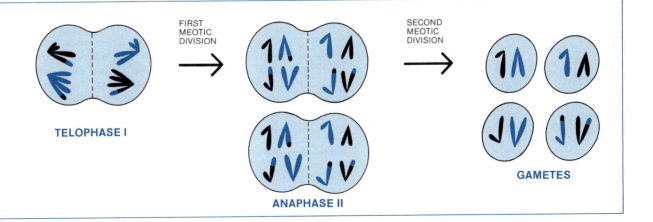

TELOPHASE I

FIRST
MEOTIC
DIVISION

ANAPHASE II

SECOND
MEOTIC
DIVISION

GAMETES

PHASE, the sister chromatids detach from each other, one going to each pole. At this point, when they are no longer connected at the centromere, they are once again called chromosomes. After this second meiotic division the number of chromosomes at each pole is half the original number, and reduction is complete.

The process of meiosis started with a single cell that had already doubled its chromosomes, so that there were four copies of each chromosome. The two meiotic cell divisions then produce four cells from one. But because there was no chromosome doubling between the two divisions, there was necessarily a reduction in chromosome number by a factor of two.

ALTERNATION OF GENERATIONS

Before meiotic cell division the cell has two copies of each chromosome; it is said to be DIPLOID (from the Greek for "double") and to have 2n chromosomes. After the meiotic divisions the cell is HAPLOID (from the Greek for "single") and has n chromosomes. Fertilization reverses the effects of meiosis, fusing two haploid cells to produce a diploid cell called a ZYGOTE. There is considerable variation in living things in the number of cell divisions separating meiosis and fertilization. In any sexual organism there are two genetic forms, the haploid and the diploid, which alternate from generation to generation. The diploid generation gives rise to the haploid generation by meiosis, and the haploid generation gives rise to the diploid generation by fertilization, the fusion of two cells. This point is illustrated by comparing the life cycles of sea lettuce, corn and man.

SEA LETTUCE

Ulva stenophylla is an edible green marine alga whose life cycle is illustrated in the chapter on reproduction (*Chapter 7, Figure 4*). It is mentioned briefly here because it exhibits ALTERNATION OF GENERATIONS in the purest sense. Sea lettuce exists in two forms, one diploid, the other haploid. The "leaves" of these two forms are indistinguishable to the unaided

eye, although one has twice as many chromosomes per cell as the other. The haploid form is called a GAMETOPHYTE because it produces gametes. Gametophytes (and the gametes they produce) can be of two different kinds, called mating types. When two gametes of different mating types meet and fuse, they form a diploid zygote that settles to the bottom and grows into a diploid "leaf" or SPOROPHYTE, whose cells can undergo meiosis to yield asexual haploid spores. After swimming about for a while, the spores settle to the bottom and grow, without fertilization, into haploid gametophyte plants of one mating type or the other.

In *Ulva* the two generations are equally prominent in the life cycle of the organism. This is unusual; most organisms emphasize either the diploid or the haploid generation. In animals and in higher plants the haploid generation is short and inconspicuous. Conversely, in mosses and fungi the diploid generation is greatly reduced.

CORN

The maize plant commonly seen growing in the cornfield belongs to the sporophyte generation. It is diploid and bisexual. By the process of meiosis, the male tassels of the plant produce haploid pollen cells and the female tissues produce haploid ova (*Figure 5*). The gametes undergo several mitotic divisions to produce the haploid gametophyte generation. The small gametophytes lie within the tissues of the sporophyte and go unnoticed by the casual observer. When pollen lands on the female flower, a long, one-celled POLLEN TUBE grows toward the ovum at the base of the silk. A male gametophyte nucleus enters the ovum and fertilizes it, leading to the formation of an EMBRYO. The whole male haploid generation consists only of small gametophytes.

The female gametes are formed in separate tissues of the diploid sporophyte. A diploid cell undergoes meiosis to produce four haploid nuclei inside a single cell membrane. Three of these degenerate and the remaining one divides three times mitotically to produce a female gametophyte with eight identical haploid nuclei. One of the eight is the egg

In animals and higher plants the haploid generation is short and inconspicuous; in mosses and fungi the diploid generation is greatly reduced.

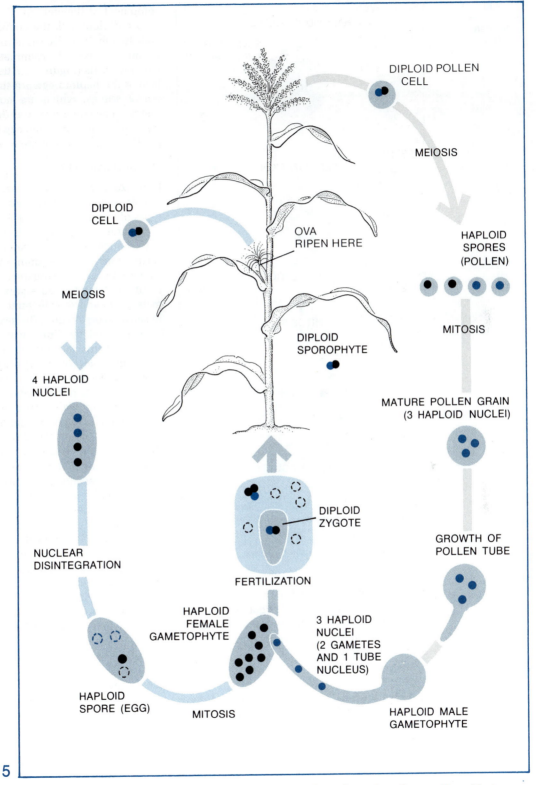

DIPLOID POLLEN
CELL

MEIOSIS

DIPLOID
CELL

OVA
RIPEN HERE

HAPLOID
SPORES
(POLLEN)

MEIOSIS

DIPLOID
SPOROPHYTE

MITOSIS

4 HAPLOID
NUCLEI

MATURE POLLEN GRAIN
(3 HAPLOID NUCLEI)

NUCLEAR
DISINTEGRATION

DIPLOID
ZYGOTE

GROWTH OF
POLLEN TUBE

FERTILIZATION

HAPLOID
FEMALE
GAMETOPHYTE

3 HAPLOID
NUCLEI
(2 GAMETES
AND 1 TUBE
NUCLEUS)

HAPLOID
SPORE (EGG)

MITOSIS

HAPLOID MALE
GAMETOPHYTE

5

LIFE CYCLE OF CORN involves alternation of generations. Most prominent phase is the diploid sporophyte (center). Haploid gametophytes are tiny structures nourished and protected by sporophyte.

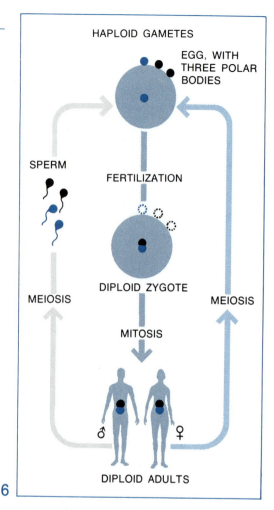

HAPLOID GAMETES

EGG, WITH
THREE POLAR
BODIES

SPERM

FERTILIZATION

DIPLOID ZYGOTE

MEIOSIS

MEIOSIS

MITOSIS

♂ ♀

DIPLOID ADULTS

6

LIFE CYCLE OF MAN is typical of most mammals. As in corn, diploid phase is most prominent. Haploid phase is reduced to a single cell (egg or sperm). Unlike corn, each mammal belongs to one sex or the other. Light- and dark-colored gametes show segregation for a specific trait.

nucleus which unites with the male gamete nucleus to form the diploid zygote of the new sporophyte. Each fertilized ovum becomes a corn kernel, and each kernel can grow into a mature sporophyte.

MAN

In higher animals, including man, the haploid stage is generally represented only by the gametes. Gametogenesis in the two sexes is different in detail. In males, a diploid spermatocyte undergoes meiosis to produce four haploid spermatids, which then mature into SPERMATOZOA. These spermatozoa are all alike in appearance. The female produces a large egg cell containing a sufficient store of nutrients to support the early stages in the de-

velopment of the embryo. The process begins with a diploid cell, the oocyte, which undergoes two meiotic divisions in which the cytoplasm is divided asymmetrically between daughter cells (*Figure 6*). Attached to the surface of the haploid egg are three tiny haploid POLAR BODIES, which are not functional gametes. They contain very little of the precious nutrient cytoplasm of the egg cell and are simply the by-products of meiosis.

SEX DETERMINATION

In maize, every diploid sporophyte gives rise to both male and female structures. These two types of tissue are genetically identical, just as roots and leaves are genetically identical. Plants such as maize and animals such as earthworms produce both male and female gametes in the same organism. In some higher plants, notably the date palm, and most animals, male and female gametes are produced in separate organisms. The sex of the offspring is determined by differences in the chromosomes, but this mechanism operates in a bewildering a variety of ways (*Figure 7*).

The sex of a honeybee depends on whether it developed from a fertilized or an unfertilized egg. A fertilized egg gives rise to a female bee — either a worker or a queen, depending on its diet during larval life. An unfertilized egg gives rise to a male drone. In many other animals, including man and the fruit fly *Drosophila*, sex is determined by a single chromosome or pair of chromosomes.

Drosophila females have two X chromosomes and males have one. But males in turn have a Y chromosome that is not found in females. The sex chromosomes of females can be represented as XX and the males as XY. Half of a male's gametes carry an X chromosome and the other half carry a Y. When an X sperm fertilizes an egg, the zygote will develop into a female; when a Y sperm fertilizes an egg, the zygote will become a male.

Humans and other mammals follow the same pattern as *Drosophila*: females are XX and males are XY. In birds, moths and butterflies, males are XX and females are XY. In these organisms it is the female that produces two type of gametes, and the sex of the offspring is hence determined by the egg, not by the sperm as in humans.

CHROMOSOME INACTIVATION

For centuries physicians were puzzled by the human disease called mongolism. It is inborn but not inherited, at least in the usual sense. Its symptoms are drastically impaired intelligence and widespread changes in body chemistry and anatomy. The structure of the eyes suggested the name mongolism to Caucasians, but because the disease has nothing to do with ethnic origins, it is better called by its proper name — DOWN'S SYNDROME — to avoid a misleading racial slur. In 1959 Jerome Lejeune and his coworkers in France made the astonishing discovery that sufferers from Down's Syndrome have 47 chromosomes instead of the normal human number of 46. The extra is a third copy of chromosome 21, one of the smallest in the genome, which presumably carries a correspondingly small portion of the genetic information. The many abnormalities in Down's Syndrome must be attributed to a 50 per cent increase in the hereditary dose of this information.

In the Lerner and Lowe musical, *My Fair Lady*, Professor Henry Higgins wonders "Why can't a woman be more like a man?" But Down's Syndrome raises the opposite question: Why are men and women so much alike? The normal female has two X chromosomes, and the normal male has one X and one Y. The Y chromosome in a male has few if any identifiable genes that are also present on the X chromosome, and appears to be genetically almost inactive. Hence there is a 100 per cent difference between women and men in the dosage of X-chromosome genes, and the X chromosome, unlike chromosome 21, is one of the largest in the human genome. Why then is not one sex or the other grossly deformed or completely inviable?

The answer was found in 1961 by two scientists working separately, Mary Lyon and Liane Russell. Lyon suggested that in any cell of a normal female one of the X chromosomes is inactivated early in embryonic life and remains inactive ever after. The choice as to which X in an XX female will remain active is random, but since the female embryo already consists of tens or hundreds of cells by the time the choice is made, virtually all women — in fact, most female mammals — contain patches of tissue in which one or the other X is active. Many diseases are known in humans which are caused by defects in an X chromosome gene, and some of these are associated with absence or abnormality of an enzyme which is easily assayed. Cells can be isolated from a woman who is known to be a heterozygous carrier of such a disease, and a separate culture can be grown from each cell. This produces two kinds of cultures: those with normal enzyme, and those with abnormal or inactive enzyme. No culture contains both normal and abnormal enzyme. The reason women who are heterozygous for such a disease — hemophilia ("bleeder's disease") for example — are not usually clinically ill is that there is enough tissue in which the normal X is turned on that the normal protein (in this case, antihemophilic globulin) can be made in sufficient amounts to supply the entire organism.

The results of cytological studies provide additional confirmation of these genetic and biochemical findings, and allow geneticists to refine the theory of active and inactive X chromosomes. It has long been clear that interphase cells of normal females have a single stainable nuclear body (called a BARR BODY, after its discoverer) which is not present in males, and which represents condensed chromatin, called heterochromatin. Heterochromatin does not produce mRNA, and is

SEX IS DETERMINED by different chromosomal mechanisms in the three types of organism depicted here. In humans two X chromosomes are characteristic of normal females, X and Y of normal males. Bees and other social insects have no sex chromosomes. Generally females, which develop from fertilized eggs, are diploid while males, which develop from unfertilized eggs, are haploid. In birds, two X chromosomes produce a male, XY a female. Human with one X is abnormal female; XXY human is an abnormal male.

metabolically inactive. Most interestingly, certain moderately abnormal women were found who had only one X chromosome and no Y chromosome. The cells of these women contained no Barr bodies at all. Other women were found who had a chromosome constitution of XXX, and their cells contained two Barr bodies. Some very abnormal men have been found with the bizarre genetic constitution, XXXXY. As might be guessed, these men had three Barr bodies in each interphase cell. The Lyon-Russell findings are now stated in the Single Active X rule: in mammals all but one of the X chromosomes in a cell are turned off. This explains why humans can survive with one X chromosome, or two, or even more than two. The rule even explains, perhaps to the dismay of everyone, why men and women are so similar.

SEX LINKAGE

In *Drosophila* and in man the Y chromosome is almost devoid of genetic content. This leads to an important deviation from the usual Mendelian laws when one examines the inheritance of genes located on the X chromosome. As noted above, any gene on that chromosome is carried in two copies by females but in only one copy by males. Therefore females may be heterozygous for genes that lie on the X chromosome, but males will always be HEMIZYGOUS for these genes. The first and still one of the best examples of sex-linked inheritance is the eye color in *Drosophila*. Their normal eye color is red, and the gene is carried on the X chromosome. In 1910, Thomas Hunt Morgan discovered a mutation that causes white eyes. When red-eyed, homozygous females were crossed to white-eyed males, all of the sons and daughters had red eyes because red is dominant over white, and all of the progeny inherited a normal X chromosome from their mother. But when a white-eyed female was mated to a red-eyed male, all of the sons were white-eyed and all of the daughters red-eyed, because the sons inherited only an X chromosome from their mother, and the Y inherited from the father does not carry any gene for eye color. The daughters, of course, get a white-bearing chromosome from their mother and a red-bearing chromosome from their father, and are therefore heterozygotes. If these heterozygous daughters are mated in turn to red-eyed males, half of their sons will have white eyes, but all of their daughters will have red eyes as shown in Figure 8.

A number of genes are located on the human X chromosome, and mutations affecting these genes are inherited in exactly the same way as white eyes in *Drosophila*. Hemophilia is a good example. A hemophilic man married to a homozygous normal woman will not produce any hemophilic children. The sons inherit a

SEX LINKAGE in Drosophila is depicted in the diagram below. The gene that governs eye color lies on the X chromosome. The wild type allele (red eyes) is dominant. Only females can be heterozygous for genes on X chromosome because Y chromosome carries almost no genes.

8

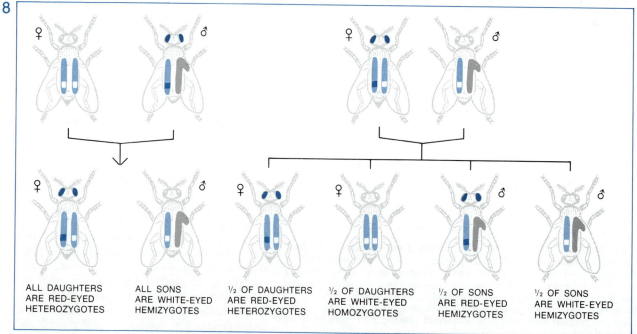

| ALL DAUGHTERS ARE RED-EYED HETEROZYGOTES | ALL SONS ARE WHITE-EYED HEMIZYGOTES | ½ OF DAUGHTERS ARE RED-EYED HETEROZYGOTES | ½ OF DAUGHTERS ARE WHITE-EYED HOMOZYGOTES | ½ OF SONS ARE RED-EYED HEMIZYGOTES | ½ OF SONS ARE WHITE-EYED HEMIZYGOTES |

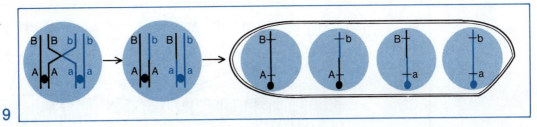

9

CROSSOVER occurs during the four-strand stage of meiosis. Diagram of meiosis in Neurospora proves this assertion. Crossover between two strands can produce only two types of spore in any one ascus. Crossover in four-strand stage can produce four different kinds of spore, as often found in nature. For simplicity, this diagram shows only one member of each pair of spores in the ascus.

single, normal X from their mother, and will neither have the disease, nor transmit it to their children. The daughters will get a normal X chromosome from their mother and a hemophilia-bearing chromosome from their father. Since hemophilia is recessive, the daughters will not be hemophilic. They will, however, be heterozygous carriers, and will transmit the disease to half their sons, and the carrier role to half their daughters. What would be needed to produce a female hemophiliac analogous to the white-eyed female fly? Her father would have to be a hemophiliac and her mother a carrier. Since hemophilia is quite rare, such a couple would be unlikely to meet. Moreover, hemophilic males rarely survive long enough to reproduce. One might expect hemophilic females to be extremely rare, and in fact very few have ever been found.

RECOMBINATION IN EUCARYOTES

It is obvious that the number of genes in a cell far exceeds the number of chromosomes, for each chromosome contains many genes. If chromosomes were not capable of pairing and recombining, the markers on a given chromosome would all segregate together, and the genes in an organism would fall into several segregation groups, one for each chromosome. The different segregation groups would assort independently, but within each group all markers would behave as a unit. A geneticist would have no way of telling that they were actually different genes. As noted earlier, Mendel's Second Law (independent assortment of alleles of different genes) applies only to genes that lie on different chromosomes.

The actual situation is more complex and therefore more interesting. Mutations located at different places on the same chromosome do separate from one another as the result of re-

combination, and the frequency with which they separate from one another depends on the distance between them on the chromosome. With eucaryotes, as with the procaryotes, geneticists can use recombination frequencies to draw genetic maps that indicate the actual arrangement of genes along the chromosome. Because meiosis in eucaryotes is a more orderly process than the pairing and recombination of the phages and bacteria, geneticists know much more about the formal details of recombination in eucaryotes than in procaryotes. But because the chromosomes of eucaryotes are much more complex than the simple DNA molecules of phages and bacteria, less is known about the molecular details of the recombination process in higher organisms than in viruses and procaryotes.

Recombination between genetic markers on the same chromosome occurs by a process called CROSSING OVER. This presumably results from the physical exchange of corresponding genetic segments between two homologous chromosomes. Crossing over occurs at the four-strand stage of meiosis, sometime before the first meiotic anaphase. The exchange event involves only two of the four chromatids in the synapsed pair of chromosomes. The points at which the chromatids break in the exchange seem to correspond perfectly, so that no genes are lost, and each chromatid ends up with the same amount of material (Figure 9).

At any point along the chromosome only two chromatids participate in crossover, although other crossovers may occur at other points. These crossovers may involve the same pair of chromatids or any other possible pair of homologous chromatids (Figure 10). Crossing over may involve two, three, or all four chromatids in a tetrad. The probability that more than one crossover will occur in the genetic segment between two mutational

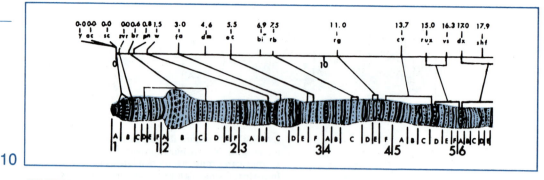

10

GENETIC AND CYTOLOGICAL MAPS of the X chromosome of Drosophila are compared in this diagram. Position of centromere is indicated by a star. Numbers below chromosome refer to the banding pattern observed under the microscope. Lines and numbers above chromosome comprise the genetic map worked out from recombination frequencies. Letters denote marker mutations.

markers depends on the distance between the markers.

NON-MENDELIAN INHERITANCE

The essence of Mendelian inheritance is that a very few copies of chromosomal information are partitioned with great precision during meiosis. But there are self-reproducing entities within eucaryote cells other than the chromosomes of the nucleus. Mitochondria, chloroplasts, and certain other cytoplasmic organelles also appear to carry some genetic information. The DNA of these organelles is subject to mutation just as is chromosomal DNA, and therefore these organelles can carry "markers." But these markers are not inherited in the same way as nuclear chromosomal markers because the amounts of cytoplasm contributed by the maternal and paternal gametes are grossly unequal. Generally speaking, all of the mitochondria (and, in eucaryotic plants, chloroplasts) in a zygote come from its mother, even though half of its chromosomes come from its father. Thus mitochondria and chloroplasts are sometimes said to be "maternally inherited."

Certain patterns of cellular architecture are inherited in a way that may not involve DNA at all. This sort of "pattern inheritance" seems to be especially important in protozoa such as *Paramecium;* but perhaps it is merely more noticeable in protozoa because they are large

cells with a distinctive external anatomy. One of the most bizarre examples of pattern inheritance is found in the protozoan, *Difflugia corona,* which was studied by H. S. Jennings in 1937. Jennings' work was eloquently summarized by David Nanney, writing in *Science:* "The organism constructs a shell by cementing sand grains together with a cellular secretion. The ventral surface of the shell possesses an opening, the 'mouth,' through which the cells communicates with the outside world. The edges of the openings are surrounded by a symmetrical array of 'teeth.' Jennings observed that the numbers of teeth varied among individuals, and he explored the question of their heredity by the only means available; he isolated individuals, allowed clones to develop, and inquired into clonal uniformity. Tooth number remained constant within a clone; differences in tooth number were hereditary. He was not able to conduct a breeding analysis, but he noted that, when the cell body divided, one of the daughter cells was extruded naked through the mouth and, while still in contact with its sister, began to construct its own sand castle, beginning in the region of contact (*Figure 11*). The new mouth structures were therefore constructed in direct contact with structures of the old mouth. This observation suggested to him that the old mouth might serve as a template to guide the organization of

Certain patterns of cellular architecture, especially in protozoa, are inherited in a way that may not involve DNA at all.

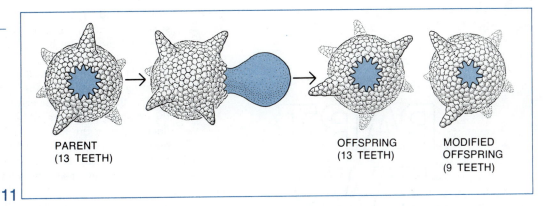

PARENT
(13 TEETH)

OFFSPRING
(13 TEETH)

MODIFIED
OFFSPRING
(9 TEETH)

11

PATTERN INHERITANCE governs the formation of "teeth" in the protozoan Difflugia corona. The teeth are made of sand. Cell with 13 teeth produces offspring with same number of teeth, because new teeth are formed while offspring is still in contact with the mouth of the parent cell (left center). If four teeth are removed by microsurgery, parent cell produces offspring with nine teeth (far right).

the new one — that new teeth were initiated in the interstices between the old teeth. This curious speculation might have remained just that had Jennings not tried his hand at oral surgery. He broke out denticles with a glass needle and examined the consequences of mutilating the parental template. Modified parents produced modified progeny, and new lineages were established with new tooth numbers. A few generations were required for symmetry to be achieved, but once that had been accomplished the tooth number stabilized and a new hereditary state was achieved. After considering these studies, one of my colleagues concluded ruefully that genetic specificity may be based on two different structural foundations — nucleic acid and sand. In view of the notorious instability of structures built upon sand, this conclusion is peculiarly disturbing."

READINGS

A.M. SRB, R.D. OWEN AND R.S. EDGAR, *General Genetics*, 2nd Edition, San Francisco, W.H. Freeman and Company, 1965. One of the favorite texts used in general genetics courses, this book is written in lucid and interesting language, and every chapter contains an instructive problem set.

C. STERN AND E.R. SHERWOOD, eds., *The Origin of Genetics: A Mendel Source Book*, San Francisco, W.H. Freeman and Company, 1966. This is a collection of original writings by the founders of genetics. It includes readable translations of Mendel's original paper, his letters to Nägeli, and papers by de Vries and Correns, who in 1900 independently rediscovered Mendel's work. In the last two articles in the book, two of the masters of mathematical genetics discuss the likelihood that Mendel fudged his data.

M.W. STRICKBERGER, *Genetics*, 2nd Edition, New York, The Macmillan Company, 1976. A fine, up-to-date text that deals in considerable depth with a broad range of genetic topics. This book also contains a number of useful problems.

A.H. STURTEVANT, *A History of Genetics*, New York, Harper & Row, 1965. An eyewitness account by one of the pioneers of genetics. The book begins with Mendel and ends at the dawning of molecular genetics. A fascinating and charming book.

A.H. STURTEVANT AND G.W. BEADLE, *An Introduction to Genetics*, New York, Dover Publications, 1962 (reprint of a book first published by W.B. Saunders Company in 1939). The text is old, but the years have treated it well. This is still one of the best introductions to "classical" genetics.

PART TWO MULTI-CELLULAR LIFE

Single-celled organisms are one of the great success stories of evolution. They probably comprise more than half the total biomass on Earth, and have successfully colonized even its harshest environments. Bacteria flourish in scalding springs and in the frozen soil of Antarctica; they survive the aridity of deserts and the crushing pressures of the ocean floor; they float freely in the atmosphere. Biochemically many microorganisms are far more versatile than man, being able to synthesize virtually everything they need from a few simple nutrients. And unlike man, many of them are potentially immortal, or at least ageless: when a bacterium reproduces by dividing in two, its daughter cells are equally young. A whole new life stretches before them. Although a bacterium can of course die of starvation or of exposure to toxic chemicals, or be eaten by another organism, it will never die of old age. It would seem that in the bacteria and their unicellular cousins nature has come close to creating the ideal organism — an observation that poses one of the profound enigmas of biology: Why, given the spectacular success of unicellular organisms, is the main thrust of evolution directed toward ever higher levels of organization?

This thrust has been marked by three great breakthroughs. The first was the invention of the eucaryotic cell, which according to one theory arose as a commune of previously free-living microorganisms. Eucaryotic cells are not only tens to hundreds of times larger than procaryotes, but they have a significant new property: the ability to aggregate into multicellular communes. The emergence of multicellular organisms was the second breakthrough, because unlike a colony of bacteria, they were more than a mere heap of cells. Their cells became specialized for a variety of functions, and interacted in a way that made the organism more than the sum of its parts. Multicells embodied a new level of organization, a whole new order of complexity. The first two breakthroughs occurred billions of years ago, but the third was relatively recent: the emergence of conscious intelligence. Although man is assembled from the same type of cells and organ systems as his biological relatives, his intelligence makes him a creature of a higher order.

The consequences of the third breakthrough are explored in Part IV; Part II focuses on the first and second. An immense gap separates the great majority of unicellular from multicellular organisms, a gap so large as to seem almost unbridgeable. Yet it was bridged long ago, and each multicellular organism, in its development from fertilized egg to mature organism, bridges the gap in a single lifetime. The basic question remains: What biological advantages favored this leap of evolutionary invention? Even with the almost explosive growth of biological knowledge in the past few decades, a really satisfying answer remains elusive. The classic argument is that multicellular organisms are somehow better adapted to flourish in certain environments, or can fill ecological niches that are totally inaccessible to unicellular forms.

One such environment is land. Life began in water, and for perhaps two billion years was confined to water, especially to the margins of the primitive seas (Chapter 16). Plants, the descendants of ancient marine algae, were probably the first organisms to colonize the land. The unicellular algae that flourish in oceans and lakes today are not very different from their ancient ancestors. Water bathes them with a solution of mineral nutrients and buoys them near the surface, where sunlight for photosynthesis is plentiful. Nevertheless, even greater rewards are available for plants on land: higher light intensity and richer concentrations of oxygen and minerals.

The first plants to invade the land enjoyed the advantage of diminished competition with other organisms. But terrestrial life imposed a whole new set of demands. Fossil remains of the first land plants have not yet been discovered, but it is reasonable to suppose that they were not unicells, or at least not procaryotic unicells like bluegreen algae. On land, water and nutrients are mostly trapped in the darkness of the soil, requiring plants to inhabit two environments at the same time: one end in the ground, the other in sunlight. No procaryotic cell is long enough to fulfill this requirement. It might seem that a long thin eucaryotic cell would serve the purpose, but as Chapter 1 pointed out, cells depend largely on diffusion for the internal transport of nutrients and other substances, and this imposes strict limits on their size. The eucaryotic marine alga *Acetabularia*, two or three centimeters high, is a giant among unicellular plants and probably lies near the upper limit of cell size. Although it is larger than many multicellular land plants, the single-celled *Acetabularia* lacks specialized physiological equipment necessary for survival on land.

CELL SPECIALIZATION

The invention of the eucaryotic cell was life's first successful attempt to exploit the advantages of greater size and complexity. Size in itself was an important new property. Not only did it help plants to colonize the land, but it enabled animals to feed on smaller organisms by overpowering and engulfing them. Even more important, the eucaryotic cell opened the way for the next breakthrough: the emergence of multicellular organisms. For reasons that are not at all clear, all multicellular organisms are comprised of eucaryotic cells. The main advantage of multicellularity is cell specialization. A multicellular organism can delegate highly specialized functions to particular groups of cells. For example, it can encase itself in a waterproof skin fashioned from one population of its cells, creating a wet internal environment favorable to the growth of other types of specialized cells. Clearly an organism capable of creating and maintaining its own internal environment is suited to pioneering new environments in a way that bacteria and other unicellular forms are not.

True, microorganisms are now found virtually everywhere on Earth, but they are there only because other organisms were there first. Parasitic bacteria, for example, rely on their hosts to carry them into new regions; many other microorganisms are saprophytes that feed on decaying material from higher organisms. Lastly, most of them depend on a supply of atmospheric oxygen created by plants. If microorganisms were forced to earn a living solely from one another they would comprise a much smaller percentage of the biomass. In the language of the ecologist, the presence of multicellular organisms increases the overall productivity of an ecosystem — that is, they not only carve out a niche for themselves, but they create new opportunities for other forms of life.

DIVERSITY

Multicellularity also carries the potential for vast diversity: millions of different shapes, specialized organ systems, and patterns of behavior. This potential greatly increased the kinds of environment that organisms could exploit, and thus constituted a powerful driving force toward the evolution of new and diverse adaptations. Once begun, the trend toward diversity reinforced itself. The existence of land plants created opportunities for land animals. Herbivores that fed on plants and carnivores that fed on herbivores are the most obvious examples, and their presence in turn created opportunities for a vast array of microorganisms. The wondrous assortment of plants and animals that inhabit the Earth today is the product of billions of years of interaction between life and Earth. The planet has provided the environment and the raw materials, and life has responded with dazzling inventiveness, fashioning organisms that can exploit every conceivable ecological niche. After glancing at the photographs of organisms in Chapters 19 and 20 one is tempted to ask, is there any limit to the potential diversity of living things?

The question applies both to whole organisms and to the systems that comprise them: digestive, sensory, reproductive and so on. The answer depends on what one means by diversity. The possible combinations of known types of cell are virtually unlimited. These combinations can be "packaged" in a vast but nonetheless limited number of sizes and shapes. The basic restrictions on the design of living creatures are imposed by the conditions on the planet, and by the laws that govern the behavior of matter. These restrictions pose a set of problems, and each species of organism represents one successful solution. For example, there are both lower and upper limits on the size of an organism.

THE LIMITS OF SIZE

The extremes of size range from viruses and bacteria at one end of the scale to the elephant, the sequoia and the blue whale at the other. The lower limit is imposed by chemistry and the upper by physics. A single-celled organism must be large enough to contain the genetic and metabolic machinery required for an independent existence — usually a few cubic microns. The elephant and the giant sequoia tree approach the theoretical upper limit to which land organisms can grow. As Galileo pointed out in the 17th century, the limitations are largely due to mechanical factors, particularly the bending moment: the gravitational effect that causes a beam supported at both ends to sag in the middle. A five-centimeter matchstick maintains its rigidity if it is held horizontally at one end, but a matchstick one meter long would droop. To maintain rigidity, the cross-sectional area of a beam must be increased in proportion to its length. This rule accounts for the difference in diameter among

the leg bones of a mouse, a deer and an elephant; the slimness of young trees, the fatness of tall ones. In the sea the buoyancy of water eases the engineering problems somewhat. The largest marine creature is the blue whale, which can attain a length of 30 meters and weigh 150 tons. Now hunted to the verge of extinction, it may be the largest animal that has ever lived on Earth.

In the arthropods (primarily insects and crustaceans) a rigid exoskeleton serves in place of the endoskeleton of our own bodies. Its size is limited by the mechanics of hollow objects. A small spherical shell can be remarkably strong, but as the diameter increases the shell becomes more fragile unless it is proportionately thickened. This is one reason for the absence of truly gigantic insects or crabs. Apparent giants (such as the Alaskan king crab, almost two meters across) achieve size not by an enormous body shell but by having long tubular legs whose strength lies in their narrow diameter. Few insects are longer than a few centimeters, and few are shorter than 0.2 millimeters; at the lower size limit the need for a rigid exoskeleton disappears.

Another factor that limits size is the disproportion between surface area and volume, which was mentioned in Chapter 1. The larger the diameter of an organism, the smaller the surface area per unit volume. The consequences are particularly clear in warm-blooded organisms. Heat loss depends on surface area. Very small animals have such a large surface area per gram of tissue that they suffer very high heat losses to the environment, and need to consume disproportionate quantities of food to maintain their body temperature. A mouse may eat its own weight of food in a day, while we eat about two percent of our weight per day. Large animals suffer the opposite problem. Their need to dispose of metabolic waste heat requires a special heat disposal mechanism, usually an elaborately controlled circulatory system that transports overheated blood to the body surfaces. The insect lacks such a controlled circulation, another reason that giant insects are an impossibility. Other solutions to the heat disposal problem among large animals include sluggish behavior, which generates less heat (elephant), or a partially aquatic lifestyle (hippopotamus).

DEATH AS A SURVIVAL STRATEGY

In the course of billions of years of evolutionary experiments, life has found successful solutions to each of the problems posed by existence on Earth. It is logical to ask why nature has not combined these solutions into an ageless superorganism. One answer is that nature is less concerned with the survival of the individual than with the survival of DNA. Paradoxically, the death of the individual is often the best strategy to ensure the survival of the species. Among multicellular organisms, the production of offspring is a better strategy for perpetuating DNA than the mere tenacious survival of individual organisms.

Agelessness has three main drawbacks as a survival strategy. First, on a changing planet the problems and opportunities for living organisms vary over the years in unpredictable ways. Climate shifts, dry land becomes flooded, new predators evolve. Such unforeseen threats can be answered only by providing opportunities for trial changes in the organism. These changes arise from sexual reproduction, which confers on the next generation a new assortment of genetic combinations and mutations. Among the offspring the failures die out, but those that succeed in the new environment survive to reproduce their better adapted genes. Second, ageless parents would compete for food and mates with their offspring, which would jeopardize the survival of future generations. A third drawback of agelessness is that living matter deteriorates.

In particular, the genes are under the constant bombardment of background radiation from space, and the resulting damage accumulates over a period of years. The probability of birth defects in humans increases with the age of the mother. Exposure to 20 extra years of radiation distinguishes the 40-year-old mother from the 20-year-old, and often results in damaged genes and inferior offspring. A population of immortal mothers would give birth to an appalling number of deformed children and genetic misfits. There are still other factors that favor the evolution of senescence and death, which we shall discuss in the more detailed treatment of evolutionary theory in Part IV.

The next 9 chapters explain the workings of multicellular organisms, beginning with development and growth: the journey from a one-celled fertilized egg to a mature organism consisting of from thousands to trillions of cells. The emphasis is on the diversity of systems that life has devised to propagate itself, to acquire raw materials and energy, to transport substances from one part of the organism to another, and to integrate and control its activities, including behavior. The strict design limitations imposed by conditions on Earth will become clear, and the reader will note how similar are the problems that each organism faces, and how different are the solutions.

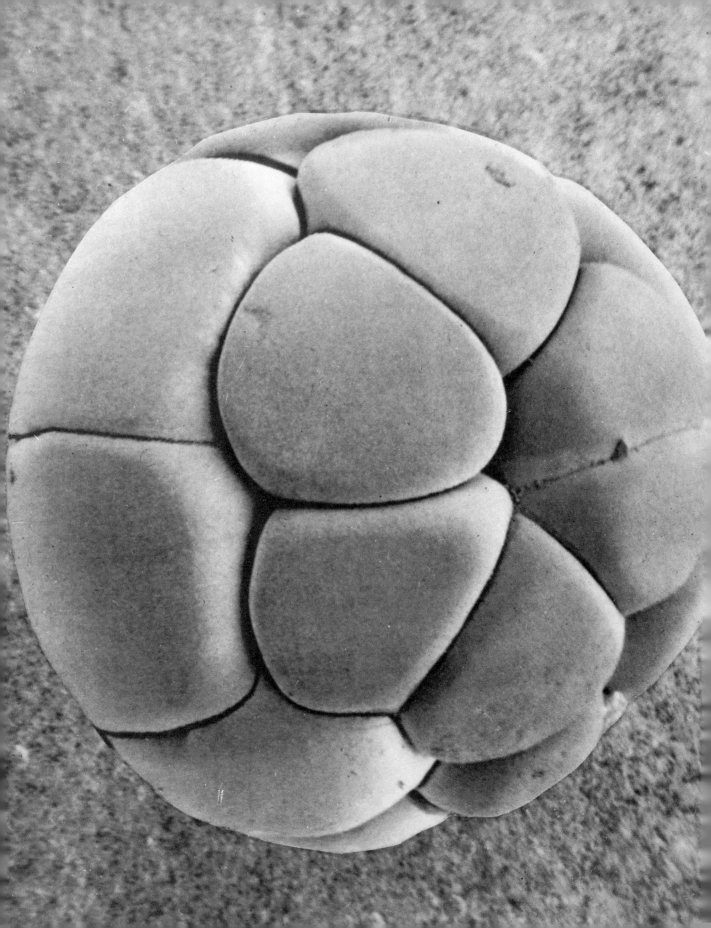

6

development

And what a glorious society we would have if men could regulate their affairs as do the millions of cells in the developing embryo.

R.M. EAKIN

Cells are not static structures; they change with time. The life cycle of an organism, from birth through maturity to death, can be described in terms of changes in its cells: the emergence of order, the maintenance of order, and finally the breakdown of order. The progressive emergence of order — the fashioning of living cells and organisms from a more or less random assortment of nonliving molecules — is called development.

At the cellular level and above there are sharp differences between the development of plants and animals, but at the molecular level the mechanisms of development are remarkably similar. Cells can be thought of as self-assembling systems. Polypeptide chains link and fold to form protein molecules; the proteins combine with other molecules to form structures like membranes and chloroplasts; and these structures in turn assemble themselves into complete cells. Some steps in the assembly process are virtually automatic, depending mainly on chemical and physical properties. Other steps are controlled directly by the DNA and RNA within the developing cell, or indirectly by hormones sent from other parts of the organism (*Chapter 12*). Like all biological systems, the cell assembly line has feedback loops and other control mechanisms that make the assembly process largely self-regulating.

SELF-ASSEMBLING MOLECULES

Many enzymes and other proteins are comprised of several polypeptide chains. In some proteins the chains are identical, while others contain two or more different chains. Once synthesized, the chains combine spontaneously into functional proteins. In the watery interior of a cell the concentration of ions, the pH of the medium, and the concentration of subunits can regulate the assembly of a protein molecule. Under the proper conditions, the whole molecule is more stable than the sum of its parts.

A good example of this is COLLAGEN, the fibrous protein that strengthens most animal connective tissue, including cartilage and bone. The fundamental protein of collagen, TROPOCOLLAGEN, consists of three polypeptide subunits coiled helically around one another to form short rods. Synthesized within the cell, the rods are transported outside the cell membrane where they assemble into collagen fibrils. The fibrils then orient themselves quite precisely, forming insoluble structures of enormous strength, such as tendons.

Tropocollagen molecules will aggregate into fiberlike structures in the test tube; the pattern in which they aggregate can be changed by altering the solution (*Figure 1*). The presence of proteins from blood serum results in one type of fibril, ATP a second, and different salt concentrations still another. Though little is known about tropocollagen aggregation within the living animal, clearly small changes in the animal's internal environment can shape the resulting structure.

ASSEMBLY OF AN ORGANELLE

A mature chloroplast is far more complex than a strand of collagen. Enclosed by a double membrane, it contains elaborate internal

ZYGOTE OF FROG depicted on opposite page has become a 16-celled sphere after four rounds of cell division. Picture was made with scanning electron microscope, and is reproduced at a magnification of about 100X.

1

SELF-ASSEMBLY OF TROPOCOLLAGEN into col-lagen rods or fibrils depends on the nature of the chemical environment. Collagen was dissolved in acetic acid, then reprecipitated in two different solutions. Electron micrograph at left shows fibrils that formed in salt solution. Banded fibrils (right) formed in acid glycoprotein solution.

membranes, precisely arranged. It is made up of proteins, a special type of DNA, and lipids. Although the self-assembly of molecules and membranes is important in chloroplast de-velopment, it is not the whole story. The synthesis of these structures requires an input of energy and information.

Chloroplasts can arise by division of pre-viously existing chloroplasts, but they can also develop from nonphotosynthetic tissue. When a seed germinates (sprouts), chloroplasts arise as small inpocketings of the cytoplasmic membrane. When exposed to light, these in-pocketings increase enormously in size to form the mature organelles. In the dark, however, one finds only very small PROPLASTIDS con-taining small crystalline structures called PRO-LAMELLAR GRANULES (Figure 2). The granules contain a few pigments, including carotenoids and minute amounts of protochlorophyll. When a seedling that sprouts in the dark is transferred to the light, the absorption of light triggers a photochemical reaction that converts protochlorophyll to chlorophyll. So long as the light shines protochlorophyll is synthesized and converted to chlorophyll. Many structural changes accompany these reactions. The pro-lamellar granules loosen up, disperse into small vesicles that migrate toward the periphery of the chloroplast and finally be-come flattened into stacks of discs. Protein and membrane synthesis begin on a massive scale. The organelle increases manyfold in size, and ultimately becomes an elaborate complex of soluble enzymes, membranes, membrane-bound proteins, and pigments.

The complete development and assembly of a chloroplast cannot occur in a test tube be-cause energy is needed to synthesize the giant molecules, and because the DNA in the pro-plastid does not contain the blueprints for all of the molecules in a mature chloroplast. The codes for many chloroplast proteins lie on the DNA in the cell nucleus. These proteins must be synthesized on ribosomes in the cytoplasm, then imported into the chloroplast. Interaction among organelles is required for the develop-ment of many components of the cell. As with the self-assembly of molecules, the assembly of organelles depends on the right environment.

CELL DEVELOPMENT

Although the developmental biologist can speak with some confidence about the assem-bly of molecules and organelles, the nature of the program that governs development re-mains a mystery. It is not enough to say that the program must somehow be encoded in the genes, because that does not explain why two embryonic cells with identical genes can fol-low two completely different pathways of de-velopment. It is clear that development is based on a precise series of changes in the pattern of protein synthesis. The synthesis of one group of proteins slows down or stops, and the synthesis of others begins. As the se-quence unfolds, the cell becomes progres-sively more specialized. At pre-set intervals the cell may divide by mitosis, and the daugh-ter cells continue the journey toward speciali-zation.

Although the developmental biologist can speak with some confidence about the assembly of molecules and organelles, the nature of the program that governs development remains a mystery.

The earliest stages in the development of eucaryotic organisms are controlled by messenger RNA already present in the egg. By some process that is not well understood, this mRNA is "unblocked" and made available for translation. Messenger RNA remains active for a very short time. Once the RNA in the egg has served its purpose, the next stages of development depend on the synthesis of new mRNA.

At different times during development, different sets of genes are active in RNA synthesis. Development must involve some mechanism for turning genes on or off — that is, for controlling transcription. Known as GENE ACTIVATION, this mechanism regulates the amount and kind of proteins that appear or disappear during development. How does it work? There is a great temptation to invoke the operon theory of Jacob and Monod, described in detail in Chapter 10. But the operon model is based on work with bacteria. The chromosomes of eucaryotes are far more complicated than the naked DNA of bacteria. They consist of a complex of DNA, RNA and protein (collectively called CHROMATIN). At this point one can say only that an operonlike mechanism could explain gene activation in higher organisms, but such a process is not yet documented in eucaryotes.

The basic problem of development is to explain how a one-celled fertilized egg becomes an adult organism comprised of many different types of cell. (Although this chapter will refer occasionally to development in unicellular organisms, the emphasis is primarily on multicellular ones, particularly the higher plants and animals.)

The story begins with the fusion of egg and sperm to form a ZYGOTE. Cell division of the zygote and its daughter cells leads to the aggregation of cells into tissues, tissues into organs, and organs into organisms. In one sense, development is best understood in terms of geometry, particularly topology, or "rubber-sheet geometry." Animal zygotes quickly develop into a hollow ball of cells. As the cells multiply and migrate, the embryo attains its final shape as a result of a complicated series of bulgings, foldings, and in- and out-pocketings of its layers of tissue. In plant embryos the rigid cell walls prevent cell migration, so the embryo attains its shape by differential growth: some parts of the embryo grow much more rapidly than others, forming new leaves and roots.

The study of development is also an exercise in the mastery of an unfamiliar vocabulary. This chapter contains many new terms, and

DEVELOPMENT OF CHLOROPLAST is depicted in four electron micrographs below. Protoplastid at far left contains crystalline prolamellar granule (top). In the presence of light the granule begins to break up into elongated vesicles (left center). Vesicles flatten into stacks of disks, seen edge-on at right center. Mature chloroplast is shown at far right.

2

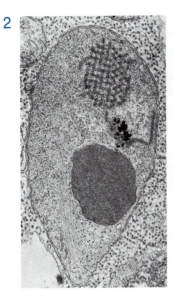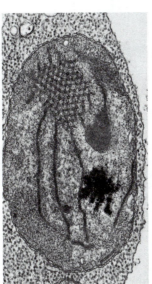

Determination in a narrow sense is the process by which a group of cells, a single cell, or even part of a cell becomes restricted to some predictable pathway of development. In certain fertilized eggs, for example, an observer can pinpoint the part of the cytoplasm that will become the epidermis or the lining of the gut.

Differentiation is the actual expression of determination. A cell differentiates both from its neighbors and its own past. A common definition of differentiation is: change leading to modification of structure and/or function.

Growth is simply irreversible increase in mass. Does growth involve differentiation? Cell division? Protein synthesis? The answer to these questions is normally yes, but there are many examples of growth in the absence of one or more of these processes.

Morphogenesis, literally the emergence of shape or form, occurs as a consequence of all the other developmental changes. It leads ultimately to the formation of a leaf, a hand, or a heart. Many differentiating systems, acting with remarkable coordination, mutual interaction, and growth, lead to morphogenesis.

---

the reader must master the definitions of four of the most important: DETERMINATION, DIFFERENTIATION, GROWTH and MORPHOGENESIS (*Box A*).

FERTILIZATION IN ANIMALS

Before looking closely at the events which follow the fusion of egg and sperm the origins of these two remarkable cells should be briefly summarized. In both males and females certain

PRIMORDIAL GERM CELLS, which will eventually give rise to the functional gametes, are recognizable in extremely young embryos. One usually thinks of the gametes as being products of the gonads: the male testis for sperm and the female ovary for eggs. But the primordial germ cells are evident long before the gonads themselves have appeared. With the eventual development of the gonads, germ cells migrate into the testis or ovary and continue their course of development.

The primordial egg cells go through a period of cell division followed by extensive growth and then enter meiosis. Of the four nuclear products of meiosis, three may simply be extruded as POLAR BODIES. The fourth will serve as the egg nucleus. Depending on species, a small or large amount of yolk is deposited within the egg. Yolk consists of droplets or platelets of protein, phospholipid, and fat. In eggs with a large amount of yolk the yolk will nourish the developing embryo for up to a few weeks. Much, if not all, of the yolk material is synthesized elsewhere in the organism, transported to the developing egg, and taken up by invagination (in-pocketing) of the egg membrane. In some insects, even ribosomes may be donated to the egg by other cells in this way.

Although the egg provides substantial biochemical assistance to the young embryo, sperm do nothing of the sort. The primordial sperm cells find their way into the testis, proliferate, and eventually go through meiosis, with each meiotic cell developing into a characteristic sperm. There is a head region,

EGGS AND SPERM. Sperm cell (top) consists of head, mitochondrial sheath and tail. Genetic material is packaged in the nucleus. Broken areas in tail indicate longer length. Diagram depicts sperm of the rabbit. Eggs of three species (bottom) contain different proportions of yolk (gray). Yolk of sea urchin and frog egg can subdivide into daughter cells by cleavage; yolk of bird egg cannot. Amount of yolk determines how long the egg can nourish developing embryo. Diagrams are not drawn to the same scale.

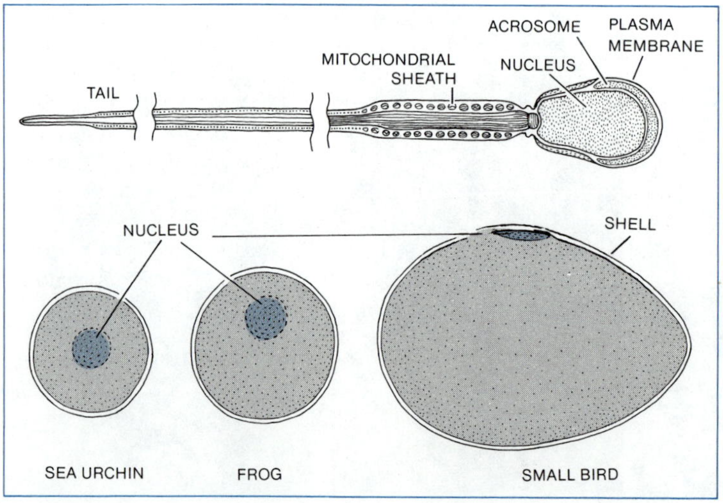

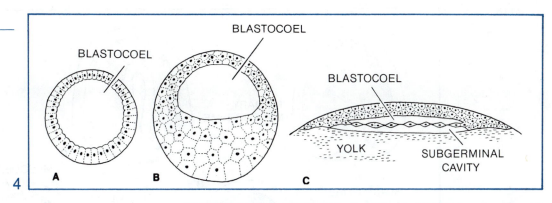

BLASTULAS of a sea urchin (A), a frog (B) and a bird (C). In sea urchin and frog, blastula is a hollow ball of cells. In birds it is a disk of cells that rests on the surface of the yolk.

consisting most anteriorly of a head cap and an underlying ACROSOME, frequently a vesicle or granule, and then the haploid nucleus. A middle section contains mitochondria, presumably to supply the power for swimming, plus a usual array of microtubules within (*Figure 3*). In some lower plant species with swimming sperm, the middle section may also contain a chloroplast. The acrosome plays an important role in fertilization. But the sperm contributes very little cytoplasm to the zygote.

The various mechanisms by which egg and sperm are brought together in different groups of animals, and the evolution of reproductive systems in the animal kingdom, are discussed in Chapter 7. As the small sperm approaches the much larger egg the acrosome makes the first physical contact. Most eggs secrete a jellylike protein-polysaccharide mixture that covers their plasma membrane, and the acrosome contains enzymes that digest this material. At the point of contact the egg plasma membrane fuses with the advancing membrane of the sperm head. Ultimately the sperm nucleus and a centriole enter the egg cytoplasm, and egg and sperm nuclei fuse. The whole process may take only a few seconds. In some unknown way the entrance of one sperm usually prevents the subsequent entry of others, probably because of rapid changes in the egg surface.

THE ANIMAL ZYGOTE

The course of events following fertilization depends largely on the amount of yolk present in the egg. This determines how long the egg can nourish the developing embryo, and thus the extent of the development that occurs within the egg before hatching.

Events are easiest to picture in the sea urchin, whose egg contains little yolk. The zygote divides, cleaving the cytoplasm into two equal portions, each with a nucleus. Cleavage is repeated until the embryo becomes a small hollow ball of cells called a BLASTULA, surrounding a central cavity or BLASTOCOEL (*Figure 4*). Much of blastula development occurs without any real increase in the size of the embryo, the original cytoplasm merely being partitioned into smaller units. Thus development occurs in the absence of growth. In amphibia such as the frog, with a large amount of yolk at one end of the egg, cleavage is somewhat different. Cells formed at the yolk end are very large, while those at the opposite end are small. The small cells contribute most to the young embryo. In birds the entire egg does not cleave; instead the zygote nucleus divides to form a flat plate of cells, the BLASTODISC, perched on the massive yolk.

The next stages of development involve the formation of a hollow three-ply embryo, most

Much of blastula development occurs without any real increase in the size of the embryo, the original cytoplasm merely being partitioned into progressively smaller units.

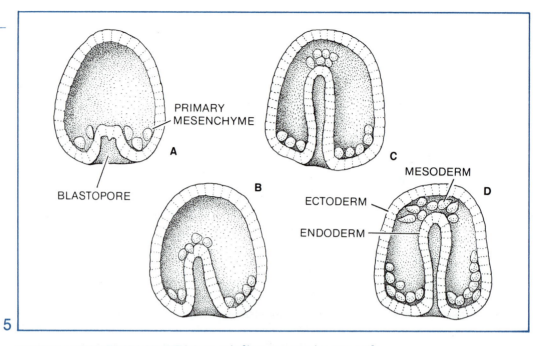

5

GASTRULATION IN SEA URCHIN. Letters indicate progressive stages of development from blastula (A) to gastrula (D). Invagination produces the primitive gut. Blastopore will become the anus.

easily visualized as three concentric tubes surrounding a central cavity. With the beginning of these stages, a major difference between plant and animal development becomes obvious. Animal cells, unrestricted by a tough cellulose wall, migrate extensively during development. The cells may move by cytoplasmic flowing, like amoebas, or by extending long filaments that attach to other cells and then contract, pulling the cells together. Within the cytoplasm minute microfilaments

GASTRULATION OF CHICK BLASTODISK. Arrows indicate direction of cell migration into and through the primitive streak to form mesoderm. Result is formation of thin three-layered embryo.

6

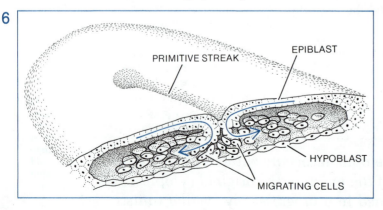

mediate the changes in cell shape that cause these movements.

The first of these MORPHOGENETIC MOVEMENTS appear in the process of GASTRULATION. Again the sea urchin is the simplest case. Though the urchin egg may look symmetrical it actually has an inherent polarity, with distinct ANIMAL and VEGETAL halves. The animal half will become the anterior (head) region of the embryo, and the vegetal half the posterior. In the blastula the cells at the vegetal end form a small pocket as they begin migration into the blastocoel (*Figure 5*). The whole embryo finally begins to increase in size. The new cavity is the ARCHENTERON or primitive gut, which opens to the outside through the BLASTOPORE, destined to become the anus. By now the zygote has become a two-ply embryo. The outermost layer of cells is the ECTODERM, and the innermost layer the ENDODERM. What is the origin of the third layer? Close to the blastopore, near the junction of ectoderm and endoderm, certain cells begin to migrate away from the two primary cell layers and move into the blastocoel. These cells are the primary MESENCHYME, and will form the middle layer or MESODERM of the three-ply embryo. Other mesenchyme cells become detached from the advancing tip of the primitive gut, and will

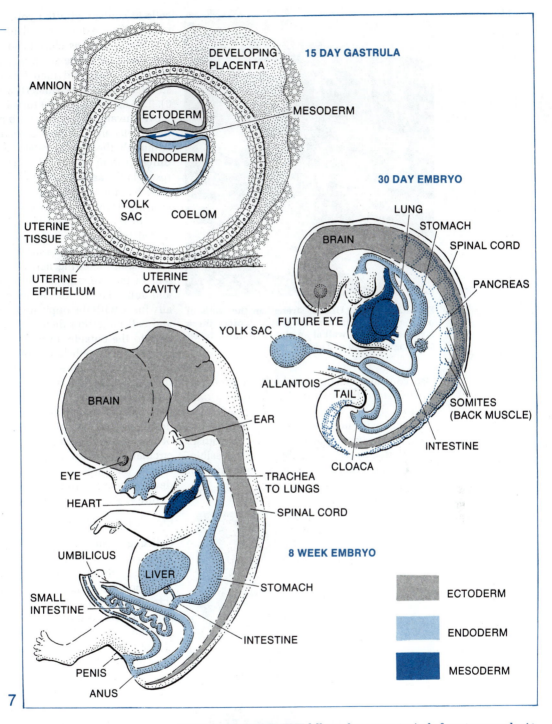

15 DAY GASTRULA

DEVELOPING PLACENTA

AMNION

ECTODERM

MESODERM

ENDODERM

YOLK SAC

COELOM

UTERINE TISSUE

UTERINE EPITHELIUM

UTERINE CAVITY

30 DAY EMBRYO

LUNG

STOMACH

SPINAL CORD

BRAIN

PANCREAS

FUTURE EYE

YOLK SAC

ALLANTOIS

TAIL

SOMITES (BACK MUSCLE)

INTESTINE

CLOACA

BRAIN

EAR

EYE

TRACHEA TO LUNGS

HEART

SPINAL CORD

8 WEEK EMBRYO

UMBILICUS

LIVER

STOMACH

SMALL INTESTINE

INTESTINE

PENIS

ANUS

7

ECTODERM

ENDODERM

MESODERM

HUMAN DEVELOPMENT follows the pattern typical of most mammals. At gastrula stage (top) embryo is a hollow ball of cells, folded in at the center. Colors indicate the three germ layers: ectoderm, endoderm and mesoderm. Mesoderm has just begun to migrate from its point of origin. Diagram at right is a longitudinal cross-section through a 30-day embryo. By this time the three germ layers have begun to differentiate into primitive tissues and organs. The major structures derived from each germ layer are indicated by the same colors in all three diagrams. By the 8th week the embryo has developed recognizably human form (left).

8

FIBRILS OF CELLULOSE strengthen the walls of plant cells. Electron micrograph depicts three layers of fibrils in wall of an algal cell, reproduced at a magnification of 24,000 ×.

also contribute to the developing mesoderm. At this stage the blastula has become a GAS-TRULA.

In eggs with massive yolks a three-ply embryo must be constructed from a flat blastodisc; gastrulation as described above is impossible. Instead, the disc rises, leaving one layer of

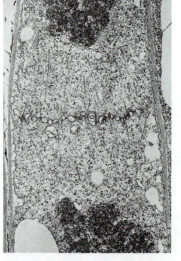

9

FORMATION OF CELL PLATE marks the last stage of cell division in plants. Electron micrograph of two cells of a seedling of soft maple (Acer sacharinum) reveals a horizontal row of vesicles against which new cellulose will be deposited to form plate that divides the two cells.

cells in contact with the yolk, forming a blastocoel. Then extensive cell migration begins. The blastodisc has clear head-to-tail polarity, and the cells at the sides begin to migrate to the center line and then forward, forming a structure called the PRIMITIVE STREAK (Figure 6). The cells toward the forward end begin to move inward and enter the blastocoel. All of these cells are destined to become mesoderm. Although the spatial relationships are quite different from those in the sea urchin, the end result is the same: a three-layered embryo.

Each of the three cell layers of the completed gastrula develops into a specific group of tissues in the adult. Ectoderm gives rise to the neural tube, which will produce the nervous system; and to the outermost layer of skin, the EPIDERMIS, including nails, hair, feathers, the lens of the eye, and the linings of the mouth and anus. The tip of the archenteron eventually fuses with the opposite layer of ectoderm, and cell migration then opens the end of the tube to the outside, forming the mouth. As a consequence teeth have both ectodermal and endodermal components. The endoderm forms the entire digestive tract and all related structures, such as the lungs and respiratory system, liver, pancreas, thyroid gland, and bladder. The mesoderm gives rise to a vast amount of tissue: bone, muscle, blood, and the entire circulatory system including the heart. The fate of the three germ layers during human development is illustrated schematically in Figure 7.

DEVELOPMENT IN PLANTS

Cell migration is absent in plants. Almost all plant cells are encased in a cell wall whose principal structural element is cellulose, a giant polymer of glucose molecules. Packaged into discrete micelles held together by hydrogen bonding, the cellulose molecules are organized into microfibrils of enormous strength, clearly visible under the electron microscope (Figure 8). The rigidity of the wall depends on an extensive network of such microfibrils embedded in a gelatinous matrix of pectins and hemicelluloses, both polymers of various sugars and sugar acids. The wall restricts the size of the plant cell and prevents the cell movement which is so characteristic of animal cells during development.

Cell division is different in plants and animals. After mitosis in animal cells, the membrane between the two daughter nuclei becomes constricted and the cytoplasmic

Lacking yolk, the plant egg is normally surrounded by cells that actively synthesize nutrients for the developing embryo. The embryo does not become independent until the dormant seed germinates, which may be years after fertilization.

connection between the two daughter cells is finally pinched shut. Once cell division is complete, the two cells are free to wander their respective ways, and frequently do. For reasons lost in evolutionary history, plant cells do not divide by constriction of the cell membrane. In plant cells after mitosis a plate forms across the center of the mitotic axis, and eventually the entire cytoplasm is cut in two (*Figure 9*).

Following cell division, the daughter cells increase in size. In animal cells extensive protein synthesis, the uptake of water, and the manufacture of more membrane material are sufficient. Small molecules readily enter the cell, and larger molecules such as proteins, and even solid particulate matter, can be taken up by invagination of the cell membrane. The small vesicle thus formed becomes pinched off inside the cell; eventually its contents are released into the cytoplasm.

REORIENTATION OF CELLULOSE FIBRILS occurs during elongation of plant cell. At the start of elongation (left) microfibrils of cellulose in the cell wall show roughly horizontal orientation. As the cell grows, the original fibers are displaced vertically (center), while new horizontal fibers are added to inner wall. At a late stage of growth (right) the original fibrils are almost vertical, but recently added innermost fibrils are still horizontal. Cross-ply orientation strengthens cell wall.

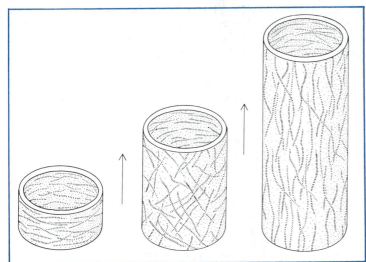

The growth of plant cells differs from that of animal cells in two important ways. First, they grow mainly by taking up water. Far less protein synthesis is involved, and much of the entering water, plus inorganic ions and small organic molecules, become concentrated in a large central vacuole. The cytoplasm shrinks to a thin layer lining the plasma membrane. Cytoplasm may comprise less than 10 percent of the total cell volume. Second, during plant cell growth there must be a substantial increase in the area of the cell wall. As the cell increases in size, the existing wall becomes stretched progressively thinner, and new wall material is laid down.

The absence of coordinated cell migration imposes a strict requirement for coordinated cell enlargement. Adjacent cells are cemented together rigidly, and connected by thin strands of cytoplasm. Two adjacent walls must increase in area in precisely the same manner. The walls normally don't slip past each other, and their cytoplasmic connections remain intact.

With cell enlargement and cell movement thus restricted in plants, what determines cell shape, and ultimately organ shape? The answer is that the cellulose microfibrils are not oriented at random. In a rapidly elongating stem, for instance, the orientation of the microfibrils is predominantly horizontal (across the long axis of the stem). As entering water forces the cell contents against the wall, the wall can bulge very little at the sides (along the axis of the microfibrils), and instead expands lengthwise separating the microfibrils from one another (*Figure 10*). The horizontal orientation is not perfect; some of the microfibrils do become tilted, but the new ones are deposited horizontally. The thickness of the wall, as determined by the number and orientation of the fibrils, regulates the rate and direction of plant cell growth.

Plants also differ from animals in their solution to the problem of forming organs of specific shape and size from undifferentiated groups of cells. In animals cell migration and

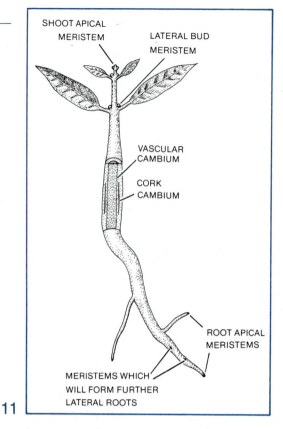

SHOOT APICAL
MERISTEM

LATERAL BUD
MERISTEM

VASCULAR
CAMBIUM

CORK
CAMBIUM

ROOT APICAL
MERISTEMS

MERISTEMS WHICH
WILL FORM FURTHER
LATERAL ROOTS

11

MERISTEMS are centers of mitotic activity and growth in plants. Diagram shows their locations in shoot, stem and roots. Girth of plant is controlled by cylindrical meristems: the vascular and cork cambium (shown in cutaway view at center).

the extension of filamentlike cell processes (as in nerve cells) serve this purpose. Cell proliferation simply provides the raw material. In plants the absence of cell movement requires that cell division be confined to specific regions of the tissue. These local centers of mitotic activity are called MERISTEMS. At the tip of each shoot or branch is a SHOOT APICAL MERISTEM, and at the tip of each root, an analogous ROOT APICAL MERISTEM (Figure 11). Flattened leaves are formed by localized lateral development of the shoot apical meristem, and lateral roots by development deep within root tissues well behind the root apex. Increase in girth of roots and shoots, when it occurs, is the result of activity of a cylindrical meristem, the CAMBIUM. The inner derivatives of the cambium becomes the woody XYLEM, the tissue through which water and minerals flow upward from the roots, and the outer derivatives become the PHLOEM, through which products

of photosynthesis are transported from leaves to other portions of the plant (Chapter 10). Outside the phloem there may be another cambium, the CORK CAMBIUM, which produces the bulk of the tissue we call bark. The shape of these meristems and the distribution of cell divisions within them determine the final shape of plant tissue and organs.

PLANT EMBRYOS

Higher plants have entirely dispensed with the swimming male gamete so characteristic of animal reproduction; instead they rely on the mechanism of POLLINATION which was discussed briefly in Chapter 5. The egg is deeply buried within the parent tissues. A pollen grain adheres to the sticky surface of a female portion of a flower and germinates, forming a POLLEN TUBE which may grow several centimeters, digesting its way through female tissue to reach the egg. The tip of the pollen tube then ruptures, releasing a sperm nucleus that fuses with the egg nucleus to form the zygote. The zygote resides in a specialized structure called the OVULE, which will eventually develop into a dormant seed. The seed serves two functions: it harbors the young plant embryo and permits dispersal.

Lacking yolk, the plant egg is normally surrounded by cells which actively synthesize nutrients for the developing embryo. The embryo does not become independent of the surrounding tissue until the dormant seed germinates, an event that may come months or years after fertilization.

Figure 12 shows the sequence of events following fertilization in the flowering plant *Capsella*, known as Shepherd's Purse. The zygote begins development by forming a short filament of cells. One end of the filament then begins cell division in three dimensions, forming a small globular embryo.

The slender remaining filament, called the SUSPENSOR, then elongates, pushing the globular embryo into the surrounding nutrient tissues. Eventually, localization of cell division produces two small lobes, making the embryo appear heart-shaped. These lobes are the so-called seed leaves or COTYLEDONS. Continued cell division and enlargement elongate the basal region of the embryo into the stemlike HYPOCOTYL (Greek for below the cotyledons) and elongate the cotyledons themselves, producing a torpedo-shaped embryo. Continued development then causes the whole embryo to be bent back on itself, as shown in Figure 12.

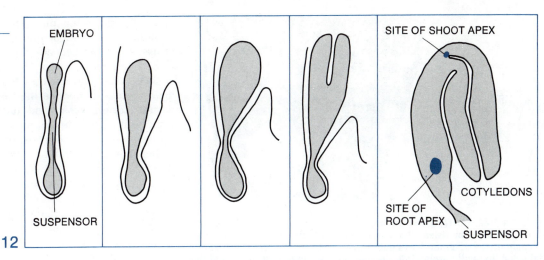

EMBRYONIC DEVELOPMENT OF PLANTS proceeds during formation of a seed. Sequence of drawings illustrates events that follow fertilization in the flowering plant Capsella, a familiar weed known as Shepherd's Purse. Suspensor elongates (far left), pushing embryo into surrounding nutrient tissues. Embryo eventually develops two small lobes (cotyledons) and below them a stemlike hypocotyl (right center). With continued differentiation embryo bends back upon itself and meristems appear (far right). Meristems are rendered in color in drawing at far right.

Meanwhile the surrounding tissues also undergo substantial development, finally producing a mature seed consisting of a seed coat, stored food material, and the embryo itself. At the same time cellular differentiation produces strands of elongated cells which will develop into the first functional xylem and phloem, tissues specialized for long-distance transport of water and nutrients (*Chapter 10*). These cells are called the PROCAMBIUM. The root apical meristem arises fairly deep within the lower end of the hypocotyl, while the shoot apical meristem appears as an insignificant mound of tissue between the cotyledons.

When the seed germinates, both root and shoot apical meristems begin the rapid and organized cell division which produces the young seedling. Here another major difference between plant and animal development emerges. Following gastrulation and neurulation, the animal embryo goes on to elaborate all of the various tissues and organs of the mature individual, at least in embryonic form. (The only exceptions are animals that undergo METAMORPHOSIS, which will be considered later.) Following this embryonic period the embryonic structures grow extensively and increase in complexity as the animal matures. By contrast the plant embryo, even at the mature seed stage, remains undifferentiated. None of

the many leaves, branches, or roots of the mature plant is represented by primordial cells or organs. Only the rather unimpressive primary root and shoot apical meristems suggest the events to come. These two meristems (or meristems derived from them) remain active throughout the life of the plant, maintaining the presence of embryonic and developing regions.

ROOT AND SHOOT MERISTEMS

At the lower end of the hypocotyl, but still beneath the embryo surface, a center of active cell division appears: the primary root apical meristem (*Figure 13*). The products of this meristem radiate out in all directions. Those below form the root cap, a structure which among other things protects the delicate root meristem as it advances through the soil. Those behind produce the root EPIDERMIS, from which absorptive root hairs will soon emerge; a CORTEX of thin walled cells; an ENDODERMIS which eventually will serve to regulate entry and exit of materials from the root vascular system; a region of very small cells celled the PERICYCLE; and finally the PROCAMBIUM which will form the vascular xylem and phloem. In cross section the xylem is frequently seen as a solid star-shaped mass, with phloem between the points of the star. Both

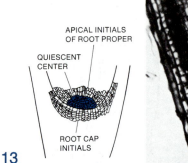

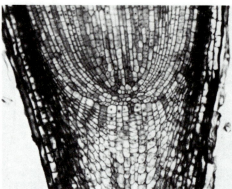

13

APEX OF PEA ROOT, shown in longitudinal section in photomicrograph above, contains meristem that gives rise to all the tissues of the root. Accompanying diagram shows undifferentiated meristem cells (called initial cells) that will develop into the root cap and into the structures of the root proper: epidermis, cortex, xylem, phloem and so on.

xylem and phloem cells are elongated and specialized for transport (*Chapter 10*). Cambium sometimes appears between the xylem and phloem, forming a complete cylinder.

Following the seedling stage of development, when the central region of the root meristem is highly active, cell division tends to occur mainly in the outer surfaces of the meristem. A quiescent zone forms in the core of the root. Here mitosis is rare, DNA synthesis very slow, and RNA and protein synthesis limited.

Figure 14 is a photomicrograph of the shoot

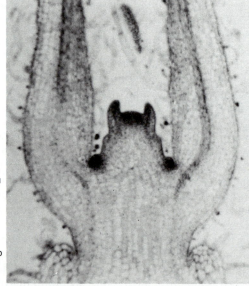

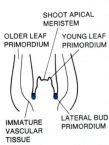

14

SHOOT APEX of the herb Coleus. Photomicrograph depicts developmental structures characteristic of flowering plants: leaf primordia, lateral bud primordia, and shoot apical meristem.

apex of a flowering plant, the herb *Coleus*. Near the tip of the shoot apex lie small mounds of tissue, the LEAF PRIMORDIA, which will develop into mature leaves. The outermost layer of shoot cells gives rise only to epidermis; the layers below it produce the rest of the stem tissues: cortex, xylem and phloem, and the central pith. Like the root meristem, the shoot meristem has a quiescent region at its core.

One other major difference between root and shoot meristems should be noted. The shoot apex produces lateral structures in a defined sequence — structures which will become leaves or flower parts. In the angle between the leaf primordium and the apex (the LEAF AXIL) a cluster of small cells forms a BUD PRIMORDIUM. Under suitable conditions, bud primordia will resume meristematic activity, become organized as shoot apical meristems, and produce lateral branches. The root apex, by contrast, produces no lateral structures except tiny root hairs. Lateral roots arise from more mature regions of the primary root. Opposite the points of the xylem star, one finds nests of pericycle cells which begin mitosis, and eventually become organized into a root apex. This root apex must digest its way through the endodermis and cortex to reach the substrate outside. Thus not only internal organization but also the origin of lateral structures are quite different in the two meristems. Yet both can remain meristematic for extremely long periods of time, in distinct contrast to the sharply limited embryonic period of animal development.

INDUCTION AND TISSUE INTERACTION

As noted earlier, the fate of an embryonic cell is determined not merely by its own genes, but also by interactions with neighboring cells. In 1924 Hans Spemann and Hilde Mangold demonstrated this in a set of definitive experiments on amphibian embryos. They removed the dorsal lip of the blastopore of a gastrulating embryo and grafted it within the blastocoel of a similar embryo. The results were startling. The dorsal lip developed into a normal neural tube, and the underlying host tissue formed a spinal cord and other associated structures (*Figure 15*). In some experiments, twin embryos developed attached belly to belly. The relationship between the grafted and host cells was clearly a partnership. Development of host tissue was induced by the grafted piece, and development of at least some tissues of the grafted piece was induced by the host.

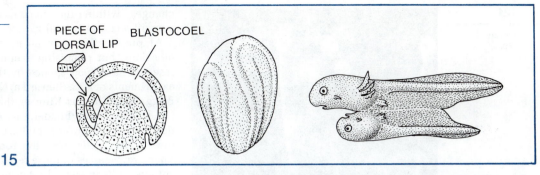

TRANSPLANTATION EXPERIMENT demonstrated the role of tissue interactions in the development of frog embryos. Part of the lip of the blastopore of one gastrula was grafted into the blastocoel of another gastrula (left). Grafted gastrulas developed two neural tubes instead of one (center), and a few of the gastrulas survived to develop into twin tadpoles joined at the belly (right).

In the study of tissue interactions, the mobility of animal cells raises an important question: What factors direct the migration of specific cell types in such a precise manner? Why, for example, are only liver cells found in liver, and in addition, why are they never found elsewhere? Shortly after the turn of the century, H. V. Wilson discovered that by pushing a sponge through a fine mesh, he could separate it into individual cells and clusters of cells. If the cells were then allowed to stand in a bath of seawater, they reaggregated themselves into a perfectly respectable sponge! Wilson and later workers then showed that if one mixed such suspensions from two different species of sponge, the cells reaggregated strictly according to species — not into hybrid sponges. The experiments suggested that a species-specific aggregation factor must be involved, but even today scientists can only guess that it must be some protein-polysaccharide complex.

In one well-known organism aggregation has been related to a single chemical substance. The cellular slime mold *Dictyostelium discoideum*, a remarkable fungus, begins its life cycle as a group of spores. The spores germinate under favorable conditions, each producing an amoeboid cell or myxamoeba which ranges about freely, feeding on bacteria and dividing in two. When the food supply dwindles, however, thousands of myxamoebae begin to aggregate into a sluglike "organism." After some aimless wandering, during which each cell retains its separate identity, the slug develops into a stalk capped by a globular fruiting body (*Figure 16*). Each cell in the globe becomes a spore, the whole structure dries, and the spores are shed to be wafted to some new favorable location for germination. It is clear that accumulation of some substance in the medium during food depletion is responsible for aggregation of the myxamoebae; the substance is the molecule adenosine 3', 5' phosphate, usually called cyclic AMP. Certain cells begin to release it, and others follow the concentration gradient to the source cells.

IS DIFFERENTIATION REVERSIBLE?

It is an axiom of genetics that virtually every cell in a multicellular organism contains a complete set of genes — all the genetic information necessary to create a complete organism. A corollary of this is that any specialized cell in a multicellular organism retains the genetic equipment necessary to carry out all the functions of any other cell in the organism. A nerve cell, for example, has the

Most botanists tend to think of differentiation as reversible while zoologists tend to think of it as permanent, but this is not a hard and fast rule. A lobster can regenerate a missing claw, but a cat cannot.

16

FRUITING BODIES of the cellular slime mold Dictyostelium discoideum are sacs of spores supported by stalks. About half an inch high, the entire structure is an aggregation of thousands of single-celled social amoebas.

same set of genes as a liver cell; they carry out different functions only because they are expressing different sets of genes. In theory it should be possible to transform a nerve cell into a liver cell, or to regenerate a missing finger from a muscle cell, or to grow an oak tree from a leaf cell. But in nature things don't work out that way, which in the past has led some biologists to speculate that differentiation is irreversible. The tentative explanation was that once development had run its course, most of the genes of the cell were permanently switched off.

In certain types of cell, differentiation is clearly irreversible. The mammalian red blood cell, which loses its nucleus during development, is one example. Another is the xylem tracheid, a water-conducting cell in higher plants. The development of the tracheid leads to the death of the cell, leaving only the pipelike cell walls that were fashioned while it was alive. In these two cases the irreversibility of differentiation can be explained by the absence of a nucleus. It is much harder to generalize about mature cells that retain normal nuclei. Most botanists tend to think of differentiation as reversible, while zoologists tend to think of it as permanent, but this is not a hard and fast rule. A lobster can regenerate a missing claw, but a cat cannot. Why is differentiation reversible in some cells but not in others? At some stage of development do

changes within the nucleus permanently commit a cell to specialization?

At present there are no answers to these questions, but promising avenues for further research have been opened by the pioneering experiments of F. C. Steward at Cornell, Robert Briggs and Thomas King of the Institute for Cancer Research, Philadelphia, and J. B. Gurdon of Oxford. Steward and his colleagues cultured callus cells from the roots of carrots. They grew masses of these relatively undifferentiated cells in special rotating flasks so that the dividing cells were constantly being tumbled and agitated in a liquid medium. When a suspension of single cells from this culture was plated on nutrient agar, the cells developed first into globular EMBRYOIDS and eventually into mature carrot plants. Clearly the nucleus of the callus cell had not differentiated irreversibly, and could still express the full genome.

The ready formation of callus by many mature plant tissues, and the capacity of these calluses to differentiate into active roots and shoots, certainly supports the argument that so long as a nucleus retains its normal complement of chromosomes, it still possesses all of the information needed to produce an entire organism. Animal cells have proved more difficult to work with. Isolated animal cells grown in tissue culture never develop into complete organisms or even into relatively simple organs. The reader will recall that organ formation in animal tissue culture requires interaction between two or more cell types. Perhaps differentiation in animals DOES involve permanent changes within the nucleus.

The first indication that this statement should be qualified came from the elegant experiments of Briggs and King. They activated a frog egg to begin development by puncturing it with a glass needle, then carefully removed the egg nucleus. Into the enucleated egg they transplanted a nucleus from one of the cells of an older embryo. If the donor cell came from a blastula, shortly before gastrulation, each recipient egg developed into a normal embryo. If the donor cells came from a somewhat later stage of development, say a late gastrula, the resulting embryos were usually defective. The experiment demonstrated that at least through blastula formation, the frog embryo nucleus contains (and can express) all of the information needed to make a whole frog.

It is not clear why Briggs and King obtained

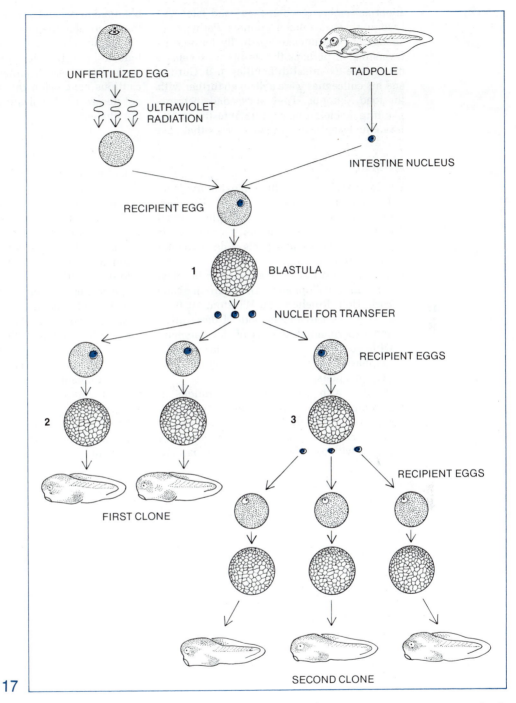

17

GURDON EXPERIMENT demonstrated that mature cells from a toad tad-pole contain all the genes necessary to guide the development of an egg into a mature adult. Nucleus from intestinal cell of Xenopus tadpole was surgically transplanted into unfertilized egg. When egg developed to blas-tula stage (1), nuclei from blastula were in turn transplanted into eggs whose nuclei had been removed. These eggs developed into normal blas-tulas (2), then into tadpoles and eventually into normal toads. Nuclei from one blastula were transplanted into new eggs, giving rise to another gener-ation of toads. Known as serial transplanting, technique can be repeated indefinitely.

abnormal embryos when they used later embryonic stages as nucleus donors. Perhaps the changes in the nucleus gradually become irreversible, or perhaps the results were caused by obscure technical difficulties. J. B. Gurdon and his colleagues were able to go further with the toad *Xenopus*. They succeeded in transplanting nuclei from mature intestinal cells into enucleated eggs. In some cases they just obtained nerve and muscle cells, but in others, they were able to grow adult frogs (*Figure 17*). The remarkable results of these experiments tend to bring the viewpoints of botanists and zoologists closer together, and prompts the cautious conclusion that differentiation does not necessarily involve permanent changes in the nucleus.

READINGS

B.I. BALINSKY, *An Introduction to Embryology*, 4th Edition, Philadelphia, W. B. Saunders, Co., 1975. An excellent and complete animal embryology text, particularly strong in its treatment of cellular aspects of development and morphogenesis.

J.D. EBERT AND I.M. SUSSEX, *Interacting Systems in Development*, 2nd Edition, New York, Holt, Rinehart and Winston, 1970. A good modern treatment both of descriptive and experimental aspects of development, with good coverage of molecular aspects. Emphasis is on animals, with only a few plant chapters.

C. FULTON AND A.O. KLEIN, *Explorations in Developmental Biology*, Cambridge, Mass., Harvard University Press, 1976. An exceptional collection of reprints of important papers with excellent text between; particular emphasis, molecular aspects of development and animal development. Contains original papers on many of the experiments described in this chapter.

J. LASH AND J.R. WHITTAKER (Editors), *Concepts of Development*, Sunderland, Mass., Sinauer Associates, 1974. A detailed, up-to-date collection of essays by various specialists on a wide range of topics in animal development.

T.A. STEEVES AND I.M. SUSSEX, *Patterns in Plant Development*, Englewood Cliffs, N.J., Prentice-Hall, 1972. A thorough and excellent treatment of cellular aspects of plant development.

F.C. STEWARD AND A.D. KRIKORIAN, *Plants, Chemicals and Growth*, New York, Academic Press, 1971. A broad treatment of plant development from molecule to organism.

NOTES

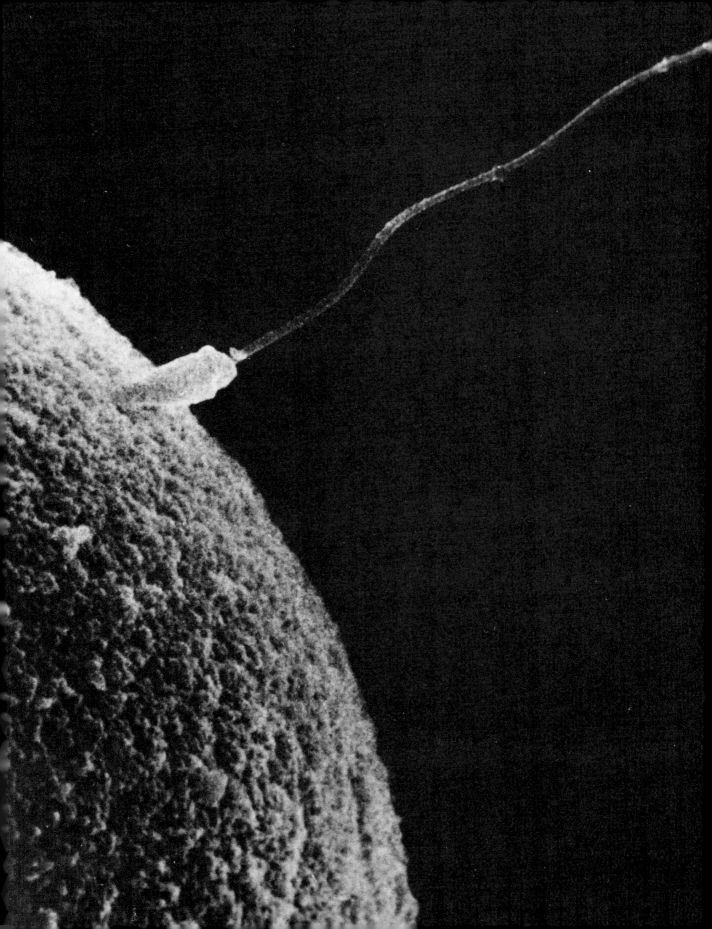

7

reproduction

The Earth, after having brought forth the first plants and animals at the beginning . . . has never since produced any kinds of plants or animals, either perfect or imperfect; and everything which we know in present times she has produced, came solely from the true seeds of the plants and animals themselves. . . .

FRANCESCO REDI (1688)

The amount of living matter on Earth remains roughly constant. If no organism ever died, there would be no room for new ones to come into existence. Reproduction is the means by which the mortal organism increases the probability that its genes will survive after it has perished. The more likely a given kind of organism is to die — in other words the shorter its average life span — the higher is its rate of reproduction. Insects have short lives and high reproductive rates; elephants have very long lives and low reproductive rates.

The fastest way to reproduce is asexually. Single-celled organisms such as *Amoeba* simply DIVIDE to produce two genetically identical daughter organisms. Another method employed by both unicellular and multicellular organisms, from parasitic protozoans to fungi, is to convert one or more of their cells into SPORES. These tiny units, each containing at least one complete set of DNA, are especially adapted for dispersal. Fungal spores, for example, are carried like dust particles for long distances — in extreme cases all the way around the Earth — in the winds of the upper atmosphere. A third method of asexual reproduction in multicellular organisms is BUDDING. The parent organism simply sprouts an offspring from part of its body (*Figure 1*). Many plants rely on an asexual process called VEGETATIVE REPRODUCTION. Special organs serve this purpose, including for example the familiar tubers of potatoes, and the runners of strawberries and certain grasses. A fifth mode of asexual reproduction is PARTHENOGENESIS, the growth of an organism from an unfertilized egg.

If asexual reproduction is so simple and fast, why has it not become universal? Sexual reproduction — the fusing of genetic material from two or more organisms — creates variety among the offspring. If an organism with the genotype AaBb reproduces asexually, all of the offspring will be AaBb unless by rare chance one or more of them mutates to a new kind of allele. But if two sexually reproducing organisms with AaBb genotypes mate with each other, the offspring could be AABB, AaBB, aaBB, AABb, AAbb, AaBb, aaBb, Aabb or aabb. In short, while the sexual organism diversifies its investments, the asexual organism puts all its eggs in one hereditary basket.

The environment of most organisms changes constantly, from hour to hour and year to year. Circumstances favor the species that faces the environment with variable offspring. This explains why sexual reproduction occurs in almost every group of organisms from bacteria to man. The exceptions are instructive. Some fungi persist by scattering huge numbers of asexually produced spores into the environment. The vast majority land where they cannot survive. But some succeed in reaching the right habitats, however scarce such places may be. Asexual reproduction is also an advantage to species which at least for a time have a uniform environment and need to reproduce rapidly to exploit it. For example, when a female aphid lands on a suitable plant in the early spring she has a very favorable home and a source of food

SPERM PENETRATES EGG of sea urchin in the photograph on opposite page. Picture was made with a scanning electron microscope.

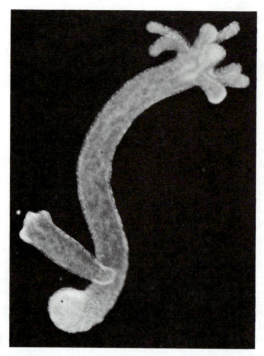

1

BUDDING is one means of asexual reproduction. Here a young hydra develops near the base of the parent animal's stalk (lower left).

two sexes. The sex cells are specialized as GA-METES whose only function is to achieve fertilization and commence the development of the new organism. In animals the female gamete, the EGG, is ordinarily the larger of the two. It is laden with nutrients that later supply the developing embryo, and incapable of movement. The male gamete, the SPERM, is usually tiny and motile, stripped down to bare essentials: a nucleus and the cilia or flagella needed to propel it to the egg, plus a mitochondrion to power the system. These sex cells are produced by a special series of cell divisions called GAMETOGENESIS: oogenesis in the female, spermatogenesis in the male (Figure 3). The crucial step is meiosis, which reduces the two sets of chromosomes in the cell to one (Chapter 5). The diploid GAMETOCYTES — the oocyte in the female and the spermatocyte in the male — undergo meiosis to create four cells. In the male, all four become sperm. But in the female, only one becomes an egg; the other three, called POLAR BODIES, shrink and eventually die. The egg grows, acquiring the extra nutrient supplies needed for future embryonic development.

In higher animals the gametes are produced

that she must exploit rapidly before predators eat her or other aphids arrive to crowd her out; she reproduces swiftly by means of parthenogenesis. Many kinds of plants and animals, including fungi and aphids, reproduce both sexually and asexually during their life cycle and thus enjoy the benefits of both kinds of reproduction.

FERTILIZATION AND SEXUALITY

In the most primitive single-celled organisms sexual union takes place in the water. Two cells conjugate, pressing closely together along their cell borders and exchanging genetic material directly through the cytoplasm (illustrated in Chapter 18). In the bacterium *Escherichia coli* this process appears in perhaps its most elementary form (Figure 2). In more evolved organisms the genes are organized into chromosomes and can be exchanged in groups.

In some single-celled organisms, such as the green alga *Chlamydomonas* and the protozoan *Paramecium*, there are often many sexes: strains of cells that can conjugate with cells belonging to other strains.

Multicellular organisms usually have only

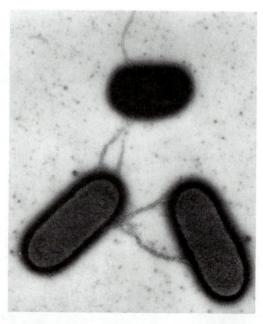

2

CONJUGATING BACTERIA. Male bacterium at lower left of electron micrograph is mating simultaneously with two females. Four long projections called pili connect male to females. Through the pili the male transfers DNA to the females. Tiny bead-like structures barely visible along the pili are bacteriophage viruses.

3

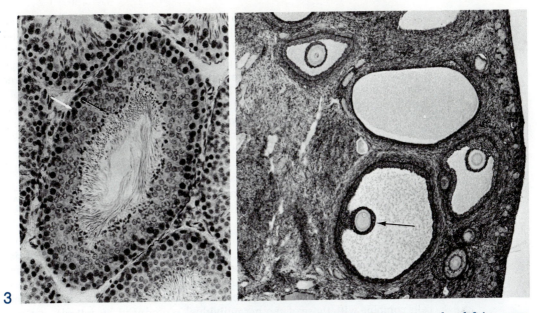

FORMATION OF SPERM AND EGGS. Photomicrograph at left is a cross-section of tubule from testis of a rat. Spermatogenesis proceeds from edge toward center of the tubule. In earliest stage primary spermatocytes (white arrow) arise by mitosis from prespermatocytes that lie near outside edge of tubule. Black arrow indicates spermatids: young tail-less sperm that arise by meiosis from primary spermatocytes. Center of tubule is lined with mature sperm ready to depart. Tails of sperm extend toward center of tubule. Photomicrograph at right depicts three eggs (arrow points to one) developing in follicles within the ovary of a cat. Photomicrographs are not reproduced to same scale.

in special organs, the female OVARIES (singular, ovary) and the male TESTES (singular, testis). These organs also include special tubes and other devices for conveying gametes to the outside. In some animals, including man and other vertebrates, these organs also produce hormones that stimulate the beginning of sexual maturity (*Chapter 12*).

Why are there only two sexes in higher organisms? The answer seems to be simply that two are enough to accomplish the purpose of genetic recombination. Three or more would create complications in the genetic basis of sex determination, in development of the complex sex organs, and in the often complicated behavior patterns that lead to fertilization.

Some animals and plants are HERMAPHRODITES, with both sex organs in the same organism. The majority of flowering plants are hermaphroditic, which in a strict botanical sense means that male and female organs occur in at least some of the same flowers. Most hermaphroditic plants have elaborate devices to insure that CROSS-POLLINATION occurs, in other words that the sperm-bearing

pollen grains fertilize the ovaries of another plant and not their own. One of the commonest devices to insure cross-pollination is the maturing and release of the pollen by the male parts of the flower before the female part of the flower fully develops.

Hermaphroditism is rarer in animals. It occurs chiefly in sessile forms, notably sponges and some mollusks, and in parasites such as tapeworms and flukes, which are unable to seek out one another. Under these circumstances it is advantageous for each individual to breed to its maximum extent. Tapeworms can fertilize themselves, but in most other instances hermaphroditism involves cross-fertilization between anatomically identical animals, as in many plants.

Why aren't all sexually reproducing organisms hermaphroditic? Why are there separate sexes? Hermaphroditism accompanied by cross-fertilization, as in the earthworms and most flowering plants, facilitates very fast reproduction. Both individuals can produce fertilized eggs, and the resulting offspring show the same genetic diversity as those created by separate sexes. The advantage to having distinct sexes in animals is that it permits a division of labor. In general, the female specializes in making eggs, finding the right places to deposit them, and (in the highest animals) in nursing and protecting the young. The male simply specializes in finding and fertilizing females. In many species he also locates and protects the territory in which the young will be born, and sometimes assists the female in rearing the young. Extreme examples of sexual division of labor are found among insects, particularly mayflies, mosquitoes and moths, in which the male is equipped to do little more than find and mate with females.

STRATEGIES OF REPRODUCTION

The familiar life cycle of the vertebrates is a very simplified one, totally committed to a sexual strategy of reproduction. Vertebrate sex cells combine with those of another individual to create a new organism. *Amoeba*, in contrast, is committed to an asexual strategy: an individual creates two individuals by dividing. Some plants and animals have committed themselves to a mixed strategy of both sexual and asexual reproduction. This usually involves ALTERNATION OF GENERATIONS: one

generation reproduces asexually and the next sexually. The period of asexual reproduction is usually devoted to multiplying rapidly and creating offspring that can spread across wide areas; the period of sexual reproduction typically produces more resistant, genetically variable offspring capable of coping with changes in the environment.

REPRODUCTION IN PLANTS

A major difference between plants and the higher animals lies in the origin of the gametes. In animals the pathway from gametocyte to gamete involves meiosis, crossover and the reduction of the chromosome number from diploid to haploid. In plants, gametes rarely arise directly from meiosis. Instead, each meiosis produces four spores that eventually germinate and divide mitotically to produce a haploid GAMETOPHYTE plant. Its gametes are produced by mitosis, not meiosis. They subsequently fuse to produce a zygote that divides mitotically to produce a diploid SPOROPHYTE plant. At some later time specialized cells of the sporophyte, called SPOROCYTES, undergo meiosis and produce haploid spores, starting the cycle anew.

This alternation of diploid-asexual with

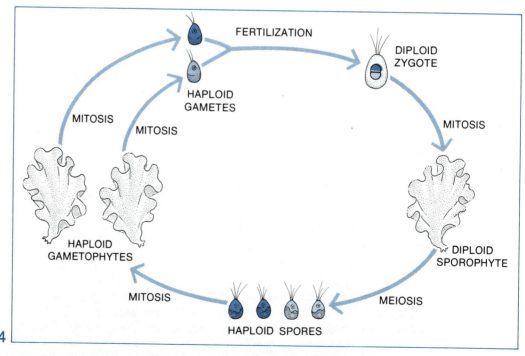

4

LIFE CYCLE OF ULVA, a green alga commonly called sea lettuce, involves diploid and haploid forms that are almost indistinguishable. Color indicates differences in mating type.

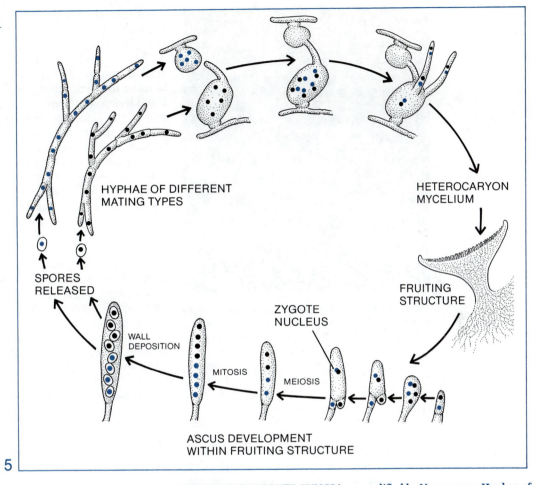

HYPHAE OF DIFFERENT
MATING TYPES

HETEROCARYON
MYCELIUM

FRUITING
STRUCTURE

SPORES
RELEASED

WALL
DEPOSITION

MITOSIS

MEIOSIS

ZYGOTE
NUCLEUS

ASCUS DEVELOPMENT
WITHIN FRUITING STRUCTURE

5

LIFE CYCLE OF HIGHER FUNGI is exemplified by Neurospora. Hyphae of different mating types grow toward one another and fuse (top). Cytoplasm and nuclei from one type invade the other. Invading nuclei divide to form a heterocaryon, which in turn forms a fruiting structure (right). Within asci of fruitimg structure paired nuclei form zygotes that develop into spores (bottom and left).

haploid-sexual generations is characteristic of multicellular plants. However, there is a broad range of variation in this basic format, particularly in the lower plants: the algae and the fungi. The life cycle of the sea lettuce *Ulva* is a classic example. The diploid sporophyte of this common alga is a "leaf" two cells thick and a few centimeters across. The motile haploid spores, propelled by four flagella, eventually find a favorable substrate. Dividing mitotically, they eventually develop into a broad thin sheet of cells that looks just like the sporophyte (*Figure 4*). A particular gametophyte can produce either male or female gametes, but never both. Male and female gametes are shed into the water. They swim together and fuse, losing their flagella when

the zygote is formed. After a brief resting period, the zygote begins mitotic division to form a new sporophyte. Any gametes that fail to find partners can settle down on a favorable substrate, undergo mitosis, and produce new gametophytes directly; in other words, the gametes can also function as spores.

The basic pattern exemplified by *Ulva* has varied in two directions during the evolution of other kinds of algae. First, some algal gametophytes produce gametes that fuse to form a zygote. The zygote, with or without a resting period, then undergoes meiosis to produce spores, which in turn produce new gametophytes. In the entire life cycle only one cell, the zygote, is diploid. Other algae go through a life cycle typical of higher animals.

6

LEAFY MOSS PLANTS belong to the gametophyte generation. Photograph shows moss growing on rocky soil in Connecticut. Pin is 2.5 cm long.

Meiosis of sporocytes produces gametes directly; they fuse to form a zygote, and the zygote divides mitotically to form a new sporophyte. In these organisms every cell except the gametes is diploid. Between these two extremes one finds algae whose gametophyte and sporophyte generations are both multicellular, but one phase (usually the sporophyte) is much larger and more prominent than the other.

Among the higher fungi a new wrinkle evolved. The cytoplasm of opposite sexes fuses long before the nuclei. This is well displayed in a group of higher fungi called ASCOMYCETES (*Chapter 19*), which includes the bread mold *Neurospora*. Its gametophyte generation is a mat of filaments called hyphae, and different gametophytes belong to different mating types. There are more than two of these "sexes" — another exception to the prevailing pattern. Hyphae of different mating types grow toward one another, join, and form a special type of filament called a HETEROCARYON ("different nuclei") containing nuclei from both parents. Branching and growing by mitosis, the hyphae then form a filamentous mass called a MYCELIUM (*Figure 5*). Finally the heterocaryon mycelium forms a fruiting structure containing many pod-shaped asci. Pairs of dissimilar nuclei enter each ascus and fuse. The resulting diploid zygote nuclei then divide to produce eight spores. The spores are

eventually shed and germinate to form new gametophyte mycelia, each of a particular mating type.

There is really no such thing as a typical algal or fungal life cycle. The array of variations on the basic alternation of generations is bewildering. But above these rather primitive plants, the plant kingdom abruptly begins to exhibit consistent life cycles. Certain evolutionary trends become apparent, including 1) progressively more protection for the developing gametes, 2) a gradual escape from water-based fertilization, 3) a dramatic shift from emphasis on the gametophyte generation to emphasis on the sporophyte, and 4) increased protection for the embryonic sporophyte.

In mosses, members of the division Bryophyta, the conspicuous plant is the small leafy gametophyte (*Figure 6*). The tips of its prominent leafy branches bear specialized sex organs: the ANTHERIDIA (singular, ANTHERIDIUM), which produce the motile male gametes, and the ARCHEGONIA (singular, ARCHEGONIUM), which produce a single immobile egg cell. The sporophyte remains attached to the gametophyte throughout its life, forming 1) an absorptive foot embedded in gametophyte tissue, 2) a stalk, and 3) a capsule in which sporocytes eventually undergo meiosis to form haploid spores. The spores are shed as in the fungi, and they germinate to form first a filamentous and then an independent new leafy gametophyte. This life cycle is monotonously consistent throughout the bryophytes (*Chapter 19*).

The most primitive plants that possess true vascular tissue for long distance transport of materials are the ferns and their relatives (*see the section on xylem and phloem in Chapter 10*). In these plants the sporophyte generation is dominant and is frequently enormous in size, as exemplified by certain tree ferns (*Figure 7*). The undersides of the fern leaves carry specialized SPORANGIA (*illustrated in Chapter 19*) in which sporocytes undergo meiosis to form haploid spores. Once shed, the spores often travel great distances and eventually germinate to form very small and inconspicuous independent gametophytes (*Figure 8*). The gametophytes produce both antheridia and archegonia, although not necessarily at the same time. Fertilization is accomplished by swimming sperm, as in the bryophytes, and the zygote develops into the new embryo sporophyte. During its early development the young sporophyte depends upon surrounding gametophyte tissue to meet its nutritional needs, but eventually it sprouts a root that gives it the capacity to grow independently. The gametophytes are small, delicate and short-lived but the sporophytes can be very large and can sometimes survive for hundreds of years.

Above the ferns on the evolutionary scale the gametophyte generation is even further reduced. The most advanced plants are the seed plants, a group comprised of GYMNOSPERMS (pines and their relatives) and ANGIOSPERMS (flowering plants, including most trees). Among seed plants the life cycle is just the reverse of that found in bryophytes: the gametophyte develops partly or entirely while attached to the sporophyte. And except in the most primitive gymnosperms, these plants have no swimming sperm — an evolutionary advance that enables them to colonize portions of the land habitat that remain closed to most lower plants. In short, not only have these higher plants escaped the aquatic environment, they have even escaped the swamp.

The colonization of dry land by the gymnosperms and angiosperms was facilitated by a new reproductive strategy. On modified leaves (such as flowers) they develop separate male and female sporangia. Within the sporangia, male and female sporocytes undergo meiosis and become spores, but the spores are not shed. Instead the gametophytes begin development within the sporangia and are dependent on them for food. In the case of the female sporangium, usually only one meiotic cell from a given sporocyte survives. Its nucleus divides and the products divide again to produce a multicellular female gametophyte — in the angiosperms normally not more than eight nuclei in all. Meanwhile, within the male sporangium, meiotic division of male sporocytes produces male spores which undergo one or a few divisions to form the male gametophytes: the familiar pollen grains. These are distributed by the wind, an insect, a hummingbird, or by a plant breeder's brush. (Instead of pollen, primitive gymnosperms release a sperm that swims to an archegonium

7

TREE FERN Cyathea contaminans grows in Java. Sporophyte shown here attains impressive size, and may survive for centuries.

and fertilizes it.) Pollen grains eventually land near the female gametophyte and send out a pollen tube. The tube reaches the egg, releasing a sperm nucleus that fuses with the egg nucleus to form the zygote. The young sporophyte develops into an embryo, and then the entire system goes dormant; the end product is a seed (Figure 9). A seed may contain tissues from three generations: the seed coat is the original sporangial wall of the parent sporophyte; within the seed coat is a layer of female gametophyte tissue (which may be fairly extensive in gymnosperms, but absent in angiosperms); and in the center of the package lies the embryo of the new sporophyte.

Compared with the delicate one-celled spores of lower plants, the multicellular seed of higher plants is well protected. Many layers

Most advanced plants have no swimming sperm —
an evolutionary advance that enables them to colonize portions
of the land habitat that remain closed to most lower plants.

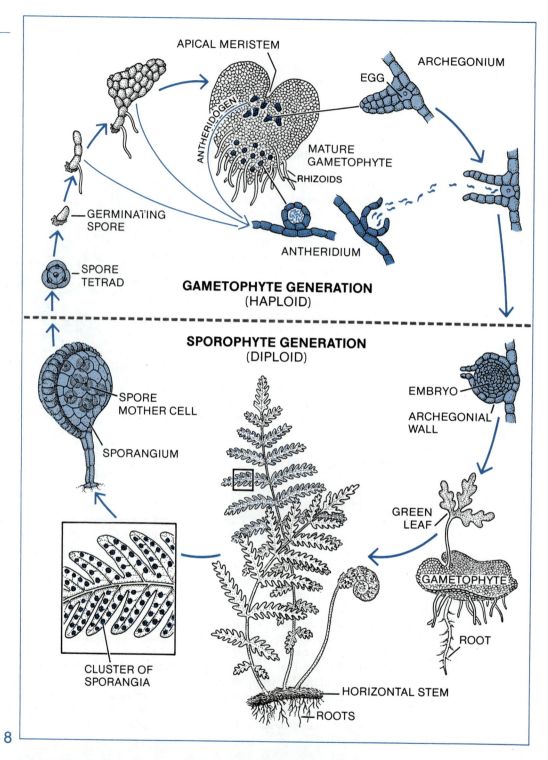

APICAL MERISTEM

ARCHEGONIUM

EGG

ANTHERIDOGEN

MATURE
GAMETOPHYTE

RHIZOIDS

GERMINATING
SPORE

ANTHERIDIUM

GAMETOPHYTE GENERATION
(HAPLOID)

SPORE
TETRAD

SPOROPHYTE GENERATION
(DIPLOID)

EMBRYO

ARCHEGONIAL
WALL

SPORE
MOTHER CELL

SPORANGIUM

GREEN
LEAF

GAMETOPHYTE

ROOT

CLUSTER OF
SPORANGIA

HORIZONTAL STEM

ROOTS

8

LIFE CYCLE OF FERN involves a large sporophyte plant and an inconspicuous gametophyte. Swimming sperm shed by antheridia of one
gametophyte fertilize eggs within archegonia of another. Zygote develops
into embryonic sporophyte within the archegonium and eventually
sprouts leaves and roots. Structure of antheridia and archegonia is illustrated in more detail in Figure 10.

of cells enclose the dormant embryo, and the seed may remain viable for years. The seed habit and the escape from aquatic fertilization are two major reasons for the enormous evolutionary success of seed plants, the dominant elements of the modern flora.

This brief survey of the reproductive cycles in plants has deliberately omitted many of the adaptations and specialized structures (such as flowers) associated with plant reproduction. The effect of day length on flowering is discussed later in this chapter and in Chapter 11; the nature of the specialized reproductive structures of both lower and higher plants is discussed in Chapter 19.

PLANT HORMONES AND PHEROMONES

In some plants reproduction is regulated by hormones; in others, it is regulated by PHEROMONES, chemicals produced by one individual and influencing another. Unlike most insect pheromones, which are released into the air as gases (*Chapter 15*), most plant pheromones are released into water.

A rather simple system is found in the water mold, *Allomyces,* a member of another major group of higher fungi, the PHYCOMYCETES. *Allomyces* is bisexual, producing distinctive motile male and female gametes, with the female gamete considerably larger than the male. Leonard Machlis of the University of California has shown that the female gamete secretes a pheromone which he named sirenin, an appropriate name for a female sex attractant. The male gamete swims toward higher concentrations of this substance, eventually reaching the female gamete and fusing with it to form the zygote.

Pheromone systems serve to get male and female gamete nuclei close enough to fuse. Ferns have a somewhat less direct but equally elegant system. As a fern spore germinates, it first forms a small gametophyte. Eventually archegonia appear near the notch of the heart-shaped gametophyte, as shown in Figure 8. At about the same time, the young gametophyte begins to produce and release a powerful

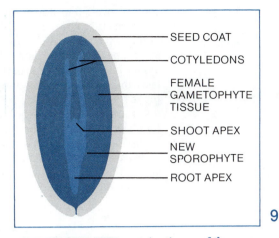

9

MATURE PINE SEED contains tissues of three generations. Embryo (light color), the new sporophyte generation, is surrounded by nutrient tissue from female gametophyte. Seed coat (gray) is from parent sporophyte. Although only two cotyledons are shown, pine embryo usually has many more.

pheromone called antheridogen. The pheromone diffuses into the environment, thus inducing nearby spores to germinate and to form antheridia (*Figure 10*). The induced antheridia mature not long after the archegonia of the original gametophyte, and shed antherizoids (swimming sperm) that fertilize the eggs in the archegonium.

The story does not end there. Germinating spores themselves produce small amounts of antheridogen, inducing neighboring spores to germinate. Thus a whole gametophyte colony on a damp log may be a highly coordinated population with respect to successive production of the male and female gametes on one individual, and their simultaneous production on adjacent individuals, insuring a measure of outbreeding in these haploid hermaphroditic plants.

In the angiosperms reproduction is controlled by hormones. Almost all studies of flowering in angiosperms have been done on plants that are sensitive to variations in photoperiod (the length of the daylight hours) because in these plants one can exert a very fine

The evolutionary story of animal reproduction, like that of plants, involves the progressive emancipation of gametes from the water, and the development of mechanisms to introduce sperm directly into the body of the female.

control over the flowering process. Some plants such as chrysanthemums bloom in autumn, when the daylight hours are short; they are called short-day plants. Long-day plants bloom in summer. What part of the plant detects the variation in day length — the stem, the leaves, or the shoot apex? Many years ago the Russian plant physiologist Mikhail K. Chailakhyan showed that it was the young leaves, and he proposed the existence of a flowering hormone. Transported to the shoot apex, the hormone (or hormones) causes changes in messenger RNA synthesis, and hence in protein synthesis, that lead to the production of the highly modified leaves that comprise the petals and other parts of the flower.

Probably at least three different kinds of substance are involved in the induction of flowering. One of these is a gibberellin, one of a class of plant growth hormones. Another is a flowering inhibitor, and the third a specific flowering substance called FLORIGEN. High levels of gibberellin and florigen and low levels of inhibitor are necessary for flowering. Any one of these three factors could limit flowering in a particular species, and the level of any one might be set by photoperiod.

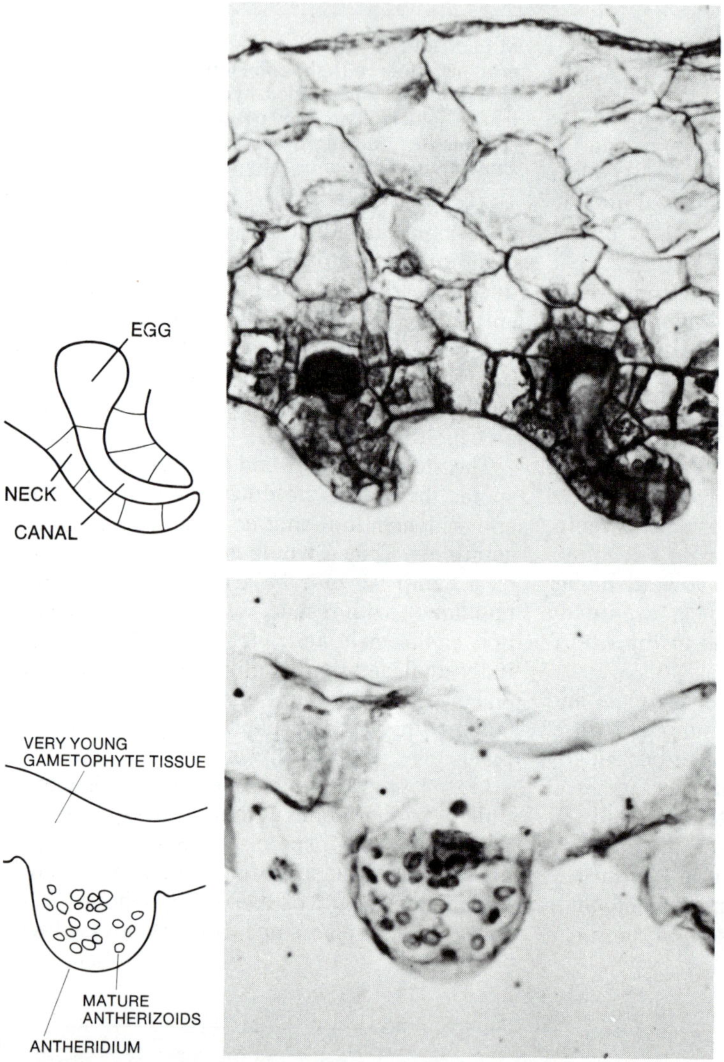

REPRODUCTION IN ANIMALS

The evolutionary story of animal reproduction, like that of plants, involves the progressive emancipation of gametes from the water, and the development of mechanisms to introduce the sperm directly into the body of the female. But throughout this lengthy sequence, which stretches from the lowest protozoans to the highest vertebrates, animal gametes remain dependent on an aqueous medium; water has simply been replaced with specially secreted body fluids.

The most elementary form of mating in multicellular animals is the release of the sex cells into the water. This method, called SPAWNING, requires an exact synchronization of the behavior of both sexes. If the timing is off, the short-lived eggs will not be fertilized. One way to achieve such synchronization is by direct communication between individuals. The common quahog (Mercenaria mercenaria) of the East Coast, for example, lives tightly packed in "clam beds". When the animals are sexually mature they hold their gametes within their bodies until they can "smell" the semen of another quahog. Then they discharge a white cloud of their own fluid and gametes. A chain reaction is set in motion among all the sexually mature quahogs in the vicinity, and the water sometimes becomes opaque with the dense concentrations of their fusing gametes. External cues from the environment can be used to achieve the same effect. The palolo worm (Leodice) lives submerged among coral reefs in the South Pacific. As the October-November moon rises for the first time in its last quarter, the worms pinch off the posterior parts of their bodies containing the gametes. These parts rise to the surface of the water in huge numbers, where spawning and fertilization im-

EGG

NECK

CANAL

VERY YOUNG
GAMETOPHYTE TISSUE

MATURE
ANTHERIZOIDS

10 ANTHERIDIUM

ARCHEGONIA AND ANTHERIDIUM OF FERN. Photomicrograph at top, a cross-section of a fern gametophyte, shows two archegonia. Accompanying diagram indicates location of eggs. Photomicrograph below, a cross-section of another gametophyte, depicts an intact antheridium with mature sperm cells visible within.

11

AMPLEXUS in frogs. Male (on top) holds female and both discharge their gametes simultaneously into water. Mass of eggs is visible at left.

mediately ensue. Meanwhile the anterior part of the worm remains alive down in the reef. During the following year it regenerates a new posterior fragment and awaits the coming of the next spawning time. Daylength is by far the commonest cue, however, with different animals, like plants, sensing increases or decreases in daylength and using the information to synchronize some reproductive activity.

A more advanced stage in evolution is spawning combined with pairing. In many kinds of fishes, one or both sexes prepare a nest into which the gametes are simultane-

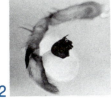

12

MALE SPIDER deposits semen on web, scoops it into a special receptacle on one of its palps and places it within reproductive organ of the female. At left, courtship of black widow spider, a ritual in which the smaller male is both the entertainment and the dessert. Picture at side depicts sperm receptacle on palp of male Micrommata, a giant crab-spider from Europe.

ously discharged. Frogs and toads engage in AMPLEXUS: the male seizes the female from above and both discharge their gametes simultaneously into the water (Figure 11). In the fishes, frogs, and toads, the pairs come together only after very particular courtship signals have been exchanged. The result is an increase in the degree of certainty that the eggs will be fertilized by males of the same species.

All of the animals mentioned to this point expose their gametes to the perils of the aquatic environment. The next logical step in evolution is INTERNAL FERTILIZATION, the direct transfer of the sperm into the reproductive tract of the female. This advance is an absolute prerequisite for living entirely on the land. Frogs and toads must return to the water to breed, but reptiles, mammals, and insects, having perfected internal fertilization, can spend their entire lives on land. The simplest method of internal fertilization is the transfer of a SPERMATOPHORE, a package of sperm which the female can pick up and insert in her own reproductive tract. Several animals have evolved variations on this technique, some of them quite bizarre. The male spider spins a small web, deposits a drop of semen (liquid containing sperm) on the web, picks the drop up in a special receptacle on one of its palps (a modified mouth part) and finally places it into the reproductive organ of the female (Figure 12). By far the most widespread form of internal fertilization involves INTROMISSION. The male penis is inserted into the female vagina (or an equivalent organ of reception) and the semen is discharged. The sperm are then able to make their way up the fluid-filled canals of the female system and fertilize the eggs internally.

THE SELF-SUFFICIENT EGG

The reader will recall that even in primitive animals the egg is typically larger than the sperm, most of the bulk being due to the yolk. This trend toward providing for offspring in advance is carried to extreme lengths by several groups of higher animals. An insect encloses its egg in a tough, waterproof covering that permits it to remain in the dry open air for weeks or months as the embryo develops inside. Insect exterminators are aware that the egg is the hardest life stage to kill. The most elaborately constructed eggs are those of birds and reptiles. The chicken's egg is a typical example. The shell of the egg is brittle but porous, permitting gases to be exchanged between the embryo and the outside air. But the shell is also waterproof, holding its liquid contents like a sealed container. Inside, the embryo is attached to the yolk sac, whose contents it gradually absorbs during development. Its body floats in sheltering fluid, enclosed by membranes that aid in respiration, the absorption of nutrients within the egg fluids, and the storage of waste materials where they will cause no harm.

NURTURE OF THE YOUNG

In the higher mammals, certain insects and other invertebrate animals, the embryo is retained within the body of the mother during part of its development. In the course of evolution, mammals have enlarged and thickened parts of the tubes that lead from the ovaries to the outside. This modified structure, the UTERUS, is especially adapted for holding the developing embryo (in man and certain other higher mammals the uterus is also called the womb). Marsupial mammals such as kangaroos and opossums have a uterus that merely protects the embryo, but does not nourish it. The young are born at a very immature stage

14

A SPIDER BALLOONING. Spider ejects a long silken thread that is caught by air currents and blown aloft like the string of a kite. The spider then releases its grip on leaf and is carried along by the wind. The thread serves the same function as a balloon carrying a human passenger. Widespread among spiders, ballooning is their single most important means of long-distance dispersal.

13

NEWBORN KANGAROO is depicted beside the urogenital opening of its mother. Baby kangaroo, born 33 days after fertilization, is about ¾ inch long. It must climb six inches to pouch, grasping mother's fur with its claws.

(Figure 13). They crawl into the marsupium, a pouch on the mother's belly, and attach themselves firmly to a nipple while they complete their development. In placental mammals, including man and most others, the membranes of the embryo come into intimate contact with the walls of the uterus (see photograph on opening page of Chapter 6). The embryonic and maternal tissues intermingle in a complex growth, the PLACENTA, through which the mother provides oxygen and dissolved foodstuffs and receives back carbon dioxide and other waste materials. This direct exchange of liquids and gases is made possible by the fact that the blood capillaries of the mother and the young are intertwined.

Animals give varying degrees of care to their offspring after they are born. Some prepare nests for the eggs or carry them on their bodies after they are laid. Others carry the young, active animals after they hatch. Feeding and grooming, which require intricate forms of communication between parents and young, have evolved in vertebrates, particularly birds and mammals, and social insects (Chapter 26).

DISPERSAL OF THE YOUNG

An important function of reproduction is to disperse the offspring. By investing in the colonization of new sites, the organism increases the likelihood that its genes will be per-

petuated in the next generation. Often the mother accomplishes the dissemination herself. The females of butterflies and moths, for example, fly from plant to plant to lay their eggs. As a result the young caterpillars that hatch later are usually not overcrowded and find ample leaves to eat. When animals disperse on their own, it is ordinarily at an early age, either soon after hatching from the egg, as in most fishes and marine invertebrates, or during early adulthood, as in many mammals, birds, and insects. The means of dispersal are extraordinarily diverse and often elaborate. Two extreme examples — a ballooning spider and an exploding fungus — are illustrated in Figures 14 and 15.

HUMAN REPRODUCTION

With the exception of certain differences in the female cycle, the reproductive biology of human beings is typical of the mammals generally. The two testes of the male are lodged outside the main body cavity in a pouch of skin, the scrotum (*Figure 16*). They contain large numbers of seminiferous tubules, the sites of spermatogenesis. The microscopic sperm are produced by the hundreds of millions and stored in the epididymis, a mass of coiled tubes at the back of the testis. Prior to ejaculation they ascend the sperm duct where they are mixed with the secretions of the seminal vesicle and prostate glands to form the SEMEN. During sexual excitement the penis becomes stiff and erect due to the engorgement of blood in the spaces of the "cavernous bodies" that make up most of its bulk. Upon insertion into the vagina, the penis is stimulated by friction until it ejaculates by a reflex action, discharging the semen into the vagina. The sperm are then able to swim under their own power upward through the uterus to the Fallopian tubes. If they encounter an egg descending from one of the ovaries, fertilization is likely to occur.

Within a woman's body, the vagina receives the semen, the ovaries produce the eggs, and the uterus (womb) carries the developing embryo. All of the cells capable of becoming eggs are already present in the ovaries of an infant girl. There are approximately 500,000 of these cells, called FOLLICLES. During the period of a woman's fertility, which ranges from puberty (12 to 15 years) to menopause (40 to 50 years), no more than 400 of the follicles are fully developed and released to attempt their rendezvous with sperm. This process of ovulation does not occur haphazardly. It is the culminating event in an ESTROUS CYCLE that is under the precise control of hormones and lasts approximately 28 days. During the first part of the cycle the wall of the uterus is greatly enlarged. It becomes engorged with blood vessels and is thus prepared to receive and nurture the embryo in the event one is created. The egg is released on about the fourteenth day of the cycle. On all but a very few occasions it is destined to make its journey without being fertilized. The "disappointed womb" then undergoes a breakdown on about the 28th day, accompanied by bleeding, which terminates the cycle and sets the stage for the beginning of the next one. If an egg is fertilized and becomes implanted in the uterine wall, the uterus maintains its functional condition steadily until the birth of the child nine months later. The hormones controlling these events are secreted by the ovary, placenta, and uterine wall.

Because the human estrous cycle is ended so conspicuously by the menses (literally, "month"), the period of uterine breakdown and bleeding, it is often referred to as the MENSTRUAL CYCLE. The human menstrual cycle is discussed in more detail in Chapter 12. In other mammals the cycle usually has a different duration, and it is seldom marked by noticeable bleeding. The most striking change comes during the time of ovulation at or near the middle of their cycle. At this time the female becomes sexually receptive and even aggressive. She is said to be IN HEAT; in fact the word estrus refers to heat. During the remainder of the cycle she is unreceptive, and seldom if ever attempts to mate. In many species ovulation occurs only during a certain breeding season. The behavior of the human female is thus atypical. With the possible exception of her menstrual period, she is sexually receptive throughout the cycle and at all seasons of the year. In other words, depending on how one looks at it, the human female is either always or never "in heat." The evolutionary significance of this trait is discussed in Chapter 27.

HUMAN SEXUALITY

The sex drive in man, as in other animals, functions as a powerful lure that entices indi-

PILOBOLUS is a fungus that grows on dung. Sac of spores (sporangium) caps the stalked sporangiophore. Fluid pressure builds up within sporangiophore and eventually bursts it, blasting the sporangium into a three- to eight-foot trajectory.

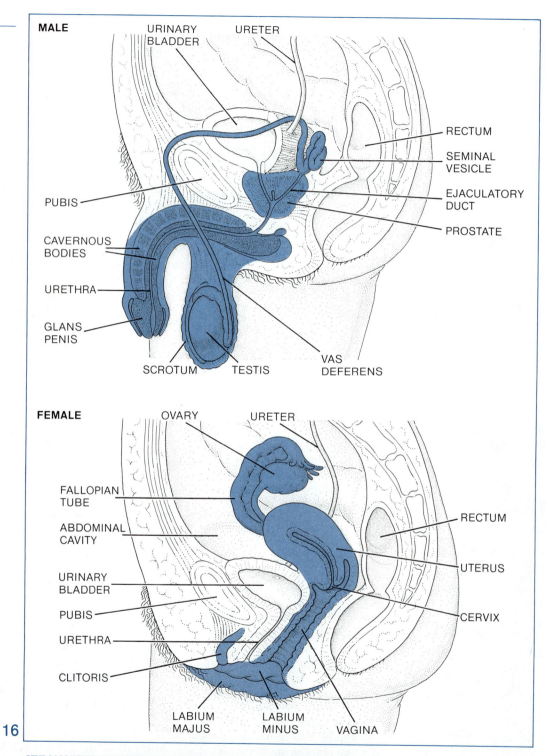

MALE

URINARY BLADDER

URETER

RECTUM

SEMINAL VESICLE

EJACULATORY DUCT

PROSTATE

PUBIS

CAVERNOUS BODIES

URETHRA

GLANS PENIS

SCROTUM

TESTIS

VAS DEFERENS

FEMALE

OVARY

URETER

FALLOPIAN TUBE

ABDOMINAL CAVITY

URINARY BLADDER

PUBIS

URETHRA

CLITORIS

RECTUM

UTERUS

CERVIX

LABIUM MAJUS

LABIUM MINUS

VAGINA

16

HUMAN REPRODUCTIVE SYSTEM. In the male, sperm produced in testis ascend the vas deferens and mix with secretions from seminal vesicle and prostate gland to form semen. Penis becomes erect during sexual excitement as cavernous bodies fill with blood. Ejaculation discharges sperm into female vagina. Sperm swim into Fallopian tubes, where they can fertilize egg descending from ovary.

viduals to reproduce. Improved contraceptive techniques have enabled man to uncouple sex from its reproductive function, and there has been an increasing tendency to view sex as an activity that is pleasurable in its own right. This view has been strongly reinforced by psychologists, who see sexual activity as an important prerequisite of mental health, and by demographers, who view the population explosion as a phenomenon that is fast approaching the proportions of a natural disaster.

Every human society has felt the need to regulate the sexual activity of its members, probably because sex has long been recognized as a force that can disrupt the fabric of family and society. Probably no aspect of human behavior has been so circumscribed by hypocrisy, myth and taboo. Compared with other cultures, Western societies have been particularly repressive in regulating sexual behavior.

In the U.S. the sexual revolution that began about a decade ago has led to a less hypocritical approach to the sexual nature of man, to a debunking of long-standing myths, and to a tendency to relax harsh moral judgments about what is "normal" in sexual behavior. But it is not easy to rewrite centuries of cultural tradition in a single generation. At present our laws and beliefs about sex are a patchwork of contradiction and inconsistency. Ideas about virtually every aspect of sexual physiology and behavior are now being reexamined, particu-

larly the sexual response patterns of men and women, homosexuality, the role of marriage, and birth control.

FEMALE SEXUAL RESPONSE

The rise of the women's liberation movement and the physiological studies of Masters and Johnson have led to increased awareness of the sexual responses of the human female. The sexual response cycle of both women and men can be divided into four phases: excitement, plateau, orgasm and resolution (*Figure 17*). In a woman, as sexual excitement begins her heart rate increases, and her nipples become erect; her clitoris and the inner lips (labia minora) of her genitals swell as they become filled with blood, and the walls of the vagina become moist with lubricating fluid. In times past the appearance of this fluid was regarded as the female equivalent of ejaculation or as evidence of orgasm, but neither is true. Victorian erotica such as Frank Harris's *My Life and Loves* abounds with accounts of women "spending" after a minute or two of genital caresses. Harris was flattering himself that he had brought his partners to orgasm when in fact he had merely begun the preliminaries.

In the plateau phase, her breathing becomes rapid and the head and shaft of the clitoris begin to retract. The sensitivity that once was restricted to the clitoris spreads over her external genitals, and the clitoris itself becomes so sensitive that touching it can be painful.

Orgasm begins with a long (two- to four-second) contraction of the outer third of the vagina, followed by shorter contractions about a second apart. Orgasm may last as long as a few minutes. Both the intensity and frequency of orgasm vary. Unlike men, some women can experience several orgasms in rapid succes-

CYCLES OF SEXUAL RESPONSE are basically similar in males and females. Differences include the presence of a refractory period in the male cycle and a greater variety of patterns in the female cycle. Female may show any of three patterns on different occasions. Commonest female pattern (curve A) resembles that of male. Alternatively, females may experience sustained multiple orgasms (curve B), or skip the plateau phase in a surge toward very intense orgasm (curve C). Dashed black lines indicate possibility of second orgasm in both sexes. Graphs are based on data of Masters and Johnson.

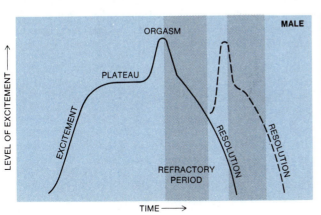

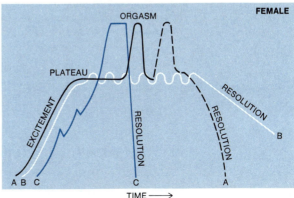

sion. Five to ten minutes after orgasm a woman's physiology has largely returned to the normal precoital level. This descent from orgasm is the resolution phase. If she does not attain orgasm, return to normal conditions may take 30 minutes or longer. Among women as among men, the descent from orgasm can be a swoop or a gentle slide. Sometimes it leaves her relaxed and ready for sleep; usually it leaves her ready for more.

Some writers in the women's movement have drawn a distinction between clitoral and vaginal orgasms; that is, between orgasms caused by stimulation of the clitoris and those caused by friction of the penis against the walls of the vagina during coitus. Masters and Johnson state that physiologically there is only one kind of female orgasm, and it can be elicited by stimulation of the clitoris or the vagina or both. It might be noted in passing that the clitoris is a remarkable and unique structure — the only organ found in nature whose sole function is to provide sexual pleasure.

The expectation that a woman should always attain orgasm during lovemaking is a relatively recent idea that has proved unsettling to both women and men: to women, because they feel that there is something wrong with them if they don't have an orgasm, and to men because they feel inadequate if they fail to bring their partner to a climax. Two thoughts should temper the current overemphasis on sheer sexual performance. First, that it is possible for a woman to enjoy sex without reaching orgasm, although of course it is considerably more pleasurable for her when she does. Second, that the full attainment of sexual pleasure depends on communication between partners. A man should be sensitive to the ways that female sexuality differs from his, particularly to the fact that a woman is generally slower to become aroused and her excitement is slower to subside. A woman in turn should not expect her partner to guess how she would like him to make love to her. Marriage manuals and erotic literature often lead partners to feel that sex is less than perfect if they fail to reach orgasm simultaneously. More experienced lovers are aware that there are no rules about that.

Cultural conditioning plays a powerful role in sexual response. Our sexual responses are generally what we expect them to be. In some cultures, such as that of Victorian England, females were taught that coitus was an unpleasant duty, and that well-bred women should show no signs of sexual passion. Women raised with that sort of conditioning rarely achieve orgasm. In sexually liberal societies such as those of the Pacific islands, both males and females are encouraged to regard lovemaking as one of life's greatest delights, and women in such cultures almost always achieve orgasm. The role of the mind in sex should not be underestimated. Most cases of frigidity in women (inability to attain orgasm) and impotence in men (inability to achieve an erection) are rooted in the mind.

THE MALE RESPONSE

The similarities between the male and female cycles of sexual response are greater than the differences. During the excitement phase the penis, engorged with blood, becomes rigid and swells to as much as double its normal length. In the plateau phase breathing becomes rapid, the diameter of the glans (the head of the penis) increases, and a clear lubricating fluid secreted by Cowper's gland oozes from the urethra. Pressure and friction against the nerve endings in the glans and in the skin along the shaft of the penis eventually trigger orgasm. Massive spasms of the muscles in the genital area, as well as contractions of the accessory reproductive organs — the prostate, seminal vesicles and ejaculatory duct — result in ejaculation: the spurting of semen from the urethra. Within a few minutes after ejaculation the penis shrinks to its normal size and body physiology drops to normal precoital levels.

Of the entire range of human sexual behavior, only two aspects appear to have no counterpart in the behavior of other animals: the human female has no period of "heat", and female orgasm — apparently a late evolutionary invention — has been demonstrated in no other species.

One of the differences between the male and female cycle is the presence of a refractory period immediately after the male orgasm, as shown in Figure 17. During this period, which may last 20 minutes or longer, a man cannot achieve a full erection or another orgasm, regardless of the intensity of sexual stimulation.

Erotic literature and folklore have propagated a number of myths about male sexuality. Foremost among them is the notion that a giant penis is much more stimulating to a woman than one of average or small size. Authors of pornography always endow their heroes with penises of truly awesome dimensions, and purveyors of sexual paraphernalia exploit male anxieties about penis size by selling weighted devices to stretch the penis to greater length. The average penis is three to four inches long when limp and about six inches long when erect, with a diameter of about 1¼ inches. Penises can be considerably larger or smaller. Although medical literature records some that measured over 13 inches long when erect, most of this length is wasted because the average woman's vagina is only about three inches deep. Contrary to popular belief, there is no relation between the size of a man's penis and his body build, race, or his ability to give and receive sexual pleasure.

SEXUAL BEHAVIOR

Standards of sexual behavior remain controversial, posing problems of what is normal, what is healthy, what is moral and what is legal. Frequently the answers to these questions are contradictory. Our society has traditionally condoned only sexual intercourse between married adults, and even in marriage certain kinds of sexual activity have long been regarded as taboo. Oral-genital contacts, for example, have been described as "unnatural acts," and they are a felony in many states. The same holds true for homosexual behavior. It is enlightening to examine the presumably more "natural" behavior of man's primate relatives. Nuzzling and licking of the female genitals has been observed among wild male chimpanzees and gorillas, and the females respond with gratitude. Homosexual behavior — males mounting males and females mounting females — is common among primates and other mammals. Of the entire range of human sexual behavior, only two aspects appear to have no counterpart in the behavior of other animals: the human female has no period of "heat" or heightened sexual receptivity, and

female orgasm — apparently a late evolutionary invention — has been conclusively demonstrated in no other species.

In the late 1940's the publication of the Kinsey Reports confronted Americans with the fact that the professed standards and laws of the society did not correspond with what people were doing in their bedrooms. Kinsey's sampling techniques had some limitations and the data are by now over two decades old, but they are still cited because they are the best available. When the reports were published many people with no contradictory evidence nevertheless denied their validity. Many of Kinsey's opponents no doubt realized that his statistics would sharply revise public attitudes about what constituted "normal" behavior, and thus would encourage more permissive attitudes toward sex. Indeed the definition of normal behavior is basically a statistical one: it is what the majority does, not what it should or ought to do. The impact of Kinsey's pioneering work no doubt helped to pave the way for the sexual revolution.

The Kinsey study showed that certain forms of behavior are rare, and therefore "abnormal," at least in the statistical sense. These included incest, pedophilia (sexual interest in children), bestiality (sexual contacts with animals), and sadomasochism (sexual satisfaction derived from inflicting or being subjected to pain). In terms of today's behavior, the Kinsey data probably understate the full extent and diversity of American sexual activity.

There is a great diversity not only in the human appetite for different kinds of sex, but also for different amounts. Beyond a certain point sexual appetite tends to diminish with age. Kinsey found that the average (median) frequency of coitus for young married couples was three times a week. This figure dropped to twice a week by age 30, once every four days by age 40, and once every 12 days by age 60. Because these figures are averages they tend to obscure the wide range of variation from couple to couple. Some couples make love several times a day, some every night, and some hardly at all. Sometimes passionate sexual activity between a married couple persists into old age, but frequently marital sex becomes an uninspired ritual. Occasionally it becomes a tedious duty, a nuisance, or worse.

BIRTH CONTROL

Birth control, also known as contraception, is practiced in one form or another by members of virtually every culture on Earth, and is approved or at least tolerated by most of the major religions. Historically the most common way to regulate family size, even in culturally advanced societies, has been infanticide. In classical Greece, for example, unwanted infants were abandoned on hillsides to die of exposure or to be eaten by animals. In China female infants, traditionally regarded as less desirable than males, were often drowned or abandoned, a practice that persisted well into the 20th century. Modern sensibility recoils at infanticide, and ethical issues raised by contraception and particularly by abortion are still the subject of controversy. On a planet already overburdened with people and afflicted by pollution and shortages of nonrenewable resources, the ethical issues of population control transcend private morality.

One of the oldest and undoubtedly simplest methods of contraception mentioned in the Old Testament is *coitus interruptus*, withdrawal of the penis before ejaculation. If ejaculation is achieved in the moments just following withdrawal, the "seed is spilled upon the ground" — a grave sin in the religion of the Israelites, who were concerned not with birth control but with a rapid increase in their own population to stave off their numerous enemies. (Throughout history many governments have discouraged birth control because a small population makes it hard to raise a large army.) A second ancient and still less pleasurable method of birth control is simple abstention from sexual relations.

In modern times a variety of sophisticated techniques of contraception have been invented that interfere slightly or not at all with the full act of sexual intercourse. These include condoms, diaphragms, spermicides and "the pill." The condom is a very thin sheath, usually made of rubber, which is pulled tightly over the penis and retains all of the semen after ejaculation. The condom seldom fails in its primary function, and in addition it greatly reduces the likelihood of transmitting venereal diseases, particularly gonorrhea and syphilis. Its chief disadvantage is the loss of some sensitivity in the penis and the necessity of interrupting foreplay in order to fit it on the penis.

The female equivalent of the condom is the diaphragm, a flexible rubber cup which before intercourse is coated on the edge and undersurface with a spermicidal (sperm-killing) jelly or cream and fitted over the cervix. It must be left in position for several hours after intercourse. The diaphragm is an especially effective device and it does not interfere with either partner's enjoyment of sexual intercourse. But because it must be inserted into the vagina and fitted exactly each time, the woman needs a prescription and special instruction from a physician. A similar and equally effective device is the cervical cap, a plastic or metal covering which is applied tightly over the cervix and must be removed only for menstruation.

A woman also has at her disposal a variety of spermicidal jellies, creams, foam tablets, aerosols, and suppositories that can be spread through the upper vagina with special applicators. Or she may use an intrauterine device, or IUD as it is usually abbreviated. The IUD is a small plastic or metallic object in the form of a spiral, an S-shaped loop, a ring, or various other designs; it is inserted inside the uterus and left in place. It operates on the general principle that any foreign body in the uterus acts to prevent pregnancy. These devices have the great advantage of being very cheap — they cost only a few cents each — and very long-lasting. On the other hand, about ten percent of women expel them spontaneously, often without realizing this has happened. Also, a small minority of women who retain the IUD's suffer minor bleeding and pelvic inflammation as side effects.

The "pill" is a radically different birth control device available to women, and one of the most effective and widely used in technologically advanced societies. It consists of a small, orally administered dose of the female hormone estrogen mixed with progestin, a synthetic chemical similar to the natural hormone progesterone, which is secreted by the ovaries. These two substances apparently act together to suppress ovulation, thus controlling the reproductive process at its very source (*Chapter 12*). When taken each day without fail, it is virtually 100 percent effective. The chief disadvantage of the pill, other than the self-discipline required to use it correctly, is its undesirable side-effects, although these have been reduced by newer, lower-dosage formulas. Some women show a few of the outward symptoms of pregnancy: swelling and tenderness of the breasts, nausea, headaches, changes in complexion, irritability, and so on. A very tiny percentage of pill users also develop increased susceptibility to thrombophlebitis (inflammation of the veins combined with blood clotting) and pulmonary embolism (development of blood clots in the lungs).

Women over 40 who use the pill run some risk of developing heart trouble. But these comparatively rare side effects must be kept in statistical perspective. The danger of complications during pregnancy and childbirth makes it much safer to use the pill than to become pregnant.

The rhythm method, or periodic abstention from sex, is the only technique of birth control now approved by the hierarchy of the Roman Catholic Church. (Surveys of U.S. Catholics reveal that laymen view things differently from their bishops: about 80 percent of Catholic couples practice some other form of birth control.) The idea of the rhythm method is simple: a woman must abstain from intercourse during the several days of each menstrual cycle in which she is capable of conceiving. More exactly, she must not be inseminated at any time from two days before ovulation to a half-day following ovulation. The rhythm method is theoretically sound, but it has some serious practical difficulties. First, the time of ovulation can only be determined by keeping careful records of the times of menstruation and changes in body temperature over a period of at least several months. On the day of ovulation the woman's temperature rises half a degree Fahrenheit, and it remains elevated until the onset of menstruation. Once this date in the monthly cycle has been calculated, the couple must add several additional days in advance of and following the estimated ovulation time to give the method an acceptable safety margin. The week or more of sexual abstention can disrupt the marital relationship. Even more serious is the fact that one woman in six has a cycle too irregular for the rhythm method to work.

Sterilization is an increasingly popular step for men and women who have completed their families and wish to enjoy a sexual life wholly free from worry. For the male, sterilization means a vasectomy, the cutting and tying off of the vas deferens. This simple operation, which is relatively painless and requires only 30 minutes to an hour in a doctor's office, ensures that sperm will no longer be included in the ejaculate. It has no effect on the sexual performance of the man or the pleasure experienced by either partner during intercourse. In fact, the only outward sign that a vasectomy has been performed is the absence of sperm in the ejaculate. The sterilization of women involves sealing the fallopian tubes, which prevents the passage of the ovum from the ovaries to the uterus. New surgical techniques have greatly simplified this operation. Sometimes called "Band-Aid surgery," the operation requires only a tiny incision near the navel. When it is performed with a local anesthetic the patient can leave the hospital in an hour or two; otherwise she goes home the next day.

ABORTION

Abortion is the most effective of all methods of population control. Almost everyone will agree that it is also the least desirable, since it requires the destruction of a fetus that is already on the developmental path toward the formation of a baby. At what point a fetus becomes a human child is a controversial biological and ethical question. The moral dilemma is further complicated by the knowledge that in certain cases a particular fetus allowed to go to term will be seriously defective, or unwanted by its parents. Born with such a handicap, a child is likely to lead a troubled life and to add a heavy burden on an already overpopulated society. Thus the choice must be made between the fetus — which may or may not be regarded as a formed human being — and the adults who must commit two decades of their lives to caring for it if it is allowed to go to term. At one extreme are those who label abortion as murder, even if the embryo consists of no more than a few microscopic cells. Opposing them are those who believe that a woman should have the right to decide for herself whether or not she will carry the fetus to term. Most biologists, including the authors of this book, concur with the 1973 decision of the U.S. Supreme Court confirming this right. We view the early fetus as a developing POTENTIAL human being, still far removed from a newborn infant. The Supreme Court decision does not say that every woman who becomes pregnant must have an abortion; it simply states that she is free to have one if she wants one. We feel it is hard to argue against that logic.

Abortion, when performed before the twelfth week of pregnancy by a physician under proper clinical conditions, is a relatively simple and safe operation. Statistically, in fact, it is much safer than childbirth. Early abortion is usually performed by SUCTION CURETTAGE. The cervix is dilated (spread open) and the contents of the uterus removed by a special tube attached to a vacuum pump. This is generally followed by scraping the uterus with a curved scalpel called a curette to make sure that no fragments of the placenta remain. The suction method simplifies and shortens the operation, and minimizes blood loss and other risks. The operation can be performed under either local or general anesthesia, according to the preference of the patient.

Pregnancies that have advanced to 16 weeks or more are usually terminated by a procedure called SALTING OUT. With a hypodermic needle some of the fluid surrounding the fetus is removed from the uterus and replaced with a salt solution that induces a miscarriage, usually within 24 to 48 hours. This type of abortion is usually performed in a hospital.

Laws against abortions did not stop them, but merely made them harder to get, riskier, and more expensive. Illegal abortionists still flourish in countries where abortion is prohibited or discouraged by law. Abortion has always been one of the most popular methods of birth control throughout the world. In Italy, where all contraception except the rhythm method is strictly forbidden by law, the abortion rate is nevertheless estimated to be almost equal to the birth rate. Such illegal abortions, usually performed by sympathetic friends or medical quacks, are of course quite dangerous. The death rate from them is higher than the combined death rate from all legal abortions and all complications arising in pregnancy and childbirth.

READINGS

V.A. GREULACH, *Plant Structure and Function*, New York, Macmillan, 1973. A good experimental botany text with an excellent treatment of plant reproduction.

H.A. KATCHADOURIAN AND D.T. LUNDE, *Fundamentals of Human Sexuality*, New York, Holt, Rinehart and Winston, 1972. A comprehensive and clearly written book developed for an undergraduate course in Human Sexuality at Stanford Univeristy.

A.C. KINSEY, W.B. POMEROY, AND C.E. MARTIN, *Sexual Behavior in the Human Male*, Philadelphia, W. B. Saunders Company, 1948. Together with the companion volume by the same authors on *Sexual Behavior in the Human Female*, published in 1953, this book is the most definitive report on the sexual practices of Americans. Heavy reading, but despite the fact that the data are more than two decades old, the studies remain a landmark.

W.H. MASTERS AND V.E. JOHNSON, *Human Sexual Response*, Boston, Little, Brown, 1966. The definitive work on the subject, based on years of careful laboratory research.

J. PEEL AND M. POTTS, *Textbook of Contraceptive Practice*, New York, Cambridge University Press, 1969. This generally excellent book includes detailed accounts of the reproductive physiology of humans and the ways it can be influenced to achieve birth control.

C.L. PROSSER, *Comparative Animal Physiology*, 3rd Edition, Philadelphia, W. B. Saunders Company, 1973. Reproductive biology presented in the setting of a full-dress review of physiological evolution.

F.B. SALISBURY, *The Biology of Flowering*, New York, Natural History Press, 1971. An excellent work on the complex process of flowering, including the role of environmental cues.

C.D. TURNER AND J.T. BAGNARA, *General Endocrinology*, 6th Edition, Philadelphia, W. B. Saunders Company, 1976. A comprehensive treatment of all aspects of the relationship of hormones to reproductive physiology in both vertebrates and invertebrates.

C.L. WILSON, W.E. LOOMIS AND T.A. STEEVES, *Botany*, 5th Edition, New York, Holt, Rinehart and Winston, 1971. A good modern botany text with a brief and clear treatment of plant life cycles.

NOTES

8

In the imperfect development of cohabitation on a crowded planet, the habit of eating one another — dead and alive — has become a general custom.

HANS ZINSSER

carbon and nitrogen sources

Throughout their lives all organisms require a source of raw materials for synthesis and a source of energy to drive their biochemical machinery. As humans we tend to take it for granted that both come from food, but not all organisms require food in the human sense of the word. Plants and many microorganisms obtain raw materials from the air and the soil, and energy from sunlight. It is more accurate to say that organisms obtain their raw materials from NUTRIENTS, a term that includes any substance used in metabolism.

Among the nutrients required by all organisms are the elements carbon, nitrogen, oxygen and hydrogen. Carbon, because it is the backbone of all organic molecules; and nitrogen, because it is an essential component of proteins and nucleic acids. Oxygen and hydrogen, as the reader will recall from Chapter 3, are essential in the energy-yielding process of respiration. Most organisms also require other elements such as phosphorus, sulfur, potassium, calcium, magnesium and iron, as well as trace quantities of molybdenum, boron, zinc and copper. Each of these elements frequently plays several different roles. The inorganic elements, sometimes loosely called minerals, may play both structural and catalytic roles in the cell, and even the simplest organisms cannot survive without some of them.

Virtually all living organisms on Earth obtain two of the four crucial elements from the same source: oxygen from the atmosphere and hydrogen from water. But the sources of carbon and nitrogen vary enormously from one type of organism to another. The diverse strategies for obtaining these two elements have shaped the anatomy, the lifestyle and the biological destiny of every type of organism on the planet.

All nutrients pass through the cell membrane by diffusion or active transport, although some cells can also ingest particulate matter by endocytosis (*Chapter 1*). The ways that nutrients arrive at the cell membranes of multicellular organisms are remarkably diverse. Inorganic elements including nitrogen enter higher plants with water through root systems; carbon in the form of CO_2 enters through the aerial parts of the plant, particularly the leaves. Oxygen may enter through both roots and leaves. All of the essential elements except oxygen enter mammals via the digestive tract; oxygen enters through the lungs. Amphibians such as frogs can take up inorganic ions and oxygen through their skin, and the gill surfaces of fish serve for exchange of dissolved gases, ion uptake and excretion of soluble waste.

The basic biochemical machinery for building, operating and replicating cells is remarkably similar from the simplest to the most complex living creatures, but lifestyles and basic chemical needs vary enormously. Autotrophs manufacture their own organic food from simple inorganic nutrients: carbon dioxide, water, nitrate or ammonium ions, and a few soluble minerals. Most autotrophs are photosynthetic, but a few are chemosynthetic, meaning that they derive energy not from light but from simple oxidizable substances available in solution around them. All other organisms are heterotrophs whose

SUNDEW PLANT (Drosera) is a small carnivore. Hairs on specialized leaves are tipped with droplets of a sticky liquid that traps insects. When insect becomes stuck on droplet other tentacles curve over and hold him fast. Plant digests insect by secreting enzymes and absorbs nutrients through leaf.

nutritional requirements include one or more organic compounds synthesized by some other organism. Versatile heterotrophs such as bacteria may require merely an oxidizable carbon source like glucose (synthesized by a plant); an exacting heterotroph like man requires an external supply of not only oxidizable carbon compounds but also eight of the 20 amino acids, certain fatty acids and an assortment of vitamins.

PHOTOSYNTHESIS

Of the four principal elements of biochemistry, only oxygen occurs abundantly in nature in a readily reactive state. Large amounts of carbon, bound in insoluble carbonates in the earth's crust, are not easily available to organisms. The carbon which is available to organisms exists as CO_2. Nitrogen, about four-fifths of the atmosphere, exists as a gas N_2, which requires considerable energy to break the triple bond linking the two atoms. Hydrogen is present in water but obtaining it or its electrons for use as reductants also requires a large amount of energy.

The photosynthetic autotrophs have evolved elegantly organized pigment systems to capture the energy of sunlight. They use the energy to raise an electron to an excited state and then trap it in such a way that a portion of its energy is extracted in usable form as the electron drops back to its ground state. In other words, light provides the energy to split water into reducing hydrogen and oxygen gas. The energy is also used to make ATP, which is required (with the hydrogen) to reduce or "fix" carbon dioxide into biologically useful compounds (Chapter 3). Photosynthesis serves two vital functions: it is the ultimate source of carbon compounds for all organisms, and it maintains a continual supply of oxygen gas in the atmosphere. Without this gas, the oxidative processes on which all aerobic organisms depend for energy would cease.

Although photosynthesis is a process unique to green plants, algae, a few unicellular flagellates and two groups of bacteria, the sheer magnitude of the process is extremely important. Though estimates vary substantially, there are roughly 10^{10} tons of carbon fixed per year on land, and up to 10 times this amount fixed in the oceans. A quantity of oxygen equivalent to that present in the atmosphere is produced by photosynthesis every two years.

A unicellular photosynthetic autotroph in water has ready access to carbon as the bicarbonate formed when CO_2 dissolves in water. Movement of water and diffusion are normally ample to satisfy the carbon requirement, as anyone who has seen a bloom of algae in tropical waters knows. Each cell of single-celled organisms such as photosynthetic bacteria or many algae is in immediate contact with the primary source of carbon. Even enormous algae such as Macrocystis or Nereocystis, kelps which on the California coast may reach lengths of 400 feet, have ready access to bicarbonate since photosynthesis occurs in sheets of cells close to the plant surface, and therefore close to the surrounding medium.

Land plants, growing in an environment far more rigorous than the ocean, have no such simple arrangement for obtaining carbon. CO_2 must be in solution to reach the chloroplasts within the cell, requiring that moist surfaces be exposed to the gaseous environment. Unfortunately a moist surface appropriate for dissolving CO_2 also enhances the evaporation of water. In all but the most humid environments plants without some kind of specialized water-conserving adaptation could hardly survive.

An herbaceous plant, one that forms little or no woody tissue, has three major organ systems — leaves, stems, and roots. The parts of the plant directly exposed to the atmosphere are normally covered with a layer of CUTIN, a waxlike substance that provides excellent waterproofing. But cutin also blocks the passage of CO_2, so the solution to one problem simply raises another.

The architecture of a leaf is a compromise solution. A cross section of a typical leaf from a temperate zone plant is shown in Figure 1. The upper surface is coated with the layer of cutin secreted by the outermost layer of cells, the UPPER EPIDERMIS. The few nonphotosynthetic epidermal cells produce the cutin. The greater portion of the photosynthesis of the leaf occurs

A quantity of oxygen equivalent to that present in the atmosphere is produced by photosynthesis every two years.

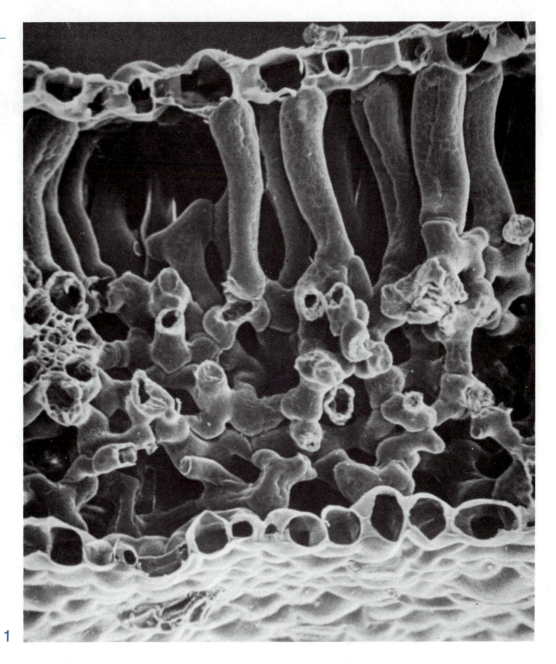

1

UPPER EPIDERMIS

PALISADE PARENCHYMA

SPONGY PARENCHYMA

INTERCELLULAR
SPACES

LOWER EPIDERMIS

STOMA

CROSS-SECTION OF BEAN LEAF (above) reveals structural compromises that adapt plants to retain water vapor while remaining permeable to the inward diffusion of carbon dioxide. A single stoma is visible at lower left. Gas diffuses readily through empty intercellular spaces. Chloroplasts lie in the parenchymal cells but not in epidermis. Drawing at left identifies structures in photomicrograph.

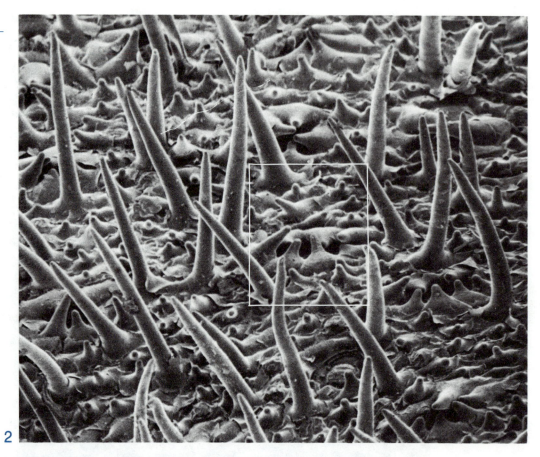

2

LEAF HAIRS create a layer of still air over the epidermis. In this scanning electron micrograph small hairs arch over two stoma (within white rectangle), restricting gas exchange and water loss.

in the PALISADE PARENCHYMA, the layer of cells beneath the epidermis. There are small intercellular spaces through which gases such as CO_2 can diffuse. Beneath this layer is a layer of irregularly shaped cells with relatively enormous intercellular spaces. This tissue, the SPONGY PARENCHYMA, is also photosynthetic. Gases can diffuse readily throughout the spongy parenchyma and the lower surfaces of the palisade parenchyma cells. Beneath the spongy parenchyma lies the LOWER EPIDERMIS which, like the upper, is nonphotosynthetic and secretes a layer of cutin.

The entrance or exit of gases in the seemingly closed system takes place between specialized pairs of epidermal cells called GUARD CELLS, each pair of which is called a stoma (plural: stomata). The guard cells, unlike the other epidermal cells, are photosynthetic. By mechanisms described in Chapter 9, the facing walls of the guard cells can open, allowing the exchange of gases between the spongy parenchyma and the outside air. Thus water loss can be regulated and enough CO_2 can still reach the photosynthesizing cells.

Plants growing in desert regions have evolved a variety of ways to conserve water while maintaining a reasonable level of photosynthesis. They may reduce the number of stomata and greatly increase the thickness of the layer of cutin. They may have dense layers of hairs which create an insulating layer of still air from which water vapor is not so rapidly swept away by air currents (*Figure 2*). They may assume shapes with minimal surface-volume ratios. The cacti are the best examples of this sort of adaptation. Leaves are reduced to nonfunctional thorns and photosynthesis is carried out by the cells in the large fleshy stems. Obviously any adaptation to reduce water loss also reduces the possible entry of CO_2. It is not surprising that such desert plants commonly show a very small increase in biomass per year in contrast to their counterparts in moister regions. Many of them possess a special photosynthetic system with an extremely high affinity for CO_2 — enabling them to do effective photosynthesis when the internal CO_2 concentration is extremely low.

In the humid tropics one finds quite different adaptations. Because potential water loss is a much less severe problem, any adaptation that increases the surface-to-volume ratio is usually advantageous. One finds enormous but relatively thin leaves (Figure 3); the leaves of Corypha, a tropical palm, may exceed 25 feet in length, and the leaves of Gleichenia, a tropical fern, may branch and cover an entire field, frequently exceeding 100 feet in length. A thinner layer of cutin and a vastly increased number of stomata per unit area of leaf surface allows more gas exchange.

A number of terrestrial plants have evolved so that they can live partially or even completely submerged in fresh water, and the leaves of these plants are appropriately specialized. For a submerged leaf, water, ions, and CO_2 are as accessible as for an alga. There is no cutin, the cells are bathed in the medium containing their elemental needs, and specialized tissue for water transport is superfluous and normally reduced.

As noted earlier, the structure of a leaf evolved as a compromise between two conflicting requirements: the need to admit carbon as the gas, CO_2, and to restrict the loss of water vapor. Somehow the compromise seems to work. In a leaf, the uptake of CO_2, the loss of water vapor, the uptake of oxygen (when a plant is respiring in the dark) or the export of oxygen (when the photosynthetic rate is vastly greater than the respiratory rate) all occur mainly through stomata. The machinery that regulates this gas exchange system will be considered in the next chapter.

CHEMOSYNTHESIS

Photosynthetic autotrophs dominate the biosphere both in terms of biomass and in terms of energy conversion. Without the light-driven

3

LEAF OF TROPICAL TREE FERN is thin and elaborately branched, adaptations that greatly increase its surface-to-volume ratio and enhance gas exchange. Such leaves are common in moist tropics.

conversion of water and CO_2 to carbohydrate, the Earth would rapidly become as sterile as the moon. There are, however, a number of bacteria, relatively small in biomass, which can extract usable energy from simple inorganic substances.

Although these chemosynthetic bacteria are autotrophs, they are really dependent on other organisms. For example, the two so-called nitrifying bacteria, Nitrosomonas and Nitrobacter, depend upon a supply of reduced nitrogen for their chemosynthetic activities. Since the biosphere is, at least at present, a strongly oxidizing environment, the nitrifying bacteria are absolutely dependent first upon their nitrogen-fixing cousins, second upon other autotrophs and heterotrophs which incorporate the fixed nitrogen, third upon the many organisms that break down the organic matter of other dead or dying organisms into relatively simple organic compounds, and finally upon

Although the list of nitrogen fixers is short, their impact on other organisms is enormous; without them no other organism could survive.

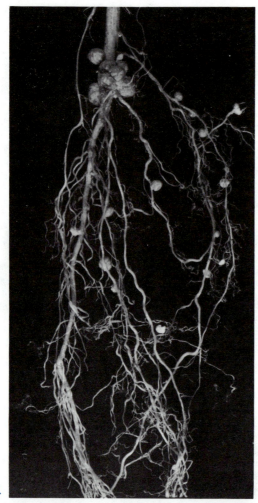

4

ROOT NODULES are abundant on this 28-day-old soybean plant. Plant was grown on nitrogen-free nutrient solution and had to rely on fixation by nodules for virtually all its nitrogen requirements.

the ammonifying bacteria and fungi which deaminate amino acids to liberate ammonia (as ammonium ion) into the environment. The nitrifying bacteria are a critical link in the nitrogen cycle.

NITROGEN FIXERS

Nitrogen gas, N≡N, is extraordinarily unreactive; the triple bond linking the two atoms must be broken to make nitrogen of any use whatsoever to organisms, and most organisms cannot perform the reaction. Fortunately there are a select few that can: the nitrogen fixers. All are procaryotic microorganisms.

Farmers in ancient Greece, Rome and China were aware that plants such as clover, peas and

alfalfa improved the soil. In 1888 a pair of German chemists, Hellriegel and Wilfarth, showed that the nodules found on the roots of these plants were sites of nitrogen fixation by bacteria (*Figure 4*). These bacteria all belong to the genus *Rhizobium*, and tend to form nodules on a particular type of legume. (The family Leguminosae includes peas, beans, alfalfa and many shrubs and trees.) Later other workers discovered the nitrogen-fixing properties of bluegreen algae (*Figure 5*) and photosynthetic bacteria. The nodule of the legume represents an excellent example of SYMBIOSIS, a situation in which two different organisms live in close association for mutual benefit (*Chapter 19*). Neither free-living *Rhizobia* nor uninfected legumes fix nitrogen; only when the two are combined in a root nodule does the reaction proceed effectively.

There is now reasonable agreement that the reduction of N≡N (N_2) proceeds through NH=NH, NH_2—NH_2, to 2 NH_3. All nitrogen-fixing systems studied to date require the same components: a powerful reductant, for instance ferredoxin; nitrogenase, a complex enzyme containing both iron and molybdenum; an oxygen-free environment (the nitrogenase can act only under anaerobic conditions); and ATP. The only unique ingredient is the nitrogenase.

About 90 million tons of atmospheric nitrogen per year are fixed by microorganisms. A very small amount of nitrogen may be fixed by nonbiological means, including lightning, volcanic eruptions and forest fires, with the resulting ammonia being dissolved in rainwater; nitrogen is also fixed commercially in chemical plants, but the total is small compared with the amount produced by fixers.

In the oceans nitrogen is fixed by photosynthetic bacteria and bluegreen algae, particularly *Anabaena* and *Nostoc*. On land the free-living soil bacteria from genera such as *Azotobacter* and *Clostridium* make some contribution, but most of the work is done in the root nodules of higher plants. Unlike the free-living procaryotes who fix what they need for their own uses and release the fixed nitrogen only upon their death, the root nodules may actually excrete some amino acids into the soil, making the nitrogen immediately available to other organisms.

Although the list of nitrogen fixers is short, their impact on other organisms is enormous. Without the nitrogen fixers, no other organism could survive. This small group of procaryotes is just as essential to the biosphere as the photosynthetic autotrophs. It is also ironic that some of these organisms, so useful to man in one sense, are so deadly in another. The genus *Clostridium* includes the dreaded botulism bacteria mentioned in Chapter 3, as well as the organisms that cause tetanus (lockjaw) and gas gangrene.

Any discussion of nitrogen fixation would be misleading and incomplete without mention of the opposite process — denitrification. There are a number of normally aerobic bacteria, mostly species of *Bacillus* and *Pseudomonas*, which can use nitrate as a terminal electron acceptor under anerobic conditions:

$$2\ NO_3^- + 10\ e^- + 12\ H^+ \rightarrow N_2 + 6\ H_2O.$$

These bacteria are extremely common and return nitrogen to the atmosphere as an inert gas. Without these organisms, atmospheric nitrogen would steadily decrease as fixation, both biological and industrial, continues on a global scale. There is some evidence that at the present time fixation exceeds denitrification, but hard figures are extremely difficult to obtain.

HETEROTROPHS

Heterotrophic organisms (all animals and parasites, as well as carnivorous and saprophytic plants) use organic materials built by others. This dependence on prefabricated goods frees an organism from the need to manufacture everything itself from simple precursors, but it also has some disadvantages. Complex materials do not usually present themselves on demand. They must be hunted or harvested, and this requires adaptive specialization, as well as the expenditure of time, energy, or both. Moreover, complex materials are usually not utilizable as such, since they are constituted largely of macromolecules that need to be reduced to smaller size before they can be absorbed. Heterotrophs depend on digestion (which is always effected by enzymes) to reduce the particles to the proper size.

FEEDING AND DIGESTION IN ANIMALS

Among animals, heterotrophic nutrition is illustrated in its simplest form by those parasites that live literally suspended in nutritious broth. The food of tapeworms (*Figure 6*) who live in the intestine of their host, is predigested, soluble, and can be absorbed without being enzymatically treated by the parasite.

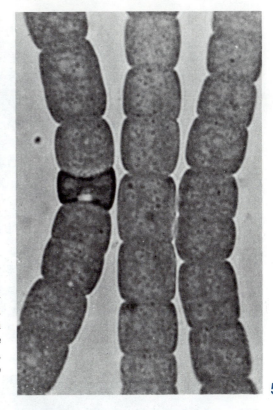

5

BLUEGREEN ALGA **Fremyella diplosiphon consists of microscopic filaments composed of many cells. Dark cell at left is dying. Filament will break at that point and new cells will proliferate from it.**

These animals need (and have) no mouth, gut or other feeding devices. Their only problem is to find the host in the first place.

Free-living animals must not only locate their food and obtain enough of it, but must also provide for subsequent digestion and absorption. How they do this differs enormously in the various groups, although the principles involved are very much the same.

Most food is taken as macromolecules, which poses two problems: the giant molecules are too large to be absorbed by the animal, and their composition is very seldom (except in cannibalism) precisely right for the feeder. Both problems are solved by digestion, that is, by reducing the macromolecules to small diffusible molecules such as amino acids, fatty acids, and monosaccharides; these can be built up later into appropriate macromolecules. This digestion makes use of a series of moderately specific enzymes; for instance there are several proteases to split protein, a few lipases to split fats, and several kinds of enzymes that split polysaccharides.

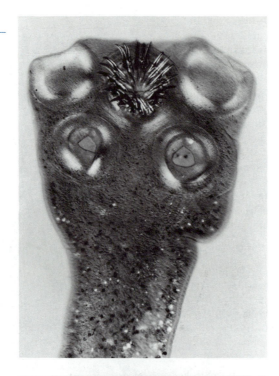

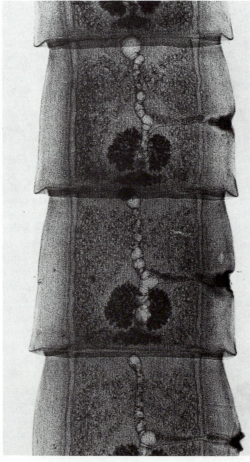

How does an organism produce enzymes to digest bodily constituents, and yet avoid digesting itself? Digestion always occurs outside the organism, and many digestive enzymes are only active when they reach the "outside". The gut is simply a tunnel through the animal, and food in the central space, the lumen, is outside the animal body in the same sense that a pebble clenched in the fist is outside the hand. Because of this external position, the food can be exposed to extreme conditions (such as high acidity or potent enzymes) which would be impossible within a cell. Many animals protect the lumen with mucus secreted by special cells which not only lubricates the lining but also insulates it against active enzymes. Insects have a different trick: they secrete a thin chitinous tube within the gut, which encloses the food and enzymes and shields the lining against abrasion and self digestion.

Digestive enzymes are invariably first produced within a cell in an inactive (zymogen) form which is passed to the outside and activated (*Figure 7*). For example, trypsin is produced as the inactive trypsinogen by the vertebrate pancreas. Within the gut it meets another enzyme called enterokinase, secreted by the gut wall, which converts it to trypsin, the active protease, by removing the protective "mask" of six amino acids.

Many other features are common in digestive-absorptive systems: chopping and churning devices to mix the food with digestive enzymes; a device to conserve the enzymes (by putting them with the food in a vacuole, or a blind sac, or a tubular gut); and suitable conditions of acidity and ionic composition for optimal digestion.

Nonparasitic protozoa have a relatively simple digestive-absorptive system. For example, an amoeba engulfs pieces of food by embracing it with protoplasmic extensions; this process, known as endocytosis, was illustrated in Chapter 1. *Paramecium* washes bacteria and other particulate matter down its oral groove (a mouth-like region) by creating a current with

TAPEWORM is one of the simplest heterotrophs. Head (top) anchors the worm to the intestinal wall of its host by means of hooks, suckers, or both. Bottom photograph depicts a few of the many body segments. Tapeworm has no trace of a digestive system. Each segment of body is little more than an envelope of reproductive organs. Worm can grow to several feet long.

the cilia that line it. Under pressure from the current, food and fluid are pushed into the cytoplasm at the base of the groove, and a food vacuole is formed. The vacuole breaks away and circulates around the cell, dwindling in size until it is finally ejected at a spot called the anal pore (*Figure 8*). As the vacuole circulates, enzymes are secreted into it and digested products are absorbed from it. The vacuole is essentially a device for intracellular digestion.

Coelenterates (from the Greek *coel*: hollow, and *enteron*: gut), simple multicellular animals comprising a large phylum that includes jellyfish, sea anemones and corals, do not have a true gut; a blind sac called the gastrovascular cavity functions as both a circulatory and a digestive system. Hydra, a half-inch-long coelenterate that clings to rocks and plants in ponds, is a carnivore that paralyzes its prey

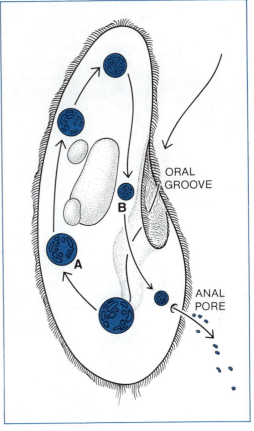

ORAL
GROOVE

ANAL
PORE

8

FOOD VACUOLE OF PARAMECIUM forms at the base of the oral groove. Vacuole circulates through the organism, growing gradually smaller as bacteria within it are digested and absorbed. Bacteria are intact at first (A), but have virtually disappeared (B) by the time that vacuole is ready for expulsion through the anal pore.

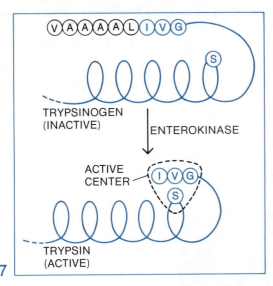

TRYPSINOGEN
(INACTIVE)

ENTEROKINASE

ACTIVE
CENTER

TRYPSIN
(ACTIVE)

7

CONVERSION OF A ZYMOGEN to an active enzyme. The hexapeptide (black circles at top) masks the essential IVG sequence from S, preventing the formation of the active site. Hexapeptide is linked by hydrogen or electrostatic bonds to a distant region of the peptide chain. In this example enterokinase removes the mask, converting inactive trypsinogen to active trypsin.

with nematocysts, tiny stingers embedded in its tentacles (*Figure 9*). The victim is then stuffed into the gastrovascular cavity, where it is partially digested by enzymes secreted by the gland cells lining the cavity. Phagocytic cells ingest the fragments and complete the digestive process by forming a vacuole, as in protozoa.

Five kinds of feeder have been recognized among animals, each with its own way of acquiring and handling food. Omnivores and herbivores together comprise the largest group; the others are carnivores, deposit feeders, filter feeders and fluid feeders.

In higher animals one finds a true gut: a tubular structure with a mouth at one end for ingestion and an anus at the other for ejection. The use of a tube instead of a vacuole or sac permits production-line processing of food. Food can be crushed in the first part, treated with an enzyme at alkaline pH, then passed to an acid area where different enzymes digest out proteins, for instance — all of which helps to reduce the particles to diffusible forms that the body can absorb. Before expelling the undigestible material as feces, the body can recover salts and water which had to be added during processing, but which are too important to waste.

All guts absorb digested matter in the lumen, whose surface is often increased to expose a greater absorptive area. The earthworm, for instance, developed a dorsal infolding or TYPHLOSOLE and the shark a spiral valve. It is also common for the wall of the gut to be folded many times, with the individual folds bearing many tiny fingerlike projections called villi. The surface of the villi may in turn be subfolded into microvilli (*Figure 10*).

Guts frequently depend on microorganisms to perform functions that the host cannot. The principal agents of digestion in ruminants (such as cows and sheep) are bacteria and protozoa living in the rumen, the first compartment of their stomachs. The microorganisms can degrade cellulose (which few animals can)

and can convert nitrogen into proteins and so contribute to the protein intake of the animal. (Cows are sometimes fed urea as a cheap nitrogen source.) Bacteria in the caecum of the guinea pig are a source of B vitamins. At night the guinea pig produces and then eats a special soft feces made up almost entirely of these bacteria. If one deprives the animal of these feces it eventually dies.

The gut is usually divided into a series of compartments, with a mouth and buccal cavity (a sort of "hallway") at one end and an anus at the other. The mouth may bear mechanical devices (teeth of vertebrates, mandibles of insects) for grasping and/or fragmenting food, or there may be a special organ further back (the gizzard of birds and some invertebrates) where this grinding is done. The churning of food and stones (small stones become lodged in the gizzard) by massive gizzard muscles breaks up the food. Some animals, of course, take in food that needs no fragmentation.

The importance, relative size, and distinctness of these compartments varies tremendously (*Figure 11*). The stomach or crop functions principally as a storage chamber that enables the animal to eat quickly and digest at leisure — a strategy that minimizes exposure to predators. Digestion and absorption may or may not occur in this storage chamber, depending on the species. If this compartment is acidic, as frequently it is, there is also a midgut or intestine. Food delivered to the intestine is well minced and mixed. This zone is usually of neutral pH, and most digestion and absorption occurs here. Enzymes derived from specialized glands (such as the pancreas of vertebrates) are commonly poured into the intestine, and the gut wall secretes other digestive enzymes. The hindgut, of which a muscular rectum may comprise all or part, recovers water and salts. These materials are especially important to terrestrial animals. The bacterial "farms" of microorganisms described above, are located here in the hindgut in man and many other animals. Sometimes there is a special intestinal diverticulum, or caecum, in which the microorganisms are housed, as for example in rodents.

Although all guts have common features, there are distinct differences from one type of animal to another. The differences depend mainly on the kind of food that its gut is called upon to process. Five kinds of feeder have been recognized, each with its own way of acquiring and handling food. OMNIVORES (eat-

9

HYDRA CAPTURES DAPHNIA and paralyzes it with stinging tentacles (left). Crustacean is then stuffed into the hydra's gastrovascular cavity (right). Cells lining cavity secrete digestive enzymes and absorb digested food. In addition, the cells engulf food vacuoles for intracellular digestion.

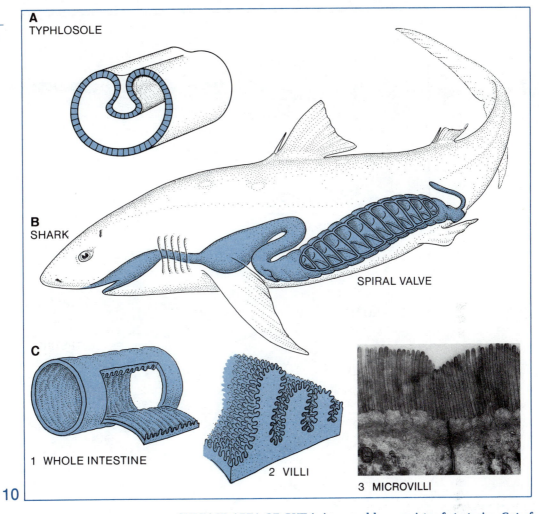

A
TYPHLOSOLE

B
SHARK

SPIRAL VALVE

C

1 WHOLE INTESTINE

2 VILLI

3 MICROVILLI

10

SURFACE AREA OF GUT is increased by a variety of strategies. Gut of earthworm (A) is infolded along its dorsal length, forming structure called the typhlosole. Intestine of shark contains spiral fold that forces food to travel a longer path, exposing it to an increased absorptive surface. In most vertebrates, including man, intestines are internally folded (C) and the folds are pitted with tiny invaginations called villi. Surfaces of villi in turn are folded into microvilli.

ers of everything) and HERBIVORES (plant-eaters) together comprise the largest group of feeders. Deer and cabbage loopers are both herbivores; men, pigs and cockroaches are all omnivores. CARNIVORES (flesh-eaters) comprise another large group, which includes species as diverse as the cheetah and the hydra. DEPOSIT FEEDERS take in great quantities of mud or similar deposits and utilize only the small fraction of organic matter present in it, which may be only four percent. The most familiar examples are the earthworm on land and lugworms and sea-cucumbers in the sea. FILTER FEEDERS, who strain out particles or

small organisms suspended in water, include most of the stationary aquatic animals such as sponges, corals, barnacles, oysters and clams. FLUID FEEDERS suck up a variety of liquids. Aphids suck the sap of plants; mosquitos, leeches and vampire bats suck blood; and spiders suck the digested contents of insect bodies.

OMNIVORES AND HERBIVORES

The digestive tract of man, so interesting to us, is a logical example of the gut of an omnivore (*Figure 12*). The teeth are used (as in all vertebrates of this group) for reducing the food to

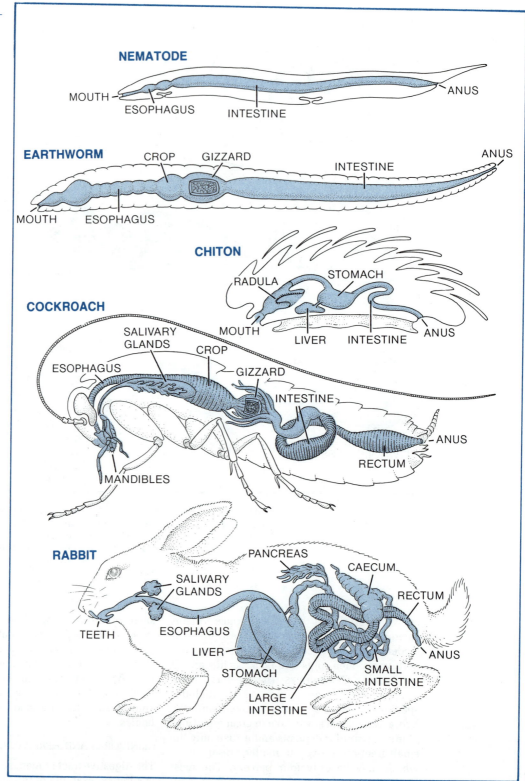

NEMATODE

MOUTH — ESOPHAGUS — INTESTINE — ANUS

EARTHWORM CROP GIZZARD INTESTINE ANUS

MOUTH ESOPHAGUS

CHITON

RADULA STOMACH

MOUTH LIVER INTESTINE ANUS

COCKROACH

SALIVARY
GLANDS CROP

ESOPHAGUS GIZZARD

INTESTINE

ANUS

RECTUM

MANDIBLES

RABBIT

PANCREAS CAECUM

SALIVARY
GLANDS RECTUM

ESOPHAGUS

TEETH ANUS

LIVER

STOMACH SMALL
INTESTINE

LARGE
INTESTINE

11

manageable proportions; digestion begins in the mouth, where a single enzyme (amylase) in the mildly alkaline saliva begins degrading starch to sugar (hence the sweet taste of well-chewed bread). A ball or BOLUS of food is transferred via the esophagus to the stomach, where 0.5 percent hydrochloric acid kills bacteria and provides optimal conditions for the activity of pepsin, a proteolytic enzyme. The food can stay for hours in the stomach, churned three times a minute by peristaltic waves — muscular contractions that sweep through the stomach and intestine mixing its contents and pushing them onwards.

From time to time a ring-shaped muscle (sphincter) at the stomach's far end (pylorus) opens and allows a portion of the partially liquified contents to be squirted into the first part of the small intestine, called the DUODENUM. Here bicarbonate makes the mass mildly alkaline (about pH 8). A digestive juice is poured in through tubes from the PANCREAS; the juice contains trypsinogen (soon to be converted to trypsin for more digestion of protein), lipase for fat digestion, and amylase for carbohydrate breakdown. The liver pours into the duodenum a bitter secretion (BILE) which it stores in the gall bladder and ejects via the bile duct. The bile contains no enzymes, but has detergents (BILE SALTS) which emulsify fats so they may be more easily digested by lipase. The bile also contains breakdown products of blood pigments degraded by the liver; thus the gut acts as a sort of sewer as well as a food pipeline.

The food is squeezed into the small intestine proper, where the cells lining the lumen secrete more juice (*succus entericus*) which contains another amylase, lipase, maltase for digesting maltose, sucrase for sucrose, lactase for lactose, and several peptidases for proteins.

Digestion is completed here and absorption begins.

The function of absorption is of course to transfer the nutrients from gut to blood, where they can circulate to needy tissues (*Chapter 11*). Tissues usually absorb small molecules such as monosaccharides, amino acids, and fatty acids. They also absorb small (0.5 μ) incompletely digested droplets of fat. (Blood from the intestine has a decidedly milky look after a fatty meal.) The concentration in the blood of any single nutrient, such as glucose,

HUMAN DIGESTIVE TRACT is suited for an omnivorous diet. Chewing reduces food to small bits and increases its surface area for attack by enzymes. Enzymatic digestion begins in the mouth and continues in the stomach and intestines. Absorption begins in the jejunum (small intestine) and continues in the large intestine and rectum.

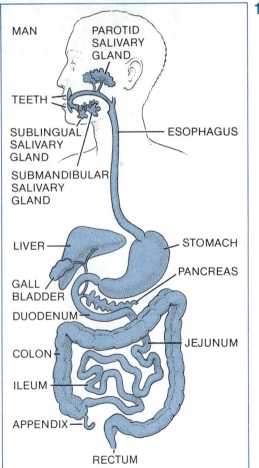

12

DIGESTIVE TRACTS OF ANIMALS are compared in drawing on opposite page. Gut of nematode worm, a fluid feeder, is little more than an unspecialized tube. Gut of earthworm contains a crop for storage and a pebble-filled gizzard for grinding soil, decayed plant matter and bodies of small animals. Gut of chiton includes a rasping tongue (radula) and a large stomach. Chiton feeds on algae. Cockroach chews food with mandibles, mixes it with saliva, and passes it first into crop, then through grinding gizzard into intestine for digestion and absorption. As in all terrestrial animals, roach's rectum is specialized for reabsorption of water. Gut of rabbit is typical of vertebrates, well equipped with digestive glands, and has a large caecum for farming symbiotic microorganisms.

13

GILL FUNGUS (Amanita muscaria) is a saprophyte that feeds on decaying plant matter. Photograph depicts the fruiting structures of two of these deadly poisonous "toadstools." Gills on underside of the fruiting structure produce numerous spores which are dispersed by the wind.

may often exceed what is in the gut. Without a special mechanism, the nutrient would not move from gut to blood, and could even be lost moving from blood to gut. This problem is solved by the evolution of active transport. Special transport systems use ATP in a sequence of chemical steps to transport key nutrients against a concentration gradient. The transport systems may be very specific, transporting glucose but not fructose. Several amino acids, however, share a common system.

Finally the residual matter from digestion is squeezed into the large intestine where the water and salts poured in during digestion are absorbed to conserve the precious resources. Here a bacterial farm manufactures our riboflavin, vitamins K, B_{12}, and nicotinic acid. A prolonged oral intake of antibiotics will poison the bacteria and can lead to deficiency in those vitamins.

PLANTS AND MICROORGANISMS

One of the themes of this chapter is the interdependence of organisms on one another for nutrients. The closest thing to an exception is a photosynthetic anaerobic bacterium, which does not need oxygen, can fix both CO_2 and N_2, and needs only a little light and a few ions to survive. At the other extreme lie the heterotrophs, whose nutritional requirements often involve compounds synthesized by several other organisms.

A second theme concerns interconversion of nitrogen compounds, starting with N_2. Plants take up ammonium ions and soluble nitrate ions (NO_3^-), produced by soil bacteria. With the expenditure of a little ATP and reducing power, and using the enzyme nitrate reductase, plants simply make ammonium again from nitrate, for use in amination reactions. The resulting proteins, nucleic acids, and so forth are already familiar to the reader. Herbivorous animals can obtain from plants the amino acids and whatever else they cannot themselves synthesize. When an organism dies many different kinds of heterotrophs complete the nitrogen cycle, simultaneously obtaining their own carbon and nitrogen. In the plant kingdom the heterotrophs include saprophytes, which live harmlessly on decaying plant and animal detritus (*Figure 13*); parasites, which infect living plants and animals; and even a few carnivores, plants that trap and digest small insects. It is the saprophytes, by and large, that complete the nitrogen cycle.

Saprophytic fungi and bacteria prevent the world from becoming one vast mass of dead organisms. These saprophytes are remarkably diverse in their appetites. They can use, for example, plastics, paint, vinegar, wax, fingerprints, paper, cloth, dung, leather, kerosene, corpses, or any of a great variety of other carbon sources. They can break these down systematically, using the carbon both for synthesis and for respiration, ultimately returning it to the air as CO_2. Many saprophytes release as ammonium ion any organic nitrogen that is surplus to their own needs, a process called ammonification. No matter how hard an organic chemist may work producing new compounds of carbon, chances are that some fungus or bacterium will have the capacity to make enzymes to use the compound as a carbon source.

The saprophytes are essential organisms in that they close both the carbon and nitrogen cycles, returning large amounts of CO_2 to the atmosphere for reuse, and returning fixed nitrogen to the soil, or reducing nitrate back to nitrogen gas (*Figures 14 and 15*). A thin line

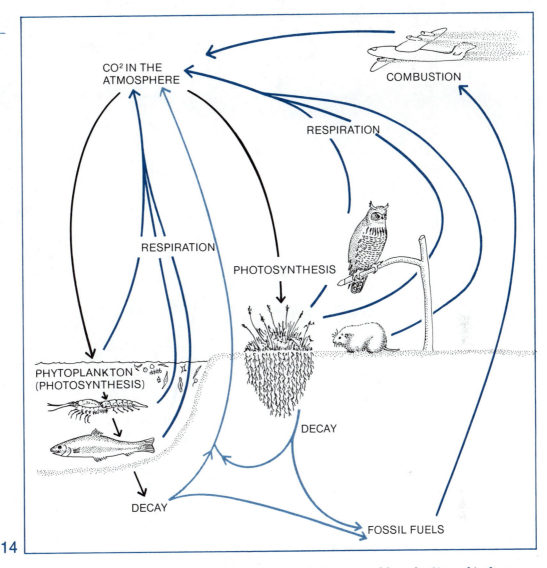

CO² IN THE ATMOSPHERE

COMBUSTION

RESPIRATION

RESPIRATION

PHOTOSYNTHESIS

PHYTOPLANKTON
(PHOTOSYNTHESIS)

DECAY

DECAY

FOSSIL FUELS

14

THE CARBON CYCLE. Carbon is removed from the CO₂ pool in the atmosphere by the process of photosynthesis. Animals obtain carbon by eating plants or other animals that feed on plants. Animal respiration, decay of organic matter, and combustion of fossil fuels return CO₂ to atmosphere.

separates the saprophytes from the parasites: the organic matter that provides carbon and nitrogen for the parasites is not yet dead. Thus parasitic bacteria and fungi are responsible for a large number of diseases in both plants and animals. Viruses represent the epitome of parasitism.

The few carnivorous plants have remarkable adaptations for allowing them to capture small insects or worms which they then use as a source of available nitrogen. The sundew *Drosera* has leaves covered with a clear sticky liquid, fairly high in sugar (*see photograph on opening page of this chapter*). An insect which touches the leaf becomes stuck, and in fact the other hairs or "tentacles" will curve over and stick to the insect too. Digestion is taken care of by the secretion of appropriate enzymes, and carbon and nitrogen are taken up.

One of the major uses of the food synthesized by plants or ingested by animals is to provide the organism with energy. The commonest way to unlock this energy is to oxidize the food to CO_2 and water. Most organisms

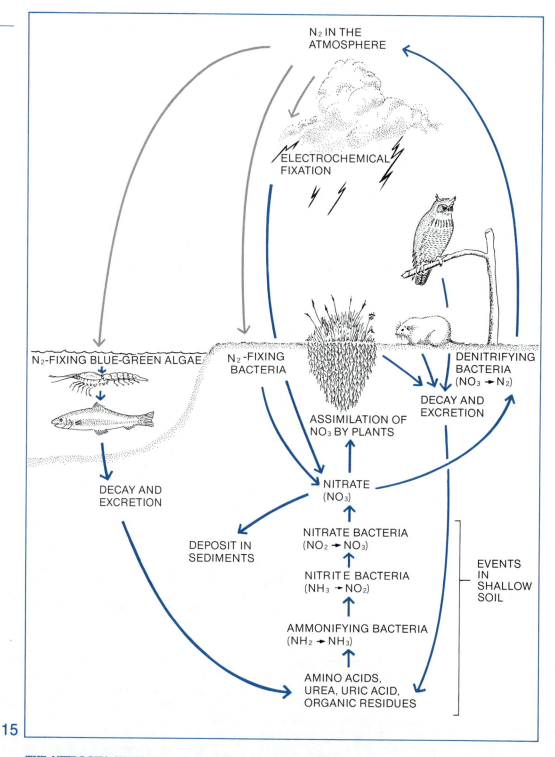

N₂ IN THE
ATMOSPHERE

ELECTROCHEMICAL
FIXATION

N₂-FIXING BLUE-GREEN ALGAE

N₂-FIXING
BACTERIA

DENITRIFYING
BACTERIA
(NO₃ → N₂)

ASSIMILATION OF
NO₃ BY PLANTS

DECAY AND
EXCRETION

DECAY AND
EXCRETION

NITRATE
(NO₃)

DEPOSIT IN
SEDIMENTS

NITRATE BACTERIA
(NO₂ → NO₃)

NITRITE BACTERIA
(NH₃ → NO₂)

EVENTS
IN
SHALLOW
SOIL

AMMONIFYING BACTERIA
(NH₂ → NH₃)

AMINO ACIDS,
UREA, URIC ACID,
ORGANIC RESIDUES

15

**THE NITROGEN CYCLE. Bacteria and bluegreen algae convert atmos-
pheric nitrogen to a form that other organisms can use. Lightning has the
same effect. Plants absorb nitrates through their roots. Animals obtain
nitrogen by eating plants or other animals. Dead organic matter and waste
products of animals return nitrogen to the soil. Denitrifying bacteria recycle it back into the atmosphere.**

have developed mechanisms for the continuous acquisition of oxygen, for transporting to the site of oxidation, and for expelling the CO_2. These mechanisms are the subject of the next chapter.

READINGS

R.G.S. BIDWELL, *Plant Physiology*, New York, Macmillan, 1974. A rather biochemical textbook with good treatment of photosynthesis, the nitrogen cycle and respiration.

T.C. CHENG, *Symbiosis*, New York, Pegasus, 1970. A wide-ranging account of the phenomenon, covering every level from invertebrates to man. Good solid reading, intended for laymen.

V.A. GREULACH, *Plant Structure and Function*, New York, Macmillan, 1973. A good general text in experimental botany with strong chapters on carbon and nitrogen.

J. MORTON, *Guts*, New York, St. Martin's Press, 1967. A paperback whose brevity is reflected by its title. In 58 pages it offers a readable and well-illustrated survey of what the author calls the first and most obtrusive of the great systems the student finds in dissecting.

G.R. NOGGLE AND G.J. FRITZ, *Plant Physiology*, Englewood Cliffs, N.J., Prentice-Hall, 1976. A new and up-to-date plant physiology text.

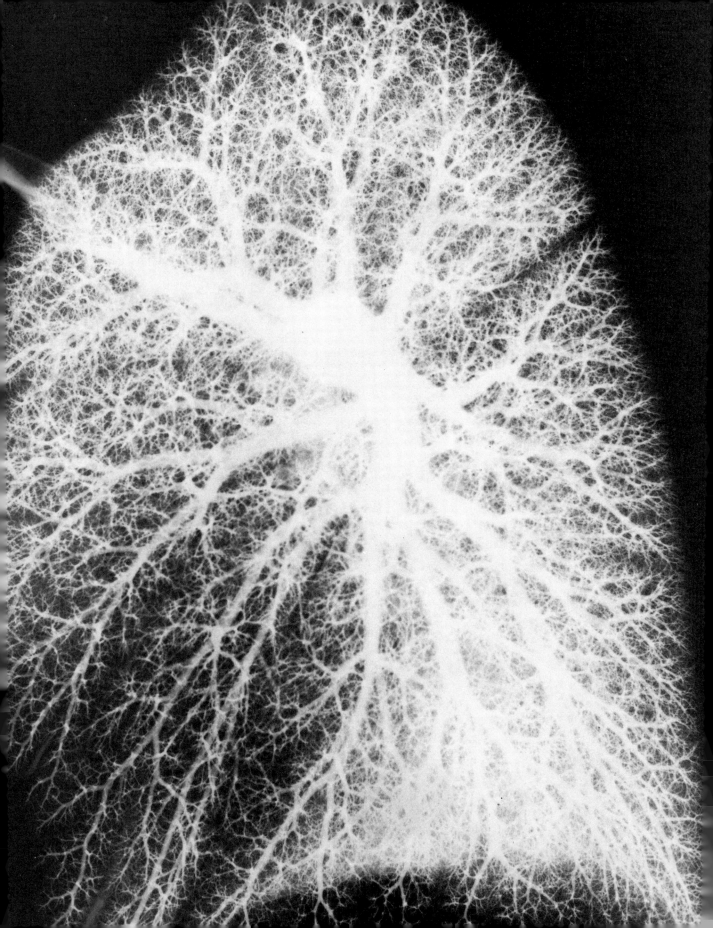

9

gas exchange

In respiration is there a sort of digestion of the air? That is to say, a kind of phenomenon that fixes the oxygen by means of a structure or tissue avid for oxygen?

CLAUDE BERNARD (1850)

Almost all organisms exchange gases with their environment. Even a saprophytic bacterium may have to rid itself of unwanted CO_2, hydrogen sulfide, or some other gaseous waste product. Because cell chemistry is aqueous chemistry, a gas must be dissolved in water to be used by an organism. Likewise if an organism is liberating a gas — be it CO_2 from respiration, O_2 from photosynthesis, or H_2 from an anaerobic photosynthetic bacterium — it is first produced in solution. Somewhere in the organism there must be a liquid-to-air interface where gases can efficiently enter or leave solution.

Unicellular aquatic organisms can cope with gas exchange very easily. So long as the water in which they grow is in motion, gaseous products are readily removed for release at this interface, and their gaseous requirements are readily met by the reverse process. Diffusion is sufficient to keep the appropriate gas available throughout the organism.

Larger aquatic organisms face two new problems. First, they may have some sort of outer tissue through which exchange of dissolved materials is severely restricted, thereby demanding that some special adaptation provide a liquid-liquid interface between the organism and the solution in which it is living. Second, increase in size means decrease in surface-volume ratio. Above a certain size, diffusion of solutes may simply be too slow for required gases or for disposal of waste gases.

For terrestrial organisms there is yet another problem. Since gases must be in solution to enter or leave the cell, there must be a solution-air interface in the organism. Too much evaporation of water from this interface can easily be fatal. Special adaptations for gas exchange have been shaped by two factors. The first is the rate of gas exchange required. A fast-living organism such as a human requires far more oxygen and evolves far more CO_2 per unit weight than a sedentary mushroom. The second factor is the function that gas exchange serves. While excessive loss of water as gaseous vapor would be harmful to a tree, the tree would be equally in trouble if no such loss occurred, because evaporation of water at the plant's interface with the air provides the driving force for raising water and essential ions from the roots to the aerial parts (*Chapter 10*). Plants and animals have evolved quite different solutions to the problem of gas exchange.

GAS EXCHANGE IN PLANTS

Aquatic plants need no special adaptations for water conservation so long as they remain submerged (the special osmotic problems of organisms in salt water are considered elsewhere). Even when they are quite large, they utilize extensive branching or make very thin organs to maximize their surface-volume ratios and keep actively metabolizing or photosynthesizing cells in close contact with the environment. Gases need never diffuse through more than a few cells. The flatworms (planaria) are an interesting parallel in the animal kingdom: mold a planarian into a sphere and its innermost cells would asphyxiate.

BLOOD VESSELS OF HUMAN LUNG appear in white in the x-ray photograph on opposite page. Large vessel at center is pulmonary artery, which arises directly from the heart.

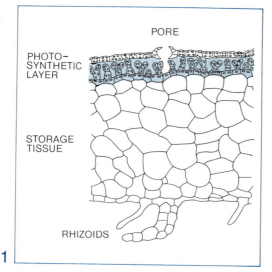

PHOTO-
SYNTHETIC
LAYER

PORE

STORAGE
TISSUE

RHIZOIDS

1

CROSS-SECTION OF LIVERWORT shows adaptations that restrict gas exchange and water loss during dry periods. Gametophyte shown here has pores that admit gases to photosynthetic layer that is only a few cells thick. Inactive storage tissue lies below it. Rhizoids attach plant to its substrate.

The lower land plants such as mosses, liverworts, and hornworts (*Chapter 19*) follow essentially the same principles. Plant parts are thin: photosynthetic organs are small and numerous; and requirements for long distance transport of any kind are minimized. But such plants are restricted severely in their distribution on land; they must grow where water is plentiful. They are usually not heavily cutinized, and they rapidly desiccate in the absence of continual moisture.

Maximizing surface area for gas exchange without elaborate modification to prevent water loss does not necessarily mean that these plants are always restricted to swamps. On a hot summer day in the dry season in the great central valley of California, one can find rocks too hot to touch. On these rocks are dried brown mosses which crumble readily in the hand. If they are crumbled onto a moist surface, however, intact cells or tissue fragments turn green and begin photosynthesizing within a few minutes. These plants become dormant when they dry out and resume growth rapidly upon rehydration. They can go

through their entire life cycle in a few weeks. Their solution to the gas exchange-desiccation dilemma is temporal, not structural.

Certain lower plants do have adaptations that permit restricted gas exchange during relatively dry periods (*Figure 1*). For example the liverwort *Marchantia*, a plant which grows flat against its substrate, has an internal structure comparable to that of a leaf. A thin layer of nonphotosynthetic cells covers the upper surface, beneath which lies an elaborate chamber system. Plates and filaments of photosynthetic cells project into these chambers and have access to the outside environment through pores in the surface layer, the epidermis. The pores open and close in response to humidity, permitting a certain degree of regulation of gas exchange. Several other lower plants produce sporophytes which project several centimeters above the substrate. They have a cutinized epidermis with the typical stomata of higher plants.

As mentioned in Chapter 7, in some higher plants such as ferns both the sporophyte and gametophyte are independent and free-living. In these cases the sporophytes, the large and obvious forms, have the typical cutinized epidermis with stomata for gas exchange. The gametophytes, however, often rely on a more primitive solution: a small flat plate of cells pressed against a moist substrate. The gametophyte phase is brief and can be completed during short periods of plentiful moisture in an otherwise arid climate.

For practical purposes, stomata are simply pores through a thin sheet of water- or gas-proof material. One is justified in asking how the gas exchange needs of a rapidly growing plant can possibly be met by what superficially appears to be a rather inefficient system. The answer lies in the physical properties of molecules. A diffusible substance in a fluid medium, whether liquid or gas, will move from a region of high concentration to one of low. If the system is closed and left undisturbed, the molecules of the substance will eventually become randomly distributed throughout the fluid. The only difference between diffusion in liquid and in gas is the rate. One can smell an overperfumed woman within seconds of the time she enters a quiet room,

Diffusion through pores, per unit area, is almost 50 times more efficient than diffusion from the uncovered surface.

because gaseous diffusion occurs at rates approaching meters per second. Diffusion in water, on the other hand, is best measured in millimeters per hour. The rate of diffusion also depends on the concentration gradient; the steeper the gradient, the greater the number of molecules transferred in a given time.

LEAVES AND STOMATA

Some 70 years ago H.T. Brown and F. Escombe showed very clearly how these principles apply to diffusion of gas through small pores such as stomata. A pan filled with water is covered with foil perforated with holes of various sizes and patterns of distribution (*Figure 2*). Brown and Escombe demonstrated that small and well-spaced holes in the foil, exposing only about 0.3 percent of the surface area

3

TWO STOMATA are visible in this scanning electron micrograph of the underside of a leaf from the spiderwort Tradescantia. The stoma at top is partly open; the other one is closed.

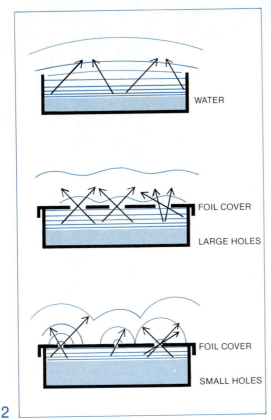

2

DIFFUSION OF GASES from open tray (top) was compared experimentally with diffusion from tray covered with foil containing large holes (middle) and with foil containing small pores (bottom). Colored contour lines indicate concentration of water vapor above trays, and arrows indicate paths of evaporating water molecules. Small widely spaced pores produce a steep hemispherical concentration gradient which leads to fast efficient diffusion.

allowed almost 15 percent as much diffusion as the uncovered pan. Diffusion through pores, per unit area, is almost 50 times more efficient than diffusion from the uncovered surface. A leaf epidermis with its neatly spaced stomata is indeed an efficient system for gas exchange (*Figure 3*).

Stomata open and close in response to the changing environment. The mechanism looks deceptively simple. The stoma can be compared to an almost empty inner tube with its opposite inner surfaces touching (*Figure 4*). If one pumps air into it, the inner surfaces are forced apart as the tube assumes its normal expanded shape. Let the air out again and it collapses. If the tube were designed so that the inner surfaces were always pressed together when it was empty, it would be a perfect mechanical model of a pair of guard cells, with each cell as half of the inner tube.

Stomata are open when water is plentiful and closed when it is scarce. Light and low CO_2 concentration cause the stomata to open; darkness causes them to close. Changes in the CO_2 concentration affect the stomata more than light. High concentrations of CO_2 usually cause stomata to close whether they are in the

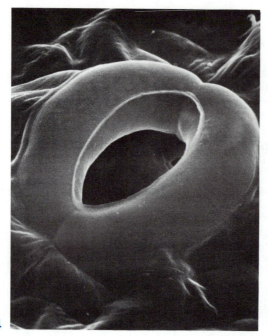

4

CLOSE-UP OF OPEN STOMA shows the two guard cells surrounding the pore in a cucumber leaf. This picture, made with the scanning electron microscope, is reproduced at a magnification of about 3100×.

light or not, and low CO_2 concentrations allow them to stay at least partially open even in the dark.

ROOTS

Roots too play a role in gas exchange. They have a large surface-volume ratio because they branch extensively, and because the surface of recently matured regions is covered with hairs. Roots can thus obtain gases directly from soil water, or from air spaces in fairly dry soil. Whether a plant will or will not grow on a particular site depends very much on the physical properties of the soil and on what else is growing there. If a soil is filled with aerobic microorganisms, for example, its oxygen content is low and its CO_2 concentration extremely high. Thus roots are exposed to somewhat

more precarious conditions than leaves. Certain large woody plants live in swampy environments in which water is extremely sluggish, and oxygen concentration extremely low because of the activities of large numbers of saprophytes. The bald cypress is a familiar example. The roots of these plants frequently develop aerial parts, or pneumatophores, which project above the water. These may be "knees" as in the cypress (*Figure 5*) or they may be the ends of specialized roots. In either case, their function seems to be to provide a path for diffusion of oxygen to the otherwise anaerobic roots.

5

PNEUMATOPHORES of the bald cypress (Taxodium distichum) project above swampy ground. Sometimes called "knees," pneumatophores rise from roots and provide path for diffusion of oxygen.

Plants are rather sluggish individuals. Their demands for gas exchange are modest for their size. Fast-living animals have requirements hundreds or thousands of times higher. A resting elephant needs 148 microliters of oxygen per gram of body weight per hour, a resting man 200, and a mouse 2500.

The stem of an herbaceous plant has an epidermis with stomata just like those of leaves. But a woody stem is another matter. The cork cells of the bark are far more effective waterproofing — and hence gasproofing — than the simple layer of cutin on the leaf. Yet within the layer of cork one finds numerous living cells, the most important of which are in the vascular cambium, laying down new xylem and phloem, and ray cells in the xylem. Thus there is at least a modest requirement for gas exchange. As with the epidermis, there are specialized regions for gas exchange, the LEN-TICELS (*Figure 6*). In the cortex beneath a stoma a group of rather spongy cells is produced. The cork cambium beneath these produces parenchyma instead of cork, and as the whole system expands with the stem, this production of spongy tissue continues. Gas diffuses through the extensive intercellular spaces of this tissue. Thus even in fairly thick bark, there are small "windows" through which oxygen can enter and CO_2 can exit.

Plants, by and large, are rather sluggish organisms. Their demands for gas exchange are modest for their size and are met locally. There is no need for a special system to move gases rapidly from one part of an organism to another, and diffusion is sufficient. As will shortly become clear, however, fast-living animals have gas exchange requirements hundreds or thousands of times higher than plants.

GAS EXCHANGE IN ANIMALS

Animals share with plants the need to provide an extensive wet surface for gas exchange, and those that live on land must, like plants, protect the exchange surface from the drying effect of air. Animals differ from plants, however, in that their gas exchange is strictly respiratory. They take in O_2 and give off CO_2, but not the reverse, as in photosynthesizing plants. Animals differ a good deal in the ways and means that they meet their respiratory demands. Two decisive factors influenced the evolution of animal respiratory mechanisms: the extent of the demand for O_2, and its availability in the environment.

OXYGEN NEED

Oxygen need depends on the activity and size of an organism, and can be expressed as microliters of oxygen per gram of body weight per hour. (A microliter, abbreviated μl, is a millionth of a liter, about the volume of a pinhead. A gram is about half the weight of a dime.)

Growth and movement make large energy demands. Slowly moving and nongrowing organisms use relatively little oxygen. To give a few examples, a python uses 6.2 μl, a carrot root cell 25 μl, a mussel 22 μl, a dry barley grain 0.06 μl. Organisms with a more active lifestyle require, even at rest, substantially more: man needs 200 μl, the butterfly *Vanessa* 600 μl, a chicken 497 μl. These figures increase greatly when such organisms become highly active: man's need climbs 20-fold to 4000 μl, *Vanessa*'s need increases an amazing 170-fold to 100,000 μl. Active forms of plants also have large needs: the male gametes of the alga *Fucus vesiculosus* need 2550 μl. A germinating barley seed uses 108 μl, which is 1800 times more than the resting seed needs. Obviously growth, like movement, imposes large energy demands.

Size influences the oxygen needs of warm-blooded animals because of the increase of relative surface area (and therefore of relative heat loss) with decreasing size. The resting elephant needs only 148 μl of oxygen per gram of body weight per hour, the resting man 200

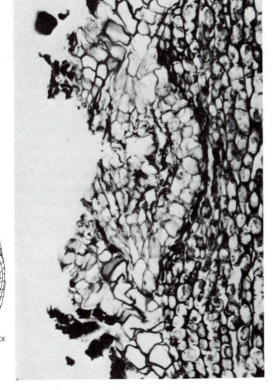

CORK

BREAK IN CORK

BARK

SPONGY TISSUE (NO WATER SEAL)

CORK

CORTEX

6

LENTICEL in the bark of a lilac tree. Gases pass through break in cork and diffuse freely through intercellular spaces in underlying spongy bark. Diagram is key to photomicrograph.

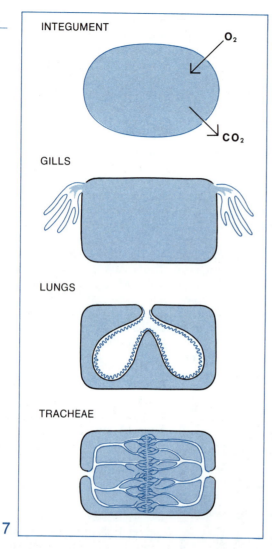

INTEGUMENT

O₂

CO₂

GILLS

LUNGS

TRACHEAE

7

GAS EXCHANGE IN ANIMALS can occur either through the integument (skin), gills, lungs or tracheae. Gills are outfolded structures that increase the exchange surface in aquatic animals. Lungs are infoldings that serve the same purpose in terrestrial animals; the exchange surface is internal and thus protected from drying out. Tracheae are branched internal respiratory tubes; exchange occurs in tiny end tubes (tracheoles). Tracheal systems are found in insects and some other arthropods.

μl, and the resting mouse 2500 μl. Simple diffusion of oxygen through the body wall is adequate only for microscopic animals or those with very modest oxygen requirements. Moreover the distance from the center of the organism to the body surface increases with increasing size, so that even if the required oxygen could be absorbed at the surface, a large organism could suffocate at its center. It was pointed out earlier in the chapter that one way to circumvent this problem is to maintain a flat body shape, so that every part of the body is near a surface, as in the flatworms.

Larger organisms have had to develop special devices to cope with the inadequacy of simple surface diffusion. All the devices provide large interfaces between the animal and its oxygen source. Some portion of the body usually becomes specialized to provide a vast surface area to speed up gas exchange, as in gills and lungs (*Figure 7*). The interface must always be wet because the permeability of dry tissues is inadequate. In terrestrial animals special precautions are needed to maintain a wet surface, yet not lose excessive amounts of water from it. Except when needs are small, the air or water in contact with the exchange surface must be continuously renewed to avoid depletion of its oxygen or unwarranted accumulation of carbon dioxide.

OXYGEN AVAILABILITY

The quantity of oxygen available in water is far less than in air, and the amount of oxygen in both can vary considerably. A liter of air contains 210 ml of oxygen, a liter of fully aerated water at 15°C contains only 7 ml (fresh water) or 5 ml (sea water). The oxygen in water falls as the temperature rises (which is why fish leave lake shores as the days warm up and seek cooler, deeper places); at 35°C, fresh water has only 5 ml per liter. These are the levels of oxygen in fully aerated water, and water is often not fully aerated. For instance, unless water movement is excellent (and it rarely is) there will be oxygen-poor zones where organic matter is being oxidized by bacteria. This is a major reason why pouring sewage into rivers can kill massive numbers of fish.

The low oxygen content of water sets severe limits on the possible activity of aquatic animals. One consequence is that all warm-blooded animals, even those whose existence is entirely aquatic (such as whales), rely upon air as an oxygen source. Additional problems arise when water is likely to be warm and stagnant. The electric eel *Electrophorus* has solved this problem by becoming an air-breather; it rises to the surface regularly to gulp air, and absorbs the oxygen through a network of blood vessels in its throat. Many other fish surface for air and some, such as lungfish, can store air in special sac-like chambers of the gut.

Oxygen insufficiency occurs less in terrestrial animals because of the relatively high oxygen content and the rapid mixing of air. But oxygen content drops proportionately to air pressure, so that at high altitudes it is much reduced. At 17,500 feet air contains about half as much oxygen as at sea level. Most animals exposed to such conditions undergo "acclimation" changes, which are well seen in man. The lung capacity, the number of red cells and the cardiac output all increase notably.

SPECIAL RESPIRATORY MECHANISMS

Life began in water, and it is there that one should look for the simplest modifications of systems that depend on free diffusion over the whole body surface, which are adequate for very small organisms. The fundamental mechanism of diffusion across a wet membrane is found at every level, but in animals of increasing complexity various specializations of the wet membranes have evolved. In the simplest type, a portion of the body wall protrudes and increases its surface area by branching to form a gill. Axolotls are among the many amphibians using this device (Figure 8). Many crustaceans also use gills of this type, but because they are covered (and so protected) by the shell, water has to be circulated over them by miniature paddles in front of the gills. Similarly, clams circulate water over their enclosed gills by the beating of millions of cilia, and the octopus actively pumps water in and out of its gill cavity.

Fishes have developed more elaborate gills, forming richly branched fringes around holes (gill slits) in the tubular gut. Water is brought continuously into the mouth and passes through these slits. Some fish actively pump the water through the gill slits by muscular contraction of the mouth cavity, coupled with a valve-like action of flaps of tissue which prevent backflow. Other fast-swimming fish such as the mackerel rely on their movement through the water to keep up water flow; they cannot stay still or they will suffocate. The efficiency of the fish gill is further improved by counter-current circulation (Figure 9). As the blood becomes progessively oxygenated in its passage through gills, it becomes progressively harder to get more oxygen into it. By having the most oxygenated blood meet with the least deoxygenated water, gas exchange is maximized. Comparable counter-current devices are used by engineers in designing heat exchangers for power plants.

LUNGS

When animals adapted to life on land, one advantage was greatly increased oxygen supply. Gill-like devices in principle would be fully adequate to exchange the oxygen and CO_2. But in practice, gills suffer from a mechanical drawback: being loose and highly

8

AXOLOTL is an amphibian with primitive gills, which can be seen protruding from its neck. Such gills are essentially outpocketings of the body wall; fringe-like branches increase their surface area.

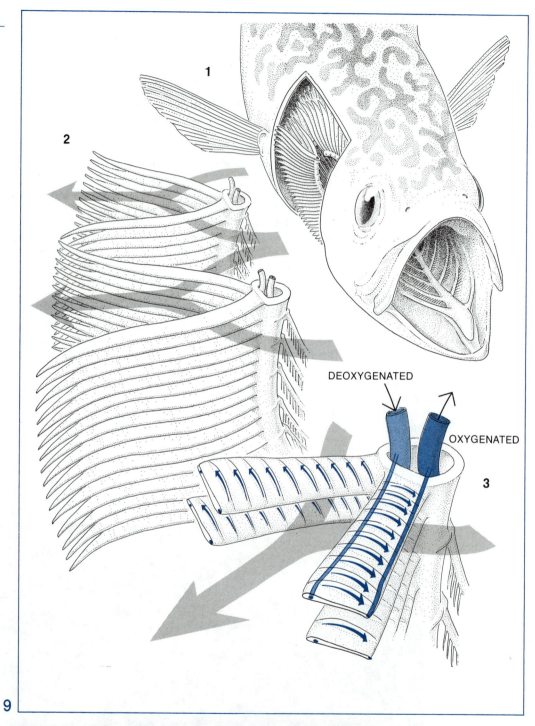

DEOXYGENATED

OXYGENATED

9

ACTION OF A GILL. Cutaway drawing of mackerel (1) shows position of gills. Broad arrows in (2) indicate flow of water past gill filaments. Diagram (3) shows circulation within filaments. Small colored arrows indicate that blood flows through transverse capillaries in opposite direction from incoming water. This countercurrent flow maximizes gas exchange.

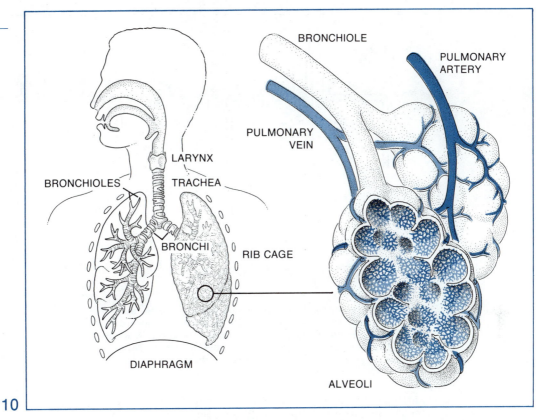

LARYNX

BRONCHIOLES   TRACHEA

BRONCHI

RIB CAGE

DIAPHRAGM

BRONCHIOLE

PULMONARY
ARTERY

PULMONARY
VEIN

ALVEOLI

10

HUMAN RESPIRATORY SYSTEM. Trachea divides into two bronchi, one for each lung. Bronchi in turn branch into bronchioles that terminate in clusters of tiny alveoli, shown in detail at right.

branched organs, they collapse when not supported by water, and their exchange surface then becomes too small. Moreover the exposure of a large and wet surface to the outside would lead to excessive water loss, one of the problems of life on land. One solution was to put the large wet surface inside as a LUNG, draw in pulses of air, and reduce one's water-loss to the amount that would saturate this relatively small pulse. The lungs of such different animals as reptiles, amphibians, mammals and birds all consist of elastic bags with linings only about 1μ thick, across which exchange occurs between blood (in a net of capillaries surrounding each chamber) and the air within the chamber.

In amphibians the skin is almost as important as the lung in gas exchange; in fact some terrestrial salamanders are lungless and use only the skin. Frogs respire through their skin and their buccal cavity as well as through their lungs. They fill their buccal cavity with air, then force their air into the lungs; this force-pump method is quite unlike the suction-pump used by most air-breathers. All reptiles have lungs, and only a few aquatic ones (such as soft-shelled turtles) also breathe through their skin.

The total gas exchange area in man may be 100 square meters, which is the surface area of a sphere 6 meters (19 feet) in diameter.

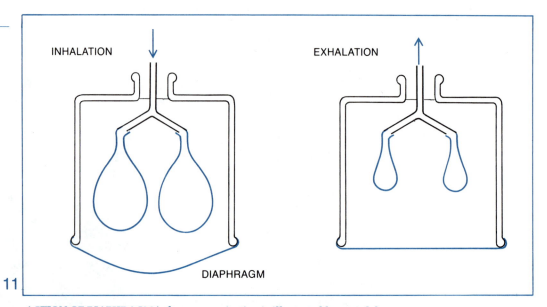

INHALATION

EXHALATION

DIAPHRAGM

11

ACTION OF DIAPHRAGM in human respiration is illustrated by a model. Colored bags represent lungs. Diaphragm is pulled downward during inhalation, expanding the lungs and sucking air into them. Relaxation of diaphragm leads to exhalation. Simplified model ignores the movements of the rib cage, which also help to fill and empty the lungs.

Mammals have elaborate lungs honeycombed with almost a billion small blind sacs called alveoli, of diameter 0.1 mm to 1 mm, in which the gas exchange occurs. The total exchange surface in man may be 100 square meters, which is the surface area of a sphere 6 meters in diameter. Air is brought in first by a sturdy trachea, dividing into two bronchi, then into about 20 bronchioles, and then into the alveoli (*Figure 10*). In this way, each ml of air becomes exposed to 300 cm² of exchange surface (in man), as compared with 20 sq. cm in

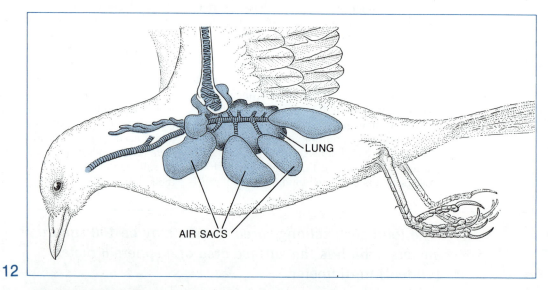

LUNG

AIR SACS

12

RESPIRATORY SYSTEM OF A BIRD includes both lungs and air sacs. The air sacs are connected to the trachea and branch throughout the body, extending even into the hollow bones.

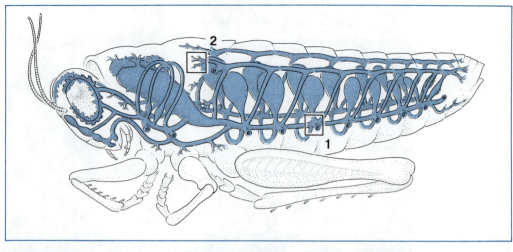

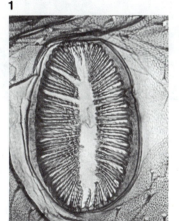

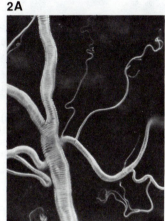

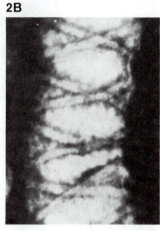

13

TRACHEAL SYSTEM OF INSECTS is rendered in color in the drawing above. Gases enter and leave through spiracles (1) on the thorax and abdomen. Spiracles are often fringed with bristles that filter the air and exclude water. Highly branched tracheae extend into virtually every part of the insect body; some are expanded into air sacs. Gas exchange occurs in thin terminal tracheoles (2) Photomicrographs at bottom depict details of spiracle (left), tracheae and tracheoles at low magnification (center), and a tracheole at a magnification of about 80,000× (right). Spiral thickenings prevent walls from collapsing.

the frog. A suction method is used to fill the lungs. The thoracic container (the pleural cavity) expands when the muscular diaphragm contracts, pulling the lungs down the cylinder formed by the rib cage. Other muscles lift the rib cage and increase its enclosed volume, also expanding the lungs. Expulsion of air is passive: the muscles relax and the elastic lungs return to their contracted volume (*Figure 11*).

Birds have even greater metabolic needs than mammals, because of the energy demands of flight, and they have developed a correspondingly more efficient respiratory ap-

paratus (*Figure 12*). In addition to lungs they have a set of communicating chambers called air sacs, connected to the tracheae and branching through the body, even into the bones. The sacs hold as much air as the lungs. Another difference from mammals is that instead of bag-like alveoli, gas exchange occurs in tubes, the air capillaries, only one hundredth of a mm wide. Yet the lung surface is relatively small; in a crow, 0.7 cm² per gram of body weight, which is $\frac{1}{10}$ that of man and $\frac{1}{100}$ that of the little brown bat. Conflicting theories exist about the way in which the air sacs help, but the general

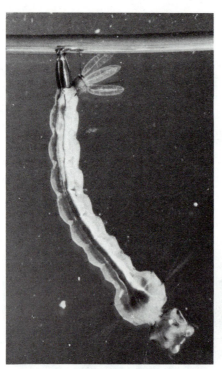

14

AQUATIC INSECTS include mosquito larva (left) and water beetle Dytiscus (right). Larva breathes by surfacing and thrusting tubelike spiracles (at
the end of its abdomen) into the atmosphere. Beetle surfaces briefly and
sucks an air bubble into space beneath its wings, where its spiracles open.

view is that they achieve a continuous air flow
through the air capillaries, in spite of the alternating cycle of inspiration and expiration.

INSECT TRACHEAE

The lungs of terrestrial vertebrates are a highly
efficient solution to the problem of gas exchange between air and blood. Another solution is to make direct use of the gaseous oxygen which terrestrial life made available. The
group that adopted this strategy was the terrestrial arthropods, including insects, spiders,
and some others. Their tissues have access to
air through a network of tubes called tracheae.
These open to the outside through holes
(spiracles) whose opening and closing are controlled by special valves, and which are protected by bristles against the entry of particles
or enemies (*Figure 13*). In the smallest insects
air simply diffuses into the trachea's increasingly fine branches, the finest of which is
called the tracheole, about 1 μ in diameter.
Where oxygen demand is high, as in muscles,
the number of tracheoles is greatly increased.
Gas exchange occurs at the tips of the

tracheoles, which are partially filled with
fluid. When the insect is active the fluid level
falls, exposing greater surface for direct exchange with tracheolar air.

In larger insects the diffusion path of the
narrow tracheae would be too long for
adequate gas flow, so the insect has taken part
of the outside atmosphere inside him in large
air-filled sacs. Air movement in and out of
these cavities is promoted by pumping
movements, especially in the abdomen. In one
locust, air flows in through anterior spiracles
and out through posterior ones at a rate
amounting to $\frac{1}{5}$ of the insect's body volume
each minute.

Some insects evolved into aquatic forms for
all or for a part of their lives. This return to the
water did not lead to a return to the conventional watery gill. Instead, special adaptations
of the tracheal system evolved. Some aquatic
insects return periodically to the surface and
thrust spiracles (commonly posterior ones)
into the atmosphere; the larvae of common
mosquitoes, such as *Aedes*, are a familiar
example. Others take the air down with them

periodically; the water beetle *Dytiscus* (Figure 14) holds a bubble under his wing-covers, and the spiracles open into the bubble.

Lungs, gills and tracheae are only part of the story of gas exchange in animals. The full story of how oxygen gets to the tissues (and CO_2 gets out) will be completed in the next chapter, which discusses the important role of blood in transporting materials to and from the animal cell.

READINGS

M.S. GORDON, G.A. BARTHOLOMEW, A.D. GRINNELL, C.B. JORGENSON AND F.N. WHITE, *Animal Function: Principles and Adaptations*, 2nd Edition, New York, Macmillan, 1971. Chapter 5 of this book (which is essentially a physiological text with a strong comparative slant) presents a fine account of respiratory mechanisms in vertebrates.

P.S. NOBEL, *Introduction to Biophysical Plant Physiology*, San Francisco, Freeman, 1974. A fairly rigorous treatment of water relationships and other topics.

G.R. NOGGLE AND G.J. FRITZ, *Plant Physiology*, Englewood Cliffs, N.J., Prentice-Hall, 1976. A new and up-to-date plant physiology text.

P.M. RAY, *The Living Plant*, 2nd Edition, New York, Holt, Rinehart and Winston, 1972. A short and excellent introduction to functional botany with a first rate treatment of water relationships, gas exchange, and stomatal physiology.

K. SCHMIDT-NIELSEN, *Animal Physiology*, 3rd Edition, Englewood Cliffs, N.J., Prentice-Hall, 1970. A brief and extremely readable book, very comparative in its approach, and excellent at explaining the adaptive significance of the various features of gas exchange in animals.

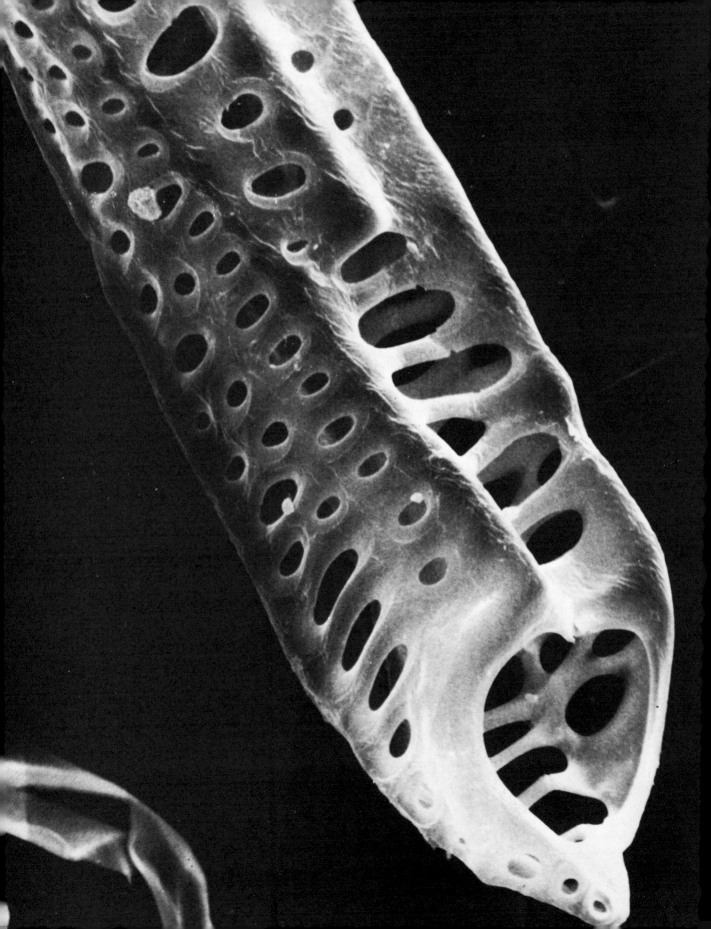

10

transport systems

It is therefore necessary to conclude that the blood in the animals is impelled in a circle, and is in a state of ceaseless movement; that this is the act or function of the heart, which it performs by means of its pulse.

WILLIAM HARVEY (1628)

All except the smallest plants and animals need a specialized internal system to deliver nutrients to every cell and to remove waste products. The design of the system depends on the size and activity of the organism. In general large organisms require more extensive systems than small ones, and active organisms require more rapid systems than inactive ones.

The differences between the lifestyles of plants and animals are reflected in the design of their transport systems. Large plants must move materials over long distances, but because they are sedentary it is not important that their biological freight be moved rapidly. Sugar synthesized in the leaves must be exported to developing flowers, fruits, roots and young expanding buds. The carbohydrates stored in bulbs, tubers and other organs must be mobilized on occasion and transported to growing regions, particularly after a period of winter dormancy. Hormones produced in one part of the plant must be moved to other regions. Ions and water entering the roots from the soil must be distributed throughout the aerial parts of the plant.

A physically active animal must have the machinery to move large quantities of material very quickly, first because working muscles consume enormous amounts of nutrients and generate corresponding quantities of toxic wastes, and second because the tissues and organs of animals do not synthesize their own food as many plant tissues do. All animal nutrients must be imported and then distributed, which imposes additional demands on the transport system.

These design requirements have favored the evolution in animals of an elaborate plumbing system complete with pumps, valves and feedback controls. And because all materials must be delivered to the cell either suspended or dissolved in water, animals have developed a fluid transport vehicle: the blood. In humans and most other higher animals blood transports not only nutrients and wastes, but also the oxygen-carrying pigment hemoglobin, which gives blood its characteristic red color. In some animals blood serves only as a food transport medium; it is unpigmented and plays no role in oxygenation.

TRANSPORT IN PLANTS

Plants have two major pathways for long-distance transport: the xylem carries water and ions and the phloem carries sugar and other carbohydrates. The force that drives water and ions through the xylem is generated within the leaf. Heated by the sun, water evaporates from the moist surfaces of the parenchymal cells and exits from the leaf through the stomata (*Chapter 9*). This evaporative water loss from leaves is called TRANSPIRATION.

When a leaf cell loses water it shrinks. Water from adjacent xylem elements — tracheids and vessel elements — enters the cell by the process of osmosis (*Chapter 1*), because the concentration of water within the cell is lower than the concentration in the xylem. The cytoplasm of the cell then swells, exerting TURGOR PRESSURE against the inside of the cell wall. Water continues to enter until the turgor

XYLEM VESSEL of a geranium plant is depicted in the photograph on opposite page. The xylem tissue has been treated to dissolve the intercellular cement, and the separated vessel was then photographed with a scanning electron microscope. Magnification, about 2300×.

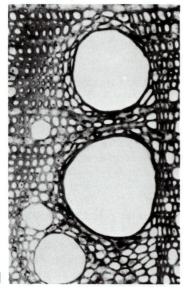
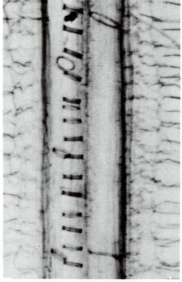

1

XYLEM ELEMENTS in plants are shown in cross-section (left) and longitudinal section (right). Photomicrograph at left depicts three large xylem vessels in oak wood. Picture at right depicts xylem strands in the stem of a squash (Cucurbita). Rings in strand at left are thickenings of xylem elements that matured while stem was still growing. Originally much closer together, they were separated as stem elongated. Strand at right consists of vessel elements that formed at the end of elongation. The two rings visible in that strand are not thickenings but the vestiges of end walls that became completely perforated on maturation of the elements, providing an open pipe for movement of water.

and 2). The reader will recall that these cells are dead at maturity, consisting only of the tubelike walls laid down when the cells were alive. A continuous ribbon of water stretches through the xylem from leaf to root. Because of the strong attractive forces between water molecules, and between water molecules and the wall material, water removed from one xylem cell is simply replaced by water from the next one. Transpirational loss of water from the leaf thus moves water upward from the underground parts of the plant.

It is incorrect to think of this process as suction, which is simply removal of air from a system. Atmospheric pressure can only raise a column of water about 32 feet; a modest redwood tree raises water at least ten times as high. A far better explanation is that of negative pressure. Consider a tree 96 feet tall. If there is no resistance to the movement of water, the equivalent of three atmospheres of pressure is needed to raise water to its crown. Such forces can readily be developed by evaporation. Transpiration can easily develop the necessary three atmospheres of negative pres-

pressure (which tends to keep water out) equals the osmotic potential generated by the dissolved solutes within the cell (which tends to pull water in). The rate of uptake is limited by the elasticity of the cell wall. A thin and highly elastic wall can be stretched far more than a thick inelastic one. A steady state is reached when the turgor pressure equals the resistance of the wall to further stretching.

If the outside temperature and humidity are moderate, the system is in dynamic equilibrium: transpirational loss is balanced by osmotic uptake. If the temperature is high and humidity low, the rate of water loss from the leaf may exceed the rate of uptake. The cell contents shrink, the turgor pressure falls to zero, and the cells collapse — the familiar phenomenon of wilting. Shrinkage beyond a certain point separates the cell membrane from the wall and breaks the delicate intercellular connections that extend through the wall. If this happens the wilted tissue cannot recover.

Xylem elements have massive walls that prevent them from collapsing even when all the water is withdrawn from them (*Figures 1*

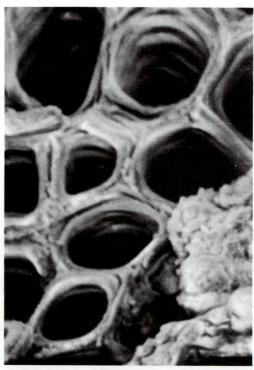

2

TRACHEIDS appear as ringed tubes in cross-section of a leaf from the Venus flytrap. Made with scanning electron microscope, picture is reproduced at a magnification of about 1700×.

*Atmospheric pressure is only sufficient to raise
a column of water about 32 feet. A modest redwood tree
raises water at least ten times as high.*

sure to do the job, and the powerful cohesive forces between molecules prevent the column of water from breaking under tension.

A water column passing through many xylem cells encounters considerable resistance to flow. This resistance depends on the number and size of the cells, the nature of the wall — or lack of it — between them, and the viscosity of water. The xylem system of a pine tree, consisting solely of tracheids, is an inefficient one. In passing from one tracheid to the next, water must pass through a double layer of walls. In the most advanced plants the vessels are far better: the vessel elements of a maple tree, for example, may be close to a millimeter in diameter, and at maturity their end walls disappear. A column of vessel elements may form a continuous pipe from root to

leaf. Nevertheless, the actual tension needed to raise water to the top of the 96-foot tree may be closer to six atmospheres than three.

Water must be moved from the soil across the root tissue to the root xylem (*Figure 3*). Careful inspection of root tissues reveals two possible pathways. First, water could enter directly into root hair cells following an osmotic potential gradient, because solutes in the soil are usually far less concentrated than solutes in the cell. So long as water is being removed at the terminus — the leaf — movement across the root from soil to xylem will continue. A second possible pathway is through the spaces between the root cells. Water can move between cells as well as through cell walls to reach the xylem. Water could cross from soil to xylem simply by following an osmotic poten-

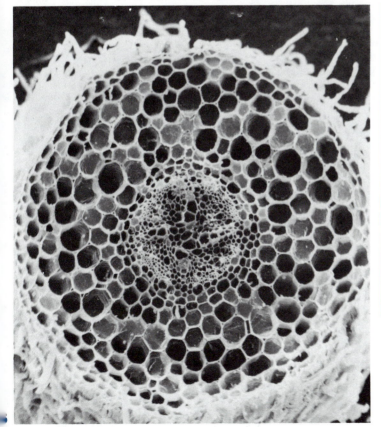

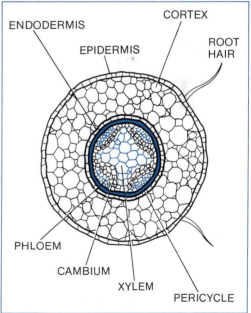

CROSS-SECTION OF BEAN ROOT shows two pathways for the entry of water. Water can enter through root hairs and follow osmotic gradient into the xylem, or it can move between the cells, through the endodermis, and into the xylem. Small diagram at right identifies the principal structures in the accompanying scanning electron micrograph, made at a magnification of 65×.

tial gradient. If the root is placed in distilled water, the water enters the xylem and can be raised to substantial heights by simple osmosis. In fact ROOT PRESSURES of several atmospheres can be developed. But the measured values are not nearly enough to move water to the heights of tall trees. Even more important, the rate of the process fails dismally to account for the movement of large volumes of water.

The root uses the energy of ATP to accumulate ions by active transport. Salts are passed from cell to cell and ultimately into the xylem. Once there they are passively carried to the upper portions of the plants, where living cells take them in by active transport.

In brief, roots show a substantial water deficit, a deficit of dual origin. First, their own solutes give them an osmotic potential which leads to water uptake. Second and probably more important, transpirational loss of water from the leaves creates a water deficit which is physically transmitted to the roots by tension on the columns of water in the xylem. Together these two forces can move water and ions to the upper branches of trees 30 stories tall.

TRANSPORT OF CARBOHYDRATES

Unlike xylem tissue, which is simply plumbing, phloem contains live active cells. Individual phloem cells can and do play a role in the movement of carbohydrates through this tissue.

It is worth considering the structure of a mature and functional phloem cell. As with xylem, the cell types in the primitive vascular plants and gymnosperms (such as conifers) are somewhat different from those in the angiosperms (flowering plants), although the basic features are similar. For simplicity this discussion is restricted to the flowering plants. The single functional phloem cell is called a SIEVE TUBE ELEMENT. A vertical column of such elements, joined by their end walls, makes a SIEVE TUBE. The end walls, called SIEVE PLATES, are specially modified with thin areas or pores leading from one sieve element to the next (Figure 4). The mature sieve element is remarkably uninteresting under the electron microscope. There is no nucleus, rarely a secondary wall, and only a very thin region of cytoplasm with at most a few mitochondria. The interior of the cell is filled with a watery solution containing a slimy protein material.

Associated with the sieve tube element is an adjacent COMPANION CELL. It has a normal nucleus and is packed with mitochondria, endoplasmic reticulum, and various other organelles. The best guess is that the companion cell provides both the energy and any needed nuclear information for itself and for the adjacent sieve tube element.

Most plant physiologists believe that the sugars produced by photosynthesis eventually reach the phloem as glucose phosphates. They enter the companion cells where they are converted to unphosphorylated sucrose and actively secreted into the sieve tubes. As the sugar concentration increases in the sieve tubes, their osmotic potential accordingly also increases, reaching values far higher than those of surrounding cells. Water, available either in the adjacent xylem or nearby parenchyma cells, therefore enters the sieve tubes, developing substantial pressure. Wall resistance prevents excessive stretching, so the sugar solution is forced out through the sieve plate into the next element. The direction of flow is determined by relative rates of sugar production and consumption in different parts of the plant. The driving force is a metabolically generated difference in osmotic potential. Movement is from sugar source to sugar sink — normally from leaves to roots, but also frequently from leaves to storage organs, developing flowers or fruits. A rapidly growing

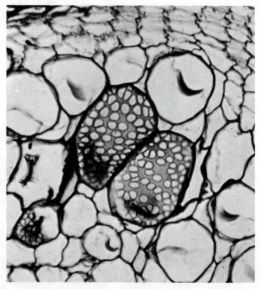

4

CROSS-SECTION OF PHLOEM of squash plant shows two sieve plates (center). The surrounding sieve tube cells were sectioned between end walls and appear clear except for distorted cytoplasm.

pumpkin gains over 40 grams of dry matter per day.

Xylem and phloem provide transport pathways for virtually everything in the plant that has to be moved over a long distance. Amino acids can be readily moved along in the phloem, as can hormones such as gibberellins (Chapter 12). Nitrogen taken up by the roots as nitrate is reduced to ammonium and bound to glutamic or aspartic acids to form glutamine or asparagine, both of which may find their way to the xylem for transport to upper parts of the plant. Xylem frequently contains substantial amounts of cytokinins, suggesting that this class of hormone can move upward with the transpiration stream. The only serious exception is found with the hormone auxin. It may move in the phloem, but in many plants it moves through other living cells too.

TRANSPORT IN ANIMALS

The metabolism of even the most rapidly growing plants proceeds at a leisurely pace compared with that of an active animal. In small invertebrates dissolved gases, nutrients and waste products are transported throughout the animal by simple diffusion. The previous chapter described how the arthropods developed a rather specialized but still essentially diffusional gas transport device (the tracheolar system), while maintaining a simpler arrangement for moving nutrients and waste products. In other invertebrates and all vertebrates, progressively more complex systems are necessary for the efficient distribution of blood to all cells. Blood is a multipurpose vehicle that carries water, oxygen, CO_2, nutrients and waste products.

As in plants, the complexity of the transport system of an animal depends on its size. When transport distances grow to a millimeter or more, free diffusion becomes inadequate, and special transport mechanisms have evolved. Exceptions include jellyfishes (up to six feet) and sea anemones (up to several inches). But 90 percent of a jellyfish is made up of a dead jelly (mesogloea) covered with a thin film of living matter, and the anemone's living tissues are thin sheets which make up a bag filled with seawater. With a few such exceptions, most metazoa have systems which serve to move the blood around the body.

The simplest transport system utilizes only the muscles of the body wall to swish blood around the organs, as in nematodes and sea-cucumbers. The next level of complexity is found in such arthropods as insects, spiders and crustaceans, in which a tubular heart is suspended in the blood; the heart fills and empties itself rhythmically, stirring the fluid in which it lies (Figure 5). Vessels and valves are sometimes associated with the heart, giving some direction and therefore greater efficiency to the stirring process.

These open systems contrast with the closed systems of more elaborate organisms, in which the blood is confined within branching tubes, through which it is pumped by one or more hearts. All vertebrates and some invertebrates (such as annelid worms) have such systems. The pumping system of some vertebrates has refinements to create greater pressure, to force blood through elaborate fine-bored vessels (capillaries), and to divert deoxygenated blood to the lungs or gills while the most richly oxygenated blood is channelled to the muscles, brain, and other oxygen-requiring organs.

THE HEART

The pumping system (heart) always has at least one thin-walled collecting bag (auricle or atrium) connected to at least one thickly-muscled pumping chamber (ventricle), and a set of valves to channel the blood in the right directions.

A sensible pattern would seem to be to collect deoxygenated blood, pump it into the lungs or gills, and then distribute it to the other tissues. The drawback is that the blood has to go through two capillary beds in series; by the time it has emerged from the respiratory bed its pressure has fallen substantially. Nevertheless fishes do precisely this, and have but one auricle and one ventricle. Cephalopod mollusks (squid, octopus) overcome the pressure-drop problem with extra booster hearts which collect deoxygenated blood, force it through the gills, and thence to the main heart which pumps it through the other tissues.

The human heart at rest pumps 284 liters of blood per hour, and in a normal lifetime it pumps 18 million barrels.

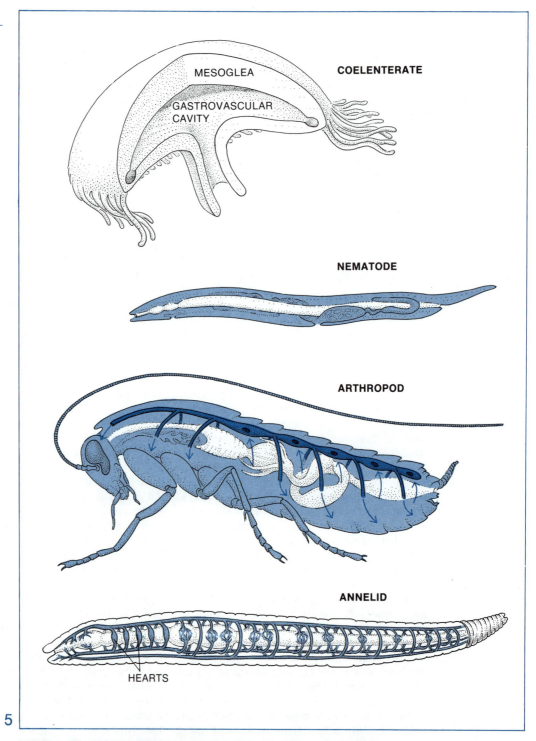

COELENTERATE

MESOGLEA

GASTROVASCULAR
CAVITY

NEMATODE

ARTHROPOD

ANNELID

HEARTS

5

ANIMAL TRANSPORT SYSTEMS. Jellyfish and related organisms (top)
have no circulatory system and rely largely on passive diffusion.
Nematode worms rely on muscles of body wall to swish around the blood
(color) that bathes their organs. Arthropods also have organs bathed in
blood, but blood is circulated by tubular heart (dark color) suspended in it.
Annelid worms have closed circulatory system typical of advanced organisms. Blood is driven through branched vessels by one or more hearts.

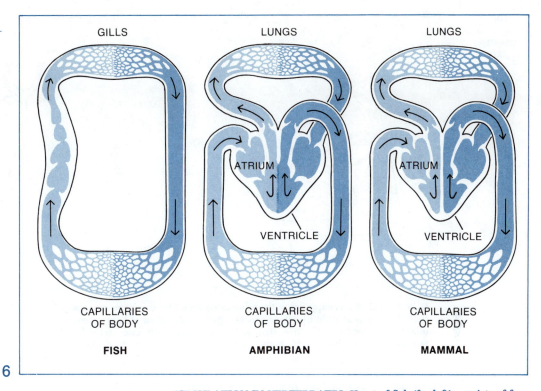

GILLS   LUNGS   LUNGS

ATRIUM   ATRIUM

VENTRICLE   VENTRICLE

CAPILLARIES OF BODY   CAPILLARIES OF BODY   CAPILLARIES OF BODY

FISH   **AMPHIBIAN**   **MAMMAL**

6

CIRCULATION IN VERTEBRATES. Heart of fish (far left) consists of four chambers in a row; respiratory circulation and systemic circulation are connected in series. Heart of amphibian has only one ventricle, but oxygenated arterial blood (dark color) and deoxygenated venous blood (light color) are partially prevented from mixing. Respiratory and systemic pathways of amphibians are connected in parallel, as in higher organisms. In mammals, birds and some reptiles, mixing of arterial and venous blood is prevented by partition that divides ventricles into left and right chambers. In effect, partition creates two hearts, one for pumping blood to lungs, the other for pumping blood to rest of body.

In amphibia the lungs are parallel with the other tissues rather than in sequence (*Figure 6*). The full force of the heart is available for these tissues, and there is no need for a booster heart. The oxygenated blood is collected in a second auricle, but there is only a single ventricle. This arrangement is advantageous only if the oxygenated blood from one auricle can be kept unmixed with that from the other. This miracle of nonmixing is achieved to a good extent by the frog although the mechanism is far from clear. Reptiles also have three-chambered hearts, but in some (crocodiles) the separation of oxygenated and deoxygenated blood in the ventricle is enhanced by a thin septum that divides the ventricle in two.

In mammals and birds the ventricles are completely separate. All the oxygenated blood is reliably distributed to the systemic circulation by the left heart (the left ventricle plus auricle), and all that received from the capillaries of the systemic circulation is reliably delivered to the lungs via the so-called pulmonary circulation by the right heart, as shown in Figure 6.

The human heart at rest pumps 284 liters of blood per hour, and in normal activity it pumps about 18 million barrels in a lifetime. Its rhythmic relaxation (diastole) sucks blood into the two atria. The blood passes from auricle to ventricle through complicated valves called the bicuspid or mitral (because it is shaped like a bishop's miter) on the left and the tricuspid on the right. The valves prevent back flow. The rhythmic contraction (systole) of the ventricles drives the blood into the arteries via the aorta (in the case of systemic circulation) as shown in Figure 7. The pressure in the arteries peaks at systole and falls during diastole, so that in reporting arterial blood

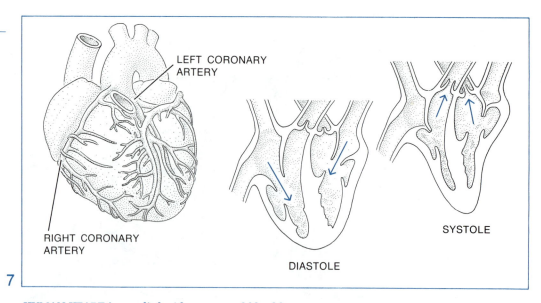

LEFT CORONARY
ARTERY

RIGHT CORONARY
ARTERY

DIASTOLE

SYSTOLE

7

HUMAN HEART is supplied with oxygenated blood by coronary arteries (left). Diagrams at right illustrate cardiac cycle. Relaxation of heart (diastole) sucks blood into the two atria at top of heart. Contraction (systole) pumps blood to lungs and body. Leaflike valves prevent back flow.

pressure physicians use two numbers, usually expressed (like the barometric pressure on a weather report) in terms of the height in mm of a mercury column which each pressure would support. A normal blood pressure in man is 120/80, in the guinea pig 77/47, and in catfish only 30/23.

The blood supply needed by the heart to support these activities is provided by the coronary artery, whose branches can be seen on the outside of the heart. Sometimes this artery becomes blocked, perhaps by a blood clot (thrombus) which lodges in it, leading to a coronary thrombosis, signalled by a ferocious pain called angina pectoris (choking in the chest). The artery may also be progressively narrowed (occluded) by development of hardened (sclerotic) tissues in its lumen, a fairly common condition known as atherosclerosis.

The heart contains a special kind of muscle, and the basic rhythm of its beat is generated by an internal pacemaker, a tiny knot of tissue called the sino-atrial node, situated in the rear wall of the right atrium. The function of heart muscle and the role of the central nervous system in speeding up and slowing down the heart rate are discussed in Chapter 14.

BLOOD VESSELS

After the blood leaves the heart and passes into the elastic aorta and its tree-like branches, it is pumped through muscular arteries, then into microscopic tubes called arterioles, and finally into the capillaries (*Figure 8*). These tiny vessels, less than 0.1 mm in diameter, permeate every tissue of the body. They bring oxygen and nutrients to the cells, and carry away carbon dioxide and breakdown products. Because of their small diameter, the surface area (across which exchange occurs) is huge: for example, each cubic cm of dog muscle has 494 sq. cm of capillary surface.

The capillary wall consists of flat cells fitted together like paving stones. The web of capillaries is not passive: it is penetrated by muscle cells, which can change the amount of blood flowing in one capillary bed as compared with another. The capillary vessel controls the flow in part by constricting or dilating the diameter of the capillary vessel itself, and in part by constricting or dilating specialized ring-shaped muscles (precapillary sphincters) located where the blood enters the capillaries. Similar control is exercised by the arteries and arterioles, but in this case the muscles are controlled by nerves that lead directly from the brain. The muscles of the capillary bed, by contrast, respond to various hormones. The response varies in different tissues; thus epinephrine constricts capillaries of the skin (the cause of turning white with anger or fright) but dilates those of the muscles, increasing the ability of the frightened animal to run or the angry animal to fight. Other factors of importance include corticosteroids (steroids from the adrenal cortex), whose absence leads to a loss of muscle tone and the collapse of circulation; and histamine, a substance released in damaged tissues, which dilates the capillaries and hence increases the blood supply available for repair processes.

After passage through capillaries, the blood collects in veins, thin walled elastic vessels that serve both as conduits back to the heart and as a blood reservoir. One-way movement through the system is aided by pocket-like

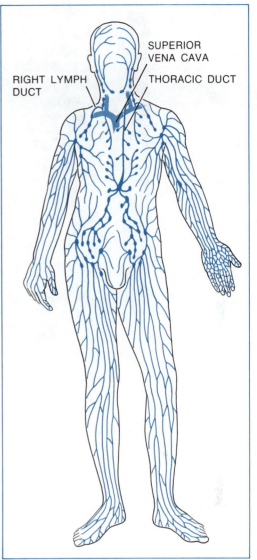

LYMPHATIC SYSTEM collects lymphatic fluid from the tissue spaces. Lymph capillaries and ducts (dark color) drain into the thoracic duct, which empties into the bloodstream via a junction with the vena cava (light color). Major lymph nodes are shown as swellings along capillaries.

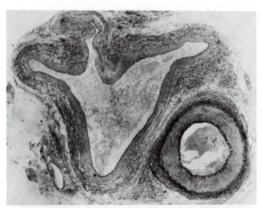

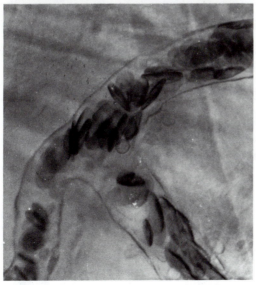

VERTEBRATE BLOOD VESSELS. Photomicrograph at top shows cross-section of a large vein and a small artery. Artery (right) can be distinguished by its firm muscular wall. Below, red blood cells squeeze through capillaries. Cross-section view of capillary appears in Chapter 11.

valves set in the veins, and by the movements of near-by skeletal muscles. The venous blood returns to the auricle and the cycle repeats itself. The venous return is reduced if there is little movement of the skeletal muscles, and blood pressure falls accordingly. This is the cause of fainting if one stands immobile for a prolonged time, and also the cause of the dizziness one feels if one springs suddenly out of bed.

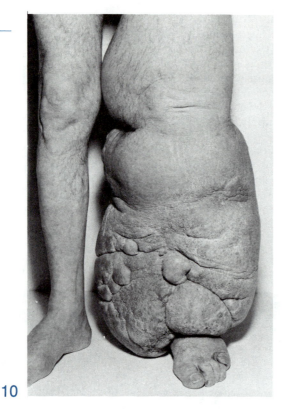

10

ELEPHANTIASIS. Lymph drainage is blocked and fluid accumulates, causing grotesque swelling of the left leg of this boy.

THE LYMPHATIC SYSTEM

The cells of the bodies of the higher vertebrates are nourished by blood flowing in capillaries. But the cells lie in a space outside the capillaries. The water and solutes that filter through the capillaries into the extracellular space are called LYMPH; they are collected by yet another set of vessels: the lymphatic system. Like the blood system, the lymphatic system has fine capillaries which collect the lymph and drain it into progressively larger vessels (lymphatic ducts) which ultimately empty into large veins in the lower part of the neck. The lymphatic vessels are fragile and hard to trace; the larger ones are equipped with valves like those of veins (Figure 9).

Along the larger lymphatic ducts are lymph nodes. These nodes (which become noticeable when they swell in the course of infection) have a triple role. They act as filters, removing particles (such as the larger kinds of bacterial cells) so that such matter is not dumped into the blood; they contain white cells which can attack and digest foreign cells; and they are secondary production sites for lymphocytes. Lymphocytes are initially made in the thymus gland, but spread to the lymph nodes, where they multiply and concentrate. Produced at 10^{10} per day in man, lymphocytes are the cells responsible for making antibodies (Chapter 11).

The filtration of blood to form lymph removes all of the red cells (and most of the others), as well as about half of the blood proteins. The concentration of small molecules (glucose, sodium, CO_2 and the like) is identical in both lymph and blood. Immersed in each cubic millimeter of lymph are about 10,000 lymphocytes and a few hundred white blood cells (leucocytes).

Local failure of the lymphatic system has severe consequences. An especially tragic one occurs in the disease which is called elephantiasis because affected body parts become swollen to elephant-like size. It can be caused by infection of the lymphatics, and also by their blockage by parasitic worms. Consequently the lymph is not drained away, but accumulates to give a great watery swelling, or lymphedema (Figure 10).

BLOOD PIGMENTS

So far this discussion has focused on the gross problems of transportation — the movement of blood solutions from place to place within the body. But the solubility of gases in the blood is also important. Given two systems of equal ability to move blood, if one of them can be made to dissolve twice as much oxygen per liter of blood, and to give it up on demand, it will be twice as effective. And if the tissues being supplied can hold on to oxygen, their ability to face emergencies will be greatly strengthened. These transport and storage functions are achieved by a few metal-containing pigments. In vertebrates and echinoderms red hemoglobin is packaged in special blood cells called erythrocytes (Figure

The job of the transport system begins and ends at the gateway to the cell: the cell membrane.

11). Many worms and crustaceans simply dissolve hemoglobin in their blood. Many mollusks and arachnids have a copper-containing pigment called hemerythrin. All these pigments behave in comparable ways.

Hemoglobin absorbs oxygen in accordance with an unexpected S-shaped curve (*Figure 12*). Examination of that curve will show that as one progressively increases the oxygen concentration, the ability to take up more oxygen is (for a while) steadily increased. In other words, the first molecules absorbed facilitate the absorption of subsequent molecules, and must therefore have modified the molecule in some favorable way. The steepness of the curve means that when working in the middle part of the curve, a modest increase of oxygen tension (movement to the right on the x axis) induces a massive increase in oxygen storage by the pigment. A subsequent modest decrease in oxygen tension induces a dramatic release of oxygen. This behavior is well matched to the body's needs. The deoxygenated blood can be loaded in the lungs with an unexpectedly large quantity of oxygen, and can subsequently be effectively released in the tissues, even though the oxygen pressure in tissues is not much lower than that in the lungs.

Oxygen is taken up automatically by the deoxygenated blood in the lungs, and is then released "on demand" in the tissues. The process is enhanced in two ways. One is the "Bohr effect:" carbon dioxide, by changing the pH of the blood, reduces the affinity of hemoglobin for oxygen. This increases the release of oxygen when the blood is in the capillaries, because carbon dioxide levels rise sharply

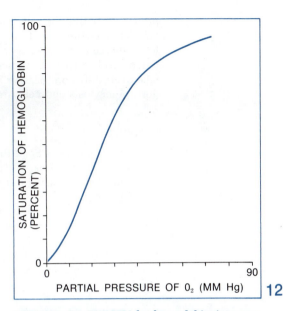

BINDING OF OXYGEN by hemoglobin increases steeply as oxygen pressure rises. This sensitivity to oxygen pressure allows hemoglobin to absorb great quantities of oxygen in the lungs and to release it in tissue, where oxygen pressure is lower.

there; and increases the uptake of oxygen in the lung, where the carbon dioxide is dumped. The second effect is caused by oxygen-combining pigments in the tissues themselves, which increase the tissues' affinity for oxygen and provide a way to store it. Especially well-known is the red myoglobin of muscle, a pigment chemically similar to hemoglobin, and also capable of storing oxygen. Very active muscles are commonly red because they contain large quantities of myoglobin. The whiteness of the meat in the breasts of domestic turkeys and chickens testifies to the low oxygen needs and hence the low myoglobin content of the flight muscles of these nonflying birds.

Carbon dioxide, the end product of oxidation, is transported without special pigments. The gas is simply dissolved in the blood plasma. The high CO_2-carrying capacity of the plasma is due to the formation of bicarbonates of blood cations, primarily sodium and potassium. In vertebrates, the rather slow hydration of carbon dioxide to form bicarbonate is speeded by an enzyme, carbonic anhydrase, present in red blood cells.

The job of the transport system begins and ends at the gateway to the cell: the cell membrane. To be of any use to an organism a nutrient must cross the membrane and move to the

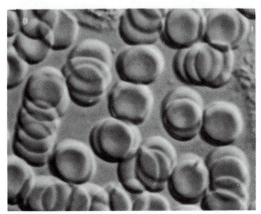

11

RED BLOOD CELLS (erythrocytes) carry hemoglobin in vertebrates. In many other animals oxygen-carrying pigments are merely dissolved in plasma.

proper site within the cell. Waste products and cellular secretions such as hormones must leave the cell the same way. Here the short-range mechanisms of intracellular transport take over: diffusion, passive and active transport, pinocytosis and cell streaming (*Chapter* 1). This chapter has emphasized the mechanisms of long-distance transport, but the reader should bear in mind that for short-distance transport the individual cells of multicellular plants and animals rely on the same mechanisms as unicellular organisms.

READINGS

K. Esau, *Anatomy of Seed Plants*, New York, John Wiley & Sons, 1960. An excellent treatment of the structural aspects of xylem and phloem.

M.S. Gordon, G.A. Bartholomew, A.D. Grinnell, C.B. Jorgenson and F.N. White, *Animal Function: Principles and Adaptations*, 2nd Edition, New York, Macmillan, 1971. Contains an excellent treatment of animal transport.

G.J. Peel, *Transport of Nutrients in Plants*, London, Butterworths, 1974. A modern approach to the transport systems of plants.

P.M. Ray, *The Living Plant*, 2nd Edition, New York, Holt, Rinehart and Winston, 1972. Contains a brief but rigorous discussion of long distance transport in plants.

M.H. Zimmerman and C.L. Brown, *Trees: Structure and Function*, New York, Springer-Verlag, 1971. Excellent chapters on transport in phloem and xylem.

NOTES

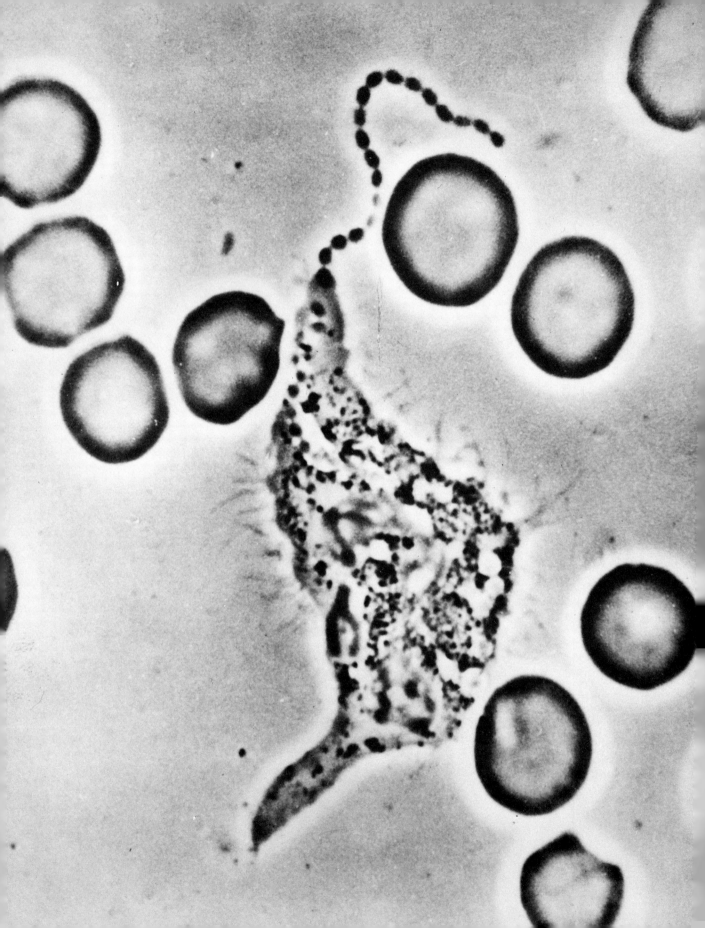

11

the internal environment

All vital mechanisms, varied as they are, have only one object, that of preserving constant the conditions of life in the internal environment.

CLAUDE BERNARD

The cells of multicellular organisms are delicate and highly specialized machines which, like high-performance engines, function at their best only under a rather narrow range of conditions. Most of these conditions are dictated by the physics and chemistry of aqueous solutions. Even the hardiest cells, for example, cannot function if the water within them is evaporated or frozen, or if their surroundings are poisonously acidic or alkaline. Cells are especially sensitive to variations in temperature. As a general rule the rate of a chemical reaction doubles for each 10°C rise in temperature. An active cell carries out thousands of reactions per second, and it is easy to see how even a small fluctuation in temperature could disrupt a network of interlocking reactions.

In the course of evolution multicellular animals have devised a portable and constant internal environment that insulates most of their cells from the rigors of the outside world. Claude Bernard, the father of physiology, introduced the concept of the internal environment more than a century ago, pointing out that most of the cells of the body live not in contact with the variable external environment, but bathed in body fluids that comprise what he called the *milieu intérieur*. He observed that a stable internal environment is a prerequisite of an independent existence. In 1930 this concept was extended by the Harvard physiologist Walter Cannon to include HOMEOSTASIS: the tendency to maintain constancy. To some extent the notion of homeostasis may have been somewhat oversold because Cannon worked with vertebrates, particularly mammals — organisms that show the highest degree of internal constancy. Other organisms, particularly plants and cold-blooded animals, have to tolerate a good bit of internal inconstancy.

Organisms maintain the physical and chemical constancy of their internal environment by four main types of mechanism: buffering, storage, feedback control and excretion. In warm-blooded animals — mammals and birds — these mechanisms provide extraordinarily stable conditions in which body cells can function at optimum levels. The virtually weatherproof internal environment of warm-blooded animals is an enormous adaptive advantage because it enables them to survive in a vast range of habitats and at a high level of physiological efficiency.

Most organisms on Earth have some sensing and timekeeping mechanism that synchronizes their internal activities with the cyclic changes outside, particularly the daily cycle of light and darkness and the seasonal cycle of changes in climate. The workings of these biological clocks are one of the most tantalizing mysteries of physiology.

Finally, all animals have some mechanism to preserve the biological constancy of their internal environment — that is, to fight off invasion by foreign organisms. Simple invertebrates have crude immune systems. Farther up the phylogenetic tree the systems become more elaborate. Physicians learned to apply the principles of immunology long before they had any basic understanding of how the immune system works. The pioneer was Edward Jenner, an English doctor born

NEUTROPHIL ENGULFS BACTERIUM in the photomicrograph of human blood on opposite page. Part of the bacterium (a chainlike streptococcus) is visible at top center. Dark disks are red blood cells.

in 1749, who invented vaccination. He noted that milkmaids who got cowpox from their contact with cows never contracted smallpox. By 1803 vaccinations with cowpox had reduced the annual smallpox deaths in London from 2018 to 622. Today the disease has been almost eliminated throughout the world. Study of the immune system is currently one of the hottest areas of research, largely because of its possible role in suppressing cancer — but that is getting a bit ahead of the story. The logical place to begin the study of the internal environment is with the four principal stabilizing mechanisms.

BUFFERING

The term BUFFER has long been used by chemists to denote a solution that is in some way resistant to change, usually a change in pH. If a drop of concentrated hydrochloric acid (HCl) is added to a beaker of water, the liquid becomes very acidic, changing perhaps from pH 7 to pH 2.5. If the beaker contained a buffer, such as a dilute mixture of sodium acetate and acetic acid, the original pH would be about 4.7. Adding a drop of concentrated HCl would lower the pH very little, perhaps to 4.65; adding a drop of concentrated NaOH would raise the pH only to about 4.75. The solution is said to be buffered against a change of pH, which of course means a change in the concentration of hydrogen ions [H^+]. All buffers are mixtures of two substances: one that can donate hydrogen ions (such as acetic acid, abbreviated HAc), and one that can accept hydrogen ions (such as acetate ions, Ac^-). The ratio of the acceptor to the donor determines the pH. If a little HCl is added, some of the acetate will react with it:

$$H^+Cl^- + Ac^- \rightarrow HAc + Cl^-$$

In the ratio [Ac^-]/[HAc] the numerator will become smaller and the denominator larger, but so long as both are much larger than the amount of acid or base added, the ratio will not change much.

Many homeostatic mechanisms, both within cells and in extracellular fluids, involve buffering. Living systems contain many different acceptor-donor pairs, including carbonates, phosphates, and groups that are present in proteins, such as carboxylate and amino groups. Blood contains a particularly subtle and complicated buffering system involving oxygen, hemoglobin and carbonate. Buffering systems function simply and automatically, preventing the lethal changes in pH that would occur within tissues whenever substances such as lactic and carbonic acids are produced faster than they can be carried away.

STORAGE

Excess glucose in the bloodstream of humans and other animals is converted to glycogen. When needed, this storage compound can be reconverted to glucose or glucose phosphate. Similarly plants convert excess glucose to starch, another storage polymer. When an animal has reached its capacity for storing glycogen, special depot cells convert the remaining excess glucose to fatty acids, which are laid down as fat. Depot cells can also convert excess fatty acids in the bloodstream to fat.

The synthesis of glycogen, starch and fat requires an input of energy, but it solves a major problem: how to store a large mass of material without raising the osmotic pressure of the internal environment (Chapter 1). One molecule of glycogen or starch raises the osmotic pressure no more than one molecule of anything else, but it contains hundreds or thousands of glucose residues. Fat molecules have much lower molecular weight, but they do not raise the osmotic pressure because they are not in solution within the cell.

A drop in the supply of glucose and fatty acids in the bloodstream triggers the release of the hormone glucagon (Chapter 12). The bloodstream carries it to the fat storage depots and to the liver, the main storage center for glycogen. Arrival of the hormone activates certain enzymes in the fat and liver cells, and the enzymes in turn catalyze reactions that release glucose, glycerol and fatty acids into the bloodstream. These reactions are not the reverse of the storage reactions, and they do not repay the energy cost of putting the nutrients into storage. Nevertheless the storage charges are a small price to pay to achieve an almost constant level of glucose and fatty acids in the bloodstream.

NEGATIVE FEEDBACK

This form of control was discussed in Part I. A room thermostat is an example of a negative-feedback mechanism. When the room gets too warm, the thermostat switches off the heat source. In more general terms, one can say that overproduction triggers mechanisms which tend to negate it. For example, an animal that becomes overheated on a hot day is in no position to switch off the heat source; instead he initiates some compensatory behavior, such as sweating or panting. In living systems several compensatory mechanisms are usually integrated to provide a relatively constant internal environment.

Mammals and birds are the only warm-blooded animals (HOMEOTHERMS); they maintain their internal temperature within quite narrow ranges of tolerance. In man an increase of a few degrees in body temperature radically changes nervous co-ordination, as one knows from the strange feeling experienced when one has a fever. Normally man's oral temperature ranges between 36 and 38°C; severe convulsions begin if the temperature goes to 41°C, and death comes at 45°C. Most other mammals have almost identical body temperatures, but birds are a trifle warmer at about 41°C. Where does the heat necessary to maintain such temperatures come from, and how is it controlled?

The individual enzymic steps of metabolism are not highly efficient; that is, much of the energy (about 70 percent) tied up in chemical bonds of nutrients is lost as heat when the chemical energy is converted into mechanical or biochemical work. Consequently every living cell generates heat. Vigorous exercise generates great amounts of heat that must be dissipated. For instance humans sweat and dogs loll their tongues; in both cases the evaporation of water provides the necessary cooling.

In mammals and birds, detectors of body-temperature appear in two locations. There are receptors in the skin, but the more important system lies in the central brain region called the hypothalamus. It consists of a heat loss center in the front section of the hypothalamus, which will trigger suitable reactions when its temperature is raised by a fraction of a degree, and a heat conservation center that acts in the opposite way. The responses that these centers evoke vary with species, and are graded according to the extent of the deviation detected. Man adjusts to mild temperature deviations by "vasomotor regulation"; that is, when the temperature increases, blood vessels under the skin open up so that more heat is lost by radiation and convection. The opposite occurs when the temperature falls. In some animals, skin areas particularly well supplied with blood vessels serve as radiators that dump excess heat; the ears of rabbits and elephants are good examples. Conversely, shivering generates extra heat if the temperature is too low for vasomotor compensation. Hair, fur and feathers are important factors in heat control, and their thickness and hence

their insulating properties changes with the season. Special muscles can adjust their effective thickness by erecting them, encasing the animal in an insulating layer of air.

In addition to all these mechanical devices, if cooling is still excessive the homeotherm resorts to hormonal control (*Chapter 12*) by producing adrenalin, which generally increases glycolysis. An even more direct control is stimulating the production of TSH or thyroid-stimulating-hormone by the anterior pituitary gland, which lies very close to the hypothalamus. The TSH is released into the blood and provokes the thyroid to release a series of hormones that elevate the basal metabolism and, by burning stored food reserves, greatly increase the production of body heat. One of the hormones, thyroxin, acts by reducing the efficiency of ATP production; therefore more material will need to be oxidized to produce a mole of ATP. In this way energy is diverted from the production of ATP to the production of heat.

Do these mechanisms fit the category of negative feedback systems? When cold provokes a bird to erect its feathers, the response might seem to be positive rather than negative. But in fact all such mechanisms have an on and off character. It is as true to say that a thermostat turns on the furnace when the room gets cold as to say that it turns off the furnace when the room gets too hot. Positive feedback is a quite different and usually disruptive process in which the product of a reaction sequence serves to activate the sequence. Autocatalytic processes are examples of positive feedback; for example, the binding of one mole of oxygen by hemoglobin facilitates the binding of additional moles. The term negative feedback applies not only to the heating or cooling mechanisms described here, but also to the mechanisms in which glands respond to a chemical signal to turn their regulatory hormones on or off.

EXCRETION

The continual intake of food by heterotrophs leads to a continuous production of byproducts that have to be ejected entirely from the organism if it is to maintain its internal stability. The foods of animals are composed primarily of carbohydrates, proteins and fats. Most of what they ingest is oxidized to produce energy in the form of ATP. The oxidation of carbohydrates and fats produces only carbon dioxide and water as end products; the carbon dioxide

NH_4^+

AMMONIUM ION
(AMMONIA IN WATER)

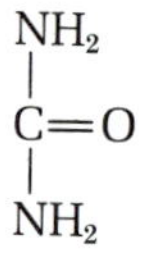

$$NH_2$$
$$|$$
$$C=O$$
$$|$$
$$NH_2$$

UREA

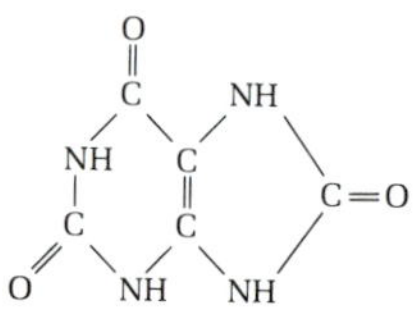

URIC ACID

is disposed of by gas exchange, and the water is a valuable acquisition. But the total oxidation of proteins produces large quantities of ammonia derived from the NH_2 groups of the amino acids. Unfortunately ammonia is very toxic and must be removed efficiently and rapidly.

In protozoans, sponges, coelenterates, and fresh water fishes there is extensive movement of water into and out of the body, and the ammonia is simply "washed out" in this water flux. In fresh-water fish the highly efficient gas-exchange mechanism of the gills is an ideal dumping device. Flatworms have developed special systems to promote water expulsion; water with its dissolved ammonia and carbon dioxide diffuses into tubules, and is expelled to the exterior by the beating of cilia

in a specialized flame cell (named after the flamelike flickering of the cilia).

Other animals do not enjoy the luxury of such an ample water flow, and have all met the problem by converting the toxic ammonia to less toxic nitrogen compounds, most often to either urea or uric acid.

The steady production of urea demands a mechanism for its steady removal from blood with only modest water loss — a problem that is complicated by the necessity to regulate water content and salt concentration. Animals that live in fresh-water habitats take up excess water through their skins by osmosis and get rid of it by urinating copiously. The frog produces urine equivalent to 25 percent of its body weight daily. Earthworms, although terrestrial, live in moist places and produce 60

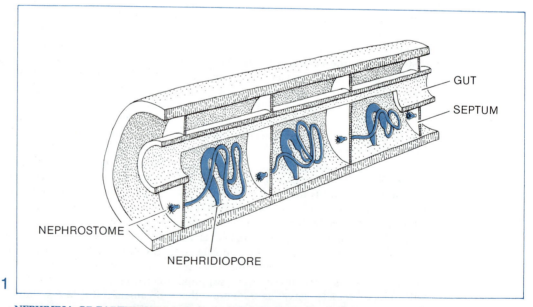

1

NEPHRIDIA OF EARTHWORM (color) are simple systems for excreting surplus water and retaining needed salts. Diagram depicts part of an earthworm sliced in half lengthwise. Liquids are drawn into each nephridium through an opening (nephrostome) in the segment ahead, and waste is expelled to the outside through the nephridiopore. Each coiled nephridial tube is associated with a capillary network (not shown) that recovers minerals, preventing their excretion in watery urine.

percent of their weight in urine daily. By contrast, man produces only two percent of his weight daily. It is clear that there are water-losing and water-conserving animals.

A simple water-losing system is the nephridium of the earthworm. It opens directly into the body cavity, from which the watery contents are drawn into the nephridium by cilia and passed down a long tubule to the exterior (*Figure 1*). Reabsorption of needed salts occurs in the tubule. The nephridium produces a watery urine that is hypo-osmotic; that is, its osmotic potential is lower than that of blood, because it is simply blood with salts removed from it by reabsorption.

In higher organisms, the simple nephridium is replaced by a more elaborate nephron; usually thousands of nephrons together comprise a kidney. A simple nephron found in fresh-water fish consists of a tubule which at one end embraces a knot of leaky blood capillaries called a glomerulus (*Figure 2*). In the glomerulus, blood filters through the capillaries to form a filtrate which lacks the cells and proteins of blood. The filtrate passes down the tubule of the nephron where reabsorption of some water and most essential solutes such as glucose and salts occurs. It is collected in a tube called the ureter, and there ejected. A copious hypo-osmotic urine results. Marine fish cannot afford this copious flow (they are always in danger of desiccation from their salty surroundings). They have reduced the flow by decreasing the size and number of glomeruli, and a few fish have abolished them completely; presumably enough filtrate leaks in through the proximal ends of the tubules.

Mammals possess a more elegant and effective kidney with a new feature added to the nephron: a highly efficient device for extracting most of the water as well as the salts from the glomerular filtrate. The crucial device in this system is the hairpin-like loop of Henle (*Figure 3*) which, like the circulatory system of a fish gill (*Chapter 9*), operates on the principle of countercurrent flow. The glomerular filtrate is collected by the nephron, and here more than half of the water and most essential substances, such as salt and glucose, are reabsorbed by active transport, just as in the simple nephron. The filtrate is then fed into the loop of Henle. Reabsorbed water and salts are returned to the bloodstream, and the remaining urine exits through a collecting duct and eventually reaches the bladder via the ureter.

Animals with even greater need to conserve

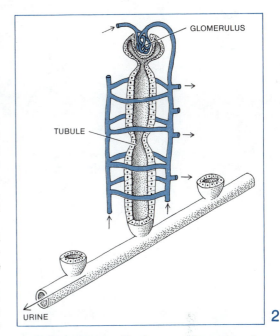

2

SIMPLE NEPHRON of fresh-water fish consists of a tubule associated with a glomerulus. Filtrate from glomerulus passes down the tubule, which reabsorbs glucose and salts. Absorbed substances are cycled back into bloodstream via the network of capillaries and veins around the tubule.

water use uric acid as their nitrogenous waste product. Uric acid is so insoluble that it is osmotically easy to abstract water from a mass of it. Birds and reptiles discharge their urine into a cloaca which also receives the feces. After the recovery of water in the cloaca both urine and feces are finally discharged as a paste, the white part of which is uric acid. Insects, whose small size increases their water-conservation problem, use a comparable approach. Their excretory apparatus is a fringe of Malpighian tubules around the midgut; these tubules have closed ends projecting into the coelom, from which they collect water and nitrogenous wastes (*Figure 4*). They discharge these tubules into the midgut, converting the nitrogenous material to uric acid. This passes through the gut and into the rectum, along with undigested matter, where the water is extracted. The waste matter is ejected as an almost dry pellet.

WATER STABILITY

The problem of water stability is a crucial one. Water makes up about two-thirds of all tissues, and is the indispensable medium for all biochemical reactions. Terrestrial organisms

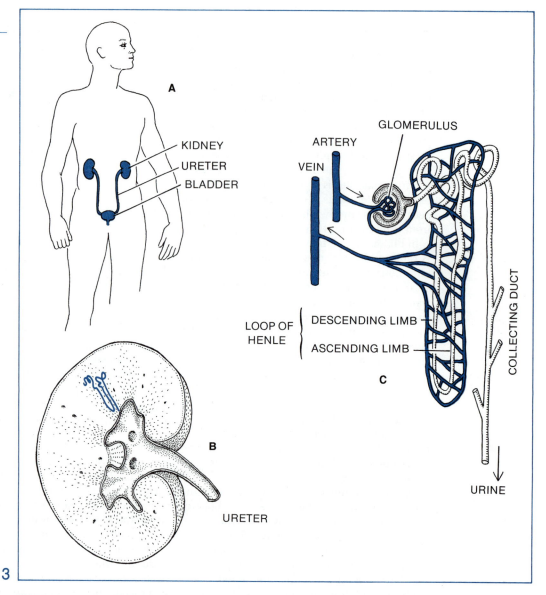

MAMMALIAN EXCRETORY SYSTEM. Each kidney contains thousands of nephrons. Cutaway view of kidney (B) shows orientation of a single nephron (color). Diagram of nephron (C) depicts glomerulus and tubules. Reabsorbed substances are returned to bloodstream via veins that drain kidney.

all share a serious problem: how to maintain enough water despite continual losses from evaporation and from the use of water to flush away the by-products of their metabolism. The threat of desiccation is the price they pay for an ample source of oxygen. Only two phyla, the arthropods and the vertebrates, have made this daring trade-off on a large scale. Many other organisms have left their ancestral seas and lakes but live in wet terrestrial habitats,

and die by desiccation if brought into the open. How do mammals and insects (for instance) avoid this fate?

Maintenance of internal wetness in a dry world demands a favorable balance between intake and output. The intake is partly as metabolic water derived from oxidation of carbohydrates and (with an even higher yield) fats. But the bulk of intake is as water in food or drink. Food can be sufficient by itself in some cases; fruit and vegetables contain 75 to 95 percent of water, meats about 50 percent. Should this source prove inadequate, most animals drink bulk water. Their drive to seek such water (which may involve a hazardous expedition from a safe place) is triggered by a reduction in water content of blood. In mammals, the trigger site has been located as the "thirst center" within the hypothalamus of the brain. If one stimulates the center electrically, the animal shows compulsive drinking, and can literally drink itself to death.

A land animal that must remain close to a water source is terribly restricted. To seek emancipation he must reduce his output to the point where he can go for long periods without access to water. Output is in three principal forms: respiratory exchange, excretion, and evaporation from the body's surface. Since respiration always occurs across a wet surface, there is inevitably some water lost in keeping it wet while passing air over it. Land vertebrates have abandoned the external-exchange route (via skin or external gills) by evolving internal lungs; the wet surface is thus exposed only to the minimum volume of air necessary. In insects exchange is also internalized in the tracheae; the external openings (spiracles) are under active control so that they are sealed off to restrict water loss when they are not in use.

Excretory losses of water are minimized by a variety of means. The copious urination of aquatic animals is reduced to a trickle in mammals. Each day a man loses a liter of water in urine and 100 ml in feces. The severity of this loss can be partly reduced in mammals if the need arises: the pituitary (a gland at the base of the brain) responds to a drop in the water content of the blood by secreting antidiuretic hormone which causes the kidney to retain more water; this can reduce the urine volume by two-thirds.

Evaporation from the body's surface is controlled by a variety of devices in fully "water-emancipated" animals such as mammals and many insects. The insect is especially vulnerable, for its small size involves a high surface to volume ratio — up to 100 times greater than in typical mammals. Insects living in dry air therefore have evolved a thick covering called cuticle, made up in large part of chitin, a polymerized amino-sugar rendered water-impermeable by being tanned with quinones and (like the cutin layer of plants) coated with wax or grease. This apolar barrier limits water loss to a rate two thousandfold less, for in-

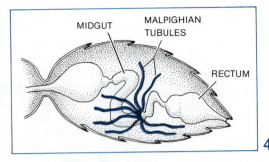

MALPIGHIAN TUBULES of insects collect water and nitrogenous wastes from coelom, convert the nitrogenous substances to uric acid, and empty into the midgut. Water is recovered in rectum.

stance, than from a red blood cell, and about a hundredfold less than from aquatic insects exposed to air.

In mammals the outermost layers of the skin are made up of dying cells (the stratum granulosum) containing a precursor of the horny protein called keratin; and dead flattened cells (the stratum corneum) thickened with keratin and lightly impregnated with wax (Figure 5). The dead and dying cells are sloughed off continuously from the outermost layer of the epidermis. The epidermis and the underlying dermis together make up the skin. Its water-proofing property depends on the strata granulosum and corneum. Studies on the guinea pig have shown that these layers appear after 54 days of fetal development; simultaneously the water permeability drops sharply.

Some mammals with unusual water-conservation problems have developed radical stratagems. Contrary to popular belief, the camel does not store water in its hump (it stores fat there); but it can tolerate extreme dehydration. It can lose 40 percent of its body water and live, whereas most mammals die if they lose 20 percent. In addition the camel can tolerate a higher body temperature than other mammals. Like them, his normal temperature is at about 34°C, but he does not turn on his water-wasting cooling system — his sweat glands — until his temperature reaches 41°C, compared with 37° for man.

Aquatic organisms might seem to have no water problems, but many of them must cope with osmotic problems. Marine invertebrates, it is true, enjoy the rare privilege of being isosmotic with their surroundings. But most modern marine fish have evolved from fresh-water ancestors, and as a result the elasmobranches like the shark are about half as salty as the sea, and the bony fishes or teleosts are about a third as salty as the sea. They are thus eternally in danger of being desiccated by water loss. By contrast all fresh-water animals, vertebrate or invertebrate, are hypertonic to their virtually salt-free environment, and are in constant danger of being flooded. Obviously all these aquatic animals survive comfortably; they must therefore have strategies to avoid such catastrophes.

The marine fish drink and absorb sea water, but excrete the surplus of salt through their gills, because their kidneys are inadequate for massive salt rejection. The fresh-water fish, by contrast, solve their water problems with their kidneys. They compensate for the steady inundation of water by urinating copiously and continuously, while jealously recovering virtually all the inorganic salts from that urine. Fresh water protozoa have also found a way to rid themselves of the excess water that inevitably invades them. They have contractile vacuoles which pump out the excess, contracting at rates that depend on the osmotic difference between their cytoplasm and the outer medium (*Figure 6*).

A few mammals and birds have returned to their ancestral home, the sea, and have had to readapt their essentially terrestrial equipment accordingly. Those that eat fish, such as seals and most whales, take in food with a salt concentration that matches their own so they do not have to drink sea water. Mammals, such as baleen whales, that eat marine invertebrates are taking in a too-salty diet, and have had to compensate by evolving special kidneys able to produce unusually concentrated urine. Marine birds (gulls, albatrosses, penguins) and reptiles (sea turtles) have adopted another tactic; the task of getting rid of surplus salt is given to a special salt gland close to the eyes. These animals can all drink sea water, after which they secrete an exceptionally salty fluid, which shows up as a salty nose-drip in the birds and salt tears in the turtles.

INTERNAL INCONSTANCY

Internal constancy is not universal, as shown by the body temperature of POIKILOTHERMS: cold-blooded animals. The terminology is misleading in Greek as well as English, for both terms imply that these animals (all animals except mammals and birds) always have cold blood. On the contrary, on a hot day their blood tends to be hot. Their special feature is that they tend to have INCONSTANT BODY TEMPERATURES that correspond to their environment, hot or cold.

In unduly hot or cold places, poikilotherms may develop behavioral adaptions to lessen the rigors of their environment, for instance by seeking shady places, or operating only at night. In other than tropical regions, such strategems do not permit them to live through the prolonged cold of winter in an active state. They therefore adopt some inactive form to last the winter, such as eggs or pupae in the case of many insects; or they may become metabolically (and therefore behaviorally) almost inert, as in reptiles and amphibia. Such adaptive transformations may be triggered not only by approaching winter, but in many cases by a variety of potentially hazardous conditions. Thus in the little crustaceans called water fleas (*Daphnia*) a specially thickened egg-case develops around a single egg, the whole package being called an ephippium. These ephippia may form in a variety of disadvantageous conditions, including drying and freezing.

Organisms that lack the ability to maintain constant body temperature must either 1) resort to special behavioral strategies such as migration, 2) resort to physiological strategies such as spending part of their lives in an inac-

STRATUM
GRANULOSUM

STRATUM
CORNEUM

EPIDERMIS

HAIR ROOT

SWEAT
GLAND

DERMIS

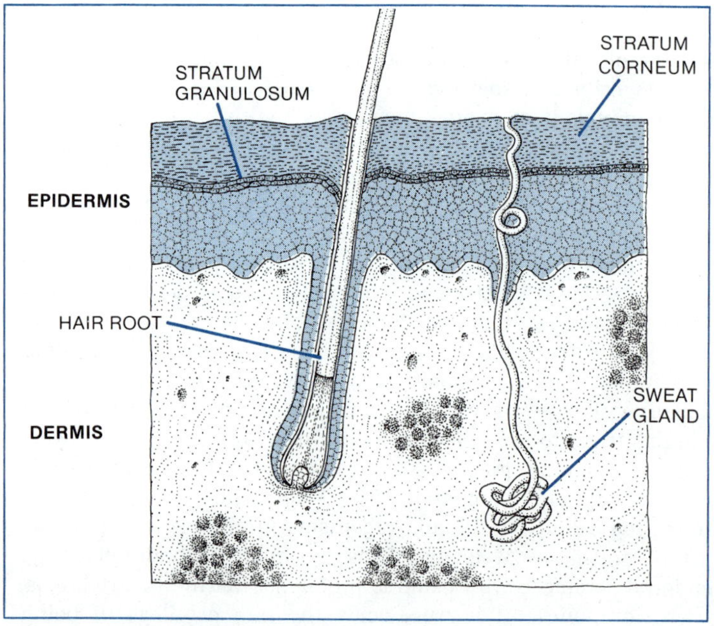

HUMAN SKIN consists of waterproofing layer of dead and dying cells (epidermis) sloughed off from the underlying layer of living cells (dermis). Sweat gland allows controlled release of water.

tive temperature-resistant state, or 3) avoid extreme environments completely. The comparatively recent evolution of sophisticated temperature-regulating mechanisms gives homeotherms some distinct biological advantages. Not only does it enhance their overall physiological efficiency, but it opens up environmental niches that are forbidden to poikilotherms: new geographic locations and new times of the year, particularly winter.

Some homeotherms have also adopted behavioral and physiological strategies to cope with the rugged winters of nontropical regions. Migratory birds, like migratory insects, escape the rigors of winter simply by going somewhere else. Many other insects, as well as many reptiles and mammals, rely on the physiological strategy of hibernation. When cold weather comes they enter a quiescent period during which their body temperature, metabolic rate and body weight drop sharply. These changes permit an organism to survive several months without food or water. In comparison with poikilotherms, hibernating homeotherms enjoy the competitive advantage of a longer active season — a little earlier in the spring, a little later in the fall — but they are at a disadvantage in year-round competition with homeotherms that possess more sophisticated all-weather temperature regulating systems.

BIOLOGICAL CLOCKS

In any part of the world in which there are pronounced seasonal changes in climate, organisms appear and disappear at particular times with uncanny precision. The swallows reappear at the Mission San Juan Capistrano in southern California within a week of March 19; the monarch butterflies settle in the trees of Monterey, California, within a few days of mid-October; noxious black flies swarm in the Maine woods in late spring and early summer, then suddenly vanish; blooms of goldenrod cover much of New England and the midwest in August; and emperor penguins breed only in the blackness of the Antarctic winter. It is obvious that these organisms can measure and predict environmental change with considerable precision, and initiate action that may lead to some important biological event many days, weeks or even months later. Biologists have long wondered how they do it, and the answer to this question is only beginning to become clear.

Response to daylength, or PHOTOPERIODISM, is well known among most groups of higher

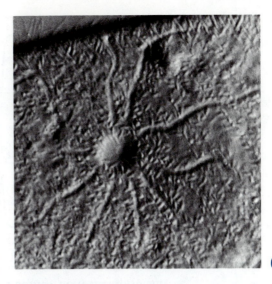

6

CONTRACTILE VACUOLE enables Paramecium to bail out excess water that enters it by osmosis. Photomicrograph depicts vacuole when full. Collecting tubules converge toward vacuole, which empties itself through a small central pore barely visible as a circular outline.

organisms. However, the most detailed studies have been done with flowering plants and insects — partly because they are readily handled in large numbers under laboratory conditions, and partly because many of them have sufficiently short life cycles that their response to photoperiod can readily be studied.

Organisms can be broadly fitted into three major groups according to their response to changing daylength. Some undergo a change only if the days become shorter than some critical number of hours. If a day is longer than 15½ hours, the cocklebur *Xanthium* grows vigorously but produces only new leaves. Only when the period of light is less than 15½ hours does it sprout flowers as well. They are obligate SHORT DAY (SD) plants. Certain aphids reproduce sexually only if daylength is less than 14½ hours. Such SD organisms do not begin sexual reproduction until the daylength is less than some critical value (Figure 7). On the other hand some plants and insects show precisely opposite behavior. Kept on short days, they fail to become sexual (or to stop growing, or to undergo seasonal changes in color). Only when the daylength is greater than some critical value does the appropriate response appear. These are the so-called LONG DAY (LD) organisms (Figure 8). Finally there are some organisms which are indifferent to daylength, and are called DAY NEUTRAL (DN);

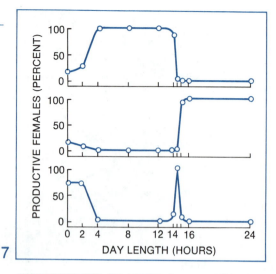

7

REPRODUCTION OF APHIDS is governed by day length. Short days (top) favor the hatching of females that reproduce asexually. Long days (middle) favor development of sexually-reproducing females. Days 14.5 hours long — the critical daylength — result in production of both types of female (bottom).

the common dandelion is a particularly obnoxious example. Some organisms show more complex patterns: those which require long and then short days in sequence (or the reverse) and those which must have a period of low temperature that precedes days of a certain length.

If organisms can tell when the light is on and when it is off, what pigment provides them with the appropriate information? In higher plants phytochrome is a primary photoreceptor in perceiving the lights-on or lights-off signal. In animals earlier workers reasonably expected vision to take care of photoperiodic perception. But the British entomologist A. D. Lees has discovered that in at least some aphids certain brain cells, and not the eyes, perceive the photoperiodic stimulus. There is also evidence that in some lizards, amphibians, and birds it is the pineal gland rather than the eye that perceives the stimulus. The pigments involved are unknown, although they may well be closely related to or identical with the visual pigments.

Why is there any response to change in daylength in the first place? It is obvious that organisms have some way of measuring time, and that they are remarkably well adapted to a 24-hour day-night cycle. The answer is that photoperiodic phenomena are tied to a

rhythmic system probably found in all eucaryotic organisms. There is without question a biological clock within such relatively advanced organisms, and the outward manifestations of this clock are known as CIRCADIAN RHYTHMS (circa: about; dies: a day). If an experimenter places certain insects in constant conditions of temperature and illumination and monitors their activity, about once every 24 hours, he will observe a peak of activity, and between these peaks, a period of relative lethargy. If he implants a thermistor, a sensitive temperature-monitoring device, into a hamster and places the hamster under constant environmental conditions with food continuously available, the hamster's temperature will continue to fluctuate up and down on an almost 24-hour cycle for as much as two years. The leaflets of a plant such as clover or the tropical tree Albizzia are normally down and folded at night and up and expanded in the daytime. Under constant dim light they continue to open and close on an approximately 24-hour cycle (Figure 9).

There are several important properties of these circadian rhythms, regardless of the organism chosen for study. First, the period (the time from peak to peak) is remarkably insensitive to temperature. Second, the rhythms are highly persistant; witness the case of the hamster. The continuous cueing provided by light-dark transition is not required for their expression. Third, they can be entrained,

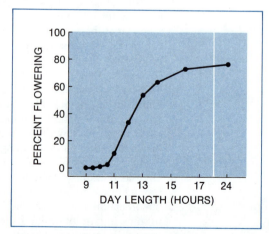

8

LONG-DAY PLANTS will flower only if exposed to a photoperiod greater than their critical daylength. Curve depicts effect of photoperiod on duckweed (Lemma gibba). Critical daylength for this plant is about 11 hours. White line at right indicates break in time scale of graph.

9

SLEEP MOVEMENTS OF PLANTS such as the tropical tree Albizzia follow a 24-hour cycle. Its leaves are normally folded at night (left) and open during daytime (right). Plant at left was kept in darkness until just before picture was made. Plant at right was left in continuous light.

within limits, by light-dark cycles. Note that the period in the above examples was noted as ABOUT 24 hours. If any of these organisms were placed on a day-night cycle totalling EXACTLY 24 hours, the rhythm expressed would show a period of exactly 24 hours. On the other hand, if the experimenter were to use a day-night cycle of, say, 22 hours, then the overt rhythms would show a 22-hour period.

There is now ample evidence that the interaction of daylength with circadian rhythms is the basis of photoperiodic behavior. Several questions, however, remain unanswered. First, what is the difference between long day and short day organisms? Second, how does the light stimulus affect the biological clock to bring about such profound changes in the growth or behavior in the organisms? Because changes in daylength are far more precise than changes in temperature, it is logical that the rhythms are insensitive to temperature; but what insulates them from the biochemical effects of temperature changes? Finally, what is the biochemical and biophysical basis for the clock? These questions are currently the subject of extensive research.

INTERNAL RHYTHMS

All these rhythmic activities have their counterparts in the metabolism of the organism.

The most obvious counterpart is in the energy use (and therefore oxygen uptake) that accompanies these varied activities. The oxygen uptake of beans, carrots, mice, crabs and beetles have all been studied, and all show circadian rhythms. It follows that all the hundreds of enzymes and intermediates involved in the energy metabolism of the body must show these rhythms.

Other enzymes and their products are geared into these cycles. In the pineal gland of the rat, for instance, several compounds involved in control of the nervous system, including the hormone norepinephrine, show circadian rhythms. The norepinephrine fluctuation in turn leads to fluctuations of an enzyme, N-acetyltransferase, whose night levels are 15 times its day levels. This enzymic fluctuation in turn causes fluctuation in the levels of serotonin (which is acetylated by the enzyme) and in melatonin (which is produced from acetylated serotonin).

No doubt these numerous internal rhythms are the consequences of detection (by an unknown mechanism) of external light rhythms, and that in turn they give rise to the fluctuation in the activity and physiological state of the whole organism. They are the intermediate steps between the times of the biological clock and its expression. But what is the timer? And

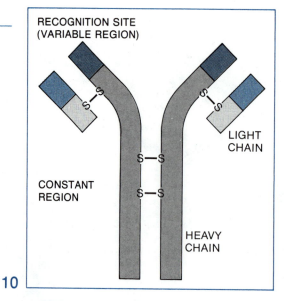

RECOGNITION SITE
(VARIABLE REGION)

LIGHT
CHAIN

CONSTANT
REGION

HEAVY
CHAIN

10

ANTIBODY MOLECULE consists of two heavy (long) polypeptide chains and two light (short) ones. The gray parts of the chains are the same in all antibodies. The variable recognition sites (color) give each antibody its unique specificity.

does the timer get its cue from external factors or is it autonomous?

F. A. Brown of Northwestern University argues that subtle geophysical factors, perhaps rhythmic changes in the earth's magnetic field or the incidence of cosmic rays, synchronize these metabolic cycles despite experimental attempts to control the external conditions. But the inability to pinpoint these factors has made many workers skeptical of their existence, and the prevailing viewpoint is that the clockwork consists of some sort of internal oscillator that is regulated by external events.

The nature of the oscillator has been much debated. The "mechanism," if there is but one, has not been located. Most experiments, though not conclusive, suggest that the body contains thousands of oscillators, and that it synchronizes them with one another by synchronizing them with a simple external oscillator.

In other words, it has been as difficult to locate the internal clock as to detect the subtle external geophysical factors. Although the mechanism of the biological clock is unclear, it is certain that many and perhaps all eucaryotic organisms possess one. The idea of the static nature of the body's composition was overturned in the 1930s and replaced by the notion of a dynamic steady state. But in the last 15 years investigators have learned that the state is not so steady. The composition of every organism is changing throughout each day, and John Doe at noon is a somewhat different organism from John Doe at midnight.

THE IMMUNE SYSTEM

Most animals have a potent defense system that protects them from invasion by foreign organisms. It consists of phagocytic cells that engulf and destroy the invaders. In addition all vertebrates, including man, have a second line of defense: an immune system that is highly specific in selecting its target organisms. In a sense the distinction between specific and nonspecific defense systems is artificial, because both systems work together, and because both depend on the white cells of the bloodstream.

In man, the three principal types of white blood cells are NEUTROPHILS and two kinds of lymphocyte, called B-CELLS (because they originate in bone marrow) and T-CELLS (because they develop in the thymus gland). Unlike red blood cells, white cells can move by themselves, either drifting with the bloodstream or swimming against it; and they can leave the capillaries, squeezing through the walls to reach damaged or infected tissues. Guided by some sort of chemical gradient, neutrophils are the first to arrive at the site of a wound. Assisted by macrophages (giant phagocytic cells found in the tissues), they attack and engulf any foreign matter that forces its way into the body, such as viruses, bacteria or the thorn of a rose. This nonspecific response is the body's first line of defense. Confronted by an infection, the white cells will literally eat themselves to death, engulfing and digesting the foreign matter until the accumulation of toxic breakdown products kills them. A neutrophil can ingest from five to 25 bacteria before it dies.

The second line of defense, specific immunity, is the job of the lymphocytes. Although B-cells and T-cells look identical under the microscope, they respond to foreign matter in different ways. A substance capable of provoking an immune response is called an ANTIGEN. Most antigens are proteins, but certain other large molecules such as polysaccharides are also antigenic. B-cells attack antigens by releasing a quantity of ANTIBODY, protein molecules that circulate in the blood and bind themselves to one specific type of antigen. The production of circulating antibodies by B-cells is sometimes called humoral immunity (from the Latin *humor* = fluid).

Antibodies are proteins consisting of long ("heavy") chains and short ("light") chains strung together by disulfide bonds, as shown in Figure 10. The so-called constant region is the same in all antibodies; the variable region, which contains the site that binds the antigen, differs from one type of antibody to another. The variable region of each type of antibody is tailored to fit one specific antigen. The sequence of amino acids in this region determines the specificity of the antibody. When the antibody encounters an unknown substance, the recognition site senses the shape of the molecule and identifies it as being either "self" (the body's own) or "non-self" (a foreign antigen). Antibodies ignore anything recognized as self, but bind themselves powerfully to foreign antigens, and thus inactivate them. Because invading cells almost always have more than one antigenic molecule on their surface, antigens can bind the cells into clumps, which are then attacked by phagocytic cells (*Figure 11*).

T-cells do not release antibodies, but react to antigens on the surfaces of cells. They recognize, engulf and destroy foreign cells, or any of the body's own cells that have been altered by virus infections, cancer, or old age. This mechanism is sometimes called cellular immunity. The important difference between the T-cells and other phagocytic cells is that a particular T-cell will destroy only one specific type of foreign cell, while neutrophils and macrophages will destroy any foreign matter they encounter.

IMMUNOLOGICAL MEMORY

The first time that an animal is exposed to a particular antigen (for example, the organism that causes whooping cough), there is a certain time lag — usually several days — before the number of antibody molecules and the number of activated T-cells circulating in the bloodstream catches up with the number of invaders. But for years afterward, sometimes for life, the immune system somehow "remembers" that particular antigen and remains tooled up to produce antibodies on short notice.

THE IMMUNE RESPONSE. Developing lymphocyte (left) differentiates into T-cells and B-cells. Specific receptors on surface of T-cell are exposed to antigen and bind it (top). Thus activated, T-cell continues to multiply (top right). B-cell exposed to antigen interacts with activated T-cell (bottom right), begins to release circulating antibodies, and continues to multiply.

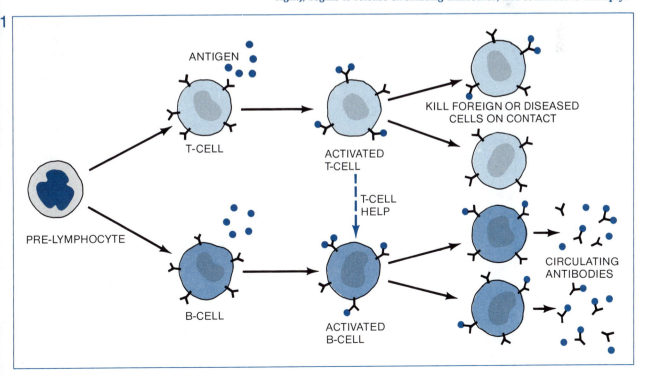

This memory effect explains why IMMUNIZATION has almost wiped out such deadly diseases as smallpox, diphtheria and polio. A very small amount of viral or bacterial protein (often treated to make it harmless) is injected into the body. Later, if very similar disease organisms should attack, the tooled-up cells recognize it and quickly overwhelm it with a massive outpouring of antibodies.

How does this memory work? And why does the system fail to attack the body's own cells and proteins? These questions are the focus of a massive research effort because the answers promise to shed light on cancer, the rejection of transplanted organs, and other medical topics. Several theories have emerged, but the one currently favored by most immunobiologists is known as the clonal selection theory.

The biosphere contains an almost endless number of potential antigens. An individual can make specific immune responses to perhaps a few million (roughly 10^6) different antigens. A cell that could make all of them would need 10^6 genes to code for them, according to the rule that one gene codes for one protein. But a human cell contains only about 10^4 genes, and they must code not only for antigens, but for all the other proteins in the body. The theory holds that an individual has a very small number of B-cells and T-cells genetically programmed (in some unknown way) to respond to only one antigen. When they come in contact with that antigen, the cells begin to multiply. After several rounds of multiplication, there is a large clone of them. (Genetically identical descendants of a single cell are known as a clone, thus the name of the theory.) In this way, immunologists believe, the immune system exhibits natural selection; the cells best able to react against a specific antigen have a reproductive advantage and soon greatly outnumber cells unable to react to the antigen.

TRANSPLANTS

If one attempts to transplant an organ or a piece of skin from one human to another, the transplant is recognized as non-self and soon provokes an immune response; the tissue is killed or "rejected". But if the transplant is done immediately after birth, or comes from a genetically identical person (an identical twin) the material is somehow recognized as self, and is not rejected. Physicians can overcome the rejection problem for a while by knocking out the immune system with drugs (immunosuppressants) that reduce the action and number of B-cells and T-cells. But this technique leaves the individual wide open to bacteria and viruses, and he must be protected from all such invaders by elaborate isolation procedures.

A slightly different problem arises when one attempts to transplant another important tissue: blood. Early attempts at blood transfusion often killed the patient. The Austrian physician Karl Landsteiner, starting in 1898, found by mixing blood cells and serum from different individuals that only certain mixes were compatible; in others the red blood cells formed clumps that would clog the blood vessels. Landsteinder deduced that there were two red cell antigens, A and B, and two corresponding serum antibodies, α and β. Mixes of A with α or B with β cause clumping. The production of these antigens and antibodies is controlled by genes. A person can be a heterozygote with AB type blood or a homozygote with AA or BB (usually designated simply as A or B). A different allele produces type O blood, which lacks both antigens. An individual who has type A blood can accept a transfusion only from type A donors. An AB individual has no serum antibody for either antigen, and can accept blood from any donor. A type O individual has antibodies for both, and needs type O blood.

The Rh factor, so named because it was first found in Rhesus monkeys, is another blood antigen. In most human populations almost 100 percent of the individuals have this antigen, and their blood is said to be Rh$^+$. Among members of the white race, however, only 87 percent are Rh$^+$; the others lack this antigen, and their blood is called Rh$^-$. In areas where the races have interbred, such as the US, the

Once exposed to a particular antigen, the lymphocytes somehow remember it for years afterward, and remain tooled up to produce the corresponding antibody on short notice.

percentage of individuals having the factor lies somewhere between 87 and 100 percent. About 95 percent of American blacks, for example, are Rh⁺. Like antigens A and B, the Rh factor is genetically determined; but the serum antibody to Rh, unlike those for blood type, is not genetically determined. It develops only after exposure to the Rh antigen. If an Rh⁻ woman becomes pregnant with an Rh⁺ child, the embryo's Rh antigen sometimes enters the mother's blood, is recognized as non-self, and the mother makes antibodies against the embryo's blood, often dooming it or the infant to an early death. The problem worsens with subsequent pregnancies. Not many years ago, when an Rh⁺ man married an Rh⁻ woman they would be advised to have only one or at most two children. In recent years this problem has been solved. A substance called Rhogam, derived from the blood of unusually Rh-reactive men and women, is injected into an Rh⁻ mother within 72 hours after her first delivery. It contains antibodies to the Rh⁺ antigens and suppresses the mother's immune attack upon her next baby's blood.

CANCER AND IMMUNITY

In a healthy body no group of cells becomes dominant. Growth stops when one group has achieved the proper size, as determined by a preset genetic program and the feedback of outside information, such as the degree of cell crowding. An effect called contact inhibition slows or stops the reproduction of cells when they are crowded together. What if a cell should mutate and escape this restriction, continuing to multiply at the expense of its neighbors? This is most likely to happen with epithelial cells (those of the skin and the linings of the body cavities) which wear out and are continuously replaced. Is it a coincidence that (based on figures from Denmark over a 25-year period) 92 percent of cancers occur in epithelial cells?

One theory of cancer argues that a principal role of the immune system is to destroy cells that become altered in any way that changes their surface antigens. A key feature of many cancerous cells appears to be a change in their surface properties, so that they become insensitive to contact inhibition. Such a change could be caused by a spontaneous mutation, by introduction of foreign DNA or RNA by a

virus, or by the disruption of genes by an environmental agent such as smoking, ultraviolet radiation or pollutants. Most cancers are diseases of old age. When the body gets old, the immune system becomes less effective. Perhaps during youth and middle age the immune system destroys any cancerous cells that might arise; but when it falters cancer cells can multiply unchecked. Humans with a naturally inadequate immune response, or who are under prolonged treatment with immunosuppressants, are particularly prone to cancer. If the immune system does indeed play a major role in the suppression of cancer, the day may come when vaccination will cause cancer to go the way of smallpox, diphtheria and polio.

READINGS

F.A. BROWN, J.W. HASTINGS AND J.D. PALMER, *The Biological Clock: Two Views*, New York, Academic Press, 1970. A delightful little paperback describing rhythmic phenomena and the different views of their control. Well illustrated and easy to read.

A.B. BURNETT AND T. EISNER, *Animal Adaptations*, New York, Holt, Rinehart and Winston, 1964. In 135 pages, this book concentrates on the diversity of the adaptations which animals have evolved to solve basic problems. The section on "Maintenance of the Internal Milieu" is particularly relevant to this chapter.

A.W. GALSTON AND P.J. DAVIES, *Control Mechanisms in Plant Development*, Englewood Cliffs, N.J., Prentice-Hall, 1970. The first chapter is an excellent discussion of photoperiodism in plants, with brief mention of animal systems.

R.A. GOOD AND D.W. FISHER, Editors, *Immunobiology*, Sunderland, Mass., Sinauer Assoc., 1971. A collection of essays on both biological and clinical aspects of immunology. Handsomely illustrated.

G.J.V. NOSSAL, *Antibodies and Immunity*, New York, Basic Books, 1969. Written for the layman by one of the authorities in the field, but by now a bit out of date.

C.L. PROSSER, *Comparative Animal Physiology*, 3rd Edition, Philadelphia, W. B. Saunders Co., 1973. A thorough treatment of regulatory mechanisms (and other matters) in animals. The chapters on water, excretion and temperature are excellent.

12

hormones

When seedlings are freely exposed to a lateral light some
influence is transmitted from the upper to the lower part,
causing the latter to bend.

CHARLES DARWIN

The remarkable constancy of the internal environment depends
mainly on hormones and nerves. Hormones are internal chemical
signals that effect relatively slow responses lasting minutes, hours or
months. Speedier responses are controlled by nerves (*Chapter 13*).
Hormones are produced by one part of a multicellular organism and
transported to another part, where they induce a change in the
biochemical functions of the responding cells. Such signals can be
compared to messages put in bottles by shipwrecked sailors. In animals, they are cast into the bloodstream, the lymphatic system or other
bodily fluids. In plants, they may enter the xylem or phloem. In their
journey through the organism, they may reach many different groups
of cells. But only certain "target" cells are genetically programmed to
detect a particular message and respond to it.

In higher plants hormones are crucial in regulating the direction
and timing of growth, including the development of flowers and fruit.
In higher animals they play much broader roles, being involved not
only in growth and reproduction but also in digestion, metabolism,
muscle tone and behavior.

Tiny amounts of hormone can produce a dramatic response in the
target cells. The plant hormone IAA (indoleacetic acid) can show
activity at concentrations of well below one part per million.
Hormone-mediated changes qualify as communication in a special
sense often attached to the word: the amount of energy involved in the
response greatly exceeds that put into the signal. Compared with the
response to nerve signals, which typically occur in a fraction of a
second, hormonal communication is slow. The swiftest effects are
produced by hormones such as adrenalin and noradrenalin, which
mediate the stress ("fight or flight") reactions of mammals, but even
this dramatic action is sluggish compared with behavioral activity
based on nerve conduction.

Hormonal communication has proved entirely adequate to coordinate the rather leisurely metabolism of plants, but mobile animals
obviously need a swifter means of internal communication. Then why
do animals still rely heavily on hormones? Why, in the course of
evolution, didn't animals discard their sluggish hormone systems
entirely, and replace them with nerves? To answer the question, one
must look at hormones in an evolutionary perspective.

When the first organisms appeared in the primeval seas billions of
years ago they probably communicated with one another during sexual reproduction by means of chemicals passed back and forth in the
water. Such chemical messages exchanged between organisms are
called PHEROMONES (*Chapters 15 and 26*). This is the prevailing mode
of communication among microorganisms and lower invertebrates
alive today, because it utilizes only the simplest sending and receiving
apparatus. When multicellular organisms evolved, there arose a need
to integrate the activities of many cells. Two kinds of internal communication evolved. The first hormones might have been modified
pheromones. When multicellular organisms closed off their body
cavities, part of the "sea" in which the cells lived became internal,
and the pheromones that they used for communication became re-

"GENERAL" TOM THUMB stands beside the
showman P.T. Barnum in photograph on opposite
page. Born Charles Stratton in 1838, Tom was exhibited as a midget by Barnum, who invented his
stage name and military rank. Because of a hormone deficiency, Tom stood less than 61 cm tall
when Barnum discovered him as a boy, and less
than 84 cm (33 inches) when fully grown.

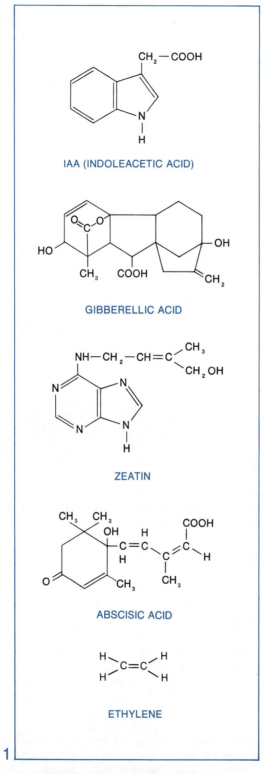

IAA (INDOLEACETIC ACID)

GIBBERELLIC ACID

ZEATIN

ABSCISIC ACID

ETHYLENE

1

PLANT HORMONES. Indoleacetic acid is an auxin, gibberellic acid a gibberellin, zeatin a cytokinin, abscisic acid a growth inhibitor. Ethylene promotes the ripening of fruit and many other processes.

defined as hormones. The first nerve cells, which evolved much later, may have been modified hormone-secreting cells. True, a nerve cell is greatly elongated and specialized for the swift propagation of electrical impulses, but its action remains partly chemical: the impulses pass from one nerve cell to another, across the synapse that separates them, when transmitter substances are released like hormones from the tips of the cells (*Chapter 13*). Hormones and nerves are intimately related and serve similar functions, under the control of the central nervous system. This point will become clearer in the later discussion of the relationship between the pituitary gland and the hypothalamus of the brain.

In animals hormones are made and released by specialized groups of cells called endocrine tissues. In higher animals the tissues are sometimes compacted into endocrine organs such as the thymus, pituitary and thyroid glands. All of the hormone-producing tissues and organs collectively comprise the ENDOCRINE SYSTEM. In plants less specialized cells from a whole region may make a particular hormone.

As one might expect, hormone molecules do not belong to any single chemical category. They have evolved in many different biological contexts, and many different organic substances have been enlisted to serve as messengers. Some are relatively small polypeptides or proteins; others are steroids (lipids containing the multiple-ring structure shown in Figure 8), amines, or amino acid derivatives. Still others, including some plant hormones, belong to none of these familiar categories. Nevertheless most hormone molecules do have one significant property in common: they are neither very large nor very small. They cannot be very large, because large molecules tend to be insoluble and difficult to transport through the body. They cannot be very small, because small molecules are not complex enough to be chemically distinctive and clearly recognizable by the target cells. The largest hormone molecules are proteins such as the human growth hormone, which consists of 188 amino acid units, and the smallest by far is the plant hormone ethylene, which contains only six atoms.

PLANT HORMONES

Hormones can be classified, according to their functions, into two broad categories: developmental and regulatory. Developmental hor-

KINETIN CONCENTRATION (MG/ML)	IAA CONCENTRATION (MG/L)				
	0	0.005	0.18	1.08	3.0
0	SMALL CALLUS	SMALL CALLUS	MEDIUM CALLUS ROOTS	MEDIUM CALLUS ROOTS	MEDIUM CALLUS ROOTS
0.2	MEDIUM CALLUS	MEDIUM CALLUS	MEDIUM CALLUS	LARGE CALLUS	LARGE CALLUS
1.0	MEDIUM CALLUS, FEW SHOOTS	MEDIUM CALLUS, MANY SHOOTS	MEDIUM CALLUS, MANY SHOOTS	MEDIUM CALLUS, FEW SHOOTS	LARGE CALLUS

INTERACTION OF AUXIN AND CYTOKININ determines the course of development in culture of tobacco pith callus. Chart summarizes effects of the auxin indoleacetic acid (IAA) and the cytokinin kinetin. Maximum tissue development requires both hormones. High ratio of auxin to cytokinin favors development of roots (upper right); low ratio favors development of shoots (bottom center) and intermediate ratio favors the multiplication of undifferentiated callus cells (middle row).

mones are found in both plants and animals; regulatory hormones, mainly in animals. Virtually all plant hormones affect development. They control the germination of seeds, the growth and differentiation of shoots and roots, and the maturation of flowers. Higher plants employ at least five classes of developmental hormones: AUXINS, GIBBERELLINS, CYTOKININS, ABSCISIC ACID (and its relatives) and ETHYLENE (Figure 1). There is only one important auxin: IAA. There are more than 30 known gibberellins, all structural modifications of gibberellic acid; they vary from species to species and even within a single plant, appearing or disappearing with such developmental changes as fruiting or flowering. Several may even operate at the same time. There are also several cytokinins, and abscisic acid is only one of a class of plant growth inhibitors.

Auxins were the first class of plant hormones to be discovered. Building upon research begun by Charles and Francis Darwin in 1880, Frits Went showed that the tips of grass seedlings contained a material which produced cell elongation. In grasses the first portion of the shoot to emerge from the seed is the COLEOPTILE, a hollow cylindrical structure that protects the shoot apex and the first true leaves as the shoot pushes upward through the soil. The tip of the coleoptile is the site of auxin synthesis. Transported to the lower parts of

the coleoptile, the auxin promotes extensive growth. More than three decades passed before the auxin in coleoptiles was conclusively identified as IAA.

In addition to its role in coleoptile elongation, auxin also promotes the elongation of stems and leaf veins, and induces differentiation of vascular tissue. It strongly inhibits the elongation of roots, but it induces the mitotic activity that leads to formation of lateral root primordia. Finally it regulates the process of ABSCISSION: the separation of fruit and old leaves from the parent plant.

The first cytokinin was isolated not from plant tissue but from herring sperm DNA. In 1955 Folke Skoog and his colleagues at the University of Wisconsin found that cytokinin had dramatic effects on callus cultures derived from the pith of tobacco stems. (Callus is a relatively undifferentiated plant tissue which develops on the cut ends of stems or roots grown on nutrient media.) Grown on a simple medium of salts and sugar, the callus tissue remained disorganized; the additition of cytokinin, however, induced the formation of small pockets of cells that acted like meristems (Chapter 6). Most surprising was the interaction between auxin and cytokinin. Figure 2 illustrates the results of varying concentrations of both of these growth regulators. Optimal growth required both substances. A high ratio of auxin to kinetin (a cytokinin) led to the formation of a large number of roots, while a low ratio produced many shoots. Intermediate ratios merely produced vigorous growth of the callus tissue.

The activation of bud meristems depends on the interaction of auxin and cytokinin. Bud meristems are tiny pockets of potentially meristematic tissue in the upper angle between leaf and stem, the leaf axil. In many plants auxin moving down the stem from the apex keeps these lateral bud meristems dormant. K. V. Thimann and his associates at Harvard University showed that cytokinins were effective auxin antagonists in this system. Small concentrations of kinetin could overcome the auxin inhibition, allowing the bud meristems to begin development. Again, hormonal interaction and balance determined the program

Developmental hormones are found in both plants and animals; regulatory hormones, mainly in animals.

of development. There are many other cyto-kinin-mediated processes, mostly involving initiation or maintenance of meristem activity.

The potency of gibberellins as growth hormones in higher plants was dramatically demonstrated in experiments on single-gene mutants of corn and peas. The stems of these mutants failed to elongate normally, and auxins had no effect whatsoever on them. But treatment with microgram quantities of gibberellins completely nullified the mutant gene. Treated plants were indistinguishable from wild-type seedlings (*Figure 3*). The mutation was eventually shown to involve a key enzyme in gibberellin synthesis. Like the auxins, gibberellins are synthesized in actively growing regions and exported to other parts of the plant.

Gibberellins play a special role in the germination of seeds of cereal grasses such as oats, wheat, and barley. Figure 4 shows a longitudinal section of an ungerminated barley seed. Within the seed the embryo lies at one end with the coleoptile extending toward the middle and the primary root toward the end. Both are attached to the scutellum, a structure adjacent to the nutritive endosperm. Just beneath the seed coat is a sheet of cells called the ALEURONE LAYER. When the seeds are soaked in water, one of the earliest events in germination is release of gibberellin by the embryo. The

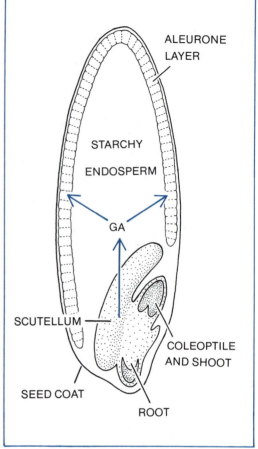

4

BARLEY SEED contains embryo at one end (bottom). Soaking seed in water triggers germination. Embryo releases gibberellic acid (GA) which reaches the aleurone cells, stimulating them to synthesize hydrolytic enzymes that digest nutrients stored in endosperm. The resulting sugars, amino acids and nucleotides become available for the growth of the embryo.

gibberellin eventually reaches the aleurone layer, where it induces these cells to begin snythesis and secretion of large amounts of hydrolytic enzymes — amylases for the digestion of starch, proteases, and nucleases. Thus sugars, amino acids, and nucleotides are made available in soluble form for the growth of the embryo. Endosperm in the seed serves the same function as yolk in eggs.

The gas ethylene is perhaps as much a pheromone as a hormone. Around the turn of the century kerosene stoves were frequently used to hasten the ripening of lemons and other citrus fruits. Substitution of more modern heating methods, however, proved to be an expensive mistake, because the fruit

3

EFFECT OF GIBBERELLIN on seedlings of mutant dwarf corn. One week before photograph was made seedlings at left were treated with gibberellin. Control seedlings at right were untreated.

failed to ripen. Evidently traces of ethylene from the stoves — not the heat — were promoting the ripening process. It was soon discovered that ripe oranges would hasten the ripening of nearby bananas; the oranges not only responded to ethylene, but eventually began producing it. Many developmental processes, including the dormancy of seeds and the falling of leaves, involve ethylene. In many cases the effects of ethylene resemble those of auxin. Indeed auxin may act in some plants by stimulating ethylene production, but in others ethylene is an auxin antagonist.

FLOWERING

A great deal of research has focused on the hormonal control of flowering in higher plants, but investigators have been repeatedly frustrated by the delicacy and complexity of the system. Chapter 7 pointed out that the usual subjects are plants that are sensitive to variations in PHOTOPERIOD (the number of hours of daylight), a trait that allows the experimenter to control precisely the timing of the flowering process. Variations in day length are detected by a photoreceptor in the young leaves: a pigment-protein complex called phytochrome. Exposure to red light transforms the receptor into a form that absorbs light of even longer wavelengths, in the far-red region of the visible spectrum. The far-red form of the receptor is biologically active. It interacts in some way with plant membranes, triggering a series of biochemical reactions that regulate the level of the hormones that control flowering, including a gibberellin and the elusive substance called florigen (page 138).

Although the evidence for the existence of florigen seems quite conclusive, florigen itself has resisted all attempts to isolate and characterize it. One might guess that different plants have entirely different florigens, compounding the difficulty, but the evidence suggests that only one substance is involved. The test is relatively simple. One takes a photoperiodically sensitive plant and induces it to flower by exposing it to the appropriate photoperiod. One then returns it to unfavorable photoperiod, removes a portion, and grafts it onto a second (uninduced) plant. The second plant will normally flower if the graft takes (Figure 5). Apparently the only limiting factor is whether or not successful grafts can be made — and they have been made between some quite distantly related species, such as the sunflower and the Jerusalem artichoke.

ANIMAL HORMONES

The hormonal control of growth and development in animals involves extensive changes, precisely coordinated. For example, the transformation of a tadpole into a frog includes destructive changes such as the elimination of the tadpole tail, gills and teeth, as well as constructive changes including the full development of limbs, tongue and middle ear. The pituitary gland, the master endocrine organ located next to the brain, begins this process by secreting a hormone that is carried by the bloodstream to the thyroid gland, which responds by secreting into the bloodstream the thyroid hormone that directly triggers metamorphosis. Experimenters have been able to induce complete metamorphosis with thyroxin, one of the thyroid hormones.

The metamorphosis of the frog also involves extensive changes at the biochemical level. A whole new set of enzymes appears in the liver, and a new kind of hemoglobin in the blood. The areas that resorb the tadpole organs contain high levels of proteases and other degradative enzymes, and a new visual pigment is formed in the retina. These changes are all inducible by thyroxin, which indicates that the specificity of the response depends not on the hormone but on the target cell. A given cell is already programmed to respond in a particular way by its physiological state and its previous development.

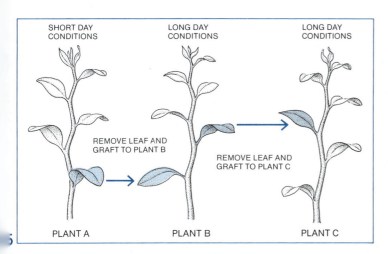

GRAFTS INDUCE FLOWERING in plants belonging to the same species. Leaf from plant that normally blooms under short-day conditions (left) is grafted onto plant under long-day conditions (center), inducing it to bloom. Grafted leaf from that plant in turn induces flowering in another (right).

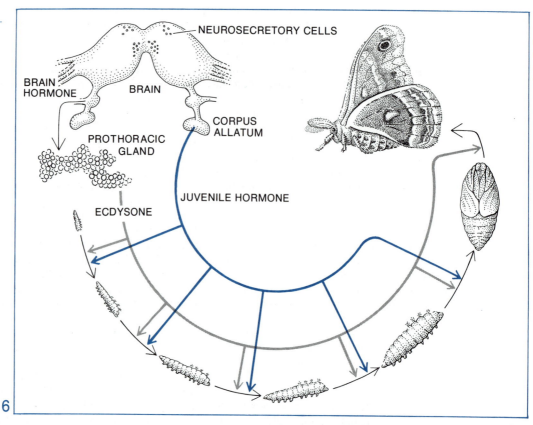

INTERACTION OF HORMONES during development of Cecropia. Molting of larva is induced by brain hormone that stimulates prothoracic gland to secrete ecdysone, a hormone that stimulates development. Ecdysone is balanced by juvenile hormone secreted by corpora allata. In pupal state the level of juvenile hormone drops. Shift in balance between two hormones allows pupa to develop into adult.

6

Metamorphosis also occurs in many invertebrate animals including sponges, starfish and insects. Carrol M. Williams and his team at Harvard University have studied metamorphosis in the silkworm moth *Cecropia*. The early embryo develops into a larva, the familiar caterpillar, which feeds and increases in size. Its growth is physically limited by an exoskeleton of chitin secreted by the epidermal cells. Eventually the entire exoskeleton is shed, a process called molting. The larva grows for a time, forms a new exoskeleton, and molts again. After the fifth molt the new exoskeleton is far more extensive; it encases the entire animal, including its appendages, forming a PUPA. Within the pupa an extraordinary reorganization begins. Small groups of larval cells, called IMAGINAL DISCS, begin extensive development, reshaping the entire animal, while the larval structures deteriorate. Finally the pupal exoskeleton is molted and the adult moth emerges.

Molting and metamorphosis involve three

Mammals probably have the most complex endocrine systems of all living organisms. During more than 100 million years of evolution, the trend in mammals has been toward bringing more physiological processes under hormonal control.

hormones. Certain brain cells secrete a hormone that causes the PROTHORACIC GLANDS to produce a mixture of two steroids called ECDYSONE. Meanwhile other glands just behind the brain, the CORPORA ALLATA (singular, corpus allatum) are producing a third substance, JUVENILE HORMONE. Throughout development ecdysone is the key substance in balance with juvenile hormone (*Figure 6*). High levels of juvenile hormone favor those synthetic activities associated with a larval molt. As the larva grows, the level of juvenile hormone drops, and ecdysone induces the transformation from larva to pupa. Finally juvenile hormone disappears, and one sees the expression of the full developmental potential of ecdysone: complete metamorphosis and the final pupal molt. But juvenile hormone is not simply an ecdysone antagonist. The British biologist Sir Vincent Wigglesworth has clearly shown that it promotes larval growth. The whole system illustrates the principle of hormonal interaction: two or more hormones acting together may bring about developmental changes quite unlike those caused by any single hormone acting alone. Compounds that inhibit these hormones, and hence prevent insects from maturing and reproducing, apparently have evolved as defense mechanisms in certain plants (*Box A*).

HORMONES IN MAMMALS

Developing mammals do not undergo the spectacular changes in form that are found among amphibians and insects, but mammals probably have the most complex endocrine systems of all living organisms. During more than 100 million years of evolution the trend in mammals has been toward bringing more and more physiological processes under hormonal control. Natural selection has accomplished this in a rather haphazard way, creating an endocrine organ in one convenient position or another, or modifying a steroid or some other available molecule to serve as a hormone. In terms of design the result is a bit of a hodgepodge, but the system as a whole works with exquisite precision.

The position of the endocrine glands of humans is shown in Figure 7, the structure of a few important hormones in Figure 8, and the sources and functions of the major hormones in Table I. In a book of this length it is not possible to consider all of the endocrine organs and hormones. The best approach is to examine a few hormones of particular importance:

A

THE PAPER FACTOR

In the 1950's Karel Slàma, a young Czech entomologist, brought a culture of European insects to the U.S. to study them in the laboratory of Carroll M. Williams at Harvard. The investigators were puzzled because instead of maturing into normal adults, the young nymphs continued to molt and to grow into still larger immature forms. In searching for the reason, Slàma and Williams discovered that the paper towels on the floor of the insects' cages contained a chemical substance that acted very much like insect juvenile hormone. Other investigators showed that this "paper factor" had a molecular structure very similar but not identical to the natural juvenile hormone of the bugs. It turned out to be present in certain kinds of paper (Scottowels, The New York Times and Scientific American) but not in others (the British scientific journal Nature). The source of the substance proved to be hemlock, balsam fir and a few other species of trees that are used to make some kinds of American paper, but not many foreign papers.

Biologists have speculated that a few plants have evolved the ability to manufacture the paper factor or similar substances as a kind of defense against the insects that attempt to feed on them. If young insects consume enough of the chemical, they either die or at least are unable to mature into reproductive adults. This discovery has led entomologists to study the possibility of controlling insects by using preparations containing specific insect hormones (or similar compounds) instead of conventional broad-spectrum insecticides, which are usually toxic both to pests and useful organisms.

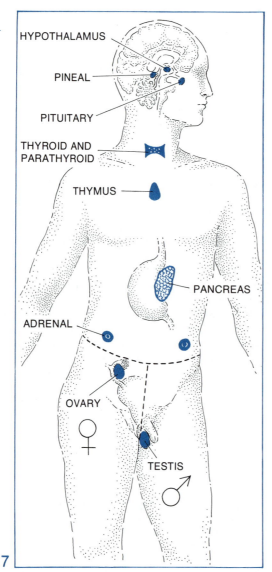

HYPOTHALAMUS

PINEAL

PITUITARY

THYROID AND
PARATHYROID

THYMUS

PANCREAS

ADRENAL

OVARY

TESTIS

7

**HUMAN ENDOCRINE ORGANS are indicated in
color. Such an endocrine system is typical of other
mammals as well.**

the growth hormone, the pituitary-hypothal-
amus system, reproductive hormones, and the
hormones that regulate stress reactions.

GROWTH HORMONE

The program of mammalian development is
under close hormonal control, and errors in
this control caused by defects in the endocrine
system can lead to tragedy. Such defects include gross errors in the final size to which the
organism grows, and disproportionate growth
of certain parts of the body.

Growth hormone is a product of the pituitary

gland (also called the hypophysis), a round
pea-sized organ enclosed in a bony cavity beneath the brain case. It consists of two lobes,
which are actually two distinct glands: the anterior and the posterior pituitary (*Figure 9*).
Excessive production of growth hormone early
in life (caused by a tumor of the anterior pituitary) leads to giantism; inadequate production
(caused by atrophy of the gland) leads to
dwarfism. The famous American dwarf, General Tom Thumb, was a victim of his pituitary.
Could physicians today prevent Tom Thumb's
unfortunate growth pattern? No, because
growth hormones from other mammals are in-
effective in humans; consequently there are
several thousand dwarfs living in the U.S. today. When biochemists learn how to synthesize the human hormone, dwarfs may disappear from the population. The reason that
many of these dwarfs now go untreated is that
treatment of a single individual would require
an extract from many human pituitaries per
day. If pituitary abnormalities occur after
adolescence, the error in growth is one of
proportion rather than of overall size. Excessive production of the hormone leads to ac-
romegaly, a fairly rare disease in which bones
of the face, hands and feet grow too large.

PITUITARY AND HYPOTHALAMUS

The anterior pituitary is often called the master
gland of the endocrine system because it secretes TROPHIC HORMONES: those that control
the development and activity of other endocrine organs, including the adrenal cortex, the
thyroid and the gonads (internal sex organs).
The control involves four different polypeptides: ACTH (adrenocorticotrophic hormone),
thyrotrophic hormone, and at least two
gonadotrophic hormones. The interaction of
the pituitary and its target glands is controlled
by both negative and positive feedback loops.
The regulation of the thyroid gland is a good
example of negative feedback. Thyroxin, mentioned earlier, speeds up oxidative metabolism. (Recall that in frogs it induces metamorphosis, an entirely different function.)
When thyrotrophic hormone from the anterior
pituitary reaches the thyroid, the production of
thyroxin increases. The rising concentration of
thyroxin in the blood in turn causes the pituitary to diminish the release of the thyrotrophic hormone. The pituitary hormone
speeds up this tightly controlled system,
thyroxin slows it down, and together they
maintain a finely balanced concentration of

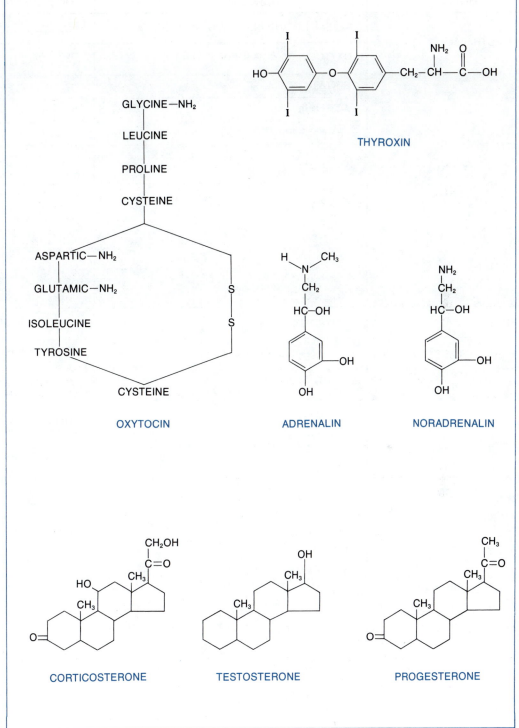

THYROXIN

OXYTOCIN

ADRENALIN

NORADRENALIN

CORTICOSTERONE

TESTOSTERONE

PROGESTERONE

8

SOME MAMMALIAN HORMONES. Thyroxin is an amino acid with four atoms of iodine. Adrenalin and noradrenalin are based on the amino acid tyrosine. Oxytocin is a simple polypeptide. Corticosterone, testosterone and progesterone are steroids, all of which have the basic ring structure shown here; they differ in the groups attached to the rings.

GLAND	HORMONE	MOST IMPORTANT EFFECTS
HYPOTHALAMUS	RELEASING AND INHIBITING FACTORS (P)	GOVERN SECRETION OF HORMONES BY THE ANTERIOR PITUITARY
	OXYTOCIN (P) & VASOPRESSIN (P)	STORED AND RELEASED BY POSTERIOR PITUITARY (SEE BELOW)
ANTERIOR PITUITARY	GROWTH HORMONE (PP)	STIMULATES GROWTH
	ACTH (PP)	STIMULATES ADRENAL CORTEX
	THYROTROPHIC HORMONE (PP)	STIMULATES THYROID
	FSH (PP)	STIMULATES GROWTH OF OVARIAN FOLLICLES (IN FEMALES) AND SPERMATOGENESIS (IN MALES)
	LH (PP)	STIMULATES CONVERSION OF OVARIAN FOLLICLE INTO CORPUS LUTEUM. STIMULATES SECRETION OF SEX HORMONES BY OVARIES AND TESTES
POSTERIOR PITUITARY	OXYTOCIN (P)	STIMULATES CONTRACTION OF UTERINE MUSCLES AND THE RELEASE OF MILK BY MAMMARY GLANDS
	VASOPRESSIN (P)	CONTROLS EXCRETION OF WATER. STIMULATES CONTRACTION OF SMOOTH MUSCLE IN WALLS OF SMALL ARTERIES
THYROID	THYROXIN (A)	STIMULATES AND MAINTAINS OXIDATIVE METABOLISM
PARATHYROIDS	PARATHORMONE (PP)	MAINTAINS NORMAL CALCIUM METABOLISM AND BONE GROWTH
THYMUS	THYMOSIN (P)	STIMULATES IMMUNE RESPONSE OF LYMPHATIC SYSTEM
PANCREAS	INSULIN (PP)	INCREASES STORAGE OF GLYCOGEN, STIMULATES OXIDATION OF CARBOHYDRATES, LOWERS BLOOD SUGAR
	GLUCAGON (PP)	STIMULATES CONVERSION OF GLYCOGEN TO GLUCOSE IN LIVER
ADRENAL MEDULLA	ADRENALIN (A)	STIMULATES FIGHT-OR-FLIGHT REACTIONS. STIMULATES CONVERSION OF GLYCOGEN TO GLUCOSE
	NORADRENALIN (A)	SUSTAINS BLOOD PRESSURE AND REGULATES REACTIONS TO STRESS
ADRENAL CORTEX	GLUCOCORTICOIDS (S) (CORTISONE, CORTICOSTERONE, HYDROCORTISONE)	CONTROL METABOLISM OF CARBOHYDRATES, PROTEINS AND LIPIDS. REDUCE INFLAMMATION OF TISSUES
	MINERALCORTICOIDS (S) (ALDOSTERONE, DEOXYCORTICOSTERONE)	CONTROL WATER BALANCE AND REGULATE METABOLISM OF SALTS (SODIUM, POTASSIUM AND OTHER MINERALS)
	CORTICAL SEX HORMONES (S)	MAINTAIN SECONDARY MALE SEXUAL CHARACTERISTICS
OVARIES	ESTROGEN (S)	STIMULATES DEVELOPMENT AND MAINTENANCE OF SECONDARY FEMALE SEXUAL CHARACTERISTICS AND BEHAVIOR
	PROGESTERONE (S)	SUSTAINS PREGNANCY AND HELPS TO MAINTAIN SECONDARY FEMALE SEXUAL CHARACTERISTICS
TESTES	TESTOSTERONE (S)	STIMULATES DEVELOPMENT AND MAINTENANCE OF SECONDARY MALE SEXUAL CHARACTERISTICS AND BEHAVIOR

I

thyroxin in the bloodstream. The overall result is a steady control of oxidative metabolism within the body.

The posterior pituitary, or neurohypophysis, is in fact an extension of the hypothalamic region of the brain (*Chapter 13*). Nerve cells in the hypothalamus produce two hormones, vasopressin and oxytocin, that flow down the stalk that connects the hypothalamus to the posterior pituitary, where they are stored. (This direct secretion of hormones by nerves, called neurosecretion, is common in both vertebrates and invertebrates.) Oxytocin induces contraction of the muscles of the uterus during childbirth. Vasopressin acts as the antidiuretic hormone, increasing the retention of water by the kidney tubules and thus decreasing the volume of urine. Vasopressin also causes blood pressure to rise by inducing constriction of the small arteries.

The intimate association between the pituitary gland and the brain raises the question: is the pituitary truly the master gland? The feedback between the trophic hormones of the anterior pituitary and their target glands is not a closed loop, but is subject to outside influences. The activities of at least some of the target glands fluctuate according to changes in the outside environment. In many mammals the concentration of sex hormones, for example, changes from season to season; there is, in other words, a breeding season and an off season. Similarly the levels of some of the hormones of the adrenal cortex, including cortisone, which assist the body in adjusting to stress, are powerfully influenced by outside events.

The explanation of the variable nature of these interactions is that the brain monitors changes in the environment and, guided by this information, directly controls the anterior pituitary. Information about changes in photoperiod, the warming and cooling of the air, the appearance of an enemy or a mate, is received from the sense organs. After being processed in the forebrain the information is transmitted to the hypothalamus, which responds by secreting special hormones of its own into a set of blood vessels connected directly to the anterior pituitary. These hormones in turn may

PITUITARY GLAND lies nestled in a bony cavity beneath the brain (top). It is intimately associated with the hypothalamus region of the brain. The two lobes of the pituitary (bottom) secrete hormones that affect many different organs of the body.

either induce or inhibit the secretion of the anterior pituitary hormones. Hypothalamic hormones that stimulate the secretion of anterior pituitary hormones are called RELEASING FACTORS; those that act in the opposite way are INHIBITORY FACTORS. Several such hypothalamic factors have now been isolated, identified, and synthesized. All are small peptides. Eventually separate inhibitory and releasing factors may be identified for all anterior pituitary hormones.

In short, the anterior pituitary is not really a master gland in the true sense of the word, but acts under the direction of the hypothalamus. This relationship too is regulated by negative and positive feedback loops. If small quantities of the female hormone estrogen are implanted directly into certain areas of the hypo-

ROLES OF PRINCIPAL HORMONES in vertebrates. Abbreviations indicate the chemical nature of each hormone: peptides (P), proteins (PP), amines or amino acids (A), and steroids (S). Table lists only the most important effects of the major hormones. Humans have many more hormones than those listed here.

thalamus, the secretion of female gonado-trophins falls sharply, limiting the further production of estrogen. The same sort of negative feedback probably prevails under natural conditions when estrogen reaches the hypothalamus through the bloodstream. The hypothalamus is probably subject to feedback effects from other hormones as well, with each hormone affecting a different group of target cells.

HUMAN REPRODUCTIVE HORMONES

The hypothalamus and pituitary play a central role in the regulation of human reproduction. At puberty the young male undergoes a series of profound physical and psychological changes. Triggered by releasing factors from the hypothalamus, the anterior pituitary secretes two gonadotrophic hormones: FSH (follicle-stimulating hormone) and LH (luteinizing hormone). LH stimulates the secretion of androgens (male hormones), including testosterone, by special endocrine cells in the testes. FSH, together with LH and testosterone, induces maturation of the testicular germ cells, and the production of spermatozoa (*Chapter 7*). The androgens are responsible for the characteristic changes of male adolescence, including the growth of hair on the face, the deepening of the voice, and development of a more muscular body.

Parallel physiological events transform the adolescent girl. The same two trophic hormones, FSH and LH, are secreted by the anterior pituitary of the female in response to releasing factors from the hypothalamus. These hormones induce the maturation of the ovaries, which then begin to secrete the two female sex hormones, estrogen and progesterone. The sex hormones in turn initiate the development of the traits characteristic of a sexually mature woman, including enlarged breasts, vagina and uterus, a broad pelvis, pubic hair, and — very suddenly — the onset of the menstrual cycle.

The endocrine control of reproductive physiology seems more complex in women than in men. Mature female mammals, including humans, are subject to an ESTROUS CYCLE. The culmination of the cycle is ovulation, the release of an unfertilized egg from the ovary. The wall of the uterus, properly primed by the effects of estrogen and progesterone, is ready to receive an embryo. In most mammals the female becomes sexually receptive only around the time of ovulation; she is then said to be "in heat" and may even become the aggressive partner. During the rest of the cycle she is unreceptive and seldom if ever attempts to mate. Some species such as sheep and deer ovulate only during a relatively short annual breeding season. In other species the cycle runs continuously; rats and mice ovulate every four or five days.

The human estrous cycle is unusual in two respects. First, as noted in Chapter 7, a woman is never in heat, but remains receptive to some degree throughout most of the cycle. Foreplay and social circumstances have a strong effect on her receptivity. Some students of human evolution ascribe this trait to the development of an unusually strong pair bond in the human species. Instead of approaching females only at the time of heat, human males remain closely affiliated at all times, perhaps because of the long and difficult period of child rearing in the human species. (The origins of human social behavior are examined more closely in Chapter 26.)

The second unusual aspect of the human estrous cycle is menstruation, the periodic flow of blood. In other mammals, as noted in Chapter 7, the cycle is seldom marked by noticeable bleeding. The human cycle lasts about 28 days, hence the term *menses*, which means a month. The hormones that control the menstrual cycle and pregnancy are secreted by the ovary, placenta and the wall of the uterus. Following menstruation, which begins the cycle, the hypothalamus induces the anterior pituitary to secrete FSH and LH, the same hormones that initiate adolescence. These gonadotrophins stimulate the growth of follicles, envelopes of cells that surround a potential egg. The growing follicle cells secrete estrogen, one of the two female sex hormones (*Figure 10*). The estrogen in turn induces the growth and thickening of the uterine wall. In a positive feedback loop, the rising estrogen concentration activates the LH-stimulating center of the hypothalamus, leading to a sharp rise in the concentration of LH. As the LH level peaks, the best-developed follicle in the ovary bursts and releases the egg, which enters the Fallopian tubes, where it awaits possible fertilization. At ovulation the follicular phase of the menstrual cycle ends and the luteal phase begins.

The luteal phase is named for the corpus luteum (Latin, yellow body), a mass of old follicle cells which have been transformed under the influence of LH into a cluster of yellowish

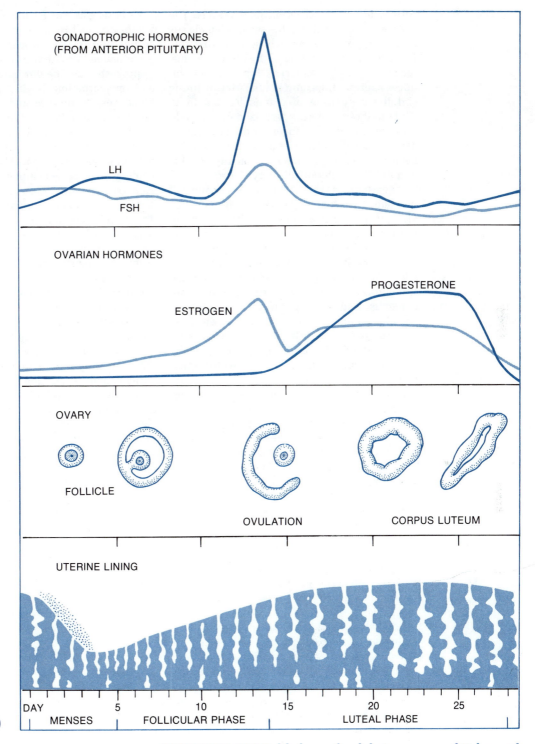

GONADOTROPHIC HORMONES
(FROM ANTERIOR PITUITARY)

LH

FSH

OVARIAN HORMONES

PROGESTERONE

ESTROGEN

OVARY

FOLLICLE

OVULATION

CORPUS LUTEUM

UTERINE LINING

DAY

5   10   15   20   25

MENSES   FOLLICULAR PHASE   LUTEAL PHASE

10

MENSTRUAL CYCLE of the human female lasts an average of 28 days, and is divided into three phases: the menses (the period of bleeding), the follicular phase and the luteal phase. The two graphs at top depict the varying levels of pituitary and ovarian hormones during the cycle. The diagram (third from top) shows the condition of the ovarian follicles and the corpus luteum. The drawing at bottom indicates the state of the lining of the uterus.

cells richly supplied with blood vessels. The corpus luteum produces the second female sex hormone, progesterone, whose principal role is the further preparation of the uterus for nurturing a possible embryo. Progesterone induces a further thickening of the uterine lining and the maturation of glands located in it. Without these changes a fertilized ovum could not implant itself. Progesterone also has a negative feedback effect on the LH-activating center of the hypothalamus, causing the LH output of the anterior pituitary to drop.

Progesterone is rightly called the hormone of pregnancy. So long as the corpus luteum survives, its progesterone prevents the ovary from releasing another egg; in other words, the menstrual cycle is suspended. This inhibition is exploited in birth-control pills, which contain synthetic compounds simular to estrogen and progesterone. They "trick" the hypothalamus into turning off the secretion of gonadotrophins by the anterior pituitary. A woman taking "the pill" does not make fully developed follicles and eggs, nor does she have a corpus luteum and a well-primed uterine wall.

What sustains the corpus luteum during pregnancy? How does it "know" that an embryo is growing in the wall of the uterus? The placenta connecting mother and embryo apparently secretes a hormone similar to LH. This hormone maintains the corpus luteum until birth, then degenerates. In other words, pregnancy sustains the corpus luteum which in turn sustains the pregnancy.

If the egg is not fertilized, no placenta develops and its LH-like hormone is lacking; the corpus luteum regresses and the level of progesterone falls. With the virtual disappearance of progesterone, the uterine wall breaks down and menstruation begins as the engorged blood vessels rupture and release their contents into the uterine cavity (*Box B*). During the first few days of the menstrual flow the body has very low concentrations of female hormones of any kind. This temporary deficiency may be accompanied by feelings of anxiety, irritability and depression, as well as abdominal cramps originating in the uterus. The drop in progesterone level also triggers the beginning of the next cycle. No longer inhibited by progesterone, the gonadotrophin-stimulating centers of the hypothalamus elicit the release of gonadotrophins from the anterior pituitary. This stimulates the development of follicles in the ovary, and a new cycle begins.

Between the ages of 40 and 50 women enter menopause, the post-reproductive period of their lives, when the menstrual cycle comes to a stop. Menopause is apparently caused by a permanent decline in the LH-stimulating activity of the hypothalamus, breaking the chain of events required to complete the menstrual cycle. The ovaries also become less sensitive to the gonad-controlling hormones, causing a decline in the output of estrogen. Although menopause is often accompanied by emotional distress, a woman's basic physical and mental powers remain virtually undiminished, and she can still enjoy a fully satisfying sexual life.

B

A STRANGE WINDOW ON THE UTERUS

How did biologists study the complex changes in the wall of the uterus during the menstrual cycle? Much of the story was worked out by surgically removing tiny bits of tissue from the uterus on different days of the cycle, then examining stained sections under the microscope. One physiologist, J. E. Markee of Duke University, used an ingenious method to observe the process directly. He transplanted bits of uterine tissue from a female rhesus monkey into the front chamber of the monkey's eye. The tissue survived and was nourished by the blood vessels that supply the eye. The hormones that govern the cycle, broadcast through the bloodstream, reached the eye as well as the uterus, and the transplant responded appropriately. By peering into the eye with a microscope, Markee was able to study the tissue transplant day by day throughout the menstrual cycle. Because the rhesus menstrual cycle is very similar to that of humans, Markee obtained valuable new insights about uterine changes in women. In particular he was able to observe what happens during the luteal phase of the cycle: the events that lead to the breakdown of the uterine lining, the rupture of blood vessels and the onset of menstrual bleeding.

One of the most important regulatory systems of the mammalian body is controlled by the adrenals, pea-sized glands perched on top of each kidney (*Figure 11*). Like the pituitary, the adrenals each consist of two glands, with a distinct set of hormones. The central part, the MEDULLA, secretes adrenalin (sometimes called epinephrine) and noradrenalin. The outer part of the adrenal, the CORTEX, produces several steroid hormones with a different set of functions.

Unlike most other endocrine glands the adrenal medulla is under direct nervous control. It is stimulated by nerves of the autonomic nervous system (*Chapter 13*) that originate in the hypothalamus. This hookup enables the adrenal medulla to respond quickly to outside stresses. When danger threatens, adrenalin and noradrenalin are promptly released into the bloodstream, affecting the body in different but complementary ways. Adrenalin acts quickly in conjunction with the autonomic nervous system to prepare the body for "fight or flight." The heart rate and blood pressure go up, veins dilate, the concentration of sugar in the bloodstream rises, and so does the number of certain white blood cells (eosinophils). The blood flow through skeletal muscle, brain and liver increases by as much as 100 percent; digestive and reproductive functions are inhibited, and one suffers a feeling of anxiety. Adrenalin is secreted whenever one is put in a stressful situation such as cold weather, hostility from others, or a narrow escape from some peril. Adrenalin also promotes the release of ACTH from the anterior pituitary, causing the release of hormones from the adrenal cortex and a more prolonged adjustment of the body to stress. Noradrenalin is released independently in stressful situations and acts mainly to sustain blood pressure. In other words, adrenalin attends to the emergency while noradrenalin plays a secondary, mainly regulatory role. A curious effect observed in humans is that participation in aggressive encounters induces the release of relatively large quantities of noradrenalin but only moderate amounts of adrenalin; conversely, the anticipation of such events, in the form of anger or fear, favors the release of adrenalin. Hockey players on the bench, for example, might secrete mostly adrenalin, while their teammates on the ice might secrete mostly noradrenalin.

The adrenal cortex has a wide variety of

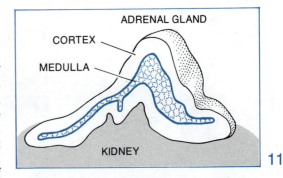

ADRENAL GLAND consists of an inner medulla and an outer cortex. Drawing depicts a cross-section through the gland. One adrenal gland is perched on top of each kidney.

roles, but in contrast with the swift-acting adrenalin and noradrenalin, its hormones generally exert relatively slow sustained effects on overall metabolism, kidney function, electrolyte balance, blood pressure and tissue inflammation. The cortical hormones are known collectively as corticoids, and all are steroid compounds. One group, the glucocorticoids, includes the well-known cortisone. They are released continuously at rates that show a daily rhythm, and promote the conversion of proteins into glucose within the liver, and they also reduce the inflammation of tissues. Another group, the mineralcorticoids (including aldosterone and deoxycorticosterone), regulates the distribution of sodium and other minerals in the tissues.

Under prolonged stress, the appropriate center in the hypothalamus induces the release of ACTH from the anterior pituitary. This substance in turn evokes an outpouring of glucocorticoids from the adrenal cortex. In general these corticoids cause responses opposite to those induced by adrenalin and noradrenalin; they serve as a brake on the body's emergency mobilization. Some quantity of corticoids is required even when the body is not under stress. Animals whose adrenals have been removed experimentally show symptoms identical to those of human patients with Addison's disease (adrenal insufficiency): lowered blood sugar, digestive disturbances, reduced blood pressure and body temperature, kidney failure, and an inability to stand stress of any kind. Without corticoids the body's condition deteriorates in the face of extreme temperatures, prolonged activity, infections, intoxication and the like.

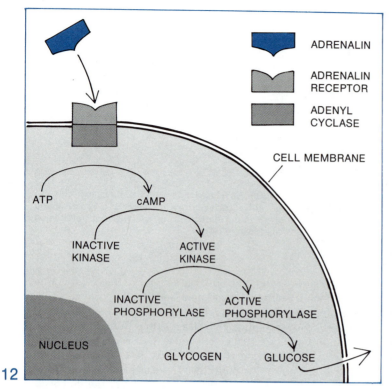

12

ADRENALIN MOBILIZES GLUCOSE in a series of steps, beginning with the binding of the hormone to a receptor in the cell membrane. Arrows show a series of reactions, beginning with the conversion of ATP to cAMP in the cytoplasm, that result in the conversion of glycogen to glucose. The glucose can be exported from the cell and carried by the bloodstream to other cells. Similar mechanisms are involved in the action of many other hormones.

In general one can say that hormones released directly by the nervous system are fast-acting and their effects are short-lived. The hypothalamus itself, for example, makes fast-acting oxytocin, and the adrenal medulla, a neural extension of the hypothalamus, produces fast-acting adrenalin. Conversely the hormones of the anterior pituitary and the adrenal cortex — both non-neural tissues — mediate effects that are relatively slow and long-lasting.

HORMONES AND CELLS

What happens at the molecular level when a hormone reaches a target cell? This is one of the key problems in modern biology. Two mechanisms have been proposed, one for protein-like hormones and the other for steroid hormones. Many of the protein, polypeptide and amine-type hormones act via a second messenger (the hormone itself being the first messenger): cyclic AMP (cAMP). Each of these hormones attaches itself to a particular receptor, probably a protein bearing a specific recognition site for the hormone. The receptor lies in the cell membrane and extends outward, making it unnecessary for the hormone to penetrate into the interior of the cell. When the hormone combines with this external receptor it activates an enzyme, adenyl cyclase, which converts ATP to cAMP within the cell. These events were first discovered by the late E. W. Sutherland of Washington University, who was awarded the Nobel Prize in 1971. Working with adrenalin, he found that the cAMP in turn activates a series of reactions that convert the enzyme phosphorylase from an inactive to an active form (*Figure 12*). The active form splits glycogen to form glucose, which becomes available not only to the target cell, but also can be transported by the bloodstream to other cells.

In the years after Sutherland's early work the list of other systems activated by cAMP seemed to multiply. It is now clear that many hormone effects in vertebrate tissues act via this second messenger. These effects include the action of adrenalin in stimulating muscle and liver cells to convert glycogen to glucose, and in stimulating the breakdown of lipids in fat cells; the action of ACTH in stimulating the production of glucocorticoids in the adrenals; the action of LH in stimulating the production of progesterone in the ovary; and many more. That such diverse effects can all be mediated by a single compound is explained not only by the specific nature of the receptor on the cell surface, but also by the fact that the target cells have secondary targets within them, targets that activate reaction systems such as those involved in the synthesis of glucocorticoids.

Steroid hormones such as estrogen, progesterone and the hormones of the adrenal cortex do not react with receptors on the cell surface,

Although hormones and nerves are described in separate chapters, one should bear in mind that they actually comprise a single response system.

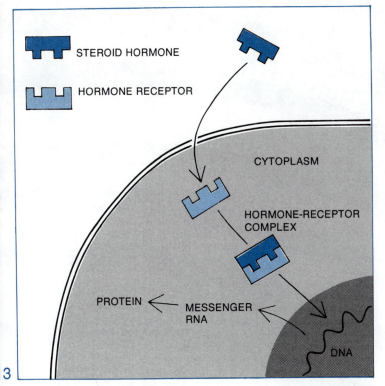

ACTION OF STEROID HORMONES involves a receptor in the interior of the cell. Hormone-receptor complex enters the nucleus and switches on the appropriate genes. Messenger RNA coded by the genes initiates protein synthesis in the cytoplasm.

but penetrate into the interior and bind to a specific receptor molecule in the cytoplasm (*Figure 13*). The hormone-receptor complex then enters the nucleus, where it somehow switches on the appropriate genes. Messenger RNA coded by these genes then moves into the cytoplasm to mediate the synthesis of specific proteins. The production of these proteins accounts for the specificity of response that is characteristic of steroid hormones.

These results are based on work with animal hormones. Much less is known about the mechanism of action of plant hormones. There is convincing evidence of the existence of membrane binding sites for only one plant hormone (auxin), and very little is known about what happens as an immediate result of auxin binding. One effect is the rapid release of hydrogen ions from treated cells, which sharply lowers their pH. Certain plant cells begin to grow when their pH is suddenly re-

duced experimentally, but the growth stops very quickly, while growth induced by auxin continues for hours. The release of hydrogen ions cannot be the sole mechanism by which auxin acts, but so far the others have not been discovered.

The next chapter describes the nervous system which, together with the endocrine system, orchestrates the responses of animals to the varying demands of their internal and external environment. Although hormones and nerves are treated separately, the reader should bear in mind that they actually comprise a single response system.

READINGS

E. FRIEDEN AND H. LIPNER, *Biochemical Endocrinology of the Vertebrates*, Englewood Cliffs, N.J., Prentice-Hall, 1971. A readable paperback providing a nice overview of vertebrate hormones, their functions, interactions, and modes of action.

A.W. GALSTON AND P.J. DAVIES, *Control Mechanisms in Plant Development*, Englewood Cliffs, N.J., Prentice-Hall, 1970. A highly readable treatment of phytochrome, plant growth hormones, and hormone-mediated phenomena.

A. LABHART, *Clinical Endocrinology*, New York, Springer Verlag, 1976. An excellent advanced level text on the physiology and biochemistry of hormone action with special emphasis on the clinical aspects of human endocrinology.

J.S. PERRY, *The Ovarian Cycle of Mammals*, Edinburgh, Oliver & Boyd, 1971. Another affordable paperback. The hormonal regulation of female reproduction in mammals is treated concisely, readably and accurately.

H.V. RICKENBERG, Editor, *Biochemistry of Hormones* (MTP International Review of Science, Volume 8), Baltimore, University Park Press, 1974. Provides scholarly articles on the molecular basis of hormone action by some of the top research practitioners in the field.

C.D. TURNER AND J.T. BAGNARA, *General Endocrinology*, 6th Edition, Philadelphia, W.B. Saunders, 1976. A comprehensive treatment of animal endocrinology, both vertebrate and invertebrate.

13

neurobiology

The period of life during which the nervous system is more or less flexible appears to vary from animal to animal and no doubt from system to system in a given species. In fact some plasticity seems to persist since many of us are able to learn new facts and skills through life.

DAVID HUBEL

For man, the nervous system is what life is about. All emotional and intellectual pleasures are functions of the brain. A man's life, when it surmounts the short-term needs for adequate food, shelter, and sleep, is devoted to modifying the contents of his brain to provide more happiness, satisfaction or fulfillment.

The nervous systems of all animals have three functions. They continuously COLLECT INFORMATION about the external and internal environments. They PROCESS AND INTEGRATE this information, evaluating it with respect to past experience and future goals. And they ACT upon this information by co-ordinating the activities of EFFECTORS — muscles and glands. In other words, the nervous system consists of three subsystems. The SENSORY SYSTEM gathers data, converts it into a form that the nervous system can handle, and feeds it to the second subsystem, the CENTRAL NERVOUS SYSTEM (CNS), which in vertebrates consists of the brain and spinal cord. It is essentially a computer containing inherited information (instincts and reflexes) and acquired information (memory and learning). The new input data may be stored or may trigger instructions to activate parts of the third system, the MOTOR SYSTEM. Motor nerves cause muscles or other organs to respond suitably to sensory data. The term "peripheral nervous system" is often used to refer collectively to the components that lie outside the brain and spinal cord.

NEURONS AND GLIA

The nervous system contains two special kinds of cell, neurons and glia. The NEURON is the only cell involved in actual transmission of the signals, called nerve impulses, which flow through the system. The GLIAL CELLS (from the Greek word for glue) appear to be merely packing material, but some workers believe that they serve other functions as well.

Neurons come in different shapes and sizes (*Figure 1*). Motor neurons are connected to muscles or glands, and sensory neurons to some kind of sensing cell. Interneurons take impulses from one neuron and pass them to another.

Most neurons are very long in comparison with their width; an extreme example is the motor neuron to the hoof of a giraffe. It is a few meters long, but its cell body (which lies in the spinal cord) is only a few microns across. As Figure 1 shows, all neurons have branches called processes. Only one of them, the AXON, can transmit impulses actively; it carries impulses away from the cell body. The other branched and knobbed processes, called DENDRITES, carry impulses (usually passively) toward the cell body.

AXONIC TRANSMISSION

ANTENNA of male silkworm moth, part of which is shown in photomicrograph on opposite page, is a sensitive chemoreceptor. Neurons at base of tiny sensory hairs can detect even a few molecules of bombykol, a female sex attractant.

Like all cells, the neuron has a polarized external membrane; there is a small potential difference between the inner and outer surface of the membrane. A common value for this potential is 70 millivolts (mV),

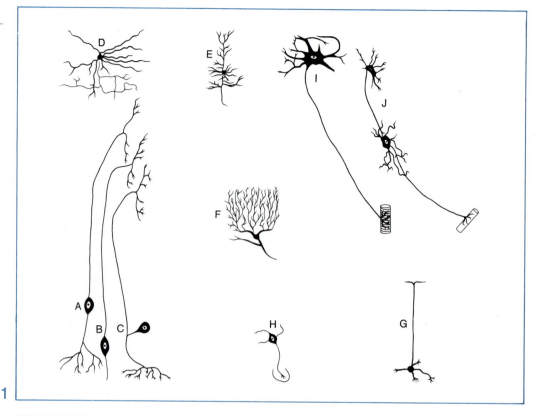

1

SHAPE OF NEURON is related to its function. Bipolar sensory neurons appear at lower left (A, B, C). D is a connecting neuron from central nervous system. Neurons at center and lower right are found in the brain: E in the cerebral cortex, F in the cerebellum, G in the optic lobe, H in the pituitary gland. I is a motor neuron, and J is a pair of neurons from the autonomic nervous system.

with the outside of the membrane being positive. If one applies the negative output (cathode) of a battery to the surface of an axon, a wave of depolarization spreads in both directions. The wave does not decrease in size regardless of the length of the axon it must traverse. This wave of depolarization is called the ACTION POTENTIAL. All data about the inner and outer worlds must be translated into action potentials if the brain is to comprehend them. Just as ATP is the indispensable currency of biological energy, the action potential is the basic unit of communication within the nervous system.

The properties of the action potential are illustrated in Figure 2. Two electrodes are connected to an oscilloscope to permit measurement of small potential differences between the electrodes. Electrode A is near (but not on) an axon, and electrode B is on the outer surface of the axon. No potential difference is seen. But if electrode B is plunged into the axon's interior, the oscilloscope shows that B is about 60 to 80 mV more negative than A. As an

Action potentials are the messages that nerves carry, the basic units of communication within the nervous system. All data about the inner and outer worlds must be translated into action potentials for the brain to comprehend them.

action potential moves, B loses its negativity with respect to A for about one millisecond, and in fact briefly becomes more positive than A. Events return to normal as the action potential passes on.

An easier procedure for watching action potentials is to hook electrode B around the axon. One then sees no resting potential, but each action potential shows up as a brief pulse (Figure 3). If the electrode were kept in any one axon, the action potential would always move at precisely the same velocity. But in general the action potential moves faster in axons with larger diameters. Moreover, if the total depolarization of the action potential in one axon were 80 mV (for instance, if the inside began at

-75 mV and at the height of the action potential reached $+5$ mV) then it would be 80 mV for all action potentials passing the axon. In other words, the amplitude and velocity along any one axon are always the same.

The action potential transmits information; for example, information about the intensity of pain or odor or light at some particular receptor cell. It cannot signal "more pain is being felt" by sending faster or larger action potentials; it can only shape the pattern of signals, for instance the number per second or the total number in a single burst. Figure 3 shows recordings from two nerves whose signals originate in a pair of auditory sensory cells of a moth's ear. All the signals in any one axon

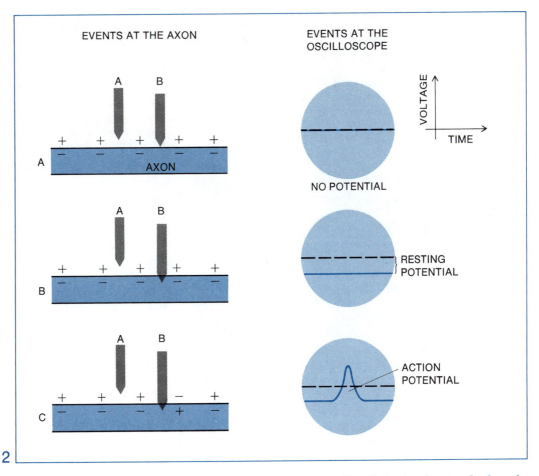

2

ACTION POTENTIAL is a wave of depolarization that travels along the nerve fiber. Oscilloscope measures the difference in electrical potential between electrodes A and B. When both electrodes are outside the axon there is no electrical potential across them (top). If electrode B penetrates the axon, oscilloscope shows 60–80 millivolt resting potential (middle). As a nerve impulse passes, oscilloscope shows positive action potential (bottom). Action potential lasts about one millisecond.

move at the same rate and have the same height, but the frequency rises as the stimulus intensity increases.

The widely (but not universally) accepted explanation of the action potential rests on the work of the English researchers Hodgkin and Huxley, who received a Nobel prize for it in 1963. Their work was based on the principle that ions, like molecules, tend to flow down a concentration gradient. The model in Figure 4 is a simplified version of a resting axon whose inside (high in K⁺) corresponds to B and whose outside (low in K⁺) corresponds to A. If one puts a 200 mM solution of KCl in compartment B, which is separated from a 10 mM solution of KCl in compartment A by a membrane permeable only to K⁺, then K⁺ will start flowing from B to A. (Cl⁻ is trapped inside because the membrane is impermeable to it.) If mixing in A or in B is negligible, the local depletion of K⁺ from the inside of the membrane and its accumulation on the outside will make the inside negative with respect to the outside; that is, it will create a membrane potential. This potential impedes the further flow of K⁺, because of the repulsion of the positive

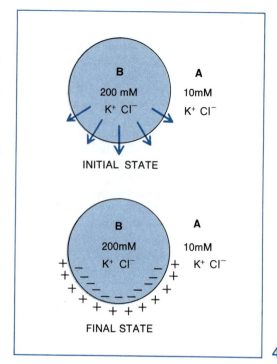

INITIAL STATE

FINAL STATE

4

MODEL OF AXON illustrates how the movement of potassium ions (K⁺) can create a resting potential across the membrane. High initial concentration of K⁺ within the membrane (top) causes K⁺ ions to diffuse outward, making the outside of the membrane positive with respect to the inside (bottom).

K⁺ by the accumulated positive charge on the inside. The potential will increase progressively until it is just enough to prevent further K⁺ flow, which will then stop. In other words the resting membrane potential of axons is a potassium potential.

How is the K⁺ concentration kept high internally? An active-transport "pump," whose molecular basis is unknown in detail, uses ATP to pump Na⁺ out of the neuron and K⁺ into it, against a concentration gradient. The pump keeps the K⁺ concentration inside 20 times higher than outside, while keeping the Na⁺ concentration inside nine times lower than outside.

When an action potential arrives, it creates a sudden leakiness in the normally sodium-impermeable membrane. Another way of saying this is that it opens an ionic "gate," permitting Na⁺ to rush in down its concentration gradient, which makes the inside temporarily positive. This reversal of polarization in turn flings open a K⁺ gate so that K⁺ suddenly pours out, reversing the effect of the Na⁺ and restoring the membrane to its former polarized state.

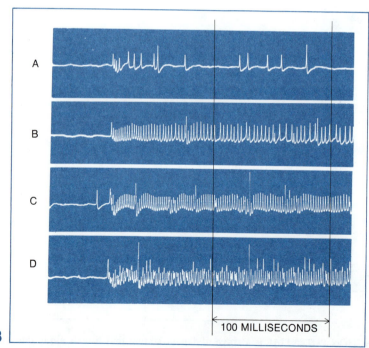

100 MILLISECONDS

3

FREQUENCY OF ACTION POTENTIAL increases as the intensity of stimulus increases. Recordings above depict the firing of two nerve cells in the ear of a moth. One cell produces a few large spikes, the other many smaller spikes. In recording A sound level is just above threshold; in B sound is seven decibels louder, in C 15 decibels, and in D 23 decibels.

As the action potential passes, each tiny segment of axon becomes briefly reverse-polarized so that the outside of the segment is briefly positive compared with the neighboring segment. A current flows to redress the imbalance, and this flow depolarizes the neighboring segment. This partial depolarization in turn triggers the sudden sodium-leakiness of the segment, releasing the whole cycle of events which occurs again. In this way the impulse is propagated from segment to segment along the axon.

The velocity of simple axonic transmission is sometimes too slow for emergencies. Two quite different ways of dealing with this problem have evolved. Invertebrates have increased the speed of transmission by increasing the diameter of their nerve fibers. The giant axon of the squid (about one millimeter in diameter) transmits at about 30 meters per second, which is almost ten times faster than most of the squid's common axons. These giant axons carry signals to the jet-propulsion device of the squid, so that the animal can move in a hurry if necessary.

Vertebrates have no giant axons, but have evolved a technique called saltatory (jumping) conduction. Many axons are coated with an insulating layer of a fatty material called MYELIN. The myelin sheath dwindles to almost nothing at intervals of a few mm, and these breaks in the sheath are called NODES OF RANVIER. This structure exploits the principle that current flows faster through a solution than along a small unmyelinated axon. When the depolarizing pulse reaches a node, the node becomes transiently negative on the outside. Current flows to it, along the axon, depolarizing the next node. The depolarization jumps from node to node along the axon, a process that permits velocities up to 120 meters per second.

THE CELLULAR BASIS OF INTEGRATION

The junction between two or more neurons is called a SYNAPSE, a term that is also applied to the active junctions between neurons and muscles or glands. At a synapse there are actual gaps between the cells, typically 200 Å wide, and the transmission of impulses across them involves special control mechanisms. Synapses are usually one-way devices; impulses can cross them in only one direction. Nearly all neurons have synapses with many other neurons. Synapses may occur between an axon and a dendrite, or an axon and a cell body, or even between an axon and an axon. Any given neuron may be an excitatory kind, which tends to pass the impulse to its neighboring cell; or it may be inhibitory, tending to prevent an excitatory impulse from passing between its neighbors. To illustrate, suppose neurons A, B, C, and D all have synapses with neuron Z in such a way that Z is always on the receiving end (*Figure 5*). By stimulating A's axon, one can cause an impulse to travel along it and pass across the synapse to Z. If Z fires (that is, an impulse is recorded in Z's axon) when A is stimulated, one says that A is an excitatory neuron, and the A-Z synapse is an excitatory synapse. If Z fails to fire when B's axon is stimulated, or when A's and B's axons are stimulated simultaneously — in other words, B is preventing A from exciting Z — then B is called an inhibitory neuron, and the B-Z synapse is called an inhibitory synapse. This ability of one neuron to prevent another from exciting its neighbor provides a basis for the integration of nervous control.

The ultimate integrating unit is the single neuron. The synapse, the point of junctional contact between neurons (or between a neuron and another cell) makes integration possible.

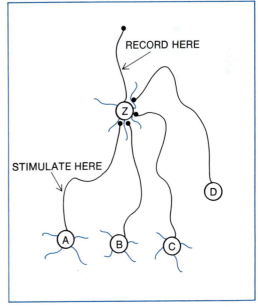

5

MODEL OF NERVE NET consists of five neurons. Axons and cell bodies are rendered in black, dendrites in color. Axons of neurons A, B, C and D synapse with cell Z. Stimulation of axon A causes impulse to travel across synapse to axon Z, where it can be recorded on an oscilloscope.

By integration we mean the ability to draw some conclusion from a variety of data. Suppose Z of Figure 5 were a motor neuron governing one's ability to reach out one's arm to pick up an apple. It might well receive excitatory inputs from higher centers (often acting via several intermediate interneurons); the hunger center might say "yes," the thirst center might say "yes," and perhaps certain limbic areas might anticipate the pleasures of apple-eating and say "yes."

But there are likely to be inhibitory inputs too. Perhaps a cortical zone says "no" because the last apple eaten led to a stomachache, and another says "no" because the apple must be purchased and money is lacking. Neuron Z thus receives both "yes" and "no" inputs, and only if the excitatory impulses are more vigorous and sustained than the inhibitory will Z be fired, and then one puts out one's arm to pick up the apple. The essence of Z's action is therefore to respond only if the excitatory influences outweigh the inhibitory.

First consider the simplest situation, in which one excitatory neuron (A of Figure 5) fires. A glance at the synapse joining A and Z shows that there is a little swelling (end-bulb) at the end of A's axon (shown in detail in Figure 6). The end bulb contains several mitochondria, and clustered around the thickened zone of membrane are hundreds of little bags called synaptic vesicles, about 200 to 400 Å across. These bags contain the chemical transmitter. Any one neuron contains only one kind of transmitter, and the commonest kinds are acetylcholine, noradrenalin and serotonin. When neuron A fires, the end bulb releases a pulse of transmitter, which diffuses across the gap approximately 200 Å wide between A and Z. For reasons that will be discussed later in the chapter, the cell body of neuron Z is normally polarized; that is, the inside of the cell membrane is negatively charged and the outside positively charged . The transmitter released by A partially depolarizes the cell body of Z. If A fires repeatedly, it may cause such a great depolarization of Z that the important area where Z's cell body joins Z's axon (an area known as the axon hillock), becomes depolarized by some critical amount. If this happens, it triggers an impulse in Z's axon; in other words, Z fires.

Like all neurons, Z does not fire continuously. It sends out one pulse at a time because the transmitter substance, once it is released into the gap, is destroyed very quickly by a specific enzyme. Acetylcholine, for example, is destroyed very quickly by the enzyme cholinesterase. Noradrenalin and related transmitters are apparently recaptured by the nerve endings that release them. In mammals this process may be a more important way to inactivate transmitters than enzymatic destruction.

Suppose A fires simultaneously with B, which is an inhibitory neuron. B releases a different transmitter, which can make it far more difficult for A to depolarize Z sufficiently to trigger the axon. This is the cellular basis of nerve inhibition.

Most neurons receive numerous excitatory and inhibitory synapses — up to several thousand in some cases. Whether the neuron fires or not depends upon the algebraic SUM OF THE STIMULI; that is, if the positive transmitters outweigh the negative transmitters, the neuron will fire. Thus the neuron can do algebra, and it can integrate different "yes" and "no" signals, in much the same way as a legislature tallies a vote to decide if it should act.

SENSORY (AFFERENT) SYSTEMS

Neurons and their associated structures collect data about the world outside — the sights,

6

SYNAPSE between body of one nerve cell and the axons of many others is depicted at progressively higher magnification. Diagram A depicts the end bulbs of many axons in contact with the cell surface. B is an enlarged view of the end bulbs. C shows mitochondria and synaptic vesicles within a single end bulb. D is a view of the synaptic cleft between the membranes of the end bulb (color) and the cell body (gray).

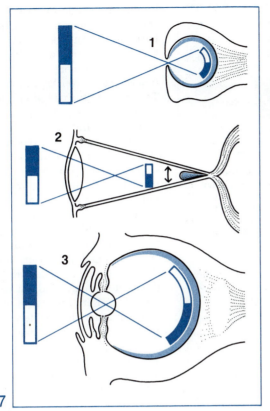

7

EYES OF INVERTEBRATES. The mollusk Nautilus has a lensless pinhole eye (1). In the eye of the arthropod Copelia (2) a single receptor scans the image projected by the lens. The eye of the squid Sepia (3) resembles that of vertebrates, with a lens whose shape can be changed for focusing on near or distant objects. Vertical bars show formation of the visual image within the eye.

VISUAL SYSTEMS

The simplest "eyes" consist of light-sensitive patches like the eye spot in the protozoan *Euglena*. They do not relay images or other patterned information, but may enable the organism to orient itself toward the light. Some organisms which have evolved true eyes have also retained simple photoreceptors of this type. Examples are the ocelli which occur in between the compound eyes of many insects, the photoreceptors found in the tails of crayfish and lobsters, and the brain area under the skull of many birds. All these organisms can perceive light even when their true eyes are removed.

Mollusks, arthropods and vertebrates have all developed true eyes which give information about contour and movement. The eyes of binocular animals such as man provide a way of estimating distances. Figure 7 shows that the mollusk *Nautilus* has a lensless eye precisely like a pinhole camera. The whole visual field is in focus, but little light penetrates the tiny opening. Another mollusk, the cuttlefish (*Sepia*) has an eye with a true lens whose shape can be changed by surrounding muscles to focus an image upon the RETINA, which contains the photoreceptors. The vertebrate has a

smells, touches, sounds, and tastes; and the world inside — including the position and the tension of muscles. These different qualities are detected by appropriate receptors which have the ability to convert (transduce) them into nerve impulses. The impulses are transmitted to the CNS, and constitute the link between the objective world, much of it outside the animal, and the subjective world within.

Not only do these sensory transducers convert sound into electrical pulses, for example, but they give information about the intensity and components of the sound. Full understanding of any sensory system requires an understanding of both the transduction mechanism and the coding system by which information is "described" in patterns of impulses relayed to the brain.

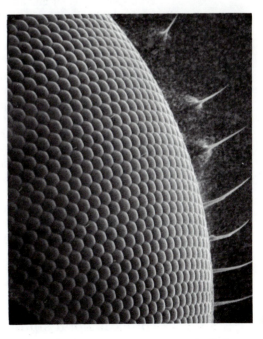

8

COMPOUND EYE OF INSECT is comprised of many ommatidia, as shown in this view of the eye of a housefly, magnified 240×.

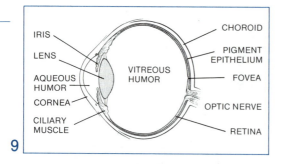

IRIS

LENS

AQUEOUS
HUMOR

CORNEA

CILIARY
MUSCLE

VITREOUS
HUMOR

CHOROID

PIGMENT
EPITHELIUM

FOVEA

OPTIC NERVE

RETINA

9

HUMAN EYE has a focusing lens and an iris diaphragm whose aperture is regulated by ciliary muscle. Interior of eye is filled with transparent gel-like material (aqueous and vitreous humor). Fovea is rodless area that affords acute vision. Point where optic nerve connects with retina is blind spot.

similar system, which differs mainly in having the receptors located at the back of the retina; *Sepia* has them in the front of the retina.

The arthropods (principally crustaceans, spiders and insects) have evolved an elaborate compound eye that consists of numerous units called OMMATIDIA; some dragonflies have 20,000 ommatidia in each eye (*Figure 8*).

This multiple-unit system provides the arthropod with a kind of mosaic image of the outside world. The image is crude and deficient in structural details, but it enables the arthropod to detect movement extraordinarily well, because the ommatidium recovers rapidly from a light impulse. A blowfly can detect the flickering of a light at a frequency of 250 flashes per second — five times the frequency threshold of the human eye.

The vertebrate eye has a lens protected by a tough cornea, and the peripheral part of the lens is covered by a retractable iris (*Figure 9*). The central part (pupil) can thus be widened in poor light by retracting the iris and letting in more light. The retina is a complicated structure whose photoreceptor cells, called RODS and CONES, lie on the outermost layers, farthest from the incoming light (*Figure 10*). The rods and cones extend the range of sensitivity. In dim light only the rods are functional; in

bright light vision shifts to the cones. The rods and cones form synapses with the dendrites of fairly simple bipolar neurons. Their axons in turn form a complex web of synapses which eventually connect with the dendrites of the innermost neuron layer, the GANGLION CELLS. The axons of the ganglion cells form the OPTIC NERVE that carries impulses to the CNS.

How does light give rise to nerve impulses? In 1967 George Wald of Harvard shared a Nobel prize for revealing the details of the first step. All land vertebrates and marine fishes (and at least some insects) have in their eyes a red material, rhodopsin, which consists of a protein (opsin) and a derivative of vitamin A called retinaldehyde. The retinaldehyde has a number of possible isomers; of these, one isomer, called the *11-cis* form, is found in rhodopsin, and light causes its conversion to a different isomer called the *all-trans* form (*Figure 11*). After this light-induced reaction, a series of other changes terminates in the splitting of opsin and retinaldehyde. This set of events leads, by a mechanism whose details are not yet clear, to an action potential being transmitted along the axons of the optic nerve.

Data about the visible world are analyzed in progressively more complex ways between its detection by photoreceptors and its final perception by the brain. Some of this analysis occurs in the retina. The ganglion cells of the retina do not simply pass on the crude data received by the receptors, but analyze and shape it before feeding it into the optic nerve. This "peripheral filtering" is common in many sensory systems.

The optic nerve enters an area on the underside of the brain called the LATERAL GENICULATE BODY, which partially processes the data before transmitting it to the VISUAL CORTEX at the rear of the brain. Various cortical cells respond to different components of the retinal image. One kind responds only to lines of light, and only responds well if the line is oriented in a particular direction. There are "horizontal line detectors" and "edge detectors" that react only to places where sizeable

The picture of the outside world presented by our visual cortex is not a simple point-by-point transcription of objects, like a television image. It is a collection of information about contours, lines and movements.

bright areas meet sizeable dark areas. Other more complex edge detectors will only respond if the bright area is to the left, and still others only if the bright area is to the right.

One can imagine the way in which man's perception of the world is stitched together from these fragments of isolated data. The picture of the outside world presented by our visual cortex is not a simple point-by-point transcription of the objects in that world, like a television image. Instead it is a collection of information about contours, lines and movements, much like short radio news bulletins which omit all but selected points of great interest.

AUDITORY SYSTEMS

Only a few invertebrates have ears, notably the cicadas, crickets, and a few moths. The moths seem to use their ears exclusively as a warning system for bat attacks (*Chapter 15*), as shown by the brilliant studies of Roeder of Tufts University. The ear of the moth employs only two sensory neurons connected to a sounding board, more properly called a tympanic membrane, on the moth's thorax. In spiders, cicadas, and crickets a similar tympanic sys-

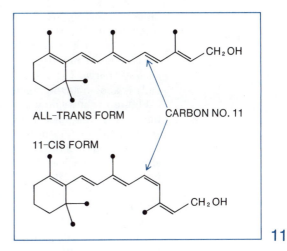

11

RETINALDEHYDE, a component of visual pigments, can be converted by light from the trans form (top) to the cis form (bottom).

tem is mounted on the legs; the number of neurons involved is seldom more than 70.

In fishes and reptiles sound plays a relatively small role; sounds above a few thousand cycles per second (or Hertz, abbreviated Hz) are not detected. In birds and mammals acoustical communication reaches tremendous complexity. In the human, for instance, detectable frequencies extend from 20 to 20,000 Hz in adults and much more in children. Elaborate structures have been developed to transmit the compression waves from a gaseous medium (air) to the almost incompressible fluid medium surrounding the sensory neurons. The energy is absorbed by a large disc (the TYMPANIC MEMBRANE or eardrum) and transmitted by a set of three levers — called the malleus (hammer), incus (anvil), and stapes (stirrup) — to a very small disc, the OVAL WINDOW of the cochlea (*Figure 12*).

In the human ear, the cochlea is a snail-shaped chamber. A flat membrane (the BASILAR MEMBRANE) is stretched tightly along the course of its canal, separating the tube into two liquid-filled canals (*Figure 13*). The pressure from the oval window is transmitted via the liquid in the vestibular canal, causing the basilar membrane to bulge; the consequent pressure in the tympanic canal is relieved by allowing outward movement of a sort of pressure-valve called the ROUND WINDOW. The basilar membrane normally presses on a group of sensitive cells (the true sound receptors) embedded in a relatively fixed plate, the tectorial membrane. When the basilar membrane

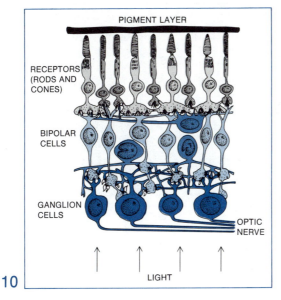

10

DIAGRAM OF RETINA of vertebrate eye depicts an area near the fovea. Rods are rendered in dark gray, cones in light gray. Horizontally oriented fibers carry information across retina at two levels, integrating the visual field. Light passes through the whole system before being absorbed by receptors and pigment epithelium.

bulges it rubs against these sensitive cells. In some unknown way, the mechanical deformation of these cells causes them to depolarize, generating an impulse in the endings of the cochlear nerve. The system of membranes and cells is known as the organ of Corti.

Figure 13 shows how the cochlea narrows at one end, at the apex of the snail shell. Consequently the width of the organ of Corti varies. Von Bekesy of the University of Hawaii has shown how the basis of pitch detection depends on this variation at ranges above 60 Hz. The basilar membrane vibrates unequally over its length; each tone gives maximal vibration at one point at which the signal is tuned to the width of the organ. Thus the brain recognizes the pitch of a sound according to the location of the sensitive cells which are firing.

The terms "outer ear" is often used for the fleshy external "cup" and the canal which terminates at the tympanic membrane; "middle ear" is used for the tympanic membrane and its enclosing chamber; and "inner ear" for the cochlea and its enclosed cells. When we hum, listen to our own voice, or to our crunching of an apple, much of the sound bypasses the outer and middle ears. This explains the big difference between one's voice as he thinks it sounds and as it sounds on a tape recorder.

CHEMORECEPTION

Chemoreceptors are neurons that detect chemical substances. The only external chemical sensations that man experiences are smell (olfaction) and taste (gustation). In most mammals, probably including man, smell is important in communicating about sex. In many animals chemical cues indicate trails to follow, dangers to avoid, and other important information (Chapters 14 and 15). There are also internal chemoreceptors, which in man monitor such things as glucose and CO_2 in the blood, and lead to compensations when necessary.

In all invertebrates and most vertebrates, chemoreceptors are simply the naked endings of specialized bipolar sensory neurons. A system which has been particularly well explored by Dethier (then at the University of Pennsylvania) is the proboscis of the blowfly. It is covered with sensory hairs, each of which contains parts of three sensory neurons. One is a touch receptor. The other two are taste receptors whose cell bodies lie at the base of the hair; dendrites pass through the hair to the tip,

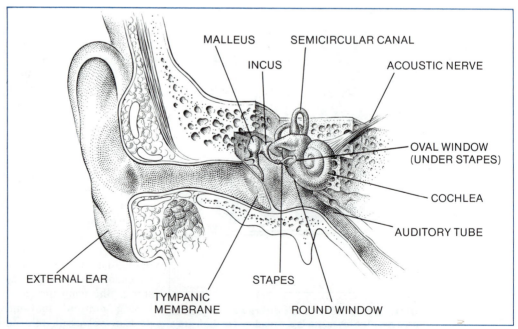

MALLEUS

SEMICIRCULAR CANAL

INCUS

ACOUSTIC NERVE

OVAL WINDOW
(UNDER STAPES)

COCHLEA

AUDITORY TUBE

EXTERNAL EAR

STAPES

TYMPANIC
MEMBRANE

ROUND WINDOW

12

HUMAN EAR. Energy of sound waves in air is absorbed by tympanic membrane and transmitted by three tiny bones (malleus, incus and stapes) to oval window of cochlea. Cochlea is connected to acoustic nerve that leads to brain. Auditory tube opens in the nasal part of the pharynx.

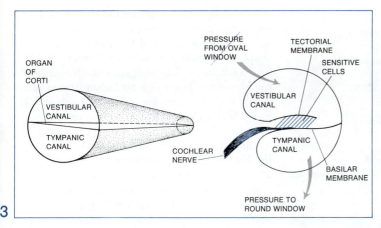

3

COCHLEA is a coiled tube, but for simplicity this schematic diagram depicts it as a straight tapered tube. Pressure from oval window causes basilar membrane to bulge, deforming sensitive cells attached to it, as shown in cross-section at right. Deformation generates a nerve impulse.

where they are exposed to the outside (*Figure 14*).

The female silkworm moth, *Bombyx mori*, attracts potential mates by releasing, from a gland at the tip of her abdomen, a chemical called bombykol. The male (*Figure 15*) detects this with incredible sensitivity, when the airborne bombykol reaches the dendrites connected to the sensory hairs of his antennae (*see photograph on opening page of this chapter*). By following a concentration gradient, a flying male can home in on a distant female.

Vertebrates have developed taste buds on their tongues, which show an important difference from the exposed-dendrite system of the blowfly. As Figure 14 shows, a group of cells buds off from the adjacent epithelium and develops hairlike processes which are exposed at the surface of the tongue. These hair cells are the actual taste receptors. A single bipolar neuron sends branches of its dendrite to the

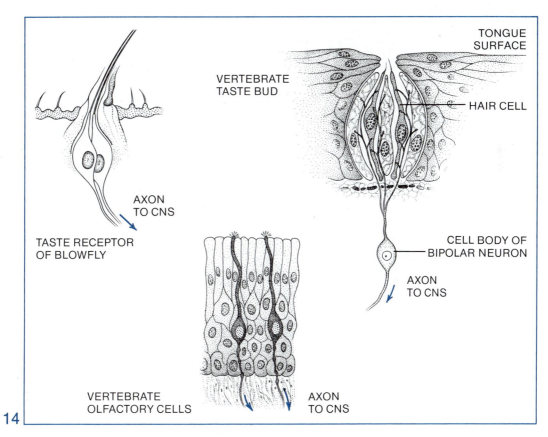

THREE CHEMORECEPTORS. Taste receptor of blowfly consists of two neurons whose dendrites extend upward along sensory hair. Vertebrate taste buds contain hair cells attached to bipolar neuron. Olfactory (smell) receptors in nose of vertebrates are modified epithelial cells.

15

MALE BOMBYX MOTH detects bombykol with his prominent antennae. A magnified view of one antenna appears on opening page of chapter.

whole group that comprises the taste bud. The number of possible tastes is still believed to be only four: sweet, sour, bitter, and salt. Specific areas of the tongue are sentitive to each: the tip for sweet, sides for sour, back for bitter, and salt all over. When we "taste" complicated and delicious flavors, we are in fact using olfactory rather than gustatory cells: we are smelling not tasting.

The olfactory endings of vertebrates lie at the back of the nose, protected by mucus, and quite densely packed; the dog has up to 40 million per square centimeter. The dendrite of the sensory neuron merges with the actual detector, usually a modified epithelial cell.

John Amoore of the U.S. Department of Agriculture argues that differences in the odor of substances are related to differences in the shape and charge of their molecules. He

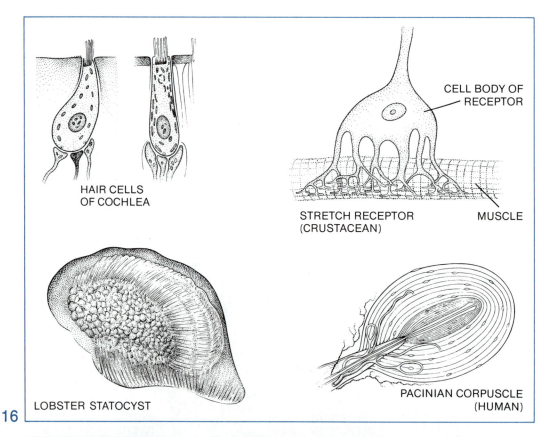

HAIR CELLS OF COCHLEA

CELL BODY OF RECEPTOR

STRETCH RECEPTOR (CRUSTACEAN)

MUSCLE

LOBSTER STATOCYST

PACINIAN CORPUSCLE (HUMAN)

16

FOUR MECHANORECEPTORS. Hair cells of cochlea are found on organ of Corti. Hair cell at left forms an end-on synapse with dendrites; cell at right, a cup-like synapse. Stretch receptor tells crustacean when a muscle is stretched or relaxed; inhibitory neuron that switches the receptor off is not shown. Statocyst is a gravity detector. Pacinian corpuscles are touch receptors in the skin of humans.

suggests that for humans, all odors are mixes of seven primary characters (ethereal, camphoraceous, musky, floral, minty, pungent, putrid), and are detected by seven kinds of receptor macromolecules that combine, with varying specificity, with a vast diversity of odorous substances. The theory may prove wrong in its details, but some kind of specific matching is almost certainly the explanation of variations in odor.

MECHANORECEPTION

A great many different purposes are served by a kind of all-purpose mechanoreceptor called a hair cell, a sensory cell with from one to 50 hairs projecting at one end (*Figure 16*). Deflection of the hairs of these cells in one direction causes the associated neuron to fire, while a reverse deflection may inhibit it. This enables the cells to indicate directionality.

One use of mechanoreceptors is to inform the organism about its own movements. The well-known SEMICIRCULAR CANAL SYSTEM of vertebrates consists of three loops at right-angles to each other, each of which provides information about angular movement in a different plane (*Figure 17*). These curved canals in the bone of the skull are filled with viscous fluid. Each canal has a gelatinous spring-loaded door called a cupula extending across it. If one rotates one's head in any plane, the fluid tends to stay in position, bending the hair cells imbedded in the cupula. The three-plane system is not universal; lampreys have two semicircular canals, and hagfish only one.

The canal system detects changes in angular (rotational) velocity. Vertebrates have a special chamber (the utriculus) below the semicircular canals in which a pebble of calcium carbonate called an OTOLITH lies on a bed of hair cells. The position of the pebble, sensed by these cells, tells the animal about the head's position with respect to the Earth's gravitational field. Invertebrates have a very similar system, a pebble-like STATOLITH lying in a hair-like sac called a STATOCYST (*Figure 16*). In 1893 Kreidl showed the importance of these organs by replacing the statoliths of a shrimp with iron filings. When he placed a magnet above them, the shrimps swam upside down.

Other types of mechanoreceptor respond to touching, stretching and twisting. A very versatile one found in the deep layers of the skin of vertebrates is called the PACINIAN CORPUSCLE (*Figure 16*). Shaped like an onion about a millimeter long, it consists of a series of con-

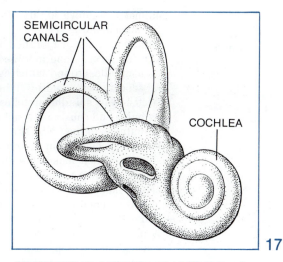

SEMICIRCULAR CANALS are fluid-filled chambers in bone of human skull. Drawing depicts a cast of the canals and the cochlea.

centric layers of connective tissue around a core that contains the naked terminal of the sensory neuron. Any kind of touch that distorts the skin and its contained "onions" can (if large enough) cause the neuron to fire.

Most animals have a variety of PROPRIOCEPTORS which indicate its posture and the stress on its muscles and joints. These receptors continuously feed information back into the CNS and permit the CNS to monitor the success of its instructions to the body. When one decides to pick up something from the floor, he follows the execution of this decision not only with his eyes, but also with information from all the joints and muscles and touch-detectors involved in the act. In fact several of these pieces of information overlap: one can do the job fairly well with his eyes closed or with anesthetized fingers.

A proprioceptor which has played an important role in neurophysiology is the stretch receptor of crustaceans. The dendrites of this neuron, which is found in the abdominal muscles, wrap around the muscle fibers as Figure 16 shows, and fire when the muscle is stretched. This neuron is contacted in turn by an inhibitory neuron which can turn off the signal from the stretch receptor. The crayfish, for instance, can tell its stretch receptors that it does not want to know about the state-of-stretch of its muscles at that particular time.

In vertebrates a more elaborate method of signalling the degree of stretch in skeletal muscle has been developed, permitting the animal to achieve highly efficient control of his musculature. At the center of each muscle bundle is a set of modified fibers called the MUSCLE SPINDLE; the modified fibers bulge in their middle, around which is wrapped a spiral sensory nerve ending. When the muscle is stretched its spindle stretches too, decreasing its bulge and causing the sensory nerve to fire an impulse to the CNS. The CNS responds by firing a specialized motor nerve to the muscle spindle, which makes the spindle relax. The overall result is that this system tells the CNS about changes in muscle status; if the muscle goes from condition A to condition B, the system only fires while the transition from A to B is occurring. If condition B is sustained, the sensory neuron stops firing — it adapts. Adaptation is a common feature of kinesthetic and many other sensory neurons. They fire when first stimulated, but then become silent unless their status changes again. In this way the CNS is not under eternal bombardment, but is only informed of changes in the status quo.

OTHER RECEPTORS

A book of this length cannot provide details of the many other kinds of neural receptor cells, but the reader should be aware that they exist. There are baroreceptors in mammals that detect changes in blood pressure by following the bulging of the carotid artery. There are thermoreceptors; in mammals and birds internal ones in the hypothalamus help to maintain body temperature within fixed limits (Chapter 11). Many invertebrates have thermoreceptors on mouth parts and legs which are useful in seeking prey and in other activities. The pit vipers also have external thermoreceptors, located in pits on their heads, which are used to locate their mammalian prey in the dark.

Electric eels, electric skates and several fresh water fish (gymnotids and mormyrids) have electroreceptors which respond to electric fields generated by the animal. More importantly they detect interruptions of those fields, and so are used to detect other fish (predators or prey) and probably to orient and navigate as well (Chapter 15).

CENTRAL INTEGRATION

How does the CNS use sensory data, integrate it with all other relevant data (incoming or stored) and when necessary undertake some suitable response? In higher animals housekeeping functions such as control of heart beat, blood sugar, body temperature, gut movement, blood flow and so on are performed by the AUTONOMIC NERVOUS SYSTEM. It operates (as its name implies) in virtual independence of the rest of the nervous system. All man's glands and smooth muscles are operated by this unobtrusive servant, so that one does not consciously have to transfer food from stomach to duodenum, nor consciously narrow one's pupils or salivate. One can preset the conditions which will be likely to set off these reactions, but the machinery is not under one's direct control.

In vertebrates the autonomic nervous system has two divisions, called SYMPATHETIC and PARASYMPATHETIC, which often have contrary actions. Thus the sympathetic system releases norepinephrine at nerve endings, which makes the heart beat faster. The parasympathetic system releases acetylcholine, which slows down the heart.

Automatic responses are not restricted to the autonomic nervous system. The well-known REFLEX ACTIONS involve a sensory and a motor neuron, with few or no interneurons. In vertebrates the apparatus for these simple reflexes lies in the spinal cord. A stimulus along a sensory neuron enters the cord, and there excites (either directly or through the mediation of a single other neuron) a motor neuron which activates the appropriate muscle. This simple set of neurons is often called a REFLEX ARC. A familiar example is the knee-jerk reflex, in which a tap on the patella provokes an almost instant kick. Similarly the eye will blink if the cornea is touched, and a finger is almost instantly withdrawn from a very hot object. These are safety reactions that require fast responses, and take priority over other bodily activities. The response is sometimes said to be wired in; that is, it is not learned. The circuitry is there at birth.

Most proprioceptive neurons fire only when their status is changed. Thus the CNS is not eternally bombarded with signals, but is informed only of changes in the status quo.

More complex activities are also often wired in, especially in invertebrates. Thus the set of muscular activities that govern flight in insects appears to be automatic; but it is turned on and off by events in the outside or inside world. Moreover, the stereotyped activity can be modified in certain limited ways. A flying locust can correct for interfering winds by biasing certain components of his complex flight program.

This simple life-style of stimulus and an automatic response is adequate for many of the simpler invertebrates. But evolution was accompanied by development of more complex behavioral repertories, demanding a kind of activity which can evaluate many factors simultaneously. A mosquito seeking its prey has a whole series of possible activities: flying, settling, walking, sucking, evading, mating, egg-laying. The appropriate behavior is displayed only in the presence of suitable prey. Each operation of landing or sucking or taking off demands an exquisite collaboration of multiple muscles, continued awareness of the outside world, and of the status of the insect's own body. Such awareness permits the animal to respond appropriately to the changing world.

The evolution from simple responses to complex behavior is accompanied by an increasing degree of CEPHALIZATION; that is, toward the formation of a brain. The brain receives and transmits data for the whole organism, largely replacing localized stimulus-and-response machinery, and confers an enormously increased potential for learning.

THE CEREBRUM

Within the vertebrates cephalization is very extensive. In the codfish the cerebrum is a relatively minor area; in the frog it has become a fairly substantial component; in the goose it is a major area but quite smooth; it has become much convoluted in the horse; and in man it has become enormous and richly convoluted, dominating the underlying regions of the brain (*Figure 18*). The convolutions add additional surface area per unit volume of cerebral hemisphere; in humans the CORTEX (outermost part) of the hemisphere is the site of associative activity, which includes all the most interesting and complex functions of the brain: the ability to put together the sights, smells, and sounds of the world outside, associate them with the learned and innate information already in the cortex, and to respond in an appropriate man-

EVOLUTION OF VERTEBRATE BRAIN shows progressive increase in size of the cerebrum (color). Cerebral cortex became convoluted, and older parts of the brain that control coordination, reflexes and instinctive behavior came to lie beneath it.

ner. Man's dominance of the planet depends on the immensely rich possibilities for associative processes conferred by this extensive cortex. The abilities to reason extensively, to speak, and to suppress short-term desires in favor of long-term goals, all stem from cortical richness.

An important question about the cortex is to what extent are its functions localized? One way to explore this has been to see what functions are lost by removing particular pieces, either by surgery or by destruction with heat, chemicals or other means. The second way is to stimulate the local area and see how pieces, either by surgery or by destruction with conscious humans, for example with epileptic problems, with much of the skull removed for exploratory surgery. If one stimulates the OC-CIPITAL CORTEX (the back of the cortex) with an electrode, flashes of light are seen by the patient. And if, by surgery or by an accident, the occipital cortex is all destroyed, the patient is blind. These effects are limited to a well-defined area, and show that this particular piece of cortex is involved in vision. In mammals the cortex also contains the final processing zones for other major sensory processes (*Figure 19*). Just as the hindmost portion is involved in visual events, portions on the side are involved in auditory events, and a strip across the top of the brain in the bodily senses. Just in front of this strip is a cortical area involved in motor activities, with any one piece connected (via numerous relays) to some particular group of muscles.

Substantial areas of the cortex, especially in the frontal portion (frontal lobes) have no simple function; they seem to be involved in complex social senses such as propriety and good and evil. In people obsessed with feelings of guilt and inadequacy, an operation used to be

performed in which these areas were destroyed or disconnected, giving a far more tranquil but overly careless personality. Nowadays this lobotomy is seldom done, because tranquillizing drugs perform the role in a less drastic and more reliable way.

Quite apart from housing the points of final analysis of sensory data, the cerebral cortex is the site where all sensory inputs and memories are integrated. But cortical integration and memory are not localized. When the smell of dead leaves evokes a wistful memory of autumns gone by, the intellectual component arises from the cortex, but the emotional component — the sad nostalgic feeling — arises from regions beneath the cortex, particularly the limbic system, which is discussed later in the chapter.

THE HYPOTHALAMUS

Lying close to the center of the brain is the area known as the HYPOTHALAMUS. The outstanding work of the Swiss neurophysiologist W. R. Hess (who won a Nobel prize in 1949) involved stimulation of this area with deep electrodes. The hypothalamus was thus shown to contain a variety of control centers for basic bodily functions and for certain drives. Chapter 11 mentioned that the hypothalamus has temperature-sensing devices that detect fluctuations from an acceptable temperature range, and trigger suitable compensating responses. Nearby centers regulate blood sugar, water balance, carbohydrate and fat metabolism, and blood pressure (*Chapter 12*). The efferent nerves from these centers, which turn on and off the appropriate machinery to accomplish control, operate the sympathetic and parasympathetic branches of the autonomic nervous system. The hypothalamus is thus the site of the systems that maintain the constancy of the internal environment.

Apparently the hypothalamus also contains a hunger center which when electrically stimulated will cause even an overfed animal to go on eating; and a satiety center, which when electrically stimulated will cause even a starved animal to refuse food. There is a thirst center, whose continuous stimulation can cause animals to drink even salty water continuously. Even more exciting (but not universally accepted) was the finding by the late James Olds, then at McGill University, that there is a pleasure center in the hypothalamus. Olds equipped rats with a lever which they could depress at will, and thus stimulate elec-

19

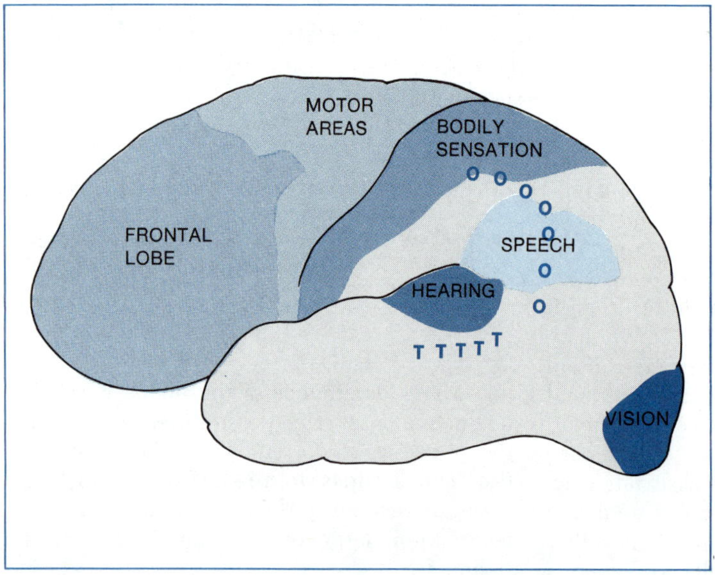

MAP OF CORTEX indicates the major sensory and motor areas on the surface of the human brain. The symbols OOO and TTT mark locations of olfactory and taste areas that lie beneath the surface.

trodes implanted in their pleasure centers. The rats lost interest in everything except pressing this lever, ignoring hunger and thirst in their orgy of self-excitement. Such electrodes have been implanted in humans suffering from severe chronic depression. Unfortunately the stimulation led to progressively less pleasure as time went on. Olds also found a punishment center whose stimulation in rats leads to apparent rage, fright and pain.

The obvious conclusion to be drawn from this work is that important drives such as hunger, thirst and perhaps pleasure are mediated through specific circuits of neurons. Perhaps each circuit is localized in a discrete hypothalamic region. If so, much of human activity may be devoted to bringing these neurons to a desirable status. Is the search for happiness merely an effort to produce a particular firing pattern in a few hundred neurons?

THE CEREBELLUM

The function of the CEREBELLUM (Figure 20) is to modify instructions to motor systems received from the cortex, in the light of the body's status as indicated by proprioceptors. For example, the primary command to pick up a piece of bread on the far side of a table originates in one's motor cortex. The execution of the command requires not only the stretching out of an arm, but literally dozens of compensating small movements in the trunk, legs, and even toes so that full balance is maintained at all times, in spite of a substantial shift in the center of gravity. These automatic compensations, involving ceaseless small adjustments in the light of ceaseless feedback from proprioceptors, are performed by the cerebellum, which is thus a kind of silent administrator of the motor system.

THE LIMBIC SYSTEM AND THE EMOTIONS

A group of brain structures collectively known as the LIMBIC SYSTEM seems to play a large role in emotion (Figure 20). The structures form a double loop approximately around the center of the brain and include part of the inner border of the cortex. In 1937 J. W. Papez of Cornell proposed that the limbic system, in coordina-

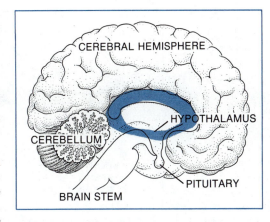

20

INTERIOR OF HUMAN BRAIN is partially exposed in the saggital section (a vertical cut from front to back). The limbic system is rendered in color.

tion with the hypothalamus and other subcortical structures, constituted the neural basis of emotion, thus giving neurological support to the popular view that thinking and feeling were fully separable functions.

Although the discussion of the brain has stressed localization of function, it would be entirely wrong to think of the brain as a clump of autonomous sub-organs, with the cortex thinking, the limbic system feeling and the cerebellum balancing. Brain areas are richly interconnected, and they function with continuous interchanges. For example, the hypothalamus contains centers of great importance to the emotions, and so the limbic system is not the sole organ of feelings. The interplay of these two subcortical systems, and their subordination to the cortex and other areas, must all be taken into account in the neural basis of emotions.

MEMORY AND LEARNING

Memory and learning are so closely related as to be difficult to distinguish experimentally. One tests the learning of an animal by testing its ability to remember particular visual or auditory or other patterns, or recall particular sequences of acts. Yet there are undoubtedly different categories of memory. Two easily demonstrated kinds are short-term and long-term. A person can readily remember a new

*Is the search for human happiness merely an effort to produce
a particular firing pattern in a few hundred neurons?*

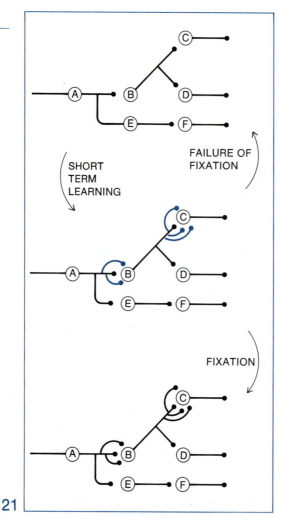

SHORT
TERM
LEARNING

FAILURE OF
FIXATION

FIXATION

21

MODEL OF MEMORY FORMATION. Before learning (top), an impulse from neuron A is equally likely to follow the path ABC or AEF. Short-term learning (middle) may lead to formation of temporary extra synapses (color) that enhance the firing of path ABC. These extra synapses may be erased or may become fixed (bottom), leading to permanent memory.

telephone number for about as long as it takes to dial it, but within a few minutes it is lost. Such short-term memory is different from the permanent memory which stores the telephone number of one's own home and a few other familiar numbers.

There seems little doubt that in mammals most or all learning is a function of the cortex. Surprisingly enough, some experiments suggest that specific memories seem not to be located in specific places in the cortex. The most celebrated experiments on this subject were those of Karl Lashley of Chicago, who showed in 1929 that a particular learned behavior in rats would not be abolished by removing any one particular piece of cortex. Instead, the general ability to remember and to learn was progressively lost, as more and more cortex was progressively removed; this was Lashley's principle of mass action. But all learning tasks in all species do not follow this principle. The ability of monkeys to learn that a red square conceals food, but a green circle does not, seems to be localized in the lateral parts of the cortex which are called the temporal lobes.

It is widely believed that memory is a consequence of pathway facilitation. For instance, when one sees a new and striking object it leads to some particular pattern of firing in the visual cortex and associated areas. If one can replay that same pattern precisely, presumably the original sensation will be experienced. The particular pathway (corresponding to the particular pattern just mentioned) is therefore replayed *in toto*, just as playing a recording involves sending the phonograph needle through precisely the original vibrations established when the recording was made. One can imagine that once some particular pathway through a complex network has been followed, that pathway can (when someone recalls the event) be followed again relatively easily. That is what is meant by pathway facilitation.

Whether or not the long-term memory trace itself is localized, the fixation site seems to be localized in a region called the hippocampus. It has long been known that a human with damage to his hippocampus acts like one who has lost his ability to fix short-term memory. Such an individual can remember a seven-digit number perfectly well for a few minutes, but cannot recall what he had for breakfast a few hours ago. An attractive possibility is that in short-term memory a particular pathway through a network of neurons becomes strengthened, perhaps by creating extra synapses (*Figure 21*). But these extra synapses might become lost in a random making-and-breaking of synapses unless a sort of fixation substance, manufactured under the control of the hippocampus, grows across the extra synapses and renders them permanent.

EFFERENT SYSTEMS

The systems responsible for carrying out the instructions of the CNS are the EFFERENTS, that is, the neurons whose impulses flow from the

CNS. Referring to them as the motor system implies that they only affect muscles; in fact, they also control glands. Moreover, certain modified neurons secrete materials such as hormones in response to instructions from the CNS. These efferent neurons do not merely control glands, in a sense they ARE glands.

In invertebrates one finds literally hundreds of NEUROSECRETORY CELLS which are clearly neuronal but produce three types of special hormones: kinetic hormones, which control the actions of muscles or glands; metabolic hormones, which control metabolic processes; and morphogenetic hormones which control growth and development. These neurosecretory cells may be set in the brain itself, as is common in arthropods. Insects have important neurosecretory cells on the dorsal surface of the brain. When the insect "decides" to moult (a decision based on day length and other data received by the CNS) these cells release a brain hormone which triggers other secretory cells outside the CNS to produce moulting hormone.

Chapter 15 discusses the behavior of the whole organism. The complexity of behavior is limited by the complexity of the neural machinery described in this chapter; and the plasticity of behavior is limited by the plasticity of the machinery. Learning about the neural machinery alone is a sterile exercise. Knowledge of behavior alone can be mere description. Knowledge of both is necessary for real understanding.

READINGS

B. KATZ, Nerve, Muscle, and Synapse, London, McGraw-Hill, 1966. An excellent little book dealing with a limited area: transmission of impulses along nerves and across synapses.

S.W. KUFFLER AND J.G. NICHOLLS, From Neuron to Brain, Sunderland, Mass., Sinauer Associates, 1976. A textbook on the nervous system, written for advanced undergraduates and graduates.

K. OATLEY, Brain Mechanisms and Mind, New York, Dutton, 1972. Somewhat popular, richly illustrated paperback which stresses the integrative functions of the brain. Good historical perspective.

T.C. RUCH, H.D. PATTON, J.W. WOODBURY AND A.L. TOWE, Neurophysiology, Philadelphia, W. B. Saunders Co., 2nd Edition, 1965. A very thorough treatment that stresses function and organization rather than anatomy or biochemistry.

D.E. WOOLRIDGE, The Machinery of the Brain, New York, McGraw-Hill, 1963. An excellent paperback which takes a bird's-eye view of the brain's function and some of its mechanisms. It may be the only neurobiology book that is easy to read.

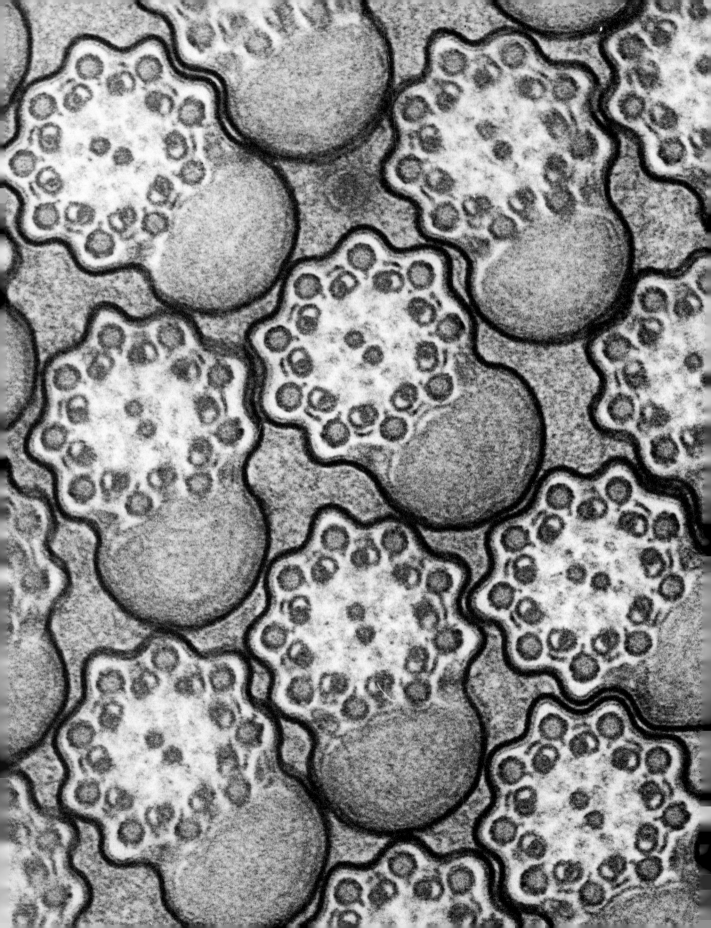

14

effectors

Four years ago, tripping over my dog in a dark corridor, I broke my kneecap. The surgeon decided to remove the broken tip of the kneecap . . . and to join the tendon of the leg muscle to what remained of the kneecap. The knee healed very well, and three years later an x-ray picture showed that the kneecap had almost regained its original shape. Under the influence of muscular pull the bone had reshaped itself to its normal form and was again as good as new.

S.E. LURIA

Multicellular organisms have evolved an astonishing array of specialized "hardware" to carry out the commands of the nervous system, as well as to perform a variety of tasks that are not under direct nervous control. Ciliated cells create mini-currents that sweep fluids from place to place; amoeboid cells engulf bacteria; other cells change the color of an organism, secrete poisons, or emit light. Special organs create sound or generate electric discharges. Such specialized structures are collectively called EFFECTORS. At best it is a very loose category because of the extreme diversity of the structures and of their evolutionary origins. From a human viewpoint perhaps the most important effector is muscle. Because muscle serves so many important functions in man, from breathing and pumping blood to walking and speaking, it has been the subject of intensive research. Recent work has elucidated the mechanism of muscle contraction at the molecular level. A logical place to begin the study of effectors is with ciliated and amoeboid cells — evolutionary carryovers from unicellular life.

CILIATED CELLS

Cilia are common among both unicellular and multicellular organisms. In many unicellular organisms, they are the only means of locomotion. Flatworms and mollusks use cilia together with the rippling motion of muscles to glide over surfaces. But most multicellular organisms use cilia not to move the cell, but to move a liquid past the surface of a stationary cell. Ciliated cells circulate water through the gills of mollusks, and through the bodies of living sponges, which feed on the tiny organisms that they filter from the water *(Chapter 20)*. Many groups of invertebrates feed by trapping microorganisms in mucus and swallowing it, or by sweeping microorganisms directly into the mouth — all with the aid of cilia. In man the respiratory passageways of the lungs and trachea are lined with mucus-coated cells equipped with hundreds of cilia *(Figure 1)*. The beating of these tiny hairlike structures helps to keep the passages clear by carrying mucus and trapped dust upward toward the throat, where it can be coughed up and spat out or swallowed. Similarly, ciliated cells lining the nasal passages carry dust and mucus downward toward the throat. (Cigarette smoke somehow inhibits normal ciliary action in the respiratory tract, and thus impairs the mechanism for removing foreign matter from the lungs.) Ciliated cells also line the female reproductive tract, creating currents that carry the egg from the ovaries to the uterus.

A cilium pushes against a liquid with the same basic motion as a swimmer's arms during the breast stroke. Cilia bend forward slowly and lash backward suddenly. Fluids exposed to the beating of cilia are

SPERM TAILS of an insect are depicted in the electron micrograph on opposite page. Cross-section view reveals the arrangement of filaments within the flagella.

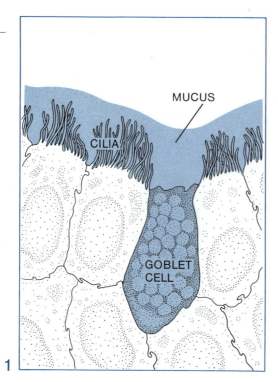

1

CILIATED CELLS in the human trachea. Dust and other foreign matter are trapped in film of mucus (color) secreted by goblet cells. Rhythmic beating of cilia sweeps mucus upward toward throat.

thus propelled in the direction of the effective stroke. Cilia typically beat in coordinated waves, which assures a steady flow past the cell surface. The mechanism that coordinates the beat remains a mystery. Suffice to say that the action of these cells appears to be automatic, and is under some form of chemical rather than nervous control. All cilia, so far as is known, are constructed like cables, with nine outer filaments surrounding two central ones. According to one theory, the cilium bends when the filaments on one side contract while the others remain relaxed. The central filaments may serve as "telegraph lines" to conduct impulses through the length of the cilium and coordinate the contraction of the outer filaments (*Figure 2*).

When a cilium occurs by itself or with only a few other cilia on the surface of a cell, it is referred to as a FLAGELLUM (plural, flagella). Flagella possess the same cable-like ultrastructure as other cilia. In some cases they are longer and employ different forms of stroke. The possession of flagella rather than cilia is a primary criterion for separating two of the

largest phyla of protozoans, the Mastigophora and Ciliophora (*Chapter 18*). Flagella also power the movement of the gametes of many kinds of organisms, including the sperm of man and other vertebrates.

AMOEBOID CELLS

The one-celled amoeba, a common denizen of pond water, moves about by extending lobe-shaped projections called PSEUDOPODS and seemingly pouring itself into them. This AMOEBOID MOVEMENT characterizes the phylum Sarcodina, to which all amoebas belong (*Chapter 18*). It also occurs widely in higher organisms, particularly in migrating cells during embryonic development, as well as in blood. Human white blood cells, for example, depend upon amoeboid movement to pursue and capture bacteria and other infectious microorganisms in the blood (*Chapter 11*). The molecular basis of amoeboid movement has yet to be demonstrated with certainty. Of several theories still in contention, all recognize that the movement is accomplished by the forward flow of a relatively liquid phase called ENDOPLASM, located in the core of the organism, followed by its spread to the periphery, where it is converted into a stiff outer layer called ECTOPLASM. The endoplasm moves the pseudopodium forward, and the ectoplasm fixes it in position. Exactly how the endoplasm moves is still unknown.

THE SKELETON

Organisms larger than amoebas need some sort of rigid or semirigid skeleton to support them against the pull of gravity. The skeleton also provides a firm body part against which muscles can pull. The simplest type of skeleton is found in the earthworm. Called a HYDROSTATIC SKELETON, it consists not of a chain of bones but rather of incompressible liquid trapped in each of its body segments. The walls of each segment are equipped with two sets of muscles, one running lengthwise and the other one encircling the segment.

Contraction of the longitudinal muscles shortens the segment, and the incompressible fluid within forces the wall of the segment to bulge outward. (Engineers apply the same principle in designing the hydraulic brakes of automobiles.) Contraction of the circular muscles has the opposite effect, squeezing the segment into a narrow elongated tube. The earthworm moves its body along first by narrowing the segments, so that they extend forward in space, then by anchoring them into the soil with bristles and contracting them, pulling forward the segments in the rear. By passing alternating waves of contraction and extension steadily along the length of its body, the earthworm is able to make fairly rapid progress through the soil or over its surface.

How can a hydrostatic skeleton be converted into a still more efficient form? One way would be to harden the walls of the segments and to add legs to them. Muscles attached to the walls could then move the legs back and forth. As odd as the idea may seem at first, this is precisely the step taken in the evolutionary origin of the phylum Arthropoda (crustaceans, insects, and related forms) from the phylum Annelida (earthworms and other segmented worms). The hardened walls of the body and appendages are referred to as the EXOSKELETON ("outer skeleton"). As mentioned in Chapter 11, in Arthropods the exoskeleton consists of CUTICLE, a distinctive layered structure comprised of a thin, wavy EPICUTICLE, which protects the body from drying, and a thicker inner layer, the ENDOCUTICLE, which forms the bulk of the structure. The endocuticle is a tough, pliable material found only in arthropods. It consists of a complex of protein and CHITIN, a complex nitrogen-containing polysaccharide. The muscles are attached to the inner surface of the arthropod exoskeleton.

THE VERTEBRATE SKELETON

The skeletal arrangement of vertebrates is the exact opposite of that in insects and other arthropods. It is an ENDOSKELETON, an inner scaffolding to which the muscles attach externally. It consists of two kinds of supportive tissue, CARTILAGE and BONE. Cartilage ("gristle" is the more familiar vernacular term) consists of cells widely separated by a rubbery matrix of mixed proteins and polysaccharides. Collagen fibers run in all directions through the matrix, adding to its strength and resiliency. Cartilage occurs in parts of the body that require both stiffness and resiliency, such as the capping material on the ends of bones, the walls of the larynx (voice box), and the protruding parts of the ear and nose. In sharks and rays, often referred to jointly as the cartilaginous fishes, the skeleton is composed entirely of cartilage. It also is the principal component of the embryonic skeleton of man and other vertebrates, being replaced gradually by bone in the course of development.

Bone, despite its dead appearance, contains large numbers of living cells. The matrix is rigid because it is impregnated with crystalline calcium phosphate and calcium carbonate in addition to collagen. It is most prominently developed in large land-dwelling animals, where its strength is required for the heavier loads moved by their powerful muscles. The architecture of bone varies with its position and function. The shafts of the long bones, for example, consist of cylinders of hard bone surrounding marrow-filled cavities (Figure 3). They are therefore both strong and light, the two prime requirements for efficient locomotion. The cavities are not idle spaces; their marrow is employed in the manufacture of red blood cells for the entire body. The more compact forms of bone are composed of structural units called HAVERSIAN SYSTEMS. Each system is a massive set of concentric bony sheets, or lamellae. Through its center runs a narrow canal containing the blood vessels (Figure 4).

The very rigidity of bones requires the existence of two supplementary forms of binding connective tissue: TENDONS and LIGAMENTS. Both contain large numbers of fibers oriented in the same direction, a property that allows them to be pulled and folded strenuously back and forth without breaking. Tendons bind the muscles tightly to the bone. As the muscle contracts, tendons absorb most of the stress along the lengths of their parallel fibers. Liga-

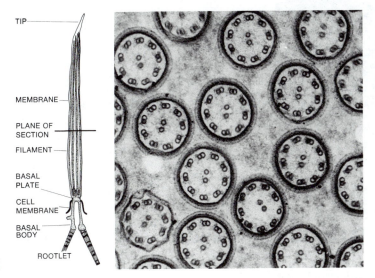

TIP

MEMBRANE

PLANE OF SECTION

FILAMENT

BASAL PLATE

CELL MEMBRANE

BASAL BODY

ROOTLET

2

STRUCTURE OF A CILIUM is depicted in the diagram at left. Right, a cross-section cut along the plane indicated in the diagram. Cross-section depicts flagella from tails of guppy sperm, and shows the typical 9 + 2 arrangement of filaments.

ments join bones together. Like leathery hinges, they hold the ends of the bones in position while allowing them sufficient play to bend freely and even to rotate to a limited extent (*Figure 5*).

MUSCLE

Multicellular animals that rely solely on the beat of cilia for locomotion are confined to a sluggish, creeping existence. The great majority of higher animals, from roundworms to man, employ muscle cells, which are contractile units specialized to move other organs in the body. Three types of muscles can be identified on the basis of differences in the structure of their cells (*Figure 6*). The simplest are SMOOTH MUSCLES, whose long and spindle-shaped cells are packed with contractile protein which is nearly indistinguishable under the light microscope from the remainder of the

4

HAVERSIAN SYSTEMS are structural components of dense bone. Cross-section shows concentric lamellae surrounding channels that contain blood vessels. Magnification, about 130×.

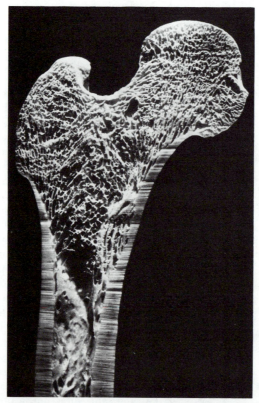

3

HUMAN FEMUR, the long bone of the upper leg, is depicted in longitudinal section. Cylinder of dense bone (bottom) surrounds the marrow-filled interior of the shaft. Marrow manufactures red blood cells. Spongy bone that comprises the head of the femur combines both strength and lightness.

cell contents. Smooth muscle cells are evolutionarily the most primitive of the three types. They occur in many of the lower animal phyla, as well as in the walls of the intestine and blood vessels of vertebrates.

In vertebrates, insects and other more complexly organized animals, muscle cells are often fused into single large units containing many nuclei. Each such unit is referred to as a MUSCLE FIBER, and it is capable of swifter and more powerful contractions than a mass of smooth muscle of comparable size. In some cases the contractile proteins of the muscle fibers are very regularly arranged, forming repetitious sequences of bands, or striations, oriented at right angles to the fiber. Such organs are called STRIATED MUSCLES. Most vertebrate muscles belong to this category, including all those that move the bones and hence provide the means of locomotion.

The third form of muscle that is most com-

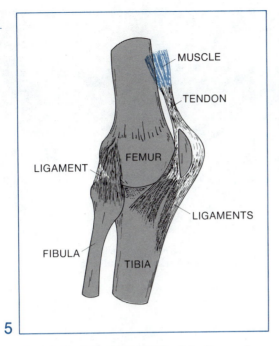

5

TENDONS AND LIGAMENTS of the human knee joint. Tendons connect muscles to bone. Ligaments are fibrous "hinges" that bind bones together.

monly recognized is CARDIAC MUSCLE, found in the hearts of vertebrates. To keep the blood circulating evenly, cardiac muscle pulses in rhythmic contractions. It is a peculiar form of striated muscle, composed of profusely branching and interconnected muscle fibers. Each fiber is divided by thin partitions representing the boundaries of the individual cells. Cardiac muscle combines the characteristics of both striated and smooth muscle tissue.

The evolution of the muscular system is a story of a gradual increase in power, speed, and complexity of action. The invention of muscle cells was a dramatic improvement over the use of cilia, and the origin of muscle fibers represented an advance beyond the exclusive use of smooth muscle cells. In the course of further evolution, as the bodies of animals became larger and more complicated, additional mobility was achieved by alterations in the

ways the muscles act. In its most elementary form, muscular movement consists of little more than an alternating contraction and relaxation of sets of muscles. By this means a roundworm can wriggle through a solid medium, a jellyfish can push through the water by expanding and then shutting the umbrella-shaped bell that makes up most of its body, and a snail can slide forward by riding upon the rippling waves of contraction in the muscles of its flattened foot. The next evolutionary step was the addition of a skeleton.

MUSCLE FUNCTION

One of the basic attributes of muscle cells and fibers is that they are able to exert force in only one direction. The complex patterns of movement required for locomotion and other actions consequently require two sets of muscles that move the body parts in opposite directions. Muscles that work against each other in this way are said to be ANTAGONISTIC. The clearest examples of antagonistic action are found in the motion of the legs and other appendages in arthropods and vertebrates. Each joint is bent ("flexed") by one or more FLEXOR MUSCLES and straightened out ("extended") by EXTENSOR MUSCLES, as shown in Figure 7.

The function of muscle is to contract at the appropriate moment, and hence either move some skeletal part to which it is attached (as when one stretches out the arm) or else cause a change in shape or size of some soft tissue (as when the heart beats or the pupil of the eye dilates). In all cases studied it appears that the contraction depends on two kinds of filaments that slide within a cell. H. E. Huxley of Cambridge University has worked extensively with vertebrate striated muscle, and much of the material that follows is the result of his work.

The skeletal striated muscle of vertebrates is under voluntary control, and ordinarily contracts when stimulated by nerves. Such muscle is made up of bundles of muscle fibers, typically 10 to 100 μ in diameter, which often run the whole length of the muscle. Each fiber contains a set of four to 20 filaments called

The invention of muscle cells for locomotion was a dramatic improvement over the use of cilia, and muscle fibers were an improvement on smooth muscle. The next big evolutionary step was the addition of a skeleton.

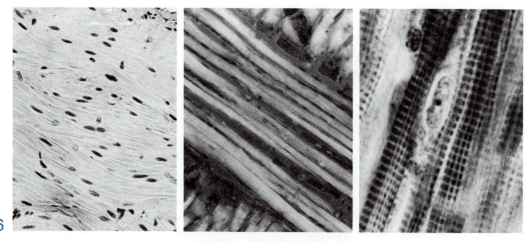

THREE TYPES OF MUSCLE. At left, smooth muscle from a snake. Center, striated muscle from a mammal. Right, cardiac muscle. (Left and right, courtesy of Carolina Biological Supply House).

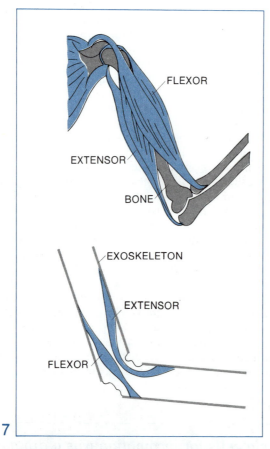

FLEXOR

EXTENSOR

BONE

EXOSKELETON

EXTENSOR

FLEXOR

FLEXOR AND EXTENSOR MUSCLES in man (top) and insects (bottom). Flexors bend the joint; extensors straighten it. Muscles of vertebrates are attached to outside of skeleton; muscles of insects, to inside of skeleton.

MYOFIBRILS, about one micron in diameter, bathed in cytoplasm. Numerous mitochondria are scattered through the cytoplasm. The banded or striped appearance of the muscle changes when the muscle contracts (*Figure 8*). Study of the changing widths of these bands led to the discovery of the contractile mechanism.

The myofibril is the actual contracting unit. In vertebrates it contains two thin filaments of about 50 Å diameter for each thick filament of about 100 Å diameter (1 μ = 10,000 Å). The thin filaments are 2 μ long and made of a protein called ACTIN; the thick ones are 1.5 μ long and made of a protein called MYOSIN. When viewed in cross section, the thin actin filaments are arrayed hexagonally around the thick myosin filaments. The two kinds of filament are connected with each other along the longitudinal axis at 60 Å intervals by bridges.

Figure 9 is a diagram of the bands in a piece of a myofibril. One distinctive set of stripes (Z lines) turns out to be the anchor point for the thin filaments. The zone between two Z lines is the basic contractile unit within the myofibril, and it is called a SARCOMERE. A broad dark band occupies much of the middle of each sarcomere; this A band is due to the thick filaments, which are not anchored at their ends, but only bound laterally (by the bridges) to the thin filaments. Because contraction is caused by thick and thin filaments sliding past each other without being shortened, the absolute length of the A band does not change in con-

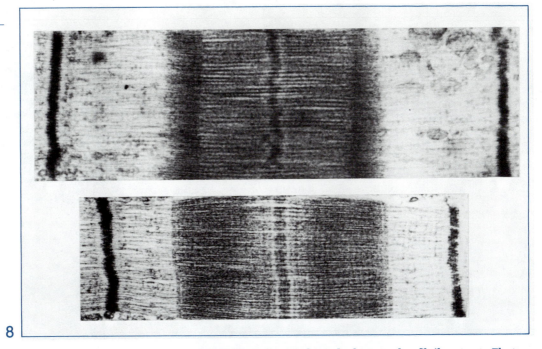

8

BAND PATTERN of striated muscle changes when fibril contracts. Electron micrograph at top depicts a single unit of a relaxed fibril. Below, a unit of a contracted fibril. Filaments of actin and myosin appear as thin horizontal lines. Light and dark bands are identified in Figure 9.

traction; but at maximal contraction the A band runs all the way from one Z band to the other. In the center of the broad A band is a thin light H band; this is the place at which the sets of thin filaments anchored to adjacent Z bands do not quite meet each other in resting fibers. When contraction occurs, the two sets of filaments approach each other, and the H band almost disappears. Finally there are pale I bands which flank the broad dark A band. These are zones where, in a relaxed muscle, one sees only thin filaments; as the muscle contracts the I bands shrink, and at maximal contraction, when the thick filaments extend right up to the Z band, the I bands disappear.

The contractile machinery is set in motion by a nerve impulse and is fueled by ATP. The last chapter described how the arrival of the impulse leads to the depolarization of the muscle membrane immediately adjacent to the nerve ending. Because each nerve fiber is branched, a single one supplies up to a hundred muscle fibers, which comprise a MOTOR UNIT. When the firing of a nerve induces a local depolarization in a fiber, the depolarization spreads throughout the fiber, being propagated in the all-or-none fashion

characteristic of axonic transmission (*Chapter 13*). Several milliseconds elapse between the initial depolarization and the contraction of the muscle. The relaxation process that returns a muscle to its resting state is essentially a reversal of the sequence that leads to contraction. When the supply of ATP is turned off, the actin and myosin are dissociated, and the elasticity of the tendons and ligaments pulls the filaments back to their original positions.

CARDIAC MUSCLE

A special kind of striated muscle is found in the hearts of vertebrates. Unlike skeletal muscle, cardiac muscle cells are branched, and mixed with noncontractile cells. Cardiac muscle can contract rhythmically even in the absence of nerve stimulation; it is "automatic" or "myogenic". Moreover, the contraction of any one part, such as the ventricle, is almost simultaneous. This synchronous contraction is essential if the ventricle is to expel its content of blood rapidly and completely.

Two special features of heart muscle explain its automatic and synchronous properties. One is that the outer membrane of each muscle fiber lies in very close contact with that of its

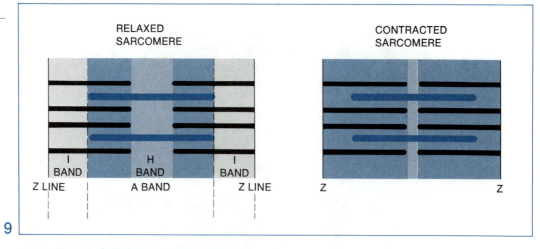

RELAXED
SARCOMERE

CONTRACTED
SARCOMERE

I BAND | H BAND | I BAND

Z LINE | A BAND | Z LINE

Z | Z

9

DIAGRAM OF SARCOMERE identifies bands seen in micrographs of striated muscle. Z lines are vertical bands at either end. Broad A band lies in middle of sarcomere, flanked by light I bands. When muscle contracts, I and H bands almost disappear. Horizontal black filaments are actin; horizontal colored filaments, myosin. Actually myosin filaments are twice as thick as those of actin.

neighbor at special regions called intercalated discs. Consequently, depolarization of any fiber sweeps rapidly throughout neighboring fibers, triggering contraction. The second special feature is the existence of a pacemaker: a small group of modified muscle cells that spontaneously and rhythmically produces a depolarization. The pacemaker of the mammalian heart is called the S-A node, or sino-atrial node, situated on the right atrium (*Figure 10*). The depolarization that it initiates spreads over the atria to another group of modified muscle cells, the A-V or atrio-ventricular node, located between the right atrium and the right ventricle. From the A-V node a bundle of special fibers carries the depolarizing wave swiftly to all parts of both ventricles. The S-A node is not unique in its ability to originate depolarization and hence contractions. If one destroys the S-A node, the A-V node may take over, but its rate is then slower. The ability to contract automatically is a property of all cardiac muscle cells, but in a normal heart the S-A pacemaker imposes its rate upon the whole organ.

The spontaneous beating of the vertebrate heart can be modified by the central nervous system. Nerve endings from the parasympathetic system release acetylcholine and slow the heart rate; and nerve endings from the sympathetic system release norepinephrine and speed up the rate. Sites that respond to

norepinephrine also respond to epinephrine (also called adrenalin) released into the blood by adrenal glands; that is why excitement or fear or surprise, all of which provoke the adrenals to secrete epinephrine, cause the heart to beat faster.

SMOOTH MUSCLE

Nonstriated or smooth muscle is made up of sets of spindle-shaped cells, each with its own nucleus, and without the banding characteristic of skeletal muscle. There are two types of smooth muscle. MULTIUNIT MUSCLES, found in the iris of the eye, and in the walls of many blood vessels, contract only when stimulated by a nerve or a hormone, and then many fibers contract quite promptly as a unit. The other type, VISCERAL MUSCLE, found in gut walls, contracts slowly and spontaneously. It is not known whether a sliding filament mechanism is involved.

Bivalve mollusks (such as clams and oysters) have developed a special kind of muscle to solve their special problem of how to hold a tremendous tension for very long periods of time, so that their shells can remain closed and protective against all but the most determined predators. The strange feature is that maintenance of this tension involves little more oxygen consumption (and therefore energy utilization) than in the resting state. This phenomenon has been considered in the past to involve some sort of latching mechanism; once the latch is set, no energy is needed to keep it closed. A recent version of this view is that the contracted muscle undergoes "setting" — a change of state, with a great increase of viscosity, rather like the setting of an epoxy glue. The muscles involved contain ribbon-like filaments of a specialized protein called paramyosin with an unusual amino acid composition (no tryptophan and little proline). But nothing is known about the molecular basis of their action.

GLANDS

A gland is a cell or an organ that is specialized to secrete a substance that is released from the cell. The two basic types are ENDOCRINE GLANDS, which secrete hormones more or less directly into the bloodstream and are therefore truly internal organs (*Chapter 12*); and EXO-CRINE GLANDS, which release their products into the digestive tract or outer surface of the body and are therefore to some degree external

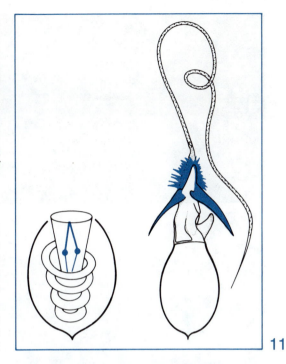

11

NEMATOCYSTS of coelenterates are barbed tubes that lie coiled within specialized cells called cnidoblasts (left). Contact with prey triggers the discharge of the nematocyst, which penetrates the tissue of the victim and injects a paralyzing poison. Discharged cnidoblast is shown at right.

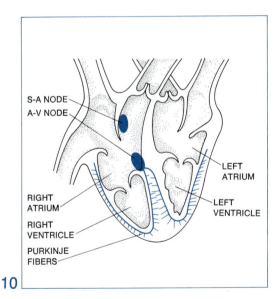

S-A NODE
A-V NODE
LEFT ATRIUM
RIGHT ATRIUM
LEFT VENTRICLE
RIGHT VENTRICLE
PURKINJE FIBERS

10

PACEMAKER OF HEART is the sino-atrial node (S-A node). Wave of depolarization spreads from S-A node to atrio-ventricular node (A-V node) and then, via the Purkinje fibers, to rest of heart.

in their activity. Exocrine glands include the digestive glands (*Chapter 8*), pheromone glands that produce odor and taste signals for use in communication (*Chapters 15 and 26*), sweat glands used in the regulation of body temperature (*Chapter 11*), and poison glands used by a great array of venomous animals, including certain mollusks, annelid worms, spiders, insects, fishes, amphibians, and even mammals (the shrew has a poisonous bite). Some plants also possess exocrine glands. The most familiar are the nectar glands found in flowers (floral nectaries) and scattered over stems and leaves (extrafloral nectaries). One remarkable plant glandular system is illustrated in Chapter 8: the carnivorous sundew plant (*Drosera*) captures insects by trapping

Electric eels generate up to 600 volts with a power output of about 100 watts — enough to light a row of bulbs or temporarily stun a man.

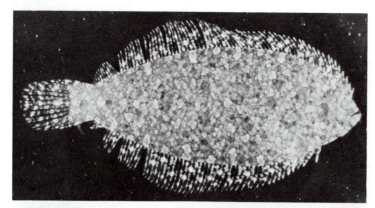

12

CAMOUFLAGE OF SOLE. Fish in top photograph has not had time to adapt its coloring to the artificial dark background. Below, sole is almost invisible after adapting its chromatophores to match sandy sea bottom. Adaptation to some backgrounds may require as much as a few hours.

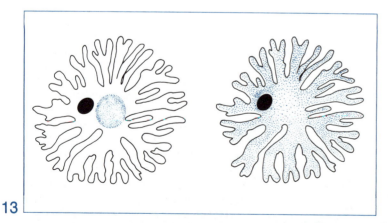

13

CHROMATOPHORE of a fish is a branched cell containing pigment (shown here as colored dots) in its cytoplasm. Animal is pale when pigment is concentrated in center (left) and turns darker when pigment fills the branches of the cell (right). Some chromatophores contain several pigments, each of which may respond to a different stimulus.

them in drops of sticky secretion produced by stalked glands that bristle from the surface of every leaf. The insects eventually die and the plant digests their decomposing remains.

TRICHOCYSTS AND NEMATOCYSTS

Some primitive animals possess highly specialized organs that are fired like miniature missiles to capture prey and repel enemies. Ciliated protozoans eject threadlike objects called TRICHOCYSTS from the surfaces of their cells (*Chapter 18*). The propulsion comes principally from the sudden elongation of the shaft of the trichocyst as it leaves the animal. Trichocysts are employed not only as weapons but also to anchor the organisms to the substratum. NEMATOCYSTS are elaborate cellular structures produced only by hydras, jellyfish and other members of the phylum Coelenterata (*Figure 11*). They are concentrated on the outer surface of the arms of the animal in huge numbers. Each nematocyst is made up of a slender thread coiled tightly within a capsule, which is armed with a spinelike trigger projecting to the outside. When the body of a potential prey organism brushes the trigger the nematocyst fires, turning the thread inside out and exposing little spines along its base. The thread either entangles or penetrates the body of the victim, and a poison is simultaneously released around the point of contact. Once the prey is subdued, it is pulled into the mouth of the coelenterate and swallowed.

CHROMATOPHORES

Chromatophores are pigment-bearing cells in the skin or at least close to the outer surface, which expand or contract to change the color of the organism. They are under nervous or hormonal control or both, and in most cases they can effect the change within minutes or even seconds. In squids, flounders, the famous chameleon (a kind of African lizard) and a few other animals, chromatophores enable the animal to blend in with the background on which it is resting and thus to escape discovery by predators (*Figures 12 and 13*). In other kinds of fishes and lizards, the color change is used as a signal to communicate with potential mates and territorial rivals belonging to the same species.

BIOLUMINESCENCE

A great many organisms can produce the strange cold light of bioluminescence: bacteria, fungi, radiolarians (a kind of protozoan),

14

LUMINOUS TOADSTOOLS of the genus Mycena were photographed by daylight (top) and by their own light (bottom). Light is emitted mainly by gills beneath the caps of the fruiting structure.

dinoflagellates (another kind of protozoan), sponges, corals, coelenterates, nemerteans (marine worms), ctenophores (jellyfish-like animals), clams, snails, centipedes, millipedes, insects, squids, and fishes. The phenomenon is especially frequent among the animals that live in the unlighted depths of the deep sea. It is also disproportionately common in animals that roam shallow water or the land at night. In a few insects, such as the fireflies (which are really beetles), bioluminescence is used in communication between the sexes. The same function is probably served in many of the deepsea fish. But in other kinds of organisms, particularly the bacteria, fungi, and protozoans, the significance of bioluminescence is still very much a mystery (Figure 14). It may in fact be no more than an incidental by-product of peculiar forms of oxidation. The light is created when a special substance, LUCIFERIN, is oxidized in the presence of the appropriate enzyme, LUCIFERASE, in specialized cells. The chemical structure of these substances varies in different kinds of animals.

ELECTRIC ORGANS

Members of at least seven families of fish can generate electricity, including the electric eel, the knife fish, the torpedo (a kind of ray), and the electric catfish. The electric organs were evolved from muscle, and rely on the same principle of creating an electric potential as nerve and muscle. The electric organs consist of very large disk-shaped cells arranged in long rows like stacks of coins. When discharged simultaneously, these organs can generate far more current than nerve and muscle. The electric eels, for example, produce up to 600 volts with an output of about 100 watts — enough to light a row of light bulbs or temporarily stun a man. In the electric catfish the discharge is apparently used only to repel enemies, but the electric eel uses it to paralyze prey. The eel also generates trains of low-voltage pulses to aid in detecting objects in the water around it, a form of orientation described in Chapter 15.

READINGS

A.L. BURNETT AND T. EISNER, *Animal Adaption*, New York, Holt, Rinehart & Winston, 1964. A brief treatment of animal behavior and physiology that describes the evolution of effector systems.

M.S. GORDON, G.A. BARTHOLOMEW, A.D. GRINNEL, C.B. JORGENSON AND F.N. WHITE, *Animal Functions: Principles and Adaptations*, 2nd Edition, New York, Macmillan, 1971. The section on muscle is a comprehensive and unusually thorough review which gives equally good descriptions of physiological, biochemical and morphological aspects.

D.R. GRIFFIN AND A. NOVICK, *Animal Structure and Function*, 2nd Edition, New York, Holt, Rinehart & Winston, 1970. This is among the best short books on animal physiology, and pays particular attention to effector systems.

H.E. HUXLEY, "The Contraction of Muscle," *Scientific American*, November 1959, and "The Mechanism of Muscular Contraction," *Scientific American*, December 1965. Two outstanding discussions on the mechanism of muscle contraction, written by the investigator who did much of the work. Together they make a lucid and quite detailed description which reads like a novel.

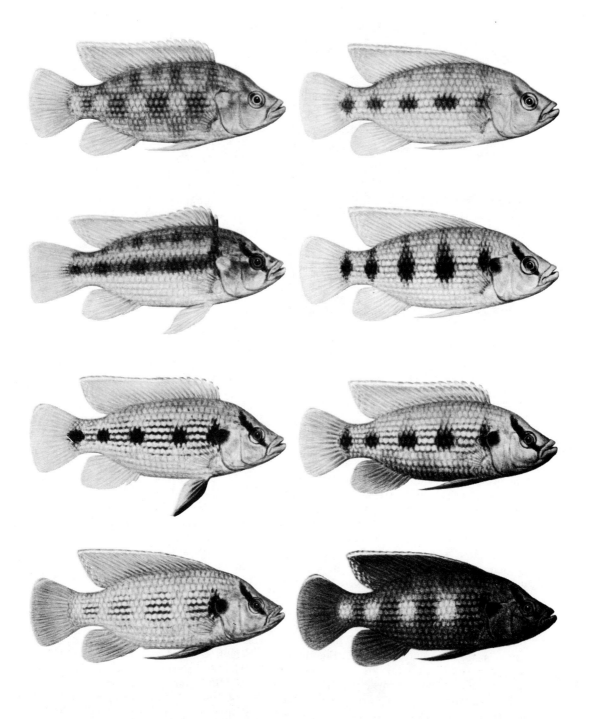

behavior

*[When the male moth gets a whiff of bombykol] it is
doubtful if he has an awareness of being caught
in an aerosol of chemical attractant. On the contrary,
he probably finds suddenly that it has become an
excellent day, the weather remarkably bracing, the time
appropriate for a bit of exercise of the old wings,
a brisk turn upwind.*

LEWIS THOMAS

Behavior consists of the actions by which organisms survive and reproduce. The more an organism is required to search through the environment to make its living, the more advanced is its behavior. A complex brain (one consisting of a greater number of neurons) permits greater movement and a more precise search of the environment; thus tigers show more complex behavior than flatworms. Nevertheless behavior is not limited to advanced animals, or even just to animals. Certain carnivorous plants are able to make sudden, directed movements to capture their prey. The Venus flytrap (*Dionaea*), for example, clamps its modified leaves around insects; the sensitive mimosas are able to fold and partially retract their leaves when they are touched. These plants accomplish movement without benefit of nerves or muscles. Instead, they rely on sudden changes in cell turgor. Within the animal kingdom, the most primitive forms of behavior are displayed by those groups which, like plants, lead a wholly sedentary existence. Adult corals and barnacles do little more than extend and retreat their bodies, move their feeding arms, and ingest the small organisms captured as prey. Truly elaborate patterns of behavior occur in animals closely related to them, such as jellyfish (which are related to corals) and shrimps and crabs (which are related to barnacles) — but obviously these relatives are mobile animals that move from place to place in search of food.

Behavior is a biological process basically like digestion and circulation. It has a genetic basis, and like any genetic trait it evolves. The potential range of an organism's behavior is wholly controlled by its DNA. The development of behavior is based primarily on the embryology of the sensory and nervous systems, and to a lesser extent on that of the endocrine system, which produces hormones that affect behavior. To an extent that varies enormously among species, development is also influenced by learning: the modification of behavior patterns by specific experiences. In a broad sense, learning can still be regarded as part of the development of an individual.

SIGN STIMULI

Animals with small brains do not have the equipment to contemplate their environment and to consult their memories before solving each problem. Often the problem is one that they have never experienced in their short lives. But to survive they must respond with elaborate and precisely timed movements. No larger-brained, more experienced animal will give them the benefit of the doubt if they make a mistake. How can such relatively simple animals be "programmed" to behave in the right way? The answer is that they respond only to a few key stimuli in the environment. These SIGN STIMULI function like special code words for the animal; they evoke the appropriate response, called a CONSUMMATORY ACT. In this world of highly abstracted behavior, the

AFRICAN FRESH-WATER FISH Hemichromis fasciatus, depicted in illustration on opposite page, can change its body coloring rapidly. The color changes are displays that express different moods.

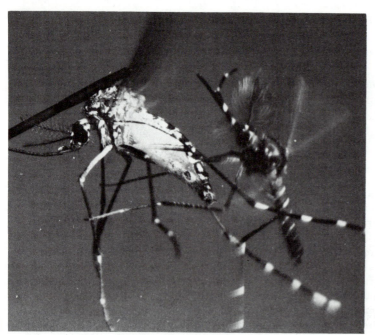

1

TETHERED FEMALE MOSQUITO (left) attracts flying male (right) who tries to mate with her. Female is glued to wire. Hum of her beating wings is an auditory stimulus that serves as a sexual lure.

buzzes her wings in attempts to fly, she attracts males from distances of 25 centimeters or more. The moment she stops, males fly past without noticing her. The great simplicity of the attractive stimulus is disclosed by the fact that tuning forks vibrating in the 300–800 cycles per second range also attract males. If the tuning forks are wrapped in soft material such as cheesecloth, a few males even try to mate with them.

Vertebrates offer equally striking examples of stereotyped behavior. When the eggs of herring gulls (and many other ground-nesting birds) accidentally fall out of the nest (or are deliberately removed by the biologist), the parents use their bills to try to roll them back into the nest. This stereotyped response reveals the sign stimuli by which the birds recognize their own eggs. Models carved from wood and painted in various colors and patterns can be presented to the parents in competition with real eggs. Such experiments have shown that the birds prefer the biggest "eggs" they can get. Any such quality that is preferred over the natural stimuli encountered in the life of an animal — in this case abnormally large size — is referred to as a SUPERNORMAL STIMULUS. Many examples of this curious phenomenon have been demonstrated experimentally. No satisfactory general explanation for supernormal stimuli has been produced, but their existence is taken as further evidence of the relative simplicity and automatic nature of much of the behavior of lower animals.

WHAT IS INSTINCT?

The word INSTINCT has had a controversial and confused history. It means too many things to different people to be valuable as a basic scientific term. Instinct will no doubt continue to be employed in popular writings to denote stereotyped behavior — behavior that is highly predictable and rigidly executed. Egg-rolling in gulls (mentioned above) could be called an instinct. The concept of instinct, however, is clouded by great variations in stereotyped behavior, even in the same species. Instinct is often believed to consist wholly of inherited responses that are built into the neuronal circuits of the brain and cannot be altered by experience. But a great deal of new information indicates that this definition is subject to many exceptions.

The complexities of "instinctive" behavior are best understood by examining some actual examples. Ground-nesting birds such as tur-

odor of a single chemical substance emitted by a female may be the sign stimulus that evokes mating behavior (the consummatory act) on the part of the male. A second chemical substance might identify a prey organism to the male and cause him to eat it. A flash of color can mean a rival male to be challenged; a single peculiar sound, the approach of a predator to be avoided. Characteristically (but not invariably) the sign stimulus consists of relatively few elements, and it is normally encountered only in the object toward which the consummatory act is directed.

These qualities of animal behavior are vividly illustrated in the sexual behavior of *Aedes aegypti*, the mosquito that carries yellow fever. The wingbeats of a flying female create a monotonous humming sound. The noise is irritating to human ears, but it is music to the male mosquito. The wingbeats generate sounds between 450 and 600 cycles per second. Males are attracted to any steady hum between 300 and 800 cycles per second, and thus are automatically drawn to the females. The effectiveness of this auditory sign stimulus can be demonstrated by introducing tethered females into cages containing males (*Figure 1*). As long as an individual female

keys, ducks, geese and pheasants sound alarm calls and crouch down when a hawk or some other bird of prey flies overhead. Experiments performed in 1937 by the pioneer behaviorists Konrad Lorenz and Niko Tinbergen indicated that the general shape of the flying bird, and not any particular anatomical feature, is crucial in recognition. When Lorenz and Tinbergen pulled cardboard silhouettes of various shapes over pens containing game birds, they discovered that a cross-shaped object whose head was shorter than its tail evoked an alarm response. When this silhouette was reversed so that the head was longer than the tail, it did not cause the alarm response. The first silhouette very roughly resembles a hawk and the second a goose (*Figure 2*). The young birds are at first afraid of all objects flying overhead, even falling leaves. Eventually they cease reacting to the ones seen repeatedly, such as the falling leaves and more common types of birds. But hawks are relatively scarce in any environment, and the birds never become accustomed to their distinctive shape. In this case learning of a special, genetically restricted kind — the identification of the dangerous stimulus by a gradual process of elimination — plays a role in the development of a stereotyped response.

Many behavioral biologists have found it impossible to make a sharp distinction between instincts and other kinds of behavior. There is a broad range of behavioral patterns, from highly stereotyped, virtually automatic responses that require no learning, to very flexible responses that are perfected only during a long learning period. The attraction of the male mosquito to the wingbeat hum of the female is an example of the first extreme, and the mastery of a particular human language is an example of the second. Many, perhaps most, patterns of animal behavior develop within strict limits imposed by heredity. But elements of learning are incorporated into the patterns before they reach their final form. A second generalization is that most behavior is GOAL-ORIENTED. This means that the final consummatory act is directed at some specific object in the presence of the animal — a food object, a mate, a territorial rival, or whatever object possesses the sign stimulus. There are typically three phases in the complete unfolding of a behavioral act.

The first is APPETITIVE BEHAVIOR: the animal searches for something it wants. A hungry animal hunts for food, for example, or a sexually mature one primed with reproductive hormones looks for a mate. The tendency to explore in a specific way, combined with a high probability of responding when the appropriate sign stimulus is encountered, is often referred to as a DRIVE. Like "instinct," the word "drive" is difficult to define, and its

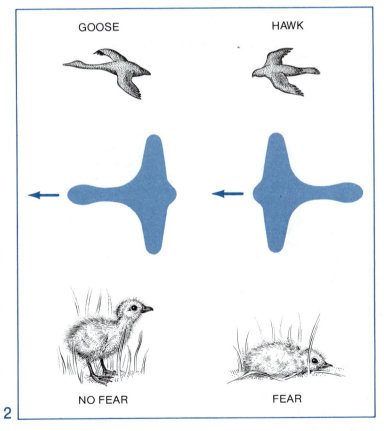

GOOSE HAWK

NO FEAR FEAR

2

DUCKLINGS SHOW FEAR and crouch in grass at the sight of a hawk flying overhead, but not at the sight of a flying goose. Same responses are evoked by sight of a cardboard silhouette (color) passing overhead. When silhouette is moved as shown at left, ducklings respond as if it were a goose. When silhouette is reversed and moved as shown at right, ducklings respond as if it were a hawk.

properties even harder to measure; most biologists avoid the term. It is important to note that appetitive behavior, whether or not one chooses to speak of it as an aspect of "drive," often has the appearance of being automatic and may require no previous experience. A young male of the mourning dove locates a place to build its nest by a slow trial-and-error process. It assumes the nest-calling position that will later be used to attract a female from within the nest. Its body is thrown forward and its head pulled down, as though its neck and breast were already fitted into the nest. During the appetitive phase, however, the body comes down in empty air. The bird now shifts its position and tries again. Eventually it finds a crevice or corner into which it fits, and the experience triggers the next round of behavior, which is the collection of straw for nest building. The appetitive behavior, in other words, brings the male dove to the point where nest building becomes automatic.

Appetitive behavior is followed by a CON-SUMMATORY ACT. The appropriate sign stimuli release the appropriate behavioral act. The posturing of the young dove brings it, almost haphazardly, to a naturally formed nest cavity, which serves as the sign stimulus to collect straw.

The final phase is QUIESCENCE. After the consummatory act, the animal normally slows or halts the appetitive behavior, and is less likely to perform additional consummatory acts when confronted with sign stimuli a sec-

ond time. Once started with its nest, the young dove no longer searches for nest cavities and does not respond to additional ones even if given the opportunity.

ORIENTATION

Most behavior is goal-oriented. For this reason it is useful to distinguish between the consummatory act itself and the orientation by which the animal directs the consummatory act at the appropriate object. Orientation is important during some forms of appetitive searching, when an animal attempts to move in a constant direction or pattern of search, and also during attempts to flee from an enemy or an unpleasant environment. Homing from the field to the nest site and navigation during migrations require the most precise orientation. Orientation movements can be separated from the consummatory acts with which they are commonly associated, and thus can be analyzed as a simpler form of behavior. Several basic types of orientation can be recognized: kineses, taxes, depth perception, and the detection of emitted energy, chiefly sound.

A KINESIS (plural: kineses) is a very elementary form of orientation in which the animal does not specifically direct its body to the stimulus. The stimulus merely causes it to move around to a greater or lesser degree, with the eventual result that it ends up either much farther away from the stimulus or else much closer to it. Consider, for example, the case of the woodlouse searching for a "home." This little isopod crustacean lives in moist places under loose stones, pieces of wood, and other objects on the ground. When a woodlouse finds itself drying out it simply begins to move around a great deal. When it encounters a moist spot it becomes less restless and may halt movement all together. As a consequence the woodlice tend to congregate in the moist spots beneath objects lying on the ground.

A TAXIS (plural: taxes), in contrast, is a movement in which the animal's body assumes a particular spatial relationship to the stimulus. The nature of the stimulus is often denoted by adding the appropriate Greek prefix to the word taxis: a phototaxis is movement guided by a light, a geotaxis is movement up or down guided by gravity, chemotaxis is movement guided by the smell or taste of some chemical substance, and so forth. A further distinction can be made between a positive taxis (movement toward the stimulus) and a negative taxis (movement away from the

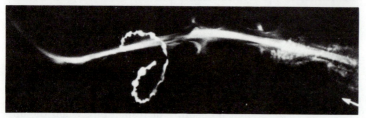

3

MOTH AVOIDS BAT in the streak photograph at top by going into a spiral dive when it detects the bat's sonar. Moth in bottom picture was less fortunate. Arrows indicate direction of bat's flight.

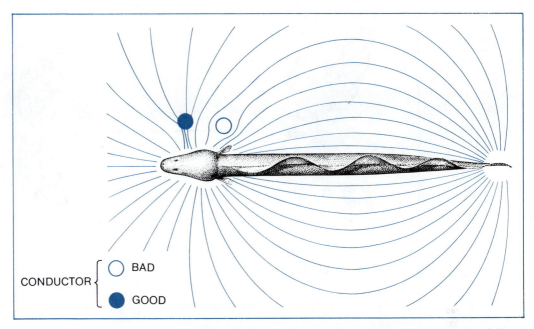

CONDUCTOR {
○ BAD
● GOOD

ELECTRIC EEL Gymnarchus generates an electric field (colored lines) around its body. Foreign objects, whether good or bad conductors, distort field. Eel detects changes with receptors in its head.

stimulus). The flight of a moth toward an electric light, for example, is a positive phototaxis; the retreat of a cockroach from the same light is a negative phototaxis. When a honeybee flies back to its hive after visiting a flower in an open field, it keeps on a straight course by moving at a constant angle to the sun — for example, straight toward the sun, straight away, 15° to the left, or whatever angle leads in the direction of the nest. This SUN COMPASS orientation, as it is called, is another form of phototaxis that is widespread in the animal kingdom. In the older biological literature, taxes used to be called tropisms, and the reader may still occasionally encounter this particular terminology, but the great majority of biologists now use the word tropism to mean only the growth of an organism toward a directional stimulus, rather than active movement.

The most sophisticated guidance systems do not depend on a signal originating from the target but instead bounce their own signals off the target. Ships at sea locate submarines and other underwater objects by means of SONAR, broadcasting a series of sound pulses into the water and detecting the echoes that come back. By noting the direction of an echo and the time required for it to return, an observer on the ship can pinpoint the position of the object in the water. Bats use the same principle to guide their flight around obstacles and to locate and capture insects in the air (Figure 3). They search for food by continuously emitting a call that consists of a rapid series of clicking sounds. The big brown bat (Eptesicus fuscus) rattles off ten clicks each second. Although the sounds are quite loud in terms of decibels, the human ear can scarcely hear them because most of the frequencies lie between 50 and 100 kilohertz (50,000 to 100,000 cycles per second), which is in the ultrasonic range, above the limit of human hearing. The bats, of

The most sophisticated guidance systems do not depend on a signal originating from the target but instead bounce their own signals off the target. Such systems are found in bats, porpoises and electric eels.

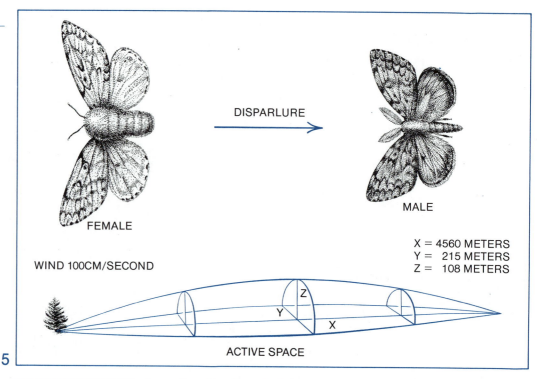

DISPARLURE

FEMALE

MALE

X = 4560 METERS
Y = 215 METERS
Z = 108 METERS

WIND 100CM/SECOND

Z

Y

X

ACTIVE SPACE

5

FEMALE GYPSY MOTH releases disparlure from glands in her abdomen. Pheromone evaporates and vapor drifts downwind, where males detect it with their antennae. Diagram below depicts a typical active space downwind from a female in tree. Males within space are attracted to her.

course, hear them very acutely. When an echo is returned from a flying insect, the bat rapidly increases the rate of emission of its clicks, permitting it to "zero in" on its prey. A dipping and swooping bat in the summer evening's sky is in the pursuit of insects by this aerial sonar. A bat can be momentarily fooled by a pebble tossed into the air near it. It will swoop at the object, which its echolocation system reports to be an "insect" but soon veer away when it realizes it has been deceived. A very similar echolocation system is used by porpoises to find their way through murky water and to hunt for fish.

A different but equally elaborate form of emitted-energy orientation, mentioned briefly in Chapter 13, is employed by the electric eels of South American and certain similar-looking fishes of Africa (*Figure 4*). When searching through the waters of their river homes, these fish generate electrical impulses from special organs located in their tails. The flow forms a dipole field around the body, with the tail negative and the head positive. The lines of flow are perceived by the fish through electric

sensory organs which are located mostly on its head. Objects in the vicinity are perceived by the distortions they create within the field.

COMMUNICATION

Much of man's esthetic experience of the living world consists of an awareness of communication between other organisms. The song of a bird, the chirp of a cricket, the flash of a butterfly's wing, the color and sweet scent of a blossom — all are signals being urgently transmitted to other plants or animals. In very general terms, BIOLOGICAL COMMUNICATION can be defined as action on the part of one organism (or cell) that alters the behavior pattern of another organism (or cell) in an adaptive way. Adaptive means that the signaling and the response contribute in some positive way to the survival or reproduction of one or both of the participants. The acts of communication include some of the most complex and fascinating forms of behavior. The best way to understand them is first to classify them according to the kind of sense organ by which they are received, and then to examine the advantages and disadvantages of each sensory system separately. Investigators have studied both communication between members of the same species and between different species, even between organisms as radically different as plants and animals. Communication involves a variety of signals including chemicals, light and color, sounds, and patterns of movement.

CHEMICAL COMMUNICATION

Chemical signals passed between different organisms of the same species are called PHEROMONES. They are the most widely used signals in both plants and animals and they were probably also the first signals put to service in the early evolution of animal behavior. Communication among single-celled organisms must antedate the origin of the higher organisms, and this primitive exchange was almost certainly chemical. Since cells in the bodies of higher animals communicate with one another by hormones, pheromones must have preceded hormones and, as noted in Chapter 12, might have given rise to them in a direct evolutionary line.

Pheromones are secreted from cells (usually special glandular cells) and transmitted either as liquids or gases. Some are normally detected at a distance by smell, others at the surface of the emitting animal by smell or taste. Still others are presented in liquid form at olfactory "signposts" to be smelled by passing animals over long periods of time. The substances function either by evoking immediate behavioral responses, called RELEASER EFFECTS, or by means of PRIMER EFFECTS that work through the endocrine system to alter the physiology and subsequent behavioral patterns of the receptor animal.

Releaser pheromones are widespread in the animal kingdom and serve a great many roles in different kinds of animals: sex attraction, simple aggregation, trail marking, territorial advertisement, individual recognition, and others. Sex attractants comprise an especially large and important category. The females of some kinds of moths, for example, secrete the pheromones from glands in the tips of their abdomens and are able to attract males over long distances (*Figure 5*). When sexually active males enter the ACTIVE SPACE in which the substance is detectable they begin to fly upwind. If they accidentally fly out of the active space, they search back and forth until they strike it again, and then continue their upwind

journey. This maneuver automatically brings them to the vicinity of the emitting female. The amounts of pheromone required to create large active spaces are often very small, because the antennae of the males are specially constructed to perceive them (*see photograph on page 242.*) In the case of the silkworm moth, for example, it has been estimated by Dietrich Schneider of Germany that only about 40 of the 40,000 hairlike receptors on the male's antennae each need be struck by a single molecule of female pheromone to start sexual activity. The pheromone is so potent that .01 microgram (one hundred millionth of a gram), the minimum average content of a single female moth, would be theoretically adequate to excite more than a billion male moths!

VISUAL COMMUNICATION

Textbooks on animal behavior lay heavy emphasis on visual communication, despite the fact that it is less common than chemical communication. The reason is simple: man has a highly developed visual sense and a relatively poorly developed chemical sense (see Chapter 27 for the evolutionary explanation of this fact). It is natural that biologists would at first pay closest attention to the forms of communication most conspicuous to them. It is also easy to see why certain kinds of animals, notably birds, fish and butterflies, have occupied center stage in studies of animal communication. These creatures, like man, rely greatly on vision. For the organisms that have well developed eyes, visual signals do indeed offer some marked advantages. By adopting a distinctive coloration or pattern of body shading, the visual animal can project a continuous message with a minimum expenditure of energy. During the breeding season a change in the color of the plumage indicates that male birds are ready for mating. Such signaling can be perceived almost instantly at long distances, and it does not even require that the sender be aware of the existence of the receiver.

Conversely, visual signals can be modified in a way that permits them to be rapidly changed. The rate of information flow is increased, and the messages are kept relevant to shifting circumstances. The illustration that opens this chapter shows an extreme example from communication in cichlid fishes. During territorial contests and sexual encounters it is important for these fishes to be able to transmit their moods, and thereby to suggest the behavior pattern likely to ensue, as each new situation demands.

AUDITORY COMMUNICATION

Sound offers certain advantages as a medium of communication. Like visual signaling it can be transmitted efficiently over relatively long distances and altered quickly to fit changing conditions. Unlike visual and chemical signals, it can be used anywhere and at any time. Auditory communication is therefore most commonly encountered among nocturnal animals or those that inhabit dense foliage where vision is limited. In the course of his evolution man selected sound as the medium of his remarkable language. The reason for this choice lies in large part in the great variety of signals that can be created by modifying 1) the intensity of the sound ("HELP!" versus "Help me, please"); 2) the frequency of each sound, measured in kilohertz (thousands of pulses) emitted per second when the sound is produced ("Heeelp" versus "Help"); and 3) the rapidity with which the sound is emitted ("Help! . . . Help!" versus "Help! Help! Help!"). Birds are like human beings, in that they have exploited all three of these qualities in sound communication. Biologists use the sound spectrogram to record and to analyze bird songs.

Intensive research on bird song over the past 20 years has revealed a number of remarkable facts about this seemingly commonplace phenomenon. The most elaborate, and to human ears the most beautiful songs, are employed by males to defend their territories and to attract females. In at least a few species each male produces its own slight variation of the song. This "signature" permits him to be recognized as an individual by his territorial neighbors. A few species transmit the song from generation to generation entirely by heredity. Others, such as the chaffinch and the sparrow, must experience the singing of other individuals to learn part or all of the song (Figure 6).

THE DANCE OF THE HONEYBEE

The waggle dance of the honeybee is the most famous of all the forms of animal behavior. First "decoded" by Karl von Frisch of Germany in 1945, it is a nearly unique symbolic language by which worker bees communicate the position of food or a new nest site to their sister bees. The waggle dance is easy to understand if one thinks of it as a ritual flight — a scaled-down version of the journey from the

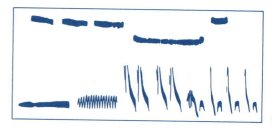

SOUND SPECTROGRAMS of the songs of two male white-crowned sparrows. At top, the rudimentary song of a bird reared in isolation; below, the song of a bird permitted to hear singing of other sparrows. Difference between the two songs reveals the role of learning. Spectrograms record frequencies of sound on the vertical axis, and the sequence in which they were emitted on the horizontal axis.

nest to the nectar. The essential element in the maneuver is the "straight run," the middle piece in the figure-eight pattern (*Figure 7*). The remainder of the figure-eight consists of doubling back to repeat the straight run. The dancing bee has just returned from several round trips to the target. The straight run is a miniaturized version of the outward flight which it invites its nestmates to try. If the bee performs on a horizontal surface with a view of the sun or open sky to guide it, the run points directly at the target. Usually, however, the bee dances on a vertical comb surface inside the darkened hive. In this case the angle toward the target is the angle between the straight run and the vertical. The length of the straight run indicates the distance to the source: the longer the straight run, the farther away the target. The straight run is emphasized by a rapid waggling motion of the body.

COMMUNICATION BETWEEN SPECIES

Communication occurs not only among members of the same species but also between radically different kinds of animals and even between animals and plants. A predator hunting for its prey, the prey seeking to detect and avoid the predator, competitors jockeying for possession of space or food, symbiotic organisms searching for each other — all communicate with one another, sometimes intentionally, sometimes not.

Many animals avoid detection through the use of CRYPTIC APPEARANCE: they resemble pieces of the environment on which they live. Some species employ only shading or color. The flounder, for example, alters its color to blend in with the patch of ocean floor on which it happens to be resting at the moment.

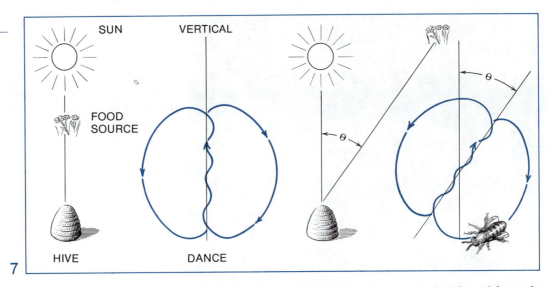

SUN   VERTICAL

FOOD
SOURCE

HIVE   DANCE

7

WAGGLE DANCE of honeybee, performed on vertical surface of the comb within the hive, indicates the location of a source of nectar. Direction of straight run (wavy arrow) relative to the vertical indicates direction to food relative to position of the sun. Dance at left indicates a food source that lies directly toward the sun. Dance at right indicates a food source that lies at an angle (Θ) to the right of the sun. The length of the bee's straight run indicates the distance to food source.

Others, such as the twig-imitating insect illustrated in Figure 8, are also shaped to resemble inedible or neutral objects. These animals typically hold their bodies in positions that increase the resemblance still further. In order to observe the phenomenon of cryptic appearance yourself, all one has to do is closely inspect a few tree trunks in warm weather. One soon encounters a variety of moths, barklice, and other insects whose color closely resembles that of the bark on which they are resting. Some will elude the observer's attention until his activity forces them to move. The great majority of examples of cryptic animals are in the act of avoiding predators. Conversely some spiders, predatory bugs and mantids resemble bark or flowers and use this deception to ambush unwary prey.

WARNING APPEARANCE is one of the most common of all biological attributes. Some examples are familiar to the layman. The boldly striped body color of the hornet, the brilliant red-and-black bands of the deadly coral snake and the fake eyespots of certain moths are cases in point (Figure 9). A good rule to remember when entering a strange habitat anywhere in the world is: leave the most conspicuous animals alone, especially if they are beautiful! Many of them are transmitting a warning to all would-be enemies. Their

appearance says in effect, "I am poisonous to eat or venomous. I don't need to hide or to run because my potential enemies learn to avoid me in ways that they will remember."

8

MIMICRY. Caterpillar of moth resembles a twig, a form of camouflage that hides it from birds. Real twigs are at top, caterpillar at bottom.

9

BIRD IS STARTLED when the sphinx moth flashes the fake eyespots on its wings. Such eyespots are found in several species of moth, and serve to discourage predators such as birds.

tion, as exemplified by the honeybee and the flowers it attends, make a marvelous story. Experiments have proven that bees can see color, but they see a visual spectrum different from ours. They can perceive ultraviolet, which man cannot, but they are far less sensitive than man to light at the opposite (red) end of the spectrum. When the human observer looks at a flower through a filter that shifts the ultraviolet energies to the visible spectrum, he sees the flower for the first time the same way as the bee. The results can be striking. What previously had seemed to be plain white or yellow blossoms now are found to possess (in the eye of the bee) strongly contrasting patterns. These ultraviolet colors permit the bees to distinguish species of plants that are superficially the same to the human eye. The patterns also provide geometric clues, called nectar guides, that direct the bee to the position in the flower where the nectar is located (*Figure 11*). No example illustrates so well the difference between the sensory world of man and that of other organisms.

MIMICRY

Perhaps the next best thing to having unpleasant traits and warning signals to advertise them is to resemble some other species that does. In this situation, called BATESIAN MIMICRY, the predator is not able to distinguish the harmless mimic from the dangerous model and leaves both alone. The mimicry typically involves color, form and behavior (*Figure 10*). A second form of mimicry is MULLERIAN MIMICRY, in which two or more unpleasant species resemble each other. Each, then, serves as both a model and a mimic. They "team up" in a way that insures more frequent exposure and quicker learning on the part of the predator than would be the case if they were advertising their repellant qualities separately.

FLOWERS AND INSECTS

The flower is the plant's signal to the insects that carry its pollen. In this special language, the plant attracts the butterfly, bee, wasp, beetle or whatever kind of insect is especially adapted to it. The shape of the flower is such that when the insect alights, it has no trouble finding the spot — usually the center of the blossom — where nectar and pollen are located. In exchange for receiving this food, the insect inadvertently carries pollen from flower to flower. Thus the plant has adopted the insect as its sperm carrier.

The details of this symbiotic communica-

AGGRESSIVE BEHAVIOR

The posturing fish in Figure 12 reveals a great deal of what behaviorists believe to be gener-

10

BATESIAN MIMICRY in butterflies. Monarch butterfly (top) is distasteful to birds, who learn to avoid it. Protection is thereby granted to the closely similar viceroy butterfly (bottom), which is quite edible.

11

BEE'S-EYE VIEW of the marsh marigold (Caltha palustris) at left is remarkably different from view seen by human eye (right) because eye of bee is sensitive to ultraviolet light. Ultraviolet photograph at left reveals pattern of nectar guides that direct bee to location of food, and incidentally ensure that she will pollinate flower. To humans, flower appears to be a bright even yellow.

ally true of animal aggression. First, except for the starkly functional act of predation, aggressive behavior is most strongly developed in encounters between members of the same species. Second, it is most frequent among males defending their territories or maintaining their status in social groupings. Third, it consists chiefly of display. Even when the opponents come into physical contact, the "fighting" is usually a ritual that results in little or no bodily harm. Fourth, the display often includes marked changes in the body, such as puffing or flattening of various body parts to make the animal look larger, and color changes that make it more conspicuous.

Aggression in animals is mostly an elaborate bluff. The goal appears to be to repel the enemy, at least to some point outside the territorial boundary, without running the risk of

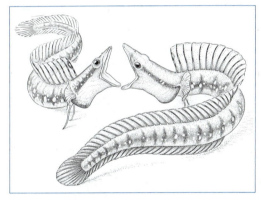

12

AGGRESSIVE DISPLAY between fishes. Male pike blennies (Chaenopsis ocellata) defend territories on sea floor with highly ritualized exchanges that include lowering of the gill case, raising the fins, gaping of the jaws, and changes of color. Winner is determined by size and aggressiveness.

Aggression in animals is mostly an elaborate bluff.
The goal appears to be to repel the enemy, at least to a point
beyond the territorial boundary, without running the risk
of dangerous combat.

dangerous combat. As Konrad Lorenz, a leading student of the subject, has expressed it, "Though occasionally, in territorial or rival fights, by some mishap a horn may penetrate an eye or a tooth an artery, we have never found that the aim of aggression was the extermination of fellow members of the species." This generalization applies less well to insects and other lower animals than it does to the vertebrates that Lorenz studied, but it is still sound enough to stand as a rule of nature.

Then is aggression in man "natural"? From a biologist's point of view it certainly seems to be. It is hard to believe that any characteristic so widespread and so easily invoked as aggressive behavior in man could be neutral or negative in the way it affects survival and reproduction. To be sure, overt aggressiveness is not characteristic of all or even most human cultures. But to be adaptive it is enough that aggressive responses be evoked only under certain conditions of stress, such as food shortages and conflicts during periodic high population densities. The lesson for man is perhaps that personal happiness has little to do with the overall pattern of aggression. Under conditions of stress, survival is purchased at the price of aggression and unhappiness. If man wishes to reduce his aggressive behavior to levels that make life more pleasant, he should consider altering population densities and social systems in such a way as to make aggression nonadaptive in the first place.

MODIFICATION OF BEHAVIOR

The examples of stereotyped behavior cited earlier in this chapter suggest certain resemblances between the actions of an animal and those of a machine. This philosophically important comparison, which is the keystone of ethology, has proved very effective in designing experiments to analyze the components of instinct. In the simplest possible case, the sign stimulus "pushes the button," and the animal responds in a machinelike way with the appropriate consummatory act. There are a few extreme examples, such as the sex attraction of male insects, in which the behavior can be characterized in approximately this fashion. But of course the great majority of animals do not really resemble robots. The total behavior of an animal follows a program of modifications that adapt it to changes in the environment. The sign stimulus still pushes the button. But sometimes the animal responds

and sometimes it does not, and often it responds with a different behavioral act. Its actions are controlled not only by the nature of the sign stimulus, but also by the season and time of day, by the level of its hormones, and by its previous experience.

CYCLICAL CHANGES IN BEHAVIOR

Few animals display uniform activity around the clock. Most are subject to circadian rhythms (Chapter 11) that affect their general levels of activity and even their "willingness" to react to sign stimuli. Sexual behavior in moths, to take one of the most clear-cut examples, is limited to certain hours of the day or night. Each species of giant silkworm moth in the United States has a particular time at which the females release sex pheromones and the males respond. In the promethea moth this is 4–6 P.M. In the cynthia moth it is 10 P.M. to 5 A.M., and in the cecropia moth around 4 A.M. These differences increase the probability that each insect will succeed in finding a mate belonging to its own species.

Reproduction tends to be strongly seasonal. The breeding season in birds, like that of other vertebrates, is heralded by changes in the structure and physiology of the ovaries and testes. These reproductive organs produce the sex hormones, which are mostly androgens in the male and estrogens in the female. The sex hormones in turn induce the development of the characteristic breeding plumage and coloration and alter the behavior of the birds so that they commence territorial and reproductive behavior. The ring dove is a species whose reproductive behavior has been analyzed by D. S. Lehrman and his associates at Rutgers University. When males and females of this small relation of the common pigeon are segregated by sex in cages, they show little evidence of reproductive activity. But if a pair are placed together in a cage, courting begins within a few minutes (Figure 13). The male is the initiator, since his testes are active and probably secreting testosterone, one of the male sex hormones. He faces the female and repeatedly bows and coos. The sight of the displaying male activates centers in the female's brain, which induce secretions from the pituitary gland. The pituitary hormones in turn stimulate the release of the two main gonadotrophic hormones, FSH and LH (Chapter 12). These hormones stimulate the growth of the female's ovaries, which begin to man-

COURTSHIP

NEST CONSTRUCTION

INCUBATION

"NURSING"

13

REPRODUCTIVE BEHAVIOR OF RING DOVE. Male initiates courtship with "bowing coos." Sight of courting male stimulates release of reproductive hormones in female. During nest-building birds copulate, and a week or so later female lays two eggs. Both male and female take turns incubating the eggs. Newly hatched squabs are fed "crop milk," a nutrient fluid regurgitated by both parents.

ufacture eggs and to release estrogen into the bloodstream. This entire chain of endocrine activity is put into motion within only a day or two. When both birds have thus been primed with hormones, they begin to construct a nest together. During nest building the male continues his courtship, and copulation follows. When the nest is finished the female becomes closely attached to it and soon lays her first egg, an event that triggers still further endocrine changes in both the female and her mate, leading to the incubation of the egg and care of the young. Step by step, as if programmed by a choreographer, hormones lead to behavioral acts; the acts elicit the release or the shutoff of other hormones, which in turn trigger further behavior, and so on until finally the young have been reared and the cycle completed. The ring dove was among the first animals whose reproductive behavior was thoroughly analyzed. By now there is evidence that many other kinds of animals, from cockroaches to higher mammals, also depend upon reciprocal relationships between hormones and behavior to guide their reproductive cycles.

The evolution of mammals shows a trend toward emancipation from hormone-directed behavior, and the substitution of learned responses. When a male rat or mouse is castrated, its sexual behavior rapidly declines and virtually disappears. Dogs and cats, which have larger brains and display less stereotyped behavior, also suffer a decline of sexual activity following castration. But the loss comes slowly over a period of months or years, and in a few individuals it does not occur at all. Man, apes, and other higher primates are not affected by castration, provided they are sexually mature and experienced before mutilation.

LEARNING

Appetites such as hunger provide a finer adjustment of behavior than the fluctuation of hormone levels. The process of learning permits the most sensitive control of all. Learning is the adaptive change of behavior as a result of experience. The organism alters its behavior to respond to specific stimuli in ways that promote its own survival or that of its offspring. Thus it closely molds its behavior pattern to adapt itself to the particular environment in which it occurs. The same learning potential can be used to fit a great many different environments.

HABITUATION is perhaps the simplest form of true learning, since it involves the loss of old responses rather than the acquisition of new ones. If an animal is repeatedly presented with a stimulus not associated with either a reward or a punishment, it will eventually cease to respond. Habituation is the basis of the "taming" of wild animals. If one carefully handles a freshly captured snake, after a few hours it will calm down and may eventually lie completely at rest in your hands. This change in behavior is simply habituation to the strange stimuli of the handling.

Most animals are able to learn to associate two or more stimuli with the same reward or punishment. The great Russian psychologist I. P. Pavlov called this conditioned reflex, but it is now more accurately called ASSOCIATIVE LEARNING. Pavlov's famous experiment still provides one of the most instructive examples. He first collected and measured the saliva released by a dog when it tasted powdered meat. Then he added a new stimulus, the ticking of a metronome, at the time the meat powder was provided. After the stimuli had been presented jointly in this fashion five or six times the dog was permitted to hear the metronome but was not given meat powder; nevertheless it still salivated.

Rather than learning to associate new stimuli with old ones that lead to reward or punishment, an animal might explore its environment until it has a particular experience, and then learn the stimuli associated with the experience. Suppose that Pavlov's hungry dog were allowed to hunt for food until it found a dish of meat powder behind the ticking metronome. Thereafter, the dog would be likely to head directly for the metronome. Learning by trial and error is sometimes called OPERANT CONDITIONING.

A student enroute to a new classroom might make casual note of such things as the location of a water fountain or alternate exit routes without discernible effort and without benefit of reward or punishment. The information is stored for potential later use. Such LATENT LEARNING also occurs in animals. Like man, they use it most frequently when placed in new surroundings. Bees and wasps, for example, often conduct orientation flights around their new nests, examining and learning enough of the terrain to find their way home on later flights.

INSIGHT LEARNING is the highest form of learning. Referred to as reasoning when it is conducted by man, it consists of recalling a previous experience that involved stimuli different from the current ones and adapting the memory to solve the current problem. Insight sometimes consists of the relatively simple process of GENERALIZATION. Honeybees can be trained to fly to a checkerboard pattern of one color in preference to all other patterns. If this particular checkerboard is now removed and the bees are limited in their choice to a checkerboard of another color, they will still select the substitute over all other patterns. They have generalized the checkerboard pattern and temporarily ignored the color. Much more complex forms of insight learning have been demonstrated in chimpanzees. Perhaps the most striking and manlike is illustrated in Figure 14. The chimp, with no previous experience of this kind, has reasoned a way to stack boxes on top of each other in order to reach a previously inaccessible banana.

The act of IMPRINTING, as the name itself implies, is a peculiar form of quick learning that is difficult to reverse, or "unlearn." It is often dramatic in effect, ordinarily occurs only during a limited time in the lifespan of an animal, and does not require a reward or punishment. The best known and most illuminating example comes from an experiment performed by Konrad Lorenz in Austria. He divided a clutch of eggs laid by a single graylag goose into two groups. When the goslings from one group hatched they were permitted to associate with the mother goose. The other goslings were hatched in an incubator, and the first living creature they saw was Konrad Lorenz. In the first few days of their lives, they were allowed to follow Lorenz as though he were their parent. Later the goslings were marked according to the group to which they belonged and placed together under a box. When released, the two groups separated from each other and ran to their respective adopted parents (Figure 15).

Similar experiments with a variety of birds, mammals, and insects have revealed that imprinting is widespread in the animal kingdom. The period of latency, during which the young animal can be trained in this special way, is often extremely short.

THE EVOLUTION OF BEHAVIOR

Circadian rhythms, appetites, hormone levels, maturation, and learning each cause major variations in the behavior of a given animal. The total pattern can be viewed as the program of its behavioral responses. This changing repertory adapts the animal on a moment-to-moment basis to its environment and permits it to survive and reproduce. Can the basic program be altered? Over long periods of time

14

INSIGHT LEARNING enables chimpanzee to stack boxes to reach a banana, but raccoon on leash fails to solve the detour problem of reaching a pan of food by first walking away from it.

15

IMPRINTED GOSLINGS follow ethologist Konrad Lorenz as if he were their parent. Lorenz was first living creature that young geese saw during latent period immediately after hatching.

each animal species must adapt by evolution to major changes in the climate and to the shifting composition of the other animals and plants that make up its living environment. The species slowly changes in the hereditary basis of its anatomical and physiological characteristics. It also changes the hereditary control of its behavior program.

For most of the first half of this century, one of the great controversies of biology and psychology raged around the role of instinct in animal behavior. Some psychologists, particularly those who belonged to the "rat psych" school of America (so named because they used white rats extensively in their experiments), believed that behavior is primarily or even exclusively learned. The young animal, they argued, is born with a brain like a *tabula rasa* — a blank tablet — on which experience leaves its mark. Behavior patterns can be developed only by learning. Most biologists regarded this theory as far too extreme. A few propounded the equally extreme theory that in lower animals behavior is usually completely controlled by heredity. The instinct versus learning controversy is but one version of the large "nature versus nurture" controversy: which is the more important determining factor, heredity or environment? Which, for example, determines eye color? The answer must be BOTH heredity and environment. Heredity dictates the color, insofar as it differs

from one individual to another, but the color itself develops only by a prolonged interaction between one's genes and the environment.

This evaluation of eye color goes to the heart of the matter. It is meaningful to speak of an "hereditary" trait only if there exist two or more such traits — such as different eye colors — which can be compared between individuals. Thus if one person has brown eyes, and another blue, the difference between the colors is an hereditary difference. The hereditary basis of this particular quantity of human variation has been established beyond doubt by standard genetic analyses. By the same token, the difference between the mating call of a ring dove and that of a rock pigeon is hereditary. The difference is also referred to as innate or "instinctive." To speak of variation in instinctive behavior is to employ a slightly different meaning from the one introduced earlier, which stated that if a behavioral pattern varied very little within a species and was therefore highly predictable, it could be referred to loosely as an instinct. Now the term "instinctive" is used for a genetically based difference between two behavioral responses. Both of these crude definitions are valid. The ambiguity of language is unfortunate, and it has been discussed here only because of the importance of the ideas that are so frequently included under the heading of "instincts."

THE ORIGIN OF SIGNALS

The many remarkable signals used in animal communication did not spring into existence suddenly. They evolved gradually, usually as modifications of behavior patterns that originally had nothing to do with communication. Consider the courtship displays of birds. Some have been derived in evolution from intention movements, the preparatory motions that animals go through as they start to run, fly, attack, or engage in some other basic functional response. One example can be seen at the beginning of the take-off leap, when a bird launches its flight. The bird pulls its head back toward the body, bends its legs, and spreads or raises its tail feathers. Many kinds, including ducks, herons, cormorants, and the turkey, have modified the take-off leap to serve as a signal during courtship (*Figure 16*). The movement resembles the true take-off from which it is derived in evolution, but it is more stereotyped and conspicuous, and no longer leads to flight. This process of the origin of a signal by modification of a behavior pattern originally

6

RITUAL TAKE-OFF LEAP of cormorant is part of courtship displays. Stereotyped body movements, including rhythmic wing-flapping, are easily distinguished from bird's true take-off leap.

having nothing to do with communication is called RITUALIZATION. Of course the animal still utilizes the ancestral forms of behavior in addition to the ritualized version. For example, the cormorant still performs a true take-off leap when it flies.

A second form of behavior that is often ritualized to create a signal is displacement activity. Animals caught in conflict situations — for example, males confronted by rivals at their territorial boundary, or thwarted in their attempts to court females — sometimes switch to behavior that is totally irrelevant to the circumstances. The animal may suddenly begin to preen itself, or go through the motions of eating or drinking. It has been seriously suggested by many authors that cigarette smoking, a form of nonfunctional "eating," is a case of a human displacement activity. It is certainly true from an anthropomorphic viewpoint that the animals engaging in displacement activities appear to "relieve nervous tension." A similar form of behavior, called a redirected activity, may also occur in

conflict situations. In this case the behavioral act is in the right context but directed at the wrong object. For example, a monkey menaced by a larger, stronger individual will often turn from this antagonist and threaten a third still smaller monkey. It is sometimes difficult to make a clear distinction between displacement and redirected activities. Consider, for instance, a rooster that turns momentarily from a rival in the barnyard and begins to peck at pebbles on the ground. Is it performing feeding movements in an inappropriate context, in other words engaging in a displacement activity? Or is it aiming its aggressive behavior at an irrelevant object, in a redirected activity?

INTELLIGENT BEHAVIOR

One of mankind's most cherished ideas about evolution is that blind instinct has gradually given way to learning and reasoned behavior, and that the trend reaches its highest (but far from perfect) expression in man. This lofty generalization is essentially true. But the evolution of behavior has not been simply the process of animal life striving upward toward a more enlightened view of the world. Behavior has evolved from highly programmed, stereotyped responses to a much more flexible form of adaptation to the environment by means of learning. Stereotyped behavior or "instinct" is the best that can be done with a small number of neurons. An animal with a very small brain resembles an automaton carefully programmed to react to certain principal features of the environment (sign stimuli) which are vital to its survival. It is guided home by one sign stimulus, it recognizes a mate by another, it seizes a food item by two more, and so on throughout its life. The large-brained animal is also programmed, but it has enough spare neurons to engage in the luxury of learning. To some extent it can learn to select from the stimuli in its own particular environment those which are the most vital to its success. It molds itself to the local envi-

An animal with a very small brain resembles an automaton programmed to react to sign stimuli vital to its survival. The large-brained animal is also programmed, but it has enough spare neurons to engage in the luxury of learning.

ronment and responds with greater precision and safety than the small-brained animal. True intelligence and enlightenment have been fortuitous by-products of this trend in human evolution.

READINGS

J. Alcock, *Animal Behavior: An Evolutionary Approach,* Sunderland, Mass., Sinauer Associates, 1975. The best short textbook on the subject. Recommended as a relatively easy next step.

V.G. Dethier and E. Stellar, *Animal Behavior: Its Evolutionary and Neurological Basis,* 3rd Edition, Englewood Cliffs, N.J., Prentice-Hall, 1970. An intermediate textbook that emphasizes the relationship between neurobiology and behavior.

I. Eibl-Eibesfeldt, *Ethology: The Biology of Behavior,* 2nd Edition, New York, Holt, Rinehart and Winston, 1975. A beautifully illustrated account of zoological contributions to the study of animal behavior, with stress on the vertebrates.

T. Eisner and E.O. Wilson, Editors, *Animal Behavior,* San Francisco, Freeman, 1975. A collection of articles from *Scientific American,* reporting the findings of behavioral research in the scientists' own words.

R.A. Hinde, *Animal Behaviour: A Synthesis of Ethology and Comparative Psychology,* 2nd Edition, New York, McGraw-Hill, 1970. A generally successful but turgidly written work that integrates the modern findings of zoology and psychology.

P.H. Klopfer and J.P. Hailman, *An Introduction to Animal Behavior: Ethology's First Century,* Englewood Cliffs, N.J., Prentice-Hall, 1967. A sound textbook on animal behavior which stresses the history of ideas.

K. Lorenz, *On Aggression,* New York, Bantam, 1967. This little classic is the source of most of the current ideas (and controversies!) on the evolutionary origin and significance of aggressive behavior in man.

P.R. Marler and W.J. Hamilton III, *Mechanics of Animal Behavior,* New York, John Wiley & Sons, 1966. An excellent full-length textbook on the physiological basis of behavior.

PART THREE
THE
ORIGIN
&
DIVERSITY
OF LIFE

The evolution of life on Earth might easily have stopped once the ancient seas were covered with a living scum of primitive green algae. These single-celled autotrophs could perpetuate life forever, capturing energy from the sun, cycling nutrients in and out of the surrounding water, and monotonously replicating themselves by endless cell division. Such organisms could persist indefinitely in an ecological steady state. Their number would remain approximately constant, with the birth rate of new cells equalling the death rate of old ones. The flow of energy too would remain constant, with the amount of radiant energy captured equalling the energy lost.

On Earth, of course, the evolution of life has followed a very different scenario. Since the appearance of the first organisms more than three billion years ago, life has relentlessly advanced and diversified. New kinds of plants, some of them comprised of many cells instead of just one, became specialized bottom-dwellers in shallow waters, while others remained floating near the surface. At a very early date the first heterotrophs appeared, including consumers that fed on living plants, predators that fed on the consumers, and scavengers that decomposed the remains of the dead. Perhaps a billion years ago the ranks of the heterotrophs gave rise to "true" animals: multicellular organisms that could move from place to place. Among the multicellular plants and animals there emerged a trend toward the evolution of new and larger organisms with ever more complex anatomy. But at the same time a diversity of small organisms with simple body plans continued to thrive. Eventually radical new "specialists" began to colonize previously uninhabited zones of the planet's surface: the deep sea, the land, and even the air. The result was a steady increase in the biomass — the amount of living matter present on Earth. At the present time the biomass is remarkably more complex and efficient than the minimal green scum that might have existed on a barren planet.

The rest of this Part will focus on the emergence of the rich diversity of life on Earth, beginning with the spontaneous appearance of life during the planet's geological youth. One chapter discusses the most venerable of the biological disciplines: taxonomy, the science of classification. Subsequent chapters reconstruct the history of life in terms of the major evolutionary inventions that produced distinctive kinds of organisms.

16

Among the impractical quests of man, none has been more alluring than that concerning the origin of life.

HANS ZINSSER

origins

In 1924 the Russian biologist A. I. Oparin published a thin monograph in Moscow entitled "The Origin of Life." Although it was never translated from Russian and had no impact on scientific thought, it contained virtually the entire blueprint for what we now believe to be the actual process of evolution of life. Five years later, the British biologist J. B. S. Haldane arrived at the same ideas independently and published them in *The Rationalist Annual,* again to little effect. It is a comment on scientific attitudes at the time that neither man felt able to propose his ideas in a recognized scientific journal. Not until Oparin expanded his ideas into a book, *"Origin of Life,"* which was published in 1936 and translated into other languages, did the problem of the appearance of life on Earth receive serious consideration.

The Introduction to this book presented essentially the Oparin-Haldane theory. Life, they argued, evolved from organic chemicals in the seas at a time when the Earth's atmosphere contained virtually no free oxygen. These organic chemicals — first simple amino acids, purine and pyrimidine bases, and sugars, and later polypeptides, polynucleic acids, and polysaccharides — were originally synthesized abiologically by the action of ultraviolet radiation, electrical discharges, heat, or radioactivity. They were not rapidly oxidized or broken down, as they would be today, because no free oxygen was available to react with them. Photosynthesis, once it appeared, changed the character of the atmosphere by liberating vast amounts of oxygen. This led to a more efficient life process: respiration. But by hastening the natural oxidation of organic matter and by screening out the Sun's ultraviolet radiation with a high-altitude layer of ozone, the oxygen atmosphere also eliminated the possibility that life could evolve a second time, or that the appearance of new life could be a continual process, as stated by the theory of spontaneous generation.

There is another good reason why the halfway stages toward living organisms could never evolve again now — they would be promptly eaten by older and more efficient competitors. Darwin recognized this in 1871: "It is often said that all the conditions for the first production of a living organism are now present, which could ever have been present. But if (and oh, what a big if!), we could conceive in some warm little pond, with all sorts of ammonia and phosphoric salts, light, heat, electricity, etc., present, that a protein compound was chemically formed ready to undergo still more complex changes, at the present day such matter would be instantly devoured or absorbed, which would not have been the case before living creatures were formed."

The evolution of life attracted the interest of relatively few biologists in the 1940's and early 1950's, but this situation changed drastically after the beginning of space exploration. With the possibility of finding life on other planets, the origin of life on our own planet became a nagging question. Paleobiology became interesting, as did exobiology: the study of life on other planets (which one skeptic has described as the only field of science with no subject matter at all).

Any reasonable theory of the origin of life must rest on three kinds

PRECAMBRIAN ANIMAL in photograph on opposite page looks somewhat like a small jellyfish but does not correspond to any living animal or known fossil species. Named Tribrachidium heraldicum ("three-armed shield") it was found in Ediacara Hills of Australia. About two centimeters in diameter, it lived on sandy sea bottom near shore.

of evidence: 1, evidence showing a common metabolic heritage in modern organisms; 2, fossil traces of Precambrian life; 3, laboratory experiments simulating presumed early Earth conditions. To these, exobiologists hope to add a fourth kind of evidence: comparisons of independent appearance of life on more than one planet.

A COMMON METABOLIC HERITAGE

The central metabolic processes of modern plants and animals are essentially the same. As explained in Chapter 3, plants and animals alike extract energy from foods by anaerobic fermentation or glycolysis

$$C_6H_{12}O_6 \rightarrow 2\ C_3H_4O_3 + 4H$$
GLUCOSE   PYRUVIC ACID

followed by aerobic respiration

$$2\ C_3H_4O_3 + 4H + 6O_2 \rightarrow 6CO_2 + 6H_2O$$

The reader will recall from Part I that green plants also synthesize their own food by the process of photosynthesis:

$$6CO_2 + 6H_2O \xrightarrow{light} C_6H_{12}O_6 + 6O_2$$

This pattern of metabolism persists all the way down the scale of life, as far as simple one-celled eucaryotes like amoebas and green algae. Many other metabolic reactions are identical or closely similar in all eucaryotes. Even where they are different, they are often merely variations on the same theme.

In terms of metabolism, the real diversity of life is much more evident in procaryotes. Bacteria and bluegreen algae (not to be confused with green algae, which are plants) display a staggering variety of mechanisms for energy storage and food synthesis that have no parallel in the eucaryotes. Procaryotes demonstrate, for example, that the combination of carbon and hydrogen with oxygen is not the only way to carry out respiration. Even the idea that respiration itself is a universal metabolic heritage is disproved by bacteria. Many bacteria manage quite well with only the energy they receive from fermentation, and do not respire at

all. Moreover several plants, including yeasts, have the option of anaerobic life under certain conditions. Photosynthesis, practically universal among plants, also illustrates the diversity of bacterial metabolism. Many bacteria use light energy to synthesize carbohydrates, but their photosynthesis does not liberate oxygen. This is because, unlike green plants, bacteria do not use water as a source of hydrogen atoms in making reduced organic compounds from carbon dioxide. Other bacteria are not photosynthetic at all, but obtain energy for carbohydrate synthesis from chemical reactions. The iron bacteria, whose ancestors date back to the Precambrian era, oxidize ferrous ions to insoluble ferric hydroxide. The nitrite bacteria oxidize ammonia to nitrite ion, and the nitrate bacteria convert nitrite to nitrate.

The important point to recognize in bacterial metabolism is that the eucaryotes, today's familiar plants and animals, have chosen one particularly efficient set of reactions from among the many possible pathways. The genuine diversity of life is obscured if one looks only at the eucaryotes. The bacteria help one to see what is not common to all organisms.

After all the variation is stripped away, the process of ANAEROBIC FERMENTATION remains. There is no living form, from bacteria to mammals, that does not break down glucose in the way that was outlined in Chapter 3. Because comparative biochemistry reveals evolutionary ancestry as clearly as does comparative anatomy, one must conclude that the original ancestor of all organisms, procaryotes and eucaryotes alike, must have absorbed its food from its surroundings and extracted energy from it without the use of molecular oxygen. In short, it must have been an ANAEROBIC HETEROTROPH.

This provides a valuable clue about the conditions under which life began. Life must have emerged when organic compounds were abundant in the seas, not only as sources of energy, but as sources of the materials from which life could evolve in the first place. There was probably no oxygen present; the evidence

The genuine diversity of life is obscured if one looks only at the eucaryotes. The bacteria help one to see the metabolic pathways that are not common to all organisms.

suggests that the atmosphere of the early Earth must have been composed mainly of hydrogen, water vapor, methane, ammonia, nitrogen and hydrogen sulfide.

THE FOSSIL RECORD

Any recognizable trace of an organic structure preserved from prehistoric times is a fossil. Some of the most dramatic examples are the mammoths that were frozen intact in the icy soil of Siberia. The mineral cast of a bone of a vertebrate and the calcified shell of a mollusk are also fossils. A fossil may consist of nothing more than the cast of the impression by the hard surface of a bone or shell on the originally soft shell. Occasionally the only trace of an organism is its footprint or the remains of a burrow that it dug in life. One of the best ways for a particular organism to become a fossil is to die in some favored spot, such as the bottom of a river or the quicksand of a swamp, where it will be quickly covered by mud, sand, peat, or some other material destined to become part of a long-lasting geological formation. As millenia pass, the mud may turn to shale, the sand to sandstone, and the peat to coal. The traces of the organism are then locked in and may remain unchanged for millions or even billions of years. When the rocks are later split open by the geologist's hammer, the fossils are exposed for inspection.

Fossils can usually be dated according to the age of the rock strata in which they lie. As a general rule the deepest strata are the oldest, provided that the formation has not been drastically folded or deformed. The age of the rock can be determined by radioisotope techniques, using trace quantities of uranium, thorium, rubidium, or potassium. Uranium-238, for example, spontaneously decays into lead at a slow but precisely known rate. By comparing the amount of U^{238} still in the rock with the amount of lead derived from its decay, geophysicists can estimate the age of a sample within an error of about 10 percent. Different kinds of fossils are used to correlate rock strata in different locations, and to delineate eras in the history of life.

Geologists divide the 4.5 billion years of the Earth's history into five major eras: the Archeozoic (primal life), Proterozoic (primitive life), Paleozoic (ancient life), Mesozoic (middle life), and Cenozoic (modern life). These eras in turn are subdivided into periods (Chapter 17). The oldest fossils discovered so far are 3.1 billion years old, which means that life was already flourishing when the Earth was only 1.4 billion years old.

Numbers in the billions are so large that they have little meaning for most readers. To convey a real sense of the relative age of fossil organisms it is useful to scale down the history of the Earth to a 30-day month. Each day on that geologic calendar represents 150 million years (Figure 1). On that calendar the Archeozoic and Proterozoic eras stretch across the first 26 days. The Cambrian period, which opened the Paleozoic era 600 million years ago, marks a great divide in the fossil record. By the dawn of the Cambrian, on the 27th day, the ancient seas teemed with life. Representatives of every modern phylum had appeared, except for vertebrates and higher plants.

PRECAMBRIAN LIFE

Below the earliest Cambrian strata there is a striking dearth of fossils — at least of the type that usually adorn museums of natural history. Various explanations have been suggested for the apparently dramatic explosion of life in the Cambrian era: changes in the radiation level of the sun, climatic changes, an increase in the oxygen content of the atmosphere past a critical point for respiration, or simplest of all, the replacement of softbodied creatures with organisms having shells, armor, and skeletons that would leave more substantial fossil remains. Moreover, in later eras most of the Precambrian rocks have been heated or deformed by geologic processes in a way that would tend to obliterate faint traces in the fossil record.

A few years ago it was thought that the record of Precambrian life was meagre and unimpressive. But the situation has changed recently because paleontologists have learned how and where to look for the microscopic

If we scale down the history of the Earth to a 30-day month, we find that fossil life does not become abundant until the dawn of the Cambrian period, on the 27th day.

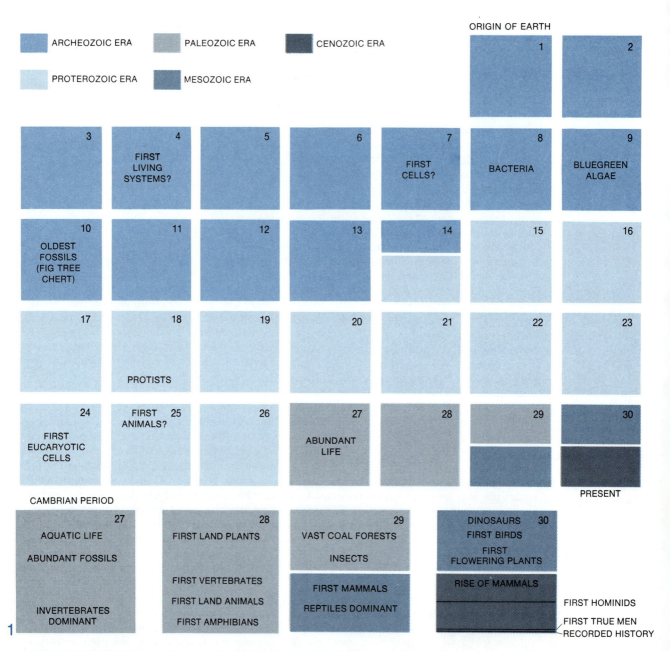

ORIGIN OF EARTH

ARCHEOZOIC ERA   PALEOZOIC ERA   CENOZOIC ERA

PROTEROZOIC ERA   MESOZOIC ERA

1 | 2

3 | 4 FIRST LIVING SYSTEMS? | 5 | 6 | 7 FIRST CELLS? | 8 BACTERIA | 9 BLUEGREEN ALGAE

10 OLDEST FOSSILS (FIG TREE CHERT) | 11 | 12 | 13 | 14 | 15 | 16

17 | 18 PROTISTS | 19 | 20 | 21 | 22 | 23

24 FIRST EUCARYOTIC CELLS | FIRST 25 ANIMALS? | 26 | 27 ABUNDANT LIFE | 28 | 29 | 30

PRESENT

CAMBRIAN PERIOD

27 AQUATIC LIFE

ABUNDANT FOSSILS

INVERTEBRATES DOMINANT

28 FIRST LAND PLANTS

FIRST VERTEBRATES

FIRST LAND ANIMALS

FIRST AMPHIBIANS

29 VAST COAL FORESTS

INSECTS

FIRST MAMMALS

REPTILES DOMINANT

DINOSAURS 30 FIRST BIRDS

FIRST FLOWERING PLANTS

RISE OF MAMMALS

FIRST HOMINIDS

FIRST TRUE MEN

RECORDED HISTORY

1

GEOLOGIC CALENDAR as shown above depicts the history of the Earth as a 30-day month. One day equals 150 million years. First living things appeared very early, perhaps four billion years ago (fourth day). Oldest known fossils lived 3.1 billion years ago, but fossil record is scanty before the opening of the Paleozoic Era (27th day). The four most recent days are depicted in greater detail at bottom. On this scale, first true men appeared only about ten minutes ago, and all recorded history, from ancient Sumer to the present, represents an interval of about 30 seconds.

Over most of the surface of the Earth, Precambrian rocks are covered by later deposits. The "shields," where large areas of Precambrian rock are exposed or at least accessible, are shown in Figure 2. The best examples of Precambrian life are listed in Table I, and the locations of these finds are marked on the map (Figure 2).

The earliest well-dated fossils are from the Fig Tree strata in eastern Transvaal, South Africa. These are flintlike black cherts, rich in carbon, and dated by rubidium/strontium methods as 3.1 billion years old. Fossil bac-

remains of one-celled organisms. If one can rid oneself of the bias that equates plants and animals with life and regards all microorganisms as pretty much alike, then one finds that the record of Precambrian life is surprisingly rich.

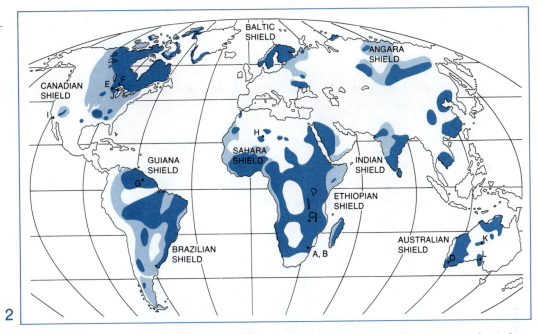

2

PRECAMBRIAN ROCKS are usually buried under later strata, but in low-lying continental shields they are relatively accessible. Exposed Precambrian strata are rendered in dark color and those overlaid with younger strata in light color. Letters A through L mark fossil sites listed in Table I. Ancient continents were shaped very differently from those of today, and occupied different positions on the face of the globe (Chapter 22). Animal illustrated on opening page of this chapter was found near location L. Bacterium in Figure 3 was found at one of two sites marked A and B.

teria were found embedded in the rock (*Figure 3*), along with spheroids and filaments of organic matter which resemble modern blue-green algae. Surprisingly, among the organic compounds extracted from the rocks are two hydrocarbons which are ordinarily considered to be degradation products of the tail of the chlorophyll molecule: pristane and phytane. The evidence of pristane and phytane is too shaky to stand alone, because these compounds have been found in living organisms that do not possess chlorophyll.

Photosynthesis represents a fairly advanced stage of chemical evolution. Did these compounds come from chlorophyll from the algae or bacteria, and were these organisms already photosynthetic 3.1 billion years ago? At present one can only speculate. Nearby, in the On-

verwacht cherts in Swaziland, lie some even older fossils: spheroids and filaments that are certainly older than 3.2 billion years. These are the oldest signs of life that have been found so far on this planet.

Modern algae that inhabit hot springs or geysers lay down limestone deposits in cone-shaped layers called STROMATOLITES. The cone grows layer by layer, with the apex of the cone downward, like a stack of nested thimbles. Limestone deposits that are almost identical with modern algal stromatolites are found in 2.7-billion-year-old limestone in southern Rhodesia. In other Precambrian deposits of roughly the same age in Australia and Minnesota investigators have found spheroidal and clustered bodies which resemble bacteria and bluegreen algae. Pristane and phytane, the

The interval from about 3.5 billion to 800 million years ago — from the 7th to 24th days of the calendar — can be described as the age of procaryotes.

I

	DEPOSIT	BIOLOGICAL REMAINS	AGE (BILLIONS OF YEARS)
ARCHEOZOIC ERA (4.5 BILLION YEARS AGO)	ONVERWACHT SEDIMENTS SWAZILAND, S. AFRICA (A)	SPHEROIDS, FILAMENTS IN CARBON-RICH CHERTS.	>3.2
	FIG TREE CHERTS TRANSVAAL, S. AFRICA (B)	FOSSIL BACTERIA (0.25 x 0.6μ), ALGAE-LIKE SPHEROIDS (PHOTOSYNTHETIC?), FILAMENTOUS ORGANIC STRUCTURES.	3.1
		COMPLEX HYDROCARBONS; PRISTANE, PHYTANE.	
	BULAWAYO LIMESTONE BULAWAYO, S. RHODESIA (C)	LIMESTONE DEPOSITS RESEMBLING STROMATOLITES LEFT BY ALGAE. LAYERS IN NESTED CONES LIKE MODERN GEYSERITE FROM YELLOWSTONE PARK.	2.7
	SOUTHERN CROSS QUARTZITE S. CROSS, WESTERN AUSTRALIA (D)	RED HEMATITE AND YELLOW GOETHITE BODIES EMBEDDED IN QUARTZITE. SPHEROIDAL (~2μ), ELONGATED, AND IN RASPBERRY-LIKE CLUSTERS.	2.7
	SOUDAN IRON FORMATION ELY, MINNESOTA (E)	CARBON-RICH ROCK CONTAINING PYRITE BALLS WITH MICROSTRUCTURES. SPHEROIDAL (0.8-1.5μ); SINGLE AND CLUSTERED. BACTERIA OR BLUEGREEN ALGAE?	>2.7
		HYDROCARBONS: PRISTANE, PHYTANE.	
LOWER PROTEROZOIC (2.5-1.7)	GUNFLINT CHERT GUNFLINT LAKE TO THUNDER BAY ALONG ONTARIO-MINNESOTA BORDER. (F)	BLUEGREEN ALGAE, FILAMENTOUS AND SPHEROIDAL, FUNGI, ORGANISMS OF NO MODERN PARALLELS. PHOTOSYNTHETIC?	1.9
		BACTERIA (0.55x1.10μ).	
		PRISTANE, PHYTANE, OTHER HYDROCARBONS.	
		STROMATOLITES (UNRELATED TO MICROFOSSILS).	
	RORAIMA CHERT BRITISH GUIANA (G)	ALGAE RESEMBLING GUNFLINT DEPOSITS.	2.0-1.7
		UNKNOWN PROTISTS, 160-400μ, SUPERFICIALLY RESEMBLE FORAMINIFERA, SPHERICAL WITH PROJECTIONS.	
MIDDLE PROTEROZOIC (1.7-1.0)	SAHARAN STROMATOLITES WESTERN SAHARA (H)	LIMESTONE ALGAL DEPOSITS. TWO GENERA, COLLENIA AND CONOPHYTON. ALTERNATE LAYERS OF $CaCO_3$ AND $Fe(OH)_3$ SUGGEST ASSOCIATION OF PHOTOSYNTHETIC BLUEGREEN ALGAE AND IRON BACTERIA.	1.6
		SAHARAN EARLIEST. LATER STROMATOLITES CONTINUING TO PRESENT. PRECAMBRIAN STROMATOLITES IN AUSTRALIA, CHINA, SIBERIA, URALS, FINLAND, N. EUROPE, CENTRAL AFRICA, TRANSVAAL, INDIA, NORTHERN U.S. AND CANADA.	
	BECK SPRINGS DOLOMITE SOUTHERN CALIFORNIA (I)	BLUEGREEN ALGAE. POSSIBLE FIRST KNOWN APPEARANCE OF GREEN ALGAE WITH NUCLEATED CELLS (EUCARYOTES).	1.4-1.2
	NONESUCH SHALES WHITE PINE, MICHIGAN (J)	FILAMENTOUS AND SPHEROIDAL BODIES THE SIZE OF ALGAE. POSSIBLY BIOLOGICAL, SOME ARTIFACTS.	1.1
		PRISTANE, PHYTANE, OTHER HYDROBARBONS.	
	ALICE SPRINGS, CENTRAL AUSTRALIA; McARTHUR FORMATIONS, AUSTRALIA (K)	POSSIBLE CASTS OF WORM TRACKS. IF VALID, EARLIEST SIGNS OF METAZOANS. POSSIBLE SPONGE SPICULES AND IMPRINTS OF JELLYFISH.	1.5-1.0
LATE PROTEROZOIC (1.0-0.6)	BITTER SPRINGS CHERT NEAR ALICE SPRINGS, AUSTRALIA (L)	BLUEGREEN ALGAE. SPHEROIDS (3-10μ), UNSEGMENTED AND SEGMENTED FILAMENTS. STRUCTURAL ANATOMY SURPRISINGLY UNDISTORTED. GREEN ALGAE.	1.0-0.9
		FIRST APPEARANCE OF ANIMALS AND MULTICELLED ORGANISMS: SILICEOUS SPONGE SPICULES—USSR, CHINA SPONGES, RADIOLARA, FORAMINIFERA—N. FRANCE SPORES—SCANDINAVIA, RUSSIA, N. FRANCE WORM BURROWS —AUSTRALIA, CHINA, USSR POLYCHAETES—GRAND CANYON ARTICULATED ANCESTORS OF TRILOBITES—SWEDEN MEDUSAS, WORMS, SEA PENS—EDIACARA HILLS, AUSTRALIA	1.0-0.6

PRECAMBRIAN FOSSILS listed in table on opposite page stretch across an interval of 3.9 billion years—the first 27 days of the calendar in Figure 1. All Precambrian organisms were aquatic. Letters in the second column refer to locations of deposits, as shown on the map of Precambrian outcroppings in Figure 2.

somewhat questionable evidence of photosynthesis, also occur in the Minnesota rocks.

One of the most dramatic and varied collections of Precambrian organisms occurs in Gunflint chert, an outcropping of 1.9-billion-year-old rock in the Canadian Shield, exposed near the Minnesota-Ontario border. Thousands of canoeists have followed the Gunflint Trail in an effort to get back to nature and turn back the clock a few years, never realizing that they could turn the clock back nearly two billion years by examining the rocks at their feet. Unmistakable bluegreen algae, fungi, and bacteria have been found, as well as organisms that have no modern parallels (Figures 4, 5, & 6). One of these unknown organisms, named *Kakabekia umbellata* (Figure 7), has since been found not to be extinct at all. It is living and thriving in special ammonia-rich soils in

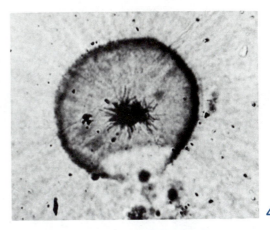

4

COLONY OF FOSSIL ALGAE found in Gunflint chert formation in Ontario resembles modern bluegreen algae of the genus Rivularia, hence its name, Paleorivularia ontarica. It apparently consists of a star of radiating algal filaments surrounded by some sort of membrane or envelope. Diameter of envelope is about 60 microns. Fossil is 1.9 billion years old.

Wales. Pristane, phytane, and other hydrocarbons are present in the Gunflint rocks, along with algal stromatolites.

The earliest porphyrins that have been found come from the Nonesuch shales of Michigan. These deposits contain algaelike bodies as well as pristane, phytane, other hy-

3

OLDEST FOSSIL ORGANISM discovered so far is the rod-shaped Eobacterium isolatum, found in the Fig-tree chert formation. During preparation of sample, fossil was moved from its original position in the rock, leaving its impression visible at bottom. "Yardstick" at far left, one micron long, indicates size of the bacterium.

SPHERICAL ORGANISM with no modern parallel was found in Gunflint chert. Named Eosphaera tyleri, it appears to be a sphere within a sphere, with the inner held away from the outer one by many tiny globules.

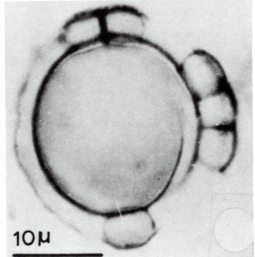

5

10μ

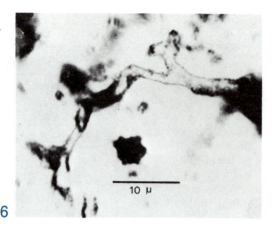

6

TUBULAR FILAMENT from Gunflint chert resembles some living types of bluegreen algae. Unlike modern algae it is branched and lacks septae (perpendicular divisions along the strand). Some tubular fossil species have segments and other structural features characteristic of modern algae.

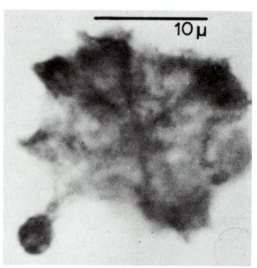

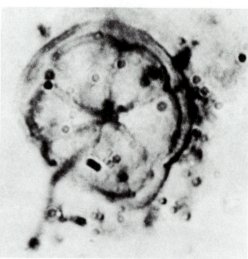

7

UMBRELLA-SHAPED MICROORGANISM from Gunflint chert (top) resembles a previously unidentified microorganism living in soil near Harlech Castle, Wales (bottom). Fossil, named Kakabekia umbellata, is about 1.9 billion years old.

drocarbons, and most significant of all, vanadium porphyrins, typical of the ring compounds that comprise the heart of the chlorophyll molecule. This may be the first solid evidence of the existence of porphyrin compounds and possibly of photosynthesis.

All of the life forms discussed so far appear to have been procaryotes — bacteria or bluegreen algae. The first true plants — green algae — may be present in the Beck Springs dolomite from California and the Bitter Springs chert from Australia. In the California specimens, 1.4 to 1.2 billion years old, can be seen what appear to be eucaryotic cells with nuclei (*Figure 8*). Both bluegreen and green algae can be seen in the Australian chert, in surprisingly undistorted form, in rocks dated between 900 and 700 million years old. If this interpretation of the fossil microorganisms is correct, then life finally arrived at the stage of green plant photosynthesis on the 25th or 26th day of our calendar, only a few hundred million years before the Cambrian era.

The first signs of multicellular life appear in the last half billion years before the Cambrian era, first as worm tracks, and then as impressions in sandstone of softbodied worms, jellyfish, and other creatures with no modern descendants (*see Figure 9 and the photograph that opens this chapter*). At this point the Precambrian fossil record blends into the Cambrian. The enormous increase in the number of fossils at the beginning of the Cambrian might be explained by a great upsurge in the population of living organisms 600 million years ago, but this population had roots in the earlier eras.

In brief, the interval from approximately 3.5 billion to 800 million years ago — from the 7th to the 24th days of our calendar — can be described as the age of the procaryotes. At the beginning of this period there flourished simple organisms morphologically similar to bluegreen algae and bacteria, which may or may not have been photosynthetic. During the course of the following 2.5 billion years, these organisms diversified and developed both

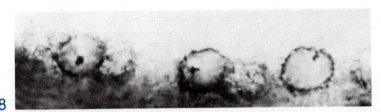

EARLIEST EUCARYOTES found so far are these globular cells from stromatolite deposits in Beck Springs, California. About 1.2 to 1.4 billion years old, fossils resemble green algae. Cells are 14 to 18 microns in diameter. Dark spot within each cell is believed to be the nucleus.

photosynthesis and respiration, leading ultimately to eucaryotes and then to multicelled creatures that one would recognize as plants and animals.

The main conclusion to emerge from the study of Precambrian microfossils is that life has been present on Earth much longer than anyone previously suspected. Many workers now estimate that the first living systems appeared sometime during the first 500 million years of the planet's existence — the first three or four days on our calendar. But even the most primitive Precambrian bacteria are complicated structures compared with nonliving matter. Their discovery sheds no light on the central question of chemical evolution: how did nonliving matter organize itself into a living system?

METEORITES AND EXTRATERRESTRIAL LIFE

One possibility is that life did not originate on Earth, but was "seeded" here by an extraterres-

trial object such as a meteor. The Earth is steadily bombarded with showers of meteors, presumably the debris of shattered asteroids, and some of this material contains molecules characteristic of living systems.

Most meteorites are stony or metallic, but a relatively small number are soft and crumbly, with a high carbon content. Called CARBONACEOUS CHONDRITES, these meteorites have been studied in the past decade by several research groups, using techniques similar to those for detecting microscopic Precambrian fossils. The results have been inconclusive, and the subject of heated debate. Hydrocarbons have been found, including pristane and phytane. Spheroids and other organized bodies of fair complexity have been reported, but contamination of the meteorite samples by airborne spores and pollen has confused the issue. In at least one instance, the organized bodies turned out to be ragweed pollen. Most of the complex organized bodies have proven to be terrestrial contaminants, and those that are unquestionably meteoric in origin are sufficiently simple that they may be natural mineral formations rather than artifacts of life. The presence of pristane, phytane, and other hydrocarbons in the meteorites indicates that at least the first step in molecular evolution — the formation of complex organic compounds — can occur spontaneously even in space. If these meteorites are not evidence for life on some shattered planet, they may be evidence for the universality of the organic-chemical-rich environment in which life could develop.

SEGMENTED WORM from the late Precambrian is about four centimeters long. Named Spriggina floundersi, it is one of many fossil metazoans found in sandstones of the Ediacara Hills in Australia.

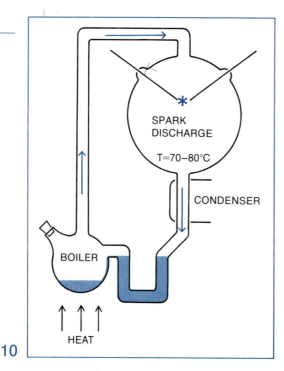

10

MILLER APPARATUS simulated conditions on the primitive Earth. The components of the primitive atmosphere—ammonia, hydrogen, methane and water—were exposed to spark discharges that simulated lightning. Condensed gases were recirculated by boiling. After running for a week, apparatus contained mixture of gases and dissolved organic compounds.

LABORATORY SIMULATION

By 1952 several scientists, including Harold Urey, were proposing that life evolved on Earth in a primitive reducing atmosphere of hydrogen, water, methane, and ammonia, from organic compounds which had been synthesized by nonbiological means. Possible energy sources for such syntheses would have been ultraviolet radiation from the sun, lightning, and to a lesser extent, radioactive decay in the Earth's crust and heat from volcanoes and hot springs. By far the most plentiful energy source would have been visible light, but it did not become significant in biochemistry until the evolution of photosynthesis.

In 1953, Stanley L. Miller, one of Urey's graduate students, made the first careful experiments to show that organic chemicals, including amino acids, could be synthesized under assumed primitive-Earth conditions. He set up a recirculating system of hydrogen, ammonia, and methane gases and water vapor, and passed these gases over a spark discharge from a high frequency tesla coil, to simulate lightning (*Figure 10*). After a week of circulation, he opened the flask and analyzed the contents. The electrical discharge had produced cyanide compounds and aldehydes from the gases. These compounds then reacted in water to yield a variety of organic compounds, including amino acids.

Since Miller's trials many other workers have experimented with electric discharges, ultraviolet light, and radioactive elements, using different gas mixtures which could represent primitive-Earth atmospheres. They have found that it would have been quite possible to produce a variety of amino acids, purines, pyrimidines, sugars such as ribose and 2-deoxyribose, nucleosides, and even complete nucleotides such as ATP, by nonbiological means. The "warm, dilute soup" in which Haldane postulated that life evolved, was quite likely to have been present on the primitive Earth. Irradiation of solutions of adenine, ribose, and phosphate compounds with ultraviolet light has led to the synthesis of AMP, ADP, and ATP. It is likely that these high energy phosphate compounds would have been present when life was evolving, and that the adenosine compounds would have been most common simply because adenine was the most easily synthesized base. The first living organisms (or to reach back one step further, perhaps self-organizing nonliving chemical systems) may have depended on ATP from the seas around them for energy. The answer to the question, Why do living organisms use ATP to store energy?, may be the same as the answer to the question of why men climbed Mt. Everest: Because it was there.

MICROSPHERES, COACERVATES, AND CELLS

It is a long way from amino acids, nucleotides, and sugars to a living cell. How is the transition to proteins, nucleic acids, and polysaccharides carried out, and how do these lead to a living organism? One idea that is not proposed is that, in some chance fashion, all of the necessary nucleotides to produce the DNA of a bacterium came together one day and started life going. As the British physicist J.D. Bernal has expressed it, "The picture of the solitary molecule of DNA on a primitive seashore generating the rest of life was put forward with slightly less plausibility than that of Adam and Eve in the Garden." Instead, investigators now picture the gradual evolution of higher and higher levels of complexity in chemical systems: first from monomers to polymers, then to collections of polymers carrying out chemical reactions and perhaps separated from the bulk of their surroundings by a membrane, and finally to isolated chemical systems with the ability to duplicate themselves by means of some template like DNA and RNA. This is the era of CHEMICAL SELECTION — of the constant formation and breakup of localized concentrations of chemicals, and of the survival of those collections of molecules that were best able to grow at the expense of their surroundings.

The spontaneous formation of polymers is not too difficult to explain. Polymerization of amino acids to protein chains is not spontaneous under standard conditions, but the reaction can become spontaneous if the concentrations of reactants are high enough. It has been proposed that polymerization took place with amino acids concentrated in drying puddles by the seashore (Haldane), adsorbed on the surface of fine-grained clays and minerals (Bernal), or enclosed in particles called coacervate drops (Oparin). Melvin Calvin has shown that in the presence of certain organic condensing agents, polymerization can even take place in dilute solutions.

Sidney Fox has found that heating a dry mixture of amino acids leads to long chain "proteinoids": polymers having a molecular weight over 10,000. He has suggested that such polymerizations took place in volcanic cinder cones, and that the proteins formed were then washed into the sea. He has also found that his proteinoids have a remarkable tendency to form MICROSPHERES about 2 microns in diameter when hot concentrated solutions are slowly cooled (*Figure 11*). These microspheres show a double-layer boundary resembling a membrane (although without lipids), and swell or shrink as the salt concentration is changed. If allowed to stand for several weeks, the microspheres absorb more proteinoid material from the solution, produce buds and sometimes divide to produce "second generation" microspheres. Cleavage or division can also be induced by changing pH or adding $MgCl_2$. These microspheres should not be taken to be the ancestors of life; rather, they show some of the remarkable self-organizing properties of relatively simple chemical systems. They illustrate how many of the functions normally associated with living organisms can be duplicated by much less complex assemblages.

Oparin has spent most of his professional career studying COACERVATE DROPS. If one shakes a mixture of a little olive oil in water, the oil will break up into a mass of tiny droplets. If instead one uses a long-chain protein such as gelatin and a polysaccharide such as gum arabic, the result is the formation of coacervate drops, which are much more stable. As was the case with the olive oil-water mixture, the coacervate preparation is divided into two separate phases: the interior of the drops and the aqueous solution around them. Coacervate drops will form in solutions of many different kinds of polymers: protein, nucleic acids, polysaccharides, and various synthetic polymers. Coacervation is simply a matter of physical chemistry, not of life. Yet coacervate drops show several properties relevant to the origin of life. Many substances, when added to a coacervate preparation, become concentrated within the droplets. Lipids can coat the boundaries of droplets with "membranes" that strengthen them against destruction. If the coacervates are prepared so as to contain enzyme molecules, they can absorb substrates, catalyze a reaction, and let the products diffuse back out into the solution once more. Coacervates containing phosphorylase, for example, will absorb glucose-1-phosphate from the surrounding solution and polymerize it into starch. If a second enzyme, amylase, is also present in the coacervate, it

MICROSPHERES prepared by adding protein-like material to water show a double-layer boundary roughly similar to the membranes of microorganisms. Microspheres are not alive, nor even direct precursors of living cells, but they illustrate the effects of physical forces that would have acted on primitive living systems as they evolved.

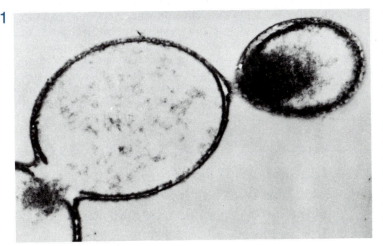

1

will break up the starch into maltose, which escapes into the solution. The coacervates containing phosphorylase and amylase are then small factories for converting the monosaccharide glucose-1-phosphate into the disaccharide maltose. The energy for the overall process, of course, comes from the high energy phosphate on the original glucose. Oparin has even made photosynthetic coacervates containing chlorophyll, which will absorb an oxidized dye from the solution, use light energy to reduce it, and return the reduced dye to the surroundings.

It is possible that the immediate precursors of living organisms were capsules of chemical reactions similar to coacervate droplets. Some coacervates would enclose reactions that led to the early breakup of the droplets; others would enclose reactions that made the droplets more stable. The more stable coacervates would survive longer, and could possibly grow at the expense of their surroundings by absorbing chemical substances derived from the remains of less stable droplets. If wave action or other mechanical forces broke a large coacervate into many small droplets, each of these might be able to absorb material and grow on its own.

Oparin postulated the existence of organized, metabolizing, but nonreproducing systems that he called PROTOBIONTS. According to this reasoning, the breakthrough that led to truly living organisms was the development of REPRODUCTION — the ability of a successful chemical system to ensure its survival by duplicating itself. The molecules in which the instructions for duplication are stored in modern living creatures are DNA or RNA. Yet the LIVING unit of life is not just the nucleic acid, but the entire cell. The organism is as helpless without its DNA as a computer without a program tape, but the DNA alone can no more live than a magnetic tape without a computer can do calculations.

What has been learned so far about the actual traces of Precambrian life, the heritage of a common chemistry in the biochemistry of modern living organisms, and possible laboratory analogues of certain facets of the evolution of life, makes it possible to construct a scenario for the origin of life. It reflects the speculations and biases of the authors in several places, and would not be accepted in its entirety by any of the scientists most concerned with such matters. But probably no two experts would argue with exactly the same points. The broad features of the scheme are almost certainly correct, but the details extend to the limits of human knowledge, and in some cases a little beyond.

THE SCENARIO

SCENE 1. The solar system condenses from a primeval dust cloud in our part of the Milky Way galaxy 4.5 to 5 billion years ago. The Earth and all of the other planets are formed by the gradual accretion of matter from the dust cloud around the new Sun.

SCENE 2. The inner planets — Mercury, Venus, Earth, and Mars — are too small and too warm to retain their original atmospheres of hydrogen, helium, and the rare gases, in contrast to the larger and colder outer planets of Jupiter and beyond. The gases of the primitive Earth atmosphere escape into space, leaving behind a solid planet.

SCENE 3. Gases escaping from the hot interior of the Earth produce a secondary reducing atmosphere composed mainly of hydrogen, water, methane, ammonia, nitrogen, and hydrogen sulfide. No free molecular oxygen is present.

SCENE 4. Water vapor condenses into seas, and a vast array of organic compounds is produced by the action of ultraviolet light, electrical discharges, radioactive decay in the crust, and volcanic heat. The oceans become a dilute "Haldane soup" of organic matter.

SCENE 5. Some polymer solutions break up into coacervate droplets, either as a result of wave action or of concentration by evaporation in tide pools. The interior composition of these droplets is often quite different from the surrounding seas. Reactions within coacervate drops differ greatly, and affect the relative

According to Oparin's line of reasoning, the breakthrough that led to truly living organisms was the invention of reproduction, which enabled a successful chemical system to ensure its survival by duplicating itself.

stabilities of drops. Droplets form and are broken up, and chemical selection operates between different types of coacervates, leading gradually to particularly stable reaction systems that could be called "protobionts."

SCENE 6. The first generally available and used energy source for thermodynamically unfavorable chemical syntheses is the supply of naturally-occurring ATP.

SCENE 7. Certain of the protobionts develop the ability to use nucleic acids as templates in the synthesis of useful proteins, which serve both as structural elements and catalysts or enzymes. When this synthetic ability progresses to the point that a protobiont can make a copy of itself, the organism can truly be called living.

SCENE 8. Organisms develop synthetic pathways to make substances that they once obtained from their surroundings, but which have been depleted. This is the beginning of what could properly be called metabolism.

SCENE 9. Anaerobic fermentation or glycolysis develops, in which high free energy compounds are used to synthesize and store ATP as the natural supply of ATP runs short. The new sources of energy are so rich, and organisms that develop glycolysis are so much at an advantage over their rivals, that all traces of the alternate metabolism vanish. Glycolysis becomes a universal heritage of all surviving life.

SCENE 10. Free oxygen from high-altitude photodissociation begins to accumulate in the atmosphere in small amounts and to poison living organisms. Hydrogen peroxide is formed. First iron, then iron porphyrins, and finally the ancestors of catalase are developed to catalyze the destruction of hydrogen peroxide.

SCENE 11. Certain organisms begin using magnesium porphyrins to absorb visible light, and to channel the energy obtained into the synthesis of ATP.

SCENE 12. Bacterial photosynthesis appears, using various donor molecules other than water as sources of hydrogen in the fixation of carbon dioxide.

SCENE 13. The bluegreen algae evolve, using water as the hydrogen source in photosynthesis. For the first time oxygen is liberated into the atmosphere from photosynthesis.

SCENE 14. Free atmospheric oxygen becomes a serious threat to life, and various organisms develop different defense mechanisms. A layer of ozone, O_3, appears in the upper atmosphere, shielding the surface of the planet from intense ultraviolet radiation, and permitting living creatures to occupy shallower waters.

SCENE 15. Respirers and chemautotrophs begin to use oxidation reactions as a source of energy for the synthesis of more ATP. The increased oxygen content of the atmosphere oxidizes the upper layers of the crust and forces the chemautotrophs to the marginal niches in deep wells and underground deposits where they are found today.

SCENE 16. Eucaryotic organisms evolve. It is probable that the chloroplasts found in eucaryotic plants are the vestigial remains of symbiotic bluegreen photosynthetic algae, and that the mitochondria found in all eucaryotes are the descendants of symbiotic non-photosynthetic bacteria. The combination of a high oxygen concentration in the atmosphere, the emergence of a really efficient system of respiration, the evolution of eucaryotes, and the appearance of multicellular life forms, all lead to an explosion of life all over the planet, and to the upsurge in number and variety of fossil remains that marks the beginning of the Cambrian era. At this stage the evolution of life as a mechanism is complete, and the further evolutionary development of plants and animals is a familiar story.

THE TIMETABLE

Scenes 1 through 8 of the scenario probably occurred within the first billion years of our planet's existence (the first week on the calendar). Scene 9, glycolysis, occurred perhaps one or two hundred million years later. The primitive spheriods and filaments in the Onverwacht sediments (*Table I*) may be the remains of such protodecomposers, with a crude anaerobic fermentation metabolism. Photosynthesis (scenes 11–13) is likely to have arisen around 3.1 to 2.7 billion years ago (between the 10th and 12th days). The Fig Tree bacteria and the Bulawayo algal stromatolites may date from this period. If the Bulawayo deposits do represent bluegreen algae, with O_2-liberating photosynthesis, then the first steps toward our present oxygen-rich atmosphere would have been taken 2.7 billion years ago. The era from that time until the late Precambrian would have been one of gradual evolutionary improvement of the photosynthetic machinery, increasing oxygen concentration, and slow development of respiratory metabolism (scenes 14 & 15). Scene 16, the appearance of eucaryotes, evidently occurred slightly less than one billion years ago (the 24th day), judging from the presence of green algae in the fossils at Bitter Springs, Australia.

The Archeozoic Era (4.5–2.5 billion years ago) would therefore be the era of the evolution of anaerobic, heterotrophic life. The Lower and Middle Proterozoic (2.5–1.0 billion years ago) would see the evolution of photosynthesis, the change in character of our atmosphere, and the development of respiration. The Late Proterozoic (1.0–0.6 billion years ago) would date approximately from the appearance of eucaryotes, include the rise of multicellular organisms, and end with the rapid rise in population brought on by an accumulation of many improvements in the machinery of life.

How much of this can be believed? Every generation needs its own creation myths, and these are ours. They are probably more accurate than any that have come before, but they are undoubtedly subject to revision as we find out more about the nature and the history of life. The best that can be said for any scientific theory is that it explains all the data at hand and has no obvious internal contradictions.

READINGS

J.D. BERNAL, The Origin of Life, Cleveland, World Publishing Company, 1967. A clear and nontechnical introduction by one of the best minds in the field. Raises philosophical as well as scientific questions. Well illustrated. The appendices of the book include two landmark papers on the origin of life, one by A. I. Oparin (1924) and one by J. B. S. Haldane. Written independently, they are so far ahead of their time that one gets the uneasy feeling that all we have done since 1924 is to fill in the technical gaps in Oparin and Haldane's schemes.

S.W. FOX (editor), The Origins of Prebiological Systems, New York, Academic Press, 1965. Proceedings of a conference held in Florida in 1963. A sequel to the Moscow symposium, with much the same cast of contributors.

A.I. OPARIN, Genesis and Evolutionary Development of Life, New York, Academic Press, 1968. Approximately every seven years from 1936, Oparin has written a book on the origin of life summarizing the state of the field at the time. This book now supersedes Life: Its Nature, Origin and Development (1961) and the various editions of The Origin of Life on Earth. This book includes a good historical introduction to what people thought about spontaneous generation of life since the time of the Greek philosophers.

A.I. OPARIN, A.G. PASYNSKII, A.E. BRAUNSHTEIN, AND T.E. PAVLOVSKAYA (editors), The Origin of Life on the Earth, New York, Pergamon Press, 1959. The proceedings of the First International Symposium of the Origin of Life, held in Moscow in 1957. Scientists from all over the world were represented. Some of the discussions and arguments (reprinted in full) are particularly interesting. Do not confuse this book with Oparin's early books with almost the same title.

I.S. SHKLOVSKII AND C. SAGAN, Intelligent Life in the Universe, San Francisco, Holden-Day, 1966. Chapters 14–17 contain a good introductory discussion of the appearance of life on Earth. Excellent color plates of Precambrian fossils.

P.H.A. SNEATH, Plants and Life, London, Thames and Hudson, 1970. More popularly written and wide-ranging than Bernal. Discussions of alternatives to terrestrial biochemistry, alien and machine intelligence, space travel and interstellar communication.

GEORGE WALD, "The Origins of Life," in The Scientific Endeavor, New York, Rockefeller University Press, 1964. This paper, given at the Centennial celebration of the National Academy of Science in 1963, is also published in Proc. Nat. Acad. Sci. (U.S.), 52:595 (1964). Ranks with the papers by Oparin and Haldane for clarity and brevity.

NOTES

"I know *who* I am but I don't know *what* I am!"

17

taxonomy and phylogeny

It must not be imagined that because we place one genus or one family before another we would consider them as more perfect, as superior to others in the system of nature. He alone has this pretension who pursues the chimerical project of placing the organisms in a single line; we have long ago renounced this scheme.

GEORGES CUVIER
NATURAL HISTORY OF FISHES (1828)

When man encounters the unknown, his first rational response is to classify. Strange objects, people, or phenomena must be compared with one another, collected into abstract groups, and named. Once classified, they can be subjected to a more sophisticated form of scrutiny. It was natural that the first efforts of biology, dating back to the time of Aristotle, were devoted to classification. We call the science of classification of organisms TAXONOMY. Other words for the same discipline are systematics and biosystematics. Since almost all taxonomists believe in evolution, a study of the similarity among organisms is normally elaborated in some fashion into evolutionary hypotheses. These interpretations are expressed in PHYLOGENETIC TREES, which represent the lines of descent from one species or other group of organisms to another. In the present chapter we shall consider first the basic rules of taxonomic procedure, including the rules of nomenclature (the assignment of technical names), and then show how biologists go about constructing classifications and phylogenetic trees.

The basic unit of classification is the SPECIES. Later, in Chapter 22, the definition and full biological meaning of this unit will be examined. For the moment, it is enough to define the species loosely as a group of organisms that closely resemble one another, either because they interbreed freely or because they have descended from common ancestors in the not too distant past. In either case, the members of the species are genetically similar; they resemble one another more than they do the members of other species.

According to the internationally accepted rules of nomenclature, each species is assigned two names, one identifying the species itself and the other the GENUS to which the species belongs. The genus is a group of species that resemble one another (the plural of the word genus is genera and its adjective form is generic). In many cases the name of the taxonomist who first proposed the species name is added on the end. Thus *Homo sapiens* Linnaeus is the name of the modern human species. *Homo* is the genus to which the species belongs, *sapiens* identifies the species, and Linnaeus was the first author to have used the species name *sapiens*. You can think of the generic name *Homo* as being equivalent to your surname and the specific name *sapiens* as being equivalent to your first name. This two-name system, referred to as binomial nomenclature, is universally employed throughout biology.

The list of names used dates back to the works of Carolus Linnaeus (1707–1778), the great Swedish biologist who first applied the binomial system on a rigorous and worldwide scale to both plants and animals. To avoid confusion, a law of priority is observed — where two or more names have been used for the same species, providing neither predates the work of Linnaeus, the oldest name is retained and the others discarded. Thus when the ant species *Lasius pallitarsus* Provancher was recently shown to be the same species as *Lasius sitkaensis* Pergande, the older name *L. pallitarsus* took precedence over the newer name *L. sitkaensis*. *L. sitkaensis*, to use the common expression, was "synonymized" or "placed in synonymy" under *L. pallitarsus*. It is the junior synonym and *L. pallitarsus* is the senior synonym, a fact that usually is expressed in taxonomic writings by the following convention: *Lasius pallitarsus* Provancher (= *Lasius sitkaensis* Pergande). Common names of animals in use for more than 50 years are exempted from the change resulting from the discovery of an obscure older name. This rule prevents taxonomists from indulging in the gamesmanship of upsetting the classification by searching through long-forgotten literature.

Notice that the generic name is capitalized and the specific name is not. In botany the specific name is sometimes capitalized, when the name of the person or place is incorporated in it, for example *Vernonia Blodgetti*, one of the ironweeds. The generic and specific names are always italicized. These scientific names, or "technical names" as they are often called, are usually derived from more or less appropriate Greek or Latin words. Thus *Homo sapiens* is Latin for "wise man," *Lasius pallitarsus* means "hairy one with pale feet," and *Lasius sitkaensis* means "hairy one from Sitka (Alaska)," the place where the first specimens were collected. If the name chosen by the taxonomist is not Greek or Latin in origin, it is usually Latinized by adding a suitable ending. Occasionally taxonomists become frivolous in their choice of names and have to be curbed. For example, the generic names *Peggichisme* (pronounced Peggy kiss me), *Marichisme*, *Pollychisme*, and so on, given originally to a group of hemipteran bugs, have been invalidated by a special ruling of the International Commission of Zoological Nomenclature; so

has the peculiar crustacean name *Cancelloidokytodermogammarus* (*Loveninuskytodermogammarus*), for obvious reasons. On the other hand, *Zyzzyx*, which was applied to a genus of wasps by an entomologist for the express purpose of contributing the last name at the end of taxonomic catalogs, has been allowed to stand. Nomenclature can sometimes be whimsical.

Suppose a taxonomist undertook to revise a group of species, meaning that he was examining the similarities among the species and attempting to classify and to name them, as well as to deduce something about their phylogeny. He would begin by deciding which species are represented and what criteria he will use to distinguish between them. The preferred criterion is whether or not the organisms interbreed freely with one another in nature. Then, by a careful study of the available information on morphology, physiology, and behavior, he would decide on a measure of resemblance among the species. There are easy-to-follow statistical procedures for taking such measures. But the procedures still contain many arbitrary features, with the result that most taxonomists still try to judge the matter subjectively. Whichever measure is used, the next problem is CLUSTERING the species into higher taxonomic categories.

Here the taxonomist is faced with a choice: he must decide the limits of each genus. A precedent he might consider, one of many available, is the placement of all of the oak species in the genus *Quercus*. The question is: Which of the various species can be clustered together to form the genera? Figure 1 shows that this is basically an arbitrary choice. By studying the position of the plants, each representing one species, one can see that it is equally valid in this particular case to combine all of the species into a single genus, or to split them into two or more genera. The lower of the two arrangements shows the species divided into three genera, but it could have been four, if we decided to put species 3 in a genus of its own, or even five, if species 6 were also sepa-

The name Zyzzyx was applied to a genus of wasps by an entomologist for the express purpose of contributing the last name at the end of taxonomic catalogs. Nomenclature can sometimes be whimsical.

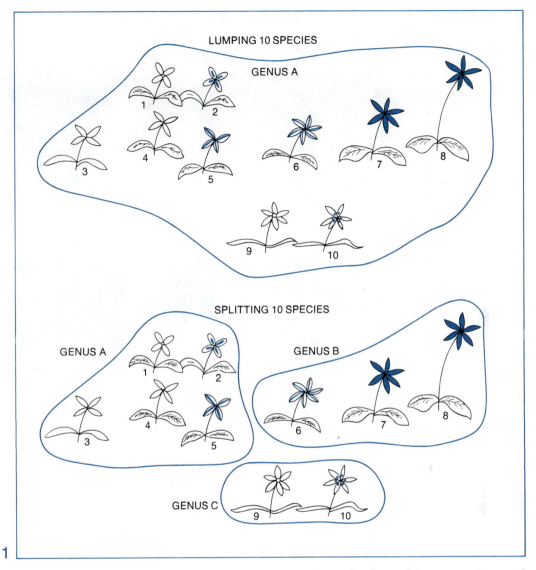

1

IMAGINARY PLANT SPECIES can be clustered into genera in several ways. Classification scheme at top might be followed by a taxonomist who is a "lumper"; scheme at bottom, by a "splitter."

rated. Taxonomists who like to combine species into a small number of large genera are often described lightly as "lumpers," while their colleagues who prefer a large number of small genera must be called, of course, the "splitters." Everyone recognizes that the choice is at least partly a matter of taste. But taxonomy is not a subject to be taken casually. The two most important criteria of good taxonomic work are first, the judgment used in defining species, or other "operational taxonomic units" such as local populations;

and second, the clarity with which the relationships are described.

Similar subjective techniques are required in the classification of categories higher than the genus. Once the limits of the genera have been decided upon, the taxonomist clusters them into FAMILIES. Figure 2 shows how the similarities among species comprising different genera form the basis for grouping particular genera into single families or setting them apart in different families. An example of a family is the Formicidae, which includes all of

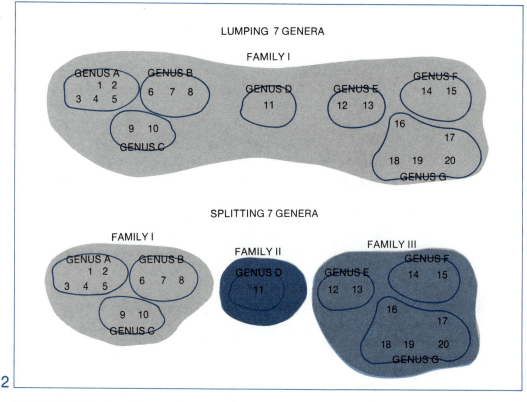

SEVEN GENERA, A to G, containing a total of 20 species (denoted by numerals) are classified into families by a lumper (top) and a splitter (bottom). Both arrangements are equally valid because a family, like a genus, is an arbitrarily defined cluster of similar species.

A
CLASSIFICATION
OF THE
HUMAN SPECIES

Kingdom: *Animalia (animals)*
  Phylum: *Chordata (chordates)*
    Subphylum: *Vertebrata (vertebrates)*
      Class: *Mammalia (mammals)*
        Order: *Primates (primates)*
          Family: *Hominidae (man and close relatives)*
            Genus: *Homo (modern man and precursors)*
              Species: *sapiens (modern man).*

B
CLASSIFICATION
OF THE
COMMON ROSE

Kingdom: *Plantae (plants)*
  Phylum: *Tracheophyta (vascular plants)*
    Subphylum: *Pteropsida (ferns and seed plants)*
      Class: *Dicotyledoneae (dicots)*
        Order: *Rosales (saxifrages, psittosporums, sweet gum, planetrees, roses, and relatives)*
          Family: *Rosaceae (cherry, plum, hawthorn, roses, and relatives)*
            Genus: *Rosa (roses)*
              Species: *gallica (domestic rose).*

To demonstrate how a taxonomic key is designed and used we have created four species of an imaginary insect called Hypotheticus.

the genera and species of ants in the world. At this point the reader should take note of an important fact about the clustering procedure that often escapes beginners: It is not necessary to include many species in each group belonging to higher categories. A genus may contain only a single species, and a family may contain only a single genus; it is possible for a family to be composed of only one species. This extreme case is exemplified by family II in Figure 2.

In animal classification the names of families are identified by -idae endings. Thus Formicidae is the family that contains all ant species, while Hominidae contains man and his fossil relatives. The family names are based on a member genus; hence Formicidae is based on *Formica* (just remove the -a and add -idae),

and Hominidae is based on *Homo* (from the base of the genitive form hominis, plus -idae). Plant classification follows the same procedure, except that the ending -aceae is used instead of -idae. Thus the Rosaceae is the family that includes the genus of roses (*Rosa*) and all its relatives. Notice that unlike the generic and species names, the name of the family is not italicized.

And so it goes on up into still higher categories. The taxonomist groups families into orders, orders into classes, classes into phyla, and phyla into kingdoms according to the same procedures. The hierarchy of units to which our own species (*Homo sapiens*) belongs is shown in Box A. Turning to the plant kingdom, a parallel set of categories can be listed for the common domestic rose, *Rosa gallica* (Box B).

Once the classification and nomenclature are settled, the taxonomist must worry about making it possible for other biologists to identify the species he has revised. He accomplishes this in three ways: by placing identified specimens in museum collections for reference purposes, by publishing illustrations and verbal descriptions in articles, and by constructing TAXONOMIC KEYS. To see how a taxonomic key is designed and used, consider the four species of the imaginary insect genus *Hypotheticus* (Figure 3). We use an arbitrary example only to be able to simplify and clarify without distorting reality. We start with four *Hypotheticus* species that can be separated easily from one another by characteristics that are obvious in the drawing. The characteristics, or CHARACTER STATES as they are usually called, are also listed in Table I and marked as either present (+) or absent (−) in each of the four species.

Using the information concerning the character states, the reader can construct a simple taxonomic key (*Table II*). By using the key to identify one or two of the species from the drawings alone, the reader can quickly see how the keys are constructed. Start with the first couplet of traits, make a choice between the two alternatives given there, and proceed on down. By examining the multiple character

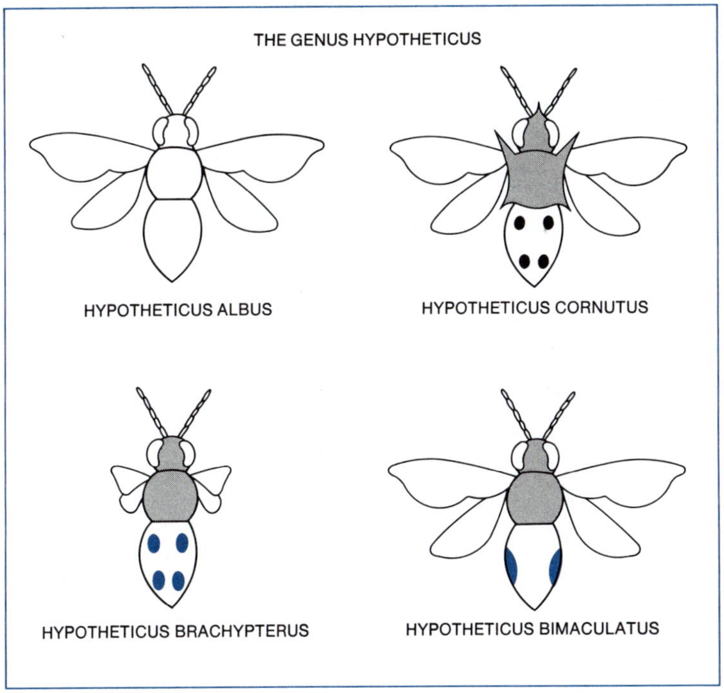

THE GENUS HYPOTHETICUS

HYPOTHETICUS ALBUS

HYPOTHETICUS CORNUTUS

HYPOTHETICUS BRACHYPTERUS

HYPOTHETICUS BIMACULATUS

IMAGINARY GENUS OF INSECTS includes four species. Early primitive species (albus) has white body, no horns, and normal wings. One advanced species (bimaculatus) has two spots on abdomen. Two other advanced species (cornutus and brachypterus) have four spots. Cornutus has horns on thorax, brachypterus has short wings. Genus was created to show principles of identification and phylogeny.

303

SPECIES	BODY ENTIRELY WHITE	HORNS PRESENT ON HEAD AND THORAX	WINGS VERY SHORT	EXACTLY TWO SPOTS ON ABDOMEN
HYPOTHETICUS ALBUS	+	−	−	−
HYPOTHETICUS CORNUTUS	−	+	−	−
HYPOTHETICUS BRACHYPTERUS	−	−	+	−
HYPOTHETICUS BIMACULATUS	−	−	−	+

I

CHARACTER STATES of the species that comprise the genus Hypotheticus. Only the presence (+) or absence (−) of each character state is indicated.

states that distinguish each of the four species, it is clear that the key to *Hypotheticus* can follow other sequences of characters than the one chosen here.

Now proceed to the construction of a hypothetical phylogeny, still using the convenient set of four *Hypotheticus* species. A first step is to deduce the BRANCHING SEQUENCES in the history of the genus. The basic, simple question is, which species evolved from which other species? Or, alternatively, which species share immediate common ancestors? Suppose we had at our disposal the following bits of information, of the sort that is often available in taxonomic studies:

All early fossils of *Hypotheticus* and similar genera are completely white and hornless and possess normally developed wings. A few fossil specimens of *Hypotheticus* from a later

A KEY TO THE SPECIES OF HYPOTHEDICUS		
1.	BODY ENTIRELY WHITE	H. ALBUS
	HEAD AND THORAX BLACK	GO TO ITEM 2
2.	HEAD AND THORAX BEARING A TOTAL OF FIVE LARGE SPINES OR HORNS	H. CORNUTUS
	HEAD AND THORAX LACKING SPINES AND HORNS	GO TO ITEM 3
3.	WINGS EACH MUCH SHORTER THAN THE WIDTH OF THE BODY	H. BRACHYPTERUS
	WINGS EACH MUCH LONGER THAN THE WIDTH OF THE BODY	H. BIMACULATUS

II

TAXONOMIC KEY of the genus Hypotheticus. Key refers to characteristics illustrated in Figure 3.

geologic period have two spots on the abdomen (like the modern species *H. bimaculatus*). The four-spotted condition, so far as is known, is found only in the modern species *H. brachypterus*.

We conclude that white bodies, hornlessness, and normal wings are the primitive (original) character states and all the other character states we listed were derived in evolution. If this is indeed the case, it is possible to study how the various character states are combined and then to come up with the simple phylogenetic diagram shown in Figure 4. The reader should try this himself to make sure he understands it. Notice that the diagram shows only the branching sequences. No attempt is made to make the time scale precise or absolute; an exact chronology can be determined only with information derived from fossil specimens.

HOMOLOGY

The phylogenetic analysis of the imaginary genus *Hypotheticus* was based on a fundamental but unspoken assumption. When two species of *Hypotheticus* were found to share similar character states, such as black coloration or horns on the body, it was implicitly assumed that these states were acquired from a similar or identical state in some ancestral species. If this assumption were true, a taxonomist would be justified in saying that possession of similar character states is evidence of a phylogenetic relationship between the two species. When the character states found in any two species owe their resemblance to a common ancestry, taxonomists say the states are HOMOLOGOUS, or are HOMOLOGUES of each other. HOMOLOGY is defined as correspondence between two structures which is due to inheritance from a common ancestor.

Homologous structures can be identical in appearance and, at least in theory, can even be based on identical genes. Conversely, such structures can diverge in evolution until they become quite different in both appearance and function. But homologous structures always retain certain basic features that betray their common ancestry. The classic example of homology combined with divergence is found in the comparative anatomy of the vertebrate forelimb. When one compares the forearm of a man with that of a monkey, for example, it is easy to make a detailed, bone-by-bone, muscle-by-muscle comparison and to con-

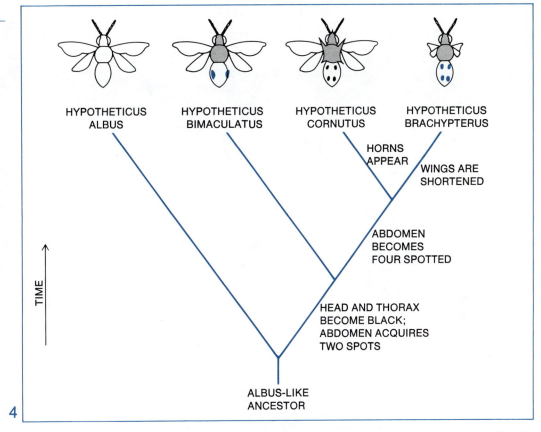

HYPOTHETICUS
ALBUS

HYPOTHETICUS
BIMACULATUS

HYPOTHETICUS
CORNUTUS

HYPOTHETICUS
BRACHYPTERUS

HORNS
APPEAR

WINGS ARE
SHORTENED

ABDOMEN
BECOMES
FOUR SPOTTED

HEAD AND THORAX
BECOME BLACK;
ABDOMEN ACQUIRES
TWO SPOTS

TIME

ALBUS-LIKE
ANCESTOR

4

ELEMENTARY PHYLOGENY of Hypotheticus is summarized in this diagram. Branching sequence is based on assumption that species having similar traits descended from a common ancestor.

clude thereby that the forearms, as well as their various parts, are homologous. But if one compares a human forearm with the foreleg of a dog, strong differences are obvious in both structure and function of the forelimb. This organ is used for locomotion by the dog but for grasping and manipulation by the man. Even so, all of the bones can still be matched. It is reasonable to conclude that these structures, and therefore the forelimbs as a whole, are homologous in man and the dog. More extreme cases exist within the vertebrate animals. The wing of a bird and the flipper of a seal at first glance appear to be radically different from each other or from the forearm of a

man, yet they too are constructed around bones that can be matched on a nearly perfect one-to-one basis.

The wing of a fly shares the same function as a bird's wing, and the two organs even resemble each other to a slight degree in external form. Yet close examination shows that the two organs are completely different in their basic structure. The entire wing of the fly is a membranous outgrowth of the external skeletal covering of the body. Its main supports are not bones but rather columns of hardened protein combined with a complex polysaccharide called chitin. The bird and fly wings are said to be ANALOGUES of each other.

The classic example of homology combined with evolutionary divergence is found in the comparative anatomy of the forelimbs of vertebrates.

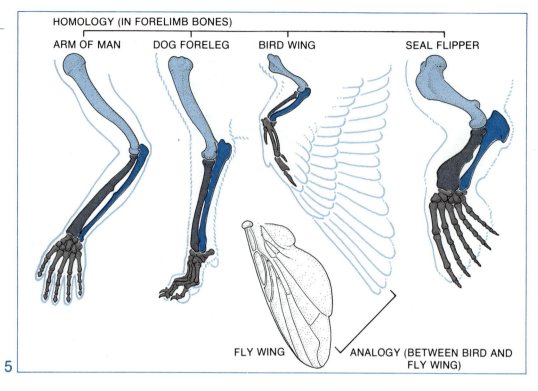

HOMOLOGY (IN FORELIMB BONES)

ARM OF MAN   DOG FORELEG   BIRD WING   SEAL FLIPPER

FLY WING   ANALOGY (BETWEEN BIRD AND FLY WING)

5

HOMOLOGY AND ANALOGY. Forelimb bones of various vertebrates are homologous, even though they sometimes serve very different functions. Homologous bones in drawing are rendered in corresponding shades of gray and color. Wing of the fly is analogous to that of a bird, meaning that it is similar in function and to some extent in form, despite its wholly different evolutionary origin.

ANALOGY is defined as a resemblance in function which is based on CONVERGENT EVOLUTION rather than on descent from a common ancestor (*Figure 5*).

FOSSILS

The taxonomist constructs his classifications from similarities and differences that he can directly observe in the relatively small number of specimens at his disposal. To add the element of phylogeny to these classifications, he must then decide whether the characteristics shared by the various groups are homologous or analogous — whether they are due to common ancestry or to convergent evolution. Such decisions are almost always subjective and difficult to make. Typically they can be put to a test only by consulting fossils, which provide the only direct record of the ancestry of living organisms.

For resolving questions of phylogeny the most useful fossils are those that preserve the features actually used by the taxonomist. Liv-

ing gastropods (snails), for example, are classified on the basis of variations in their soft bodies. Fossil gastropods are known mainly on the basis of their shells; the soft parts are rarely preserved. Thus there is an element of uncertainty in establishing lines of inheritance of the important taxonomic features.

Few fossils are composed of the actual material of the once-living animal or plant. More commonly the fossil vertebrate bone or invertebrate shell is a mineral replacement of the original (*Figure 6*). Frequently, as with petrified wood, the original structure of the organism is preserved — replaced molecule by molecule as the original material is removed by decomposition.

Sometimes images of the soft parts are preserved as casts retaining some of the original molding of the organism. The finest of the smaller fossils are those preserved in amber, the fossilized gum of trees. When a tree is injured or for some other reason exudes gum, the material serves as a natural flypaper that traps pollen, bits of leaves, and the smallest animals, especially insects. In some of the amber fossils the anatomical details are nearly as complete as those in specimens freshly fixed by the biologist and embedded in clear balsam for microscopic examination. Every hair and wrinkle can still be studied under magnification. A photograph of one of the oldest and most significant amber specimens appears in Figure 7. This worker ant, discovered in 1967, was the earliest social insect of any kind known at the time, and it proved that social life originated as long as 100 million years ago (a termite of similar age has been uncovered since then). It also constitutes one of the missing links of evolution, which connects the

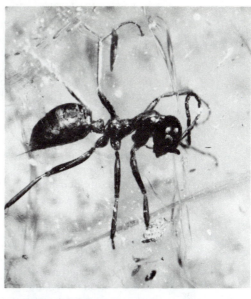

7

100-MILLION-YEAR-OLD ANT depicted in photograph above is exquisitely preserved in amber. Ant was trapped in the gum of a sequoia tree in an ancient forest near the Atlantic shore of New Jersey. This specimen is an evolutionary link between ants and their ancestors, the nonsocial wasps.

6

FOSSIL CAST OF HORSESHOE CRAB was exposed when rock containing the fossil was split open. Rock also shows impression of tracks made by crab just before it died.

ants with the nonsocial wasps that gave rise to them. With the new information provided by this fossil it was possible to state with some assurance that the ants originated from one particular family of wasps, the Tiphiidae. Accordingly, the new genus to which the fossil ant belongs was named *Sphecomyrma*, which means "wasp ant."

THE EVOLUTIONARY TIME SCALE

In the early 19th century paleontologists began the arduous task of reconstructing the history of life on Earth by the study of fossils. By the latter part of the century it was clear that the sequence of fossils reflected many of the evolutionary changes predicted by Darwin. But an urgent question remained. What was the time scale of the fossil-dated geological ages? How many thousands or millions of years did major evolution require? Geologists, by various indirect methods such as estimating the time required to lay down the total deposits of rocks, judged that the Earth must be at least hundreds of millions of years old. This, the biologists agreed, was consistent with the time required for the sweeping changes that evolution has wrought among living organisms. In the last two decades an absolute time scale was finally made possible by the radioactive dating techniques discussed in Chapter 16. Those techniques are based on decay processes that take millions of years, and therefore serve to date the older geological deposits. For the dating of younger deposits, a powerful tool is RADIOCARBON DATING, which is based on the decay of carbon-14, the isotope familiar from tracer experiments (*Chapter 2*). The C^{14} atoms in nature are eventually oxidized to CO_2 and are incorporated into plants during photosynthesis. So long as the C^{14} circulates through living plants and animals and back into the atmosphere or water, the proportion of C^{14} to C^{12} (the much more abundant nonradioactive isotope) remains constant. But as soon as the organism dies the C^{14} begins to decay into nitrogen at a steady rate. Measurement of the residual radioactivity can be used to date samples up to 50,000 years old, which makes the technique particularly useful in obtaining precise dates of prehistoric man and other relatively recent fossils.

Chapter 16 covered the history of the Earth to the beginning of the Cambrian period, 600 million years ago. Table III covers the interval from then to the present — the last four days on the 30-day calendar in Chapter 16. The curve that accompanies Table III indicates a trend toward increasing diversity among living things. Between the earliest recorded beginnings of life and the first true animal fossils, which occur in the late Precambrian, stretches an interval of no less than two billion years — about 13 days on our geological calendar. Once evolution entered its multicellular phase, 600 million years ago, the number of different organisms increased steadily to the present remarkable level. Right now life on Earth is in a period of unfolding. But this expansion of diversity has not always been a smooth upward progression. Widespread extinctions, accompanied by rapid evolutionary change and important shifts in the composition of life, took place around the close of the Cambrian, Ordovician, Devonian, Permian, Triassic and Cretaceous periods. The most serious perturbation occurred at or near the close of the Permian period, and its effect is clearly evident in the dip in the curve of animal diversity of Table III. At this time no fewer than 24 orders, and about half of the families of animals throughout the world, became extinct. Several theories have been advanced to explain these crises in the history of life. Some geologists have blamed them on sudden

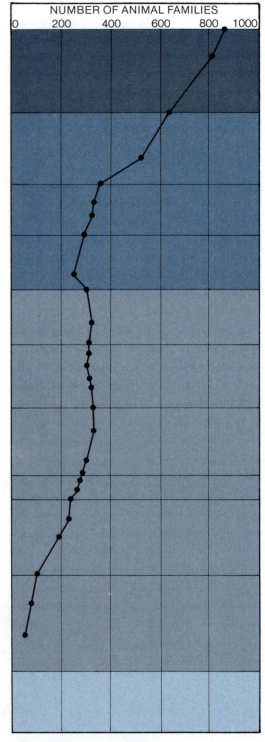

NUMBER OF ANIMAL FAMILIES

ERA	PERIOD	BIOLOGICAL EVENTS	YEARS AGO
CENOZOIC	QUATERNARY	MODERN MAN	11 THOUSAND
		EARLY MAN; NORTHERN GLACIATION	5 TO 3 MILLION
	TERTIARY	LARGE CARNIVORES	13 ± 1 MILLION
		FIRST ABUNDANT GRAZING MAMMALS	25 ± 1 MILLION
		LARGE RUNNING MAMMALS	36 ± 2 MILLION
		MANY MODERN TYPES OF MAMMALS	58 ± 2 MILLION
		FIRST PLACENTAL MAMMALS	63 ± 2 MILLION
MESOZOIC	CRETACEOUS	FIRST FLOWERING PLANTS; CLIMAX OF DINOSAURS AND AMMONITES FOLLOWED BY EXTINCTION	135 ± 5 MILLION
	JURASSIC	FIRST BIRDS, FIRST MAMMALS; DINOSAURS AND AMMONITES ABUNDANT	180 ± 5 MILLION
	TRIASSIC	FIRST DINOSAURS; ABUNDANT CYCADS AND CONIFERS	230 ± 10 MILLION
PALEOZOIC	PERMIAN	EXTINCTION OF MANY KINDS OF MARINE ANIMALS, INCLUDING TRILOBITES. GLACIATION AT LOW LATITUDES	280 ± 10 MILLION
	CARBONIFEROUS	GREAT COAL FORESTS, CONIFERS FIRST REPTILES SHARKS AND AMPHIBIANS ABUNDANT. LARGE AND NUMEROUS SCALE TREES AND SEED FERNS	345 ± 10 MILLION
	DEVONIAN	FIRST AMPHIBIANS AND AMMONITES; FISHES ABUNDANT	405 ± 10 MILLION
	SILURIAN	FIRST TERRESTRIAL PLANTS AND ANIMALS	425 ± 10 MILLION
	ORDOVICIAN	FIRST FISHES; INVERTEBRATES DOMINANT	500 ± 10 MILLION
	CAMBRIAN	FIRST ABUNDANT RECORD OF MARINE LIFE. TRILOBITES DOMINANT, FOLLOWED BY EXTINCTION OF ABOUT TWO-THIRDS OF TRILOBITE FAMILIES	600 ± 50 MILLION
	PRECAMBRIAN	FOSSILS EXTREMELY RARE, CONSISTING OF PRIMITIVE AQUATIC PLANTS AND SOME SIMPLE ANIMALS. EVIDENCE OF GLACIATION. OLDEST DATED ALGAE AND BACTERIA, 3.1 BILLION YEARS	

GEOLOGICAL ERAS since the dawn of the Cambrian period. Table corresponds to the last four days of the geological calendar in Chapter 16, and lists the major biological events of each period. Curve at left indicates the evolutionary trend toward a greater diversity of organisms. Dip in middle of curve reflects the widespread extinctions around the Permian period. Dates were obtained from measurements of radioactivity in rocks.

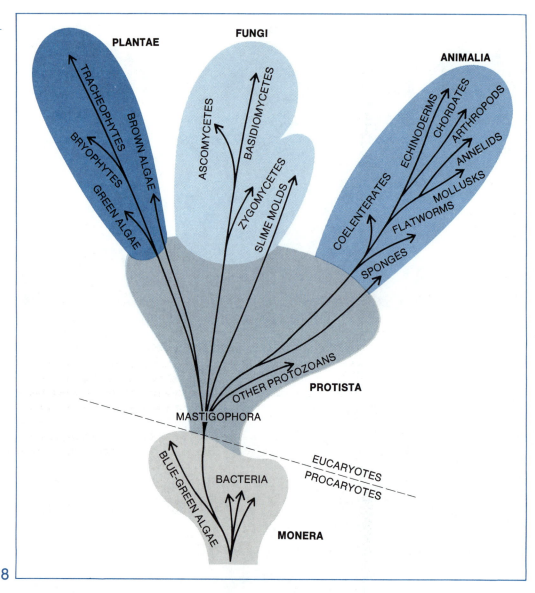

**FIVE KINGDOMS OF ORGANISMS and their relationships are repre-
sented by this phylogenetic tree. For simplicity, lower categories such as
orders and families have been omitted.**

episodes of mountain building, associated
with a general uplift of the land. Others blame
drastic changes in climate, resulting in cooling
and even glaciation of much of the Earth's surface. Still others have blamed massive out-
bursts of cosmic radiation from outer space.
The truth is that no single kind of cataclysmic
event has been consistently associated with all
of the crises. Possibly more than one cause was
at work and there may have been others that
remain undiscovered.

THE CLASSIFICATION OF ORGANISMS

This chapter closes with a table of the major
groups of organisms, including the five kingdoms (monerans, protists, fungi, plants, and
animals), most of the phyla, and a number
of the most important classes (*Box C*). The
phylogenetic relationships of the kingdoms
are roughly indicated in Figure 8. For simplicity, most groups belonging to lower categories,
such as orders and families, are omitted. There

	KINGDOM	PHYLUM	SUBPHYLUM	CLASS	SUBCLASS	ORDER
THE PROCARYOTES (CHROMOSOMES NOT LOCATED WITHIN A NUCLEUS)	**MONERA**	**SCHIZOPHYTA OR SCHIZOMYCETES:** BACTERIA (UNKNOWN NUMBER OF SPECIES)				
		CYANOPHYTA: BLUE-GREEN ALGAE (2500 SPECIES)				
THE EUCARYOTES (CHROMOSOMES LOCATED WITHIN A NUCLEUS)	**PROTISTA**	**MASTIGOPHORA OR FLAGELLATA:** FLAGELLATES		**PHYTOMASTIGINA:** PLANT-LIKE FLAGELLATES		
				ZOOMASTIGINA: ANIMAL-LIKE FLAGELLATES		
		SARCODINA OR RHIZOPODA: AMOEBAS, FORAMINIFERANS, HELIOZOANS, RADIOLARIANS				
		SPOROZOA: SPOROZOANS, INCLUDING MALARIAL PARASITES				
		CILIOPHORA OR CILIATA: CILIATES (6000 SPECIES)				
		MYXOMYCOTA: SLIME MOLDS (450 SPECIES)				
	FUNGI	**ZYGOMYCOTA** (500 SPECIES): TUBE FUNGI, INCLUDING SOME RUSTS, BREAD MOLDS, WATER MOLDS, AND OTHERS				
		ASCOMYCOTA (35,000 SPECIES): SAC FUNGI, INCLUDING YEASTS, POWDERY MILDEWS, CUP FUNGI, BLUE AND GREEN MOLDS, SOME BREAD MOLDS, AND OTHERS				
		BASIDIOMYCOTA (25,000 SPECIES): CLUB FUNGI, INCLUDING MOST MUSHROOMS, TOADSTOOLS, BRACKET FUNGI, SMUTS, AND RUSTS				
		FUNGI IMPERFECTI: UNDER THIS LOOSE CATEGORY ARE INCLUDED A WIDE VARIETY OF FUNGUS SPECIES WHICH CANNOT BE PLACED TO ONE OF THE CLASSES ABOVE, BECAUSE THE SEXUAL PART OF THE LIFE CYCLE HAS BEEN LOST IN				

	KINGDOM	PHYLUM	SUBPHYLUM	CLASS	SUBCLASS	ORDER
		EVOLUTION OR ELSE EXISTS BUT SIMPLY IS NOT YET ELUCIDATED. MOST FUNGI IMPERFECTI PROBABLY BELONG TO EITHER THE ASCOMYCOTA OR THE BASIDIOMYCOTA.—LICHENS: THESE ARE COMPOUND ORGANISMS, EACH OF WHICH CONSISTS OF ALGAE LIVING WITHIN THE BODY OF A FUNGUS, USUALLY A MEMBER OF THE ASCOMYCOTA OR BASIDIOMYCOTA.				
	PLANTAE	**PYRROPHYTA:** DINOFLAGELLATES (1100 SPECIES)				
		CHRYSOPHYTA: DIATOMS AND RELATED ALGAE (10,000 SPECIES)		**XANTHOPHYCEAE:** YELLOW-GREEN ALGAE		
				CHRYSOPHYCEAE: GOLDEN-BROWN ALGAE		
				BACILLARIOPHYCEAE: DIATOMS		
		PHAEOPHYTA: BROWN ALGAE (1000 SPECIES)				
		RHODOPHYTA: RED ALGAE (3000 SPECIES)				
		CHLOROPHYTA: GREEN ALGAE (6000 SPECIES)				
		BRYOPHYTA: LIVERWORTS, MOSSES, AND RELATED FORMS (250,000 SPECIES)		**ANTHOCEROPSIDA:** HORNWORTS		
				HEPATICAE, OR HEPATICOPSIDA: LIVERWORTS		
				BRYOPSIDA OR MUSCI: MOSSES		
		TRACHEOPHYTA: VASCULAR PLANTS (250,000 SPECIES)	**PSILOPSIDA:** PSILOTUM AND OTHERS	**PSILOPHYTALES:** PSILOPHYTON AND OTHER PRIMITIVE EXTINCT PLANTS		
				PSILOTALES: PSILOTUM AND RELATED FORMS		
			LYCOPSIDA: LYCOPODS			
			SPHENOPSIDA: HORSETAILS (EQUISETUM) AND RELATED FORMS			
			PTEROPSIDA: FERNS AND SEED PLANTS	**FILICINEAE:** FERNS		
				GYMNOSPERMAE: PINES, FIRS, GINKGOS, AND OTHER GYMNOSPERMS		
				ANGIOSPERMAE: ANGIOSPERMS OR FLOWERING PLANTS. THIS GROUP CONSISTS OF THE MONOCOTS (ORDER MONOCOTYLEDONAE), WHICH INCLUDE THE GRASSES, SEDGES, LILIES, PALMS, AND ORCHIDS; AND THE DICOTS (ORDER DICOTYLEDONAE), WHICH CONSISTS OF THE BULK OF THE REMAINING HIGHER PLANTS		

THE EUCARYOTES (CHROMOSOMES LOCATED WITHIN A NUCLEUS)

THE CLASSIFICATION OF ORGANISMS (CONTINUED)

THE EUCARYOTES (CHROMOSOMES LOCATED WITHIN A NUCLEUS)

KINGDOM	PHYLUM	SUBPHYLUM	CLASS	SUBCLASS	ORDER
ANIMALIA	PORIFERA: SPONGES (10,000 SPECIES)				
	SNIDARIA OR COELENTERATA: HYDRAS, JELLYFISH, AND RELATED FORMS (9000 SPECIES)		HYDROZOA: HYDRAS AND HYDROIDS		
			SCYPHOZOA: JELLYFISH		
			ANTHOZOA: SEA ANEMONES, CORALS		
	CTENOPHORA: COMB JELLIES, SEA GOOSEBERRIES (90 SPECIES)				
	PLATYHELMINTHES: FLATWORMS (13,000 SPECIES)		TURBELLARIA: "PLANARIANS" AND OTHER FREE-LIVING FLATWORMS		
			TREMATODA: FLUKES (ALL PARASITIC FORMS)		
	MESOSOMA: MESOZOANS (SMALL, EXTREMELY SIMPLIFIED PARASITES OF MARINE INVERTEBRATES: (50 SPECIES)		CESTODA: TAPEWORMS (ALL PARASITIC FORMS)		
	NEMERTINEA OR RHYNCHOCOELA: NEMERTEANS OR RIBBON WORMS (750 SPECIES)				
	ROTIFERA: ROTIFERS (1500 SPECIES)				
	GASTROTRICHA: GASTROTRICHS (175 SPECIES)				
	KINORHYNCHA: KINORHYNCHANS (64 SPECIES)				
	NEMATODA: ROUNDWORMS OR NEMATODES (10,000 SPECIES)				
	NEMATOMORPHA: HAIR WORMS (100 SPECIES)				
	ACANTHOCEPHALA: ACANTHOCEPHALAN WORMS (ALL PARASITES OF ARTHROPODS: 300 SPECIES)				
	ENTOPROCTA: MOSS ANIMALS OR ENTOPROCTS (60 SPECIES)				
	PRIAPULIDA: PRIAPULIDS (CUCUMBER-SHAPED MARINE ANIMALS: 8 SPECIES)				
	SIPUNCULA: SIPUNCULIDS (CYLINDRICAL MARINE WORMS: 250 SPECIES)				
	MOLLUSCA: MOLLUSKS (80,000 SPECIES)		AMPHINEURA: CHITONS		
			MONOPLACOPHORA: MONOPLACO-PHORANS		
			GASTROPODA: SNAILS, CONCHS, SLUGS		
			BIVALVIA, OR PELECYPODA: CLAMS, OYSTERS, AND OTHER BIVALVE MOLLUSKS		
			SCAPHOPODA: SCAPHOPODS OR TUSK SHELLS		
			CEPHALOPODA: SQUIDS, OCTOPODS		

C

THE CLASSIFICATION OF ORGANISMS (CONTINUED)

THE EUCARYOTES
(CHROMOSOMES LOCATED WITHIN A NUCLEUS)

KINGDOM	PHYLUM	SUBPHYLUM	CLASS	SUBCLASS	ORDER
ANIMALIA CONTINUED	**ECHIURA:** ECHIURIDS (CYLINDRICAL MARINE WORMS; 60 SPECIES)				
	ANNELIDA: SEGMENTED WORMS (7000 SPECIES)		**POLYCHAETA:** POLYCHAETES (EXCLUSIVELY MARINE WORMS)		
			OLIGOCHAETA: OLIGOCHAETES, INCLUDING EARTHWORMS AND FRESHWATER WORMS		
			HIRUDINEA: LEECHES		
	TARDIGRADA: WATER BEARS (180 SPECIES)				
	PENTASTOMIDA: TONGUE WORMS (65 SPECIES)				
	ONYCHOPHORA: ONYCHOPHORANS (65 SPECIES)				
	ARTHROPODA: ARTHROPODS (900,000 SPECIES)		**TRILOBITA:** TRILOBITES (PRIMITIVE EXTINCT FORMS)		
			CHELICERATA: SCORPIONS, SPIDERS, MITES, HORSESHOE CRABS		
			PYCNOGONIDA: SEA SPIDERS		
			CRUSTACEA: CRUSTACEANS, INCLUDING WATER FLEAS, CRABS, SHRIMPS, AND LOBSTERS		
			INSECTA OR HEXAPODA: INSECTS		**ODONATA:** DRAGONFLIES
					BLATTARIA: COCKROACHES
					ISOPTERA: TERMITES
					ORTHOPTERA: GRASSHOPPERS, CRICKETS, WALKING STICKS
					HEMIPTERA: STINK BUGS, ASSASSIN BUGS, BEDBUGS, WATER BOATMEN
					HOMOPTERA: APHIDS, SCALE INSECTS, CICADAS
					ANOPLURA: LICE
					COLEOPTERA: BEETLES
					LEPIDOPTERA: BUTTERFLIES, MOTHS
					DIPTERA: FLIES, MOSQUITOES, GNATS
					HYMENOPTERA: SAWFLIES, WASPS, BEES, ANTS
					SIPHONAPTERA: FLEAS
			DIPLOPODA: MILLIPEDES		
			CHILOPODA: CENTIPEDES		
			PAUROPODA: PAUROPODS		
			SYMPHYLA: SYMPHYLANS		
	PHORONIDA: PHORONIDS (WORMLIKE MARINE ANIMALS; 15 SPECIES)				

C

THE CLASSIFICATION OF ORGANISMS (CONTINUED)

THE EUCARYOTES
(CHROMOSOMES LOCATED WITHIN A NUCLEUS)

KINGDOM	PHYLUM	SUBPHYLUM	CLASS	SUBCLASS	ORDER
ANIMALIA CONTINUED	**BRYOZOA OR ECTOPROCTA:** BRYOZOANS OR MOSS ANIMALS (4000 SPECIES)				
	BRACHIOPODA: BRACHIOPODS OR LAMP SHELLS (260 SPECIES)				
	CHAETOGNATHA: ARROW WORMS (50 SPECIES)				
	ECHINODERMATA: ECHINODERMS (5300 SPECIES)		**CRINOIDEA:** SEA LILIES AND FEATHER STARS		
			ASTEROIDEA: STARFISH OR SEA STARS		
			OPHIUROIDEA: BRITTLE STARS, BASKET STARS		
			ECHINOIDEA: SEA URCHINS, SAND DOLLARS		
			HOLOTHUROIDEA: SEA CUCUMBERS		
	POGONOPHORA: POGONOPHORANS (80 SPECIES)				
	HEMICHORDATA: ACORN WORMS (80 SPECIES)				
	CHORDATA: CHORDATES (39,000 SPECIES)	**VERTEBRATA:** VERTEBRATES	**AGNATHA:** AGNATHS OR JAWLESS FISHES, INCLUDING LAMPREYS		
			PLACODERMI: PLACODERMS (EXTINCT ARMORED FISHES)		
			CHRONDRICHTHYES: ELASMOBRANCHS, INCLUDING SHARKS, RAYS, AND CHIMAERANS		
			OSTEICHTHYES: BONY FISHES (MOST LIVING FORMS OF FISHES)		
			AMPHIBIA: AMPHIBIANS, INCLUDING FROGS, TOADS, SALAMANDERS, AND CAECILIANS		
			REPTILIA: REPTILES		
			AVES: BIRDS		
			MAMMALIA: MAMMALS	**PROTOTHERIA:** MONOTREMES (EGG-LAYING MAMMALS, INCLUDING THE DUCKBILL PLATYPUS AND SPINY ANTEATERS, LIMITED TO AUSTRALIA AND NEW GUINEA)	
				METATHERIA: MARSUPIALS (POUCHED MAMMALS)	
				EUTHERIA: PLACENTAL MAMMALS	**INSECTIVORA:** SHREWS, MOLES, TENRECS
					CHIROPTERA: BATS
					CARNIVORA: DOGS, CATS, BEARS, SEALS, SEA LIONS
					RODENTIA: MICE, RATS, SQUIRRELS, PORCUPINES, BEAVERS
					LAGOMORPHA: RABBITS, HARES
					PRIMATES: LEMURS, LORISES, TARSIERS, MONKEYS, APES, MAN

	KINGDOM	PHYLUM	SUBPHYLUM	CLASS	SUBCLASS	ORDER
C **THE CLASSIFICATION OF ORGANISMS (CONTINUED)**	ANIMALIA CONTINUED					**ARTIODACTYLA:** EVEN-TOED UNGULATES— CATTLE, DEER, SHEEP, PIGS, CAMELS, HIPPOPOTAMUSES **PERISSODACTYLA:** ODD-TOED UNGULATES— HORSES, ASSES, ZEBRAS, RHINOCEROS **PROBOSCIDEA:** ELEPHANTS **CETACEA:** WHALES, PORPOISES **SIRENIA:** SEA COWS (MANATEES)

THE EUCARYOTES (CHROMOSOMES LOCATED WITHIN A NUCLEUS)

is no need to try to memorize these names. They are presented here so that the reader can scan them and obtain a preliminary view of the total diversity of life. For later chapters that deal with the individual phyla, the table will provide a convenient reference for comparisons among the groups.

READINGS

B. KUMMEL, *History of the Earth: An Introduction to Historical Geology*, San Francisco, W.H. Freeman & Co., 2nd Edition, 1970. A fine introduction to paleontology, strong in both geological and biological aspects.

E. MAYR, *Principles of Systematic Zoology*, New York, McGraw-Hill, 1969. This is the best available introduction to the principles of classification and is recommended as the next book to read for the student interested in exploring this growing branch of biology. Although intended for zoologists, most of the ideas and techniques are equally useful in botany.

R.H. WHITTAKER, "New Concepts of Kingdoms of Organisms," *Science*, 163: 150–160 (1969). An authoritative account of the higher categories of organisms, their evolutionary relationships, and the history of the classification of kingdoms and phyla.

NOTES

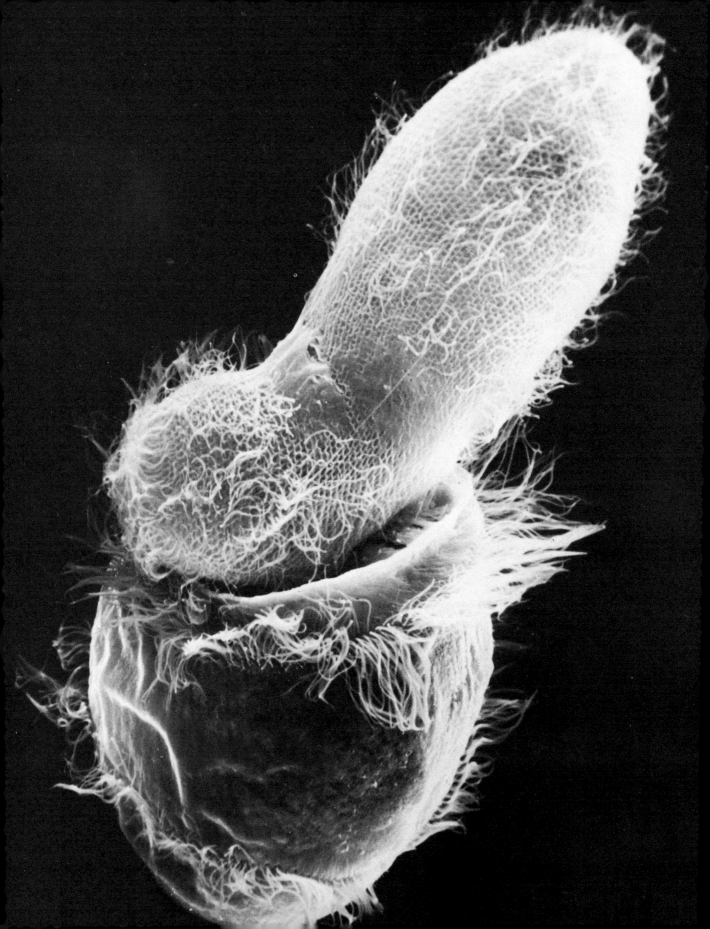

18

microorganisms

What is a microorganism? There is no simple answer to this question. The word 'microorganism' is not the name of a group of related organisms, as are the words 'plants' or 'invertebrates' or 'frogs'. The use of the word does, however, indicate that there is something special about SMALL organisms; we use no special word to denote large animals or medium-sized ones.

W.R. SISTROM

The human being, a gigantic organism, is inordinately aware of other large organisms. Although, as noted earlier, microorganisms probably comprise more than half the total biomass on Earth, we are only dimly aware of the vast and strikingly diverse world of viruses, bacteria, protozoans, and other creatures that are too small to be seen without a microscope. This bias is reflected vividly in most classification schemes, including the one outlined in the previous chapter. With considerable confidence taxonomists draw limits around such phyla of big organisms as the Mollusca (snails, clams, squids, and their relatives) and Tracheophyta (flowering plants), certain in their knowledge that the species enclosed are of common ancestry and share many basic anatomical features.

When sorting out microorganisms, however, taxonomists lack this sure touch. Until very recently all of the single-celled eucaryotes except algae were lumped into the phylum Protozoa. Now it is recognized that the differences among the "classes" of protozoans, namely the Mastigophora, the Sarcodina, the Sporozoa, and the Ciliophora, are at least as great as those that distinguish most phyla of the higher animals and plants. The Mastigophora, for example, include certain organisms that are animal-like in structure and nutrition and others that are plant-like. Consequently, many (but not all) zoologists prefer to raise the classes to the rank of full phyla, as in Chapter 17. An even more serious area of uncertainty is the division of microorganisms into kingdoms, the highest of the taxonomic categories. Bacteria and bluegreen algae are procaryotes, and a strong case can be made for distinguishing them from eucaryotes as a separate kingdom, the MONERA (pronounced mon-eé-ra). The one-celled eucaryotes comprise the other kingdom, the PROTISTA. Currently the classification of microorganisms is in a state of flux. There is a great amount of new information, and the classification scheme is likely to undergo some major changes before it is finally stabilized.

THE KINGDOM MONERA (PROCARYOTES)

Although considered the simplest of living organisms, the procaryotes are enormously more complex than single proteins or other nonliving molecules. The basic unit of these organisms is the procaryotic cell, which contains a complete complement of genetic and protein-synthesizing machinery, including DNA, RNA, and all of the enzymes needed to translate the code into protein. The cell also contains at least one system for generating the ATP needed to keep the machinery running. In earlier chapters it was pointed out that the procaryotic cell differs from the eucaryotic cell in four important ways: First, its genetic material is not organized within a distinctive nucleus, and the DNA is not part of a DNA-protein complex called chromatin. Second, there are none of the familiar membrane-bound organelles —

BATTLE BETWEEN CILIATES. Scanning electron micrograph on opposite page shows a one-celled Didinium (bottom) about to devour a Paramecium which has been paralyzed by the predator's dartlike trichocysts. Elastic cell membrane enables Didinium to engulf an organism larger than itself.

319

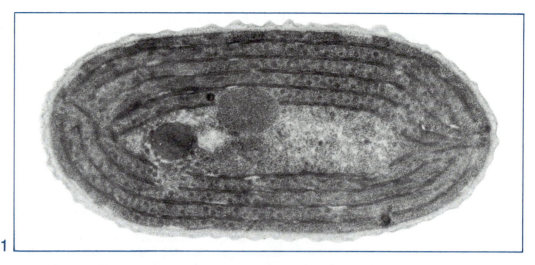

1

BLUEGREEN ALGAE are procaryotes that contain photosynthetic pigments bound to elaborate internal membranes. Electron micrograph depicts Spirulina, a bluegreen alga found in the warm sulfur springs of Yellowstone National Park. The genetic material in center of the cell is ringed by four or five layers of photosynthetic membranes. Magnification, 40,000×.

mitochondria, chloroplasts, golgi apparatus, endoplasmic reticulum — that occur in the cells of higher organisms. Third, there is no subcellular structure that can carry out the entire process of aerobic respiration, or, in the case of the photosynthetic procaryotes, photosynthesis. The cell itself is the smallest entity that can carry out these reactions. Finally, almost all procaryotes have a cell wall whose composition is unique. It consists of a backbone molecule of polymerized amino sugars, together with characteristic amino acids that cross-link the adjacent chains. Al-though this wall material, called the MUCOCOMPLEX SUBSTANCE, may in fact account for only a fraction of the total wall substance (proteins, lipids, and perhaps true polysaccharides make up the rest), it is thought to be responsible for the structural integrity of the wall. This substance is not found in the cells of any other organisms. The architecture of procaryotic and eucaryotic cells was illustrated in Chapter 1.

Absence of characteristic eucaryote organelles should not be construed as a total absence of any membranous structures within

FRUITING BODY of a myxobacterium (genus Chondromyces). This series of photomicrographs shows the development of the branched reproductive structure from an aggregated mass of individual bacterial cells. The tree-like structure is usually a little less than one millimeter high. Tips of the branches develop into microscopic cysts containing hundreds or thousands of rod-shaped cells.

2

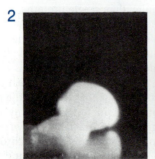  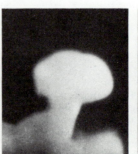 

*The eubacteria encompass an enormous number of species,
a wide variety of structural forms, and a truly incredible
array of biochemical tricks for obtaining energy.*

procaryotic cells. The procaryotic cell is bounded by a membrane. Membranous structures called mesosomes are frequently associated with formation of new cell walls during cell division. Many aerobic bacteria have rather elaborate internal membrane systems to which respiratory enzymes are bound, as in mitochondria. Similarly, the bluegreen algae and the photosynthetic bacteria may have elaborate internal membranes to which photosynthetic pigment systems are bound (*Figure 1*).

In the difficult task of classifying the procaryotes, we follow the arbitrary practice of dividing them into two phyla: the SCHIZOPHYTA, which includes the myxobacteria, spirochetes and true bacteria; and the CYANOPHYTA, which includes only the bluegreen algae. Any self-respecting botanist inevitably includes the bluegreen algae in his survey of the plant kingdom, and he may even include most or all of the bacteria among the lower fungi. Nevertheless, man's compulsion to classify requires a choice, and our scheme makes at least as much sense as calling bluegreen algae plants because they are studied primarily by botanists, and bacteria fungi because they are frequently included in surveys of the lower plants.

THE BACTERIA (PHYLUM SCHIZOPHYTA)

Several major groups of bacteria can be identified on the basis of striking characteristics. Cells of the MYXOBACTERIA are short rods that glide from place to place in some unknown way. They form remarkable fruiting structures similar to those of the cellular slime molds. A group of cells aggregates, and from the cell mass there arise characteristic fruiting bodies. These may be simple globes over a millimeter in diameter, or they may be somewhat more complex branched structures (*Figure 2*). Then either single cells transform themselves into thick-walled spores, or whole clusters of cells form an environment-resistant cyst. Both spores and cysts can germinate under favorable conditions to yield the typical gliding vegetative cells.

Like the myxobacteria, the SPIROCHETES also have rather thin and flexible walls. They possess a unique structure called the AXIAL FILAMENT. The cells themselves are long rods coiled helically around the filament, which is thought to be responsible for motility of these organisms, but the actual mechanism by which it works is unknown. Many spirochetes are parasites in man, including the organism that causes syphilis.

The EUBACTERIA, sometimes called true bacteria, unlike the other two groups possess a thick and relatively stiff cell wall. Though many are not motile, some can move by means of flagella, whiplike filaments that extend singly or in tufts from one or both ends of the cell, or all around it (*Figure 3*). The flagella of higher organisms — plant, animal, or protozoan — are monotonously alike: they consist of a hollow cylinder of nine pairs of fibrils surrounding two central fibrils (*Chapter 14*). By contrast, the bacterial flagellum consists of a single fibril of contractile protein. Thus flagellum architecture provides another major dif-

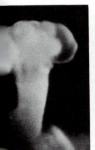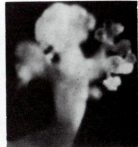

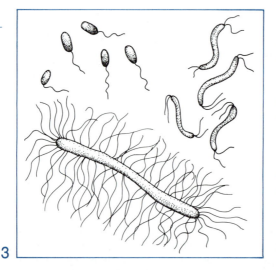

3

FLAGELLATED BACTERIA include Pseudomonas (upper left), which has a single flagellum; a spirillum (upper right) with tufts of flagella at both ends; and Proteus vulgaris, which has many flagella. Bacteria are not drawn to same scale.

ference between procaryotic and eucaryotic cells.

The eubacteria encompass an enormous number of species, a wide variety of structural forms, and a truly incredible array of biochemical tricks for obtaining energy. Some of these tricks have been mentioned in previous discussions of autotrophy and photosynthesis, and it is necessary here merely to note the extraordinary diversity of organic molecules that can serve both as carbon and energy sources for some hungry bacterium. At one time bacteriologists smugly stated that there was perhaps no carbon compound — natural or man-made — that could not be metabolized by at least one strain of bacteria. Only recently

have they found, to everyone's distress, that many detergents and pesticides are in fact relatively poor carbon and energy sources for either bacteria or fungi, and thus can accumulate in noxious amounts in the landscape. About the only important energy-related biochemical process one cannot find among these organisms is aerobic photosynthesis.

Eubacteria characteristically exist in one or another of three different shapes (Figure 4). There are spheres (coccus, plural cocci), rods (bacillus, plural bacilli), and helical forms (spirillum, plural spirilla). The cocci can occur singly, or in two or three dimensional arrays of chains, plates, or blocks of cells. Bacilli and spirilla occur singly or in chains, but the chains do not really signify multicellularity; each cell is fully viable and independent. Most bacteria reproduce simply by the fission of one cell into two. Chains arise merely by the accidental adhesion of cells after fission. A few bacteria reproduce by a process known as budding. Instead of simple fission to produce two daughter cells of equal size, these bacteria sprout a small protuberance at one end. This protuberance gradually increases in size, and ultimately separates as a new and independent cell.

The nonphotosynthetic eubacteria also include the RICKETTSIAE (singular, rickettsia) and related organisms. These extremely small parasites have never been cultured outside of living cells. Rickettsiae are agents of several serious diseases in man, including Rocky Mountain spotted fever and typhus, and are frequently carried by arthropods, particularly ticks and fleas.

This survey of the eubacteria appropriately closes with two groups of PHOTOSYNTHETIC BACTERIA: green and purple. Unlike the higher photosynthetic organisms and bluegreen algae, they lack chlorophyll *a* as their key photosynthetic pigment, having instead either bacteriochlorophyll (purple bacteria) or chlorobium chlorophyll (green bacteria). They have no difficulty in growing in stagnant water beneath fairly dense layers of algae, because the wavelengths of light that they need are much longer than those absorbed by the algae. Since these organisms can fix nitrogen, they may be very important ecological components of stagnant water.

BLUEGREEN ALGAE (PHYLUM CYANOPHYTA)

Of all known organisms, these procaryotes are without question the most independent nutri-

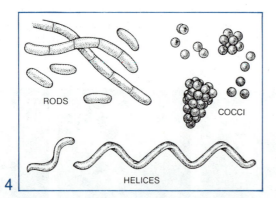

RODS

COCCI

HELICES

4

SHAPES OF BACTERIA. Bacterial cells occur in three basic forms, either singly or in groups. Rod shape is most common.

tionally. They perform aerobic photosynthesis (using chlorophyll *a*), they can respire aerobically, and many can fix nitrogen on a large scale. They require only a few mineral elements, water, nitrogen gas, oxygen, and carbon dioxide.

Bluegreen algae are either free-living, colonial, or filamentous. Some filamentous forms show differentiation into at least three different cell types: vegetative cells, spores, and heterocysts (*Figure 5*). The heterocysts are the site of nitrogen fixation. All of the known bluegreen algae with heterocysts fix nitrogen, and the forms that lack heterocysts do not.

Filamentous forms reproduce by fragmentation (breaking at a vegetative cell or at a heterocyst) to form shorter strands of cells. In spore formers, the spores are simply released, remain dormant for a while, and eventually germinate to produce new strands of cells. Although sexual reproduction is encountered among bacteria, it is absent in the bluegreen algae. Recently viruses have been discovered that can infect bluegreen algae and then transfer genetic material from one alga to another by transduction, but true sexuality has never been observed.

THE VIRUSES

As pointed out in Chapter 1, this group of very important "organisms" brings up the most difficult problems in the entire art of classification, and challenges the best ideas about the dividing line between the living and nonliving. Chapter 1 pointed out that viruses do not carry out the life processes characteristic of living organisms. It is perhaps best to consider viruses not as organisms, but as interesting and frequently dangerous biochemicals. They can even be crystallized like common table salt.

What, then, is the "life history" of a virus? Viruses are obligate parasites, if indeed a complex biochemical can really be called a parasite. The most widely studied and perhaps most complex viruses known are those that infect bacterial cells: the bacteriophages, or phages for short. The cycle of phage infection

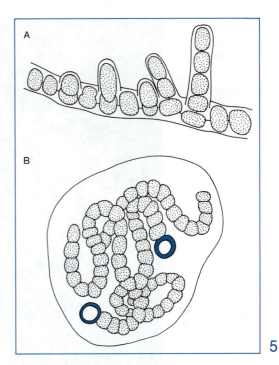

5

FILAMENTOUS BLUEGREEN ALGAE are often embedded in a slimy sheath, shown here in outline. Drawings depict Haplosiphon (A) and Nostoc (B). Nostoc has two types of cell; the colored ones are heterocysts, thought to be site of nitrogen fixation.

and reproduction has been mentioned in previous chapters.

The fundamental particle of the virus, perhaps analogous to the cell, is the VIRION, which consists of a central core either of DNA or RNA surrounded by a coat composed of one or at most a few different kinds of proteins. The reader will recall that the bacteriophage T4 consists of a head region of nucleic acid (DNA) surrounded by protein, and a hollow and rather complex tail.

Viruses that infect higher plants and animals, including those that cause poliomyelitis, influenza, and other diseases in man, do not show the clear differentiation into head and tail regions. They are rod-shaped, or perhaps cuboidal, like the geometrically symmetrical heads of the phages (*Figure 6*). Frequently the

It is perhaps best to consider viruses not as organisms, but as interesting and frequently dangerous biochemicals. They can even be crystallized like common table salt.

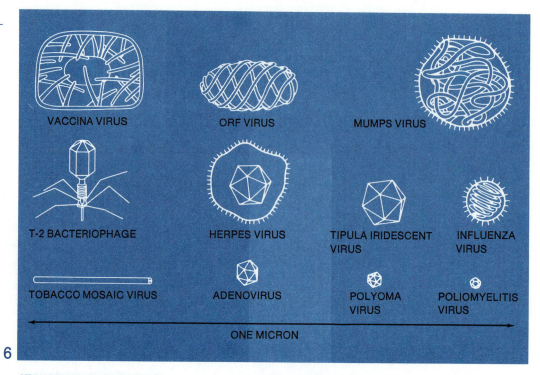

VACCINA VIRUS   ORF VIRUS   MUMPS VIRUS

T-2 BACTERIOPHAGE   HERPES VIRUS   TIPULA IRIDESCENT   INFLUENZA
                                          VIRUS   VIRUS

TOBACCO MOSAIC VIRUS   ADENOVIRUS   POLYOMA   POLIOMYELITIS
                                          VIRUS   VIRUS

ONE MICRON

6

STRUCTURE OF VIRUSES shows a symmetry that reflects their chemical composition. Coats of polyhedral viruses are molecular "boxes" fashioned from identical protein subunits. Rod-shaped tobacco mosaic virus is a helical rod (helix not shown in drawing). Internal components of mumps and influenze viruses also show helical symmetry. Other viruses display a more complex symmetry. Spiked outlines of mumps, herpes and influenza viruses indicate their assymetrical protein coats.

entire virion enters the host cell, including the viral protein. Although the coat protein may play a role in virus replication, it is not essential because in both plant and animal viruses the isolated nucleic acid is fully capable of initiating infection.

The distribution of viruses in terms of host organisms is odd. Viral diseases of flowering plants (angiosperms) are very common, but they are rare in the cone-bearing seed plants (gymnosperms), lower vascular plants (such as ferns), as well as algae and fungi. Almost all vertebrates are susceptible to viral infection, but among invertebrates such infections are common only in the arthropods. A possible explanation of this curious selectivity is that viruses originated from cells as detached pieces of genetic material. These pieces, while still within the cells, somehow acquired the facility to replicate faster than the rest of the genetic material. Upon release, perhaps at cell death, such pieces might enter or be engulfed by neighboring cells, there to repeat their rep-

licative cycle. If this hypothesis is correct the biological specificity of viral infection is no longer a mystery; only those organisms whose detached genetic material became viruses would be subject to reinfection by a nucleic acid fragment bearing some resemblance to their own. Until a new virus can be "created" by this method in the laboratory, however, such reasoning must be regarded as highly speculative.

THE KINGDOM PROTISTA

In an important sense the phylum MAS-TIGOPHORA, the flagellates, can be regarded as the most fundamental of all the eucaryotic phyla. If every trace of life on Earth were removed except the members of this one group, a large percentage of the species would probably survive — or to be more precise, they would survive so long as a supply of fixed nitrogen was available (the phylum contains no nitrogen-fixers). Not only would they sustain life entirely by themselves, but they would

Because of their extraordinary versatility in nutrition, the flagellates are said to bridge the gap between plants and animals at the unicellular level. And because some of them form large and well-organized colonies, they also span the gap between unicellular and multicellular organisms.

provide a favorable starting point from which both plants and animals could re-evolve.

Figure 7 depicts a *Euglena*. Like most other members of its phylum, this common freshwater form possesses a relatively elementary cell plan and a well-formed nucleus. It propels itself through the water with one of its two flagella, which sometimes doubles as an anchor to hold the organism in place. *Euglena* reproduces by mitosis — the simplest and most direct way possible. It has very flexible nutritional requirements. In sunlight it is fully autotrophic, using its chloroplasts to synthesize organic compounds from either organic or inorganic sources. When kept in the dark, the organism loses its photosynthetic pigment and begins to feed exclusively on dead organic material floating in the water around it.

Because of their extraordinary versatility in nutrition, the flagellates are said to bridge the gap between plants and animals at the unicel-

lular level. Some relatively large green colonial flagellates appear quite similar to some of the higher plants. Other flagellates resemble tiny animals. An impressive diversity of animal-like "zooflagellates" (a label that distinguishes them from the plantlike "phytoflagellates") live as internal parasites of larger animals, including man. Some idea of the diversity of this phylum is conveyed by Figure 8.

The flagellates also span the gap between single-celled and many-celled organisms. Surprisingly large and well-organized colonies of cells are formed in such fresh-water groups as the genus *Volvox*. The cells are not differentiated into tissues and organs as in higher plants and animals, but the colonies show vividly how the preliminary step to this great evolutionary advance might have been taken.

PHYLUM SARCODINA, the amoebas and their relatives, have for generations been portrayed

EUGLENA GRACILIS, a photosynthetic heterotroph, propels itself through the water with the longer of its two flagella. Rudimentary second flagellum does not extend beyond canal at rear of organism. Starch manufactured by photosynthesis is stored as paramylon granules.

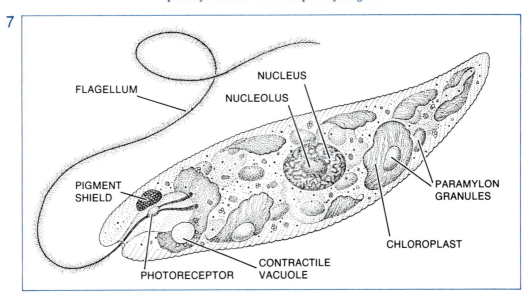

7

FLAGELLUM

NUCLEUS

NUCLEOLUS

PIGMENT SHIELD

PARAMYLON GRANULES

CHLOROPLAST

PHOTORECEPTOR

CONTRACTILE VACUOLE

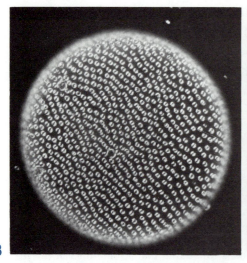
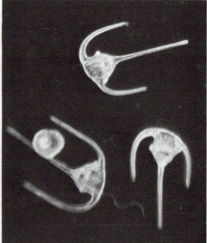

8

DIVERSITY OF FLAGELLATES is illustrated by organisms pictured above. Left: Volvox, a colonial flagellate consisting of a hollow ball of cells. Center: Ceratium tripos, a marine dinoflagellate. Right: cellulose-digesting flagellate from hindgut of woodeating cockroach Cryptocercus. Photomicrographs are not reproduced to the same scale.

in popular writing as blobs of glop — the simplest form of animal life imaginable. The animal consists of a single cell with no definite shape (*Figure 9*). Despite its apparently primitive characteristics, the amoeba is probably not a primeval organism but a rather advanced form of protozoan specialized for life on the bottom of lakes, ponds and other bodies of water. Its simplicity probably arose during the evolution of the Sarcodina from ancestors within the Mastigophora. Some intermediate forms still exist. *Mastigamoeba aspera*, shown

in Figure 10, could be classified in either phylum.

The remaining Sarcodina form an astonishing array of even more specialized forms (*Figure 11*). All are animal-like, existing as predators, parasites, or scavengers. There are SHELLED AMOEBAS that live in casings of sand grains glued together, or in spiny or scaly shells secreted by the animal itself. FORAMINIFERANS are marine creatures that secrete shells of calcium carbonate. The shells of individual species have distinctive shapes, and they are easily preserved as fossils in marine sediments. Each geological period had its own distinctive species. For this reason, plus the fact that they are so abundant, foraminiferan fossils are especially valuable as indicators in the classification and dating of

9

AN AMOEBA. Shape of this organism constantly changes as it extends pseudopods (right) to move from place to place or to engulf food.

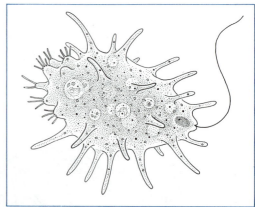

10

MASTIGAMOEBA has pseudopods like the Sarcodina and a flagellum like the Mastigophora. It can legitimately be assigned to either phylum.

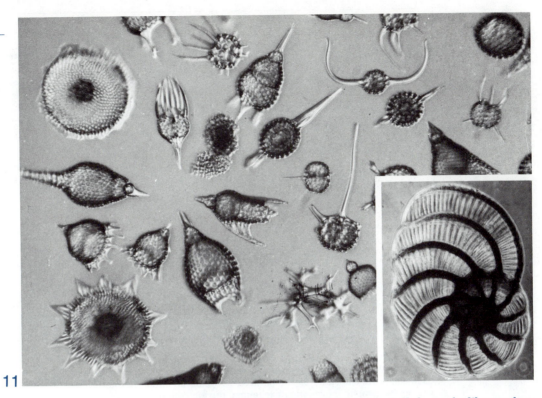

11

RADIOLARIA AND FORAMINIFERA are shelled, amoeba-like members of the phylum Sarcodina. Siliceous skeletons of radiolarians (large photo) display an extraordinary diversity of architecture. Tiny pores in surface are openings through which the living organisms extended pseudopods. Magnification, about 150×. Foraminifera secrete chalky shells (inset). Magnification, about 75×.

sedimentary rocks, and they also serve as indicators in oil prospecting. HELIOZOANS are fresh-water forms surrounded by a bristling array of long pseudopods. Like the foraminiferans, they drift in the water and use their pseudopods to trap smaller organisms. A third group with the same feeding technique are the RADIOLARIANS. Found exclusively in the sea, they are perhaps the most beautiful of all microorganisms.

The SPOROZOANS (Phylum Sporozoa) are exclusively parasitic protozoans whose name derives from the fact that some of them produce sporelike infective stages. Sporozoans generally have an amoeboid body form, but this in no way indicates a relationship to the Sarcodina. The development of the trait is common among parasitic protozoans. The sporozoans, like many parasitic forms among the higher animals, display elaborate life cycles featuring asexual and sexual reproduction by a series of very dissimilar life stages. The best

BLOOD OF MALARIA VICTIM shows presence of plasmodia in human red cells (far right and lower left). Crescent-shaped cells are gametes of the parasite, which infect Anopheles mosquito.

12

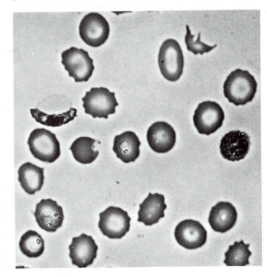

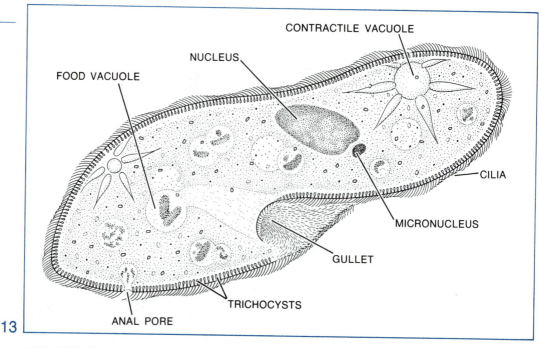

ANATOMY OF PARAMECIUM, as seen under the light microscope, exhibits the complexity of structure characteristic of ciliates. Undischarged trichocysts are embedded in the pellicle, between the cilia. Micronucleus carries the genetic information exchanged during conjugation.

known sporozoan is the malaria parasite *Plasmodium*, a highly specialized organism that spends part of its life cycle within vertebrate red blood cells (*Figure 12*).

PHYLUM CILIOPHORA, the ciliates, ranks with the flagellates as the most diverse and ecologically important of the protozoan phyla. Ciliates are all animal-like in nutrition, and are much more specialized in body form than most flagellates and other protists. They are characterized by the possession of hairlike CILIA which have the same cablelike ultrastructure as flagella and are believed to have evolved from them.

DISCHARGED TRICHOCYST consists of needlelike tip and a banded filament. In the laboratory, Paramecium ejects trichocysts when disturbed. In nature, Paramecium probably uses these "harpoons" to defend itself against predators and to anchor itself in place. Magnification, about 20,000×.

The ciliates represent the zenith of evolution among unicellular organisms. Their astonishing complexity of structure and behavior is exemplified by *Paramecium*, the most famous and thoroughly studied member of the phylum (*Figure 13*). Its slipper-shaped body is covered by an elaborate pellicle, a skin composed principally of an outer membrane and an inner layer of closely packed, kidney-shaped structures called alveoli that embrace the cilia. Also present as a layer of the pellicle are the unique defensive organelles called trichocysts. Expelled by a microscopic explosion in a few thousandths of a second, the trichocysts emerge as sharpened darts driven forward at the tip of a long shaft (*Figure 14*).

The cilia provide a form of locomotion that is generally superior to that made possible by flagella and pseudopods. (Some of the larger

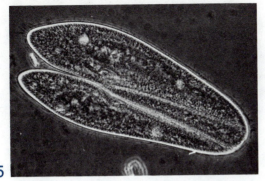

15

CONJUGATING PARAMECIA align themselves tightly side by side, and their oral grooves fuse. After exchanging genetic material (micronuclei), the two organisms separate and depart.

ciliates hold the speed record for protists — over 2 millimeters per second.) Paramecia usually reproduce by cell division, but they also reproduce sexually. During CONJUGATION two paramecia line up tightly against each other and their oral regions fuse (*Figure 15*). During the next several hours there is an extensive reorganization and exchange of nuclear material. The exchange is perfectly reciprocal — each of the two paramecia gives and receives an equal amount of genetic material. The two organisms then pull away and depart, having been genetically refreshed. As a rule, each familial line of paramecia must periodically go through the process of conjugation. It has been shown by laborious experimentation that if some species are not permitted to conjugate, the asexual lines can live through no more than about 350 cell divisions before they die out.

Most ciliates possess all of the traits just described for *Paramecium*. Some, however, are especially notable for the exceptional degree of development of individual organelle systems. Certain of the hypotrich ciliates, for example, have the equivalent of legs. Fused cilia called CIRRI move in an independent but coordinated fashion, enabling the animal to walk over surfaces. This degree of coordination is made possible by nervelike neurofibrils that lead to individual cirri. When these are cut in laboratory experiments, the coordination is lost. Many kinds of ciliates possess myonemes, musclelike fibers within the cytoplasm. Possibly the ultimate in cytoplasmic

organization is displayed by highly specialized ciliates that live in the digestive tracts of cows and many other hooved animals. They possess not only myonemes, neurofibrils and elaborately fused cilia, but also a cytoplasmic "skeleton" and a "gut" complete with mouth, esophagus, and anus. When examining the intricate structure of one of these organisms, the observer may have to stop and remind himself that he is looking at only one cell.

READINGS

R.D. BARNES, *Invertebrate Zoology*, Philadelphia, Saunders, 3rd Edition, 1974. One of the best textbooks in the field, containing an excellent brief review of the protozoans.

T.D. BROCK, *Biology of Microorganisms*, Englewood Cliffs, N.J., Prentice-Hall, 2nd Edition, 1974. An attractively produced and very readable account of microorganisms, with emphasis on the bacteria.

H. CURTIS, *The Marvelous Animals: An Introduction to the Protozoa*, Garden City, New York, The Natural History Press, 1968. An excellent popular account of the protozoans, with enough detail to satisfy almost anyone but a research specialist.

A. JURAND, AND G.G. SELMAN, *The Anatomy of Paramecium Aurelia*, New York, St. Martin's, 1969. A modern treatise of the anatomy of a protist. Remarkable photographs of the ultrastructure of organelles offer a view of the phenomenal complexity that prevails even at this microscopical level of biological organization.

G.F. LEEDALE, *Euglenoid Flagellates*, Englewood Cliffs, N.J., Prentice Hall, 1967. An authoritative treatise of this fascinating group of plant-animals, including summaries of much contemporary experimental work.

W.R. SISTROM, *Microbial Life*, New York, Holt, Rinehart, & Winston, 2nd Edition, 1969. A brief and well-written survey of bacteria and viruses.

R.Y. STANIER, E.A. ADELBERG, AND J.L. INGRAHAM, *The Microbial World*, Englewood Cliffs, N.J., Prentice-Hall, 4th Edition, 1976. Perhaps the best introductory textbook on microbiology, containing a review of the broad principles of classification and the biology of each of the major groups.

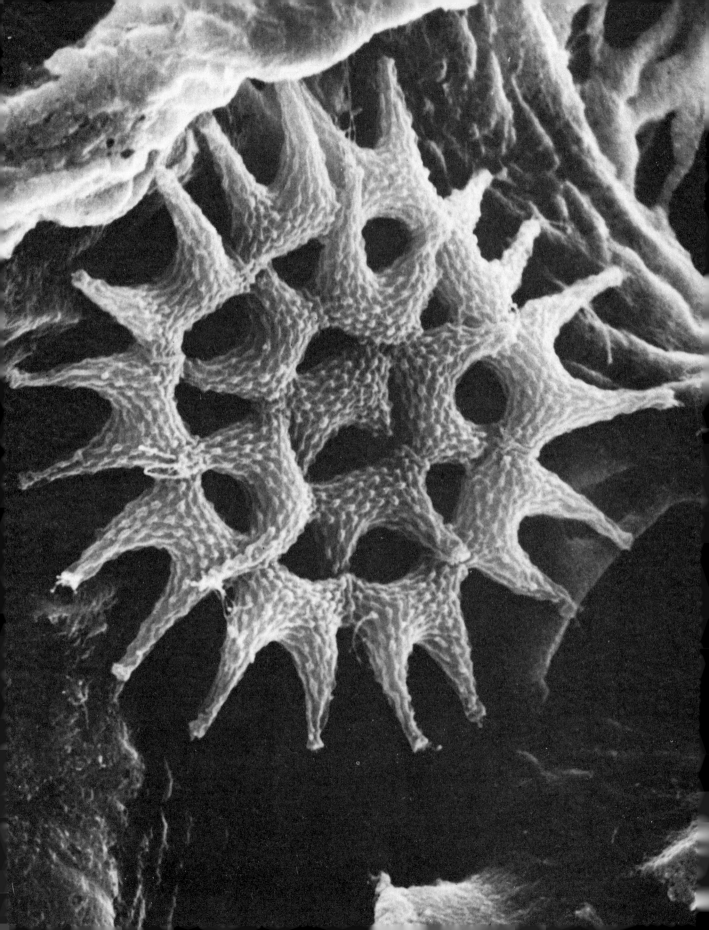

19

fungi and plants

We have not to deal with a directly ascending series of organic forms which advance from the lower and simpler to the higher and more complex, but we must rather conceive of the Vegetable Kingdom . . . as a copiously branching tree, whose boughs have their origin in a common stem, but stand in no direct communication with each other.

K. GOEBEL (1882)

The evolutionary steps that led from single-celled organisms to multicellular fungi and plants produced new strategies of existence. Monerans and protistans diversified to exploit the opportunities open to microscopic organisms. Conversely the fungi and plants diversified to exploit the opportunities open to large organisms. The fungi are important degraders of dead organic matter; they probably never possessed the ability to photosynthesize or to live autotrophically in any other way. The higher plants, including the algae, stand in sharp contrast. Virtually all are capable of photosynthesis, and they comprise the overwhelming bulk of autotrophic organisms on the planet today.

THE KINGDOM FUNGI

Most people think of fungi as sprawling globs of living matter that occupy a very low place in the evolutionary order of things. The truth is nearly at the opposite extreme. Most fungi are exquisitely constructed, and their life cycles are among the most complex to be found anywhere in nature. Their great diversity makes them difficult to define. Some workers consider the fungi simply as a category of nonphotosynthetic plants. But we feel that a group that includes slime molds, molds, yeasts, rusts, and mushrooms is sufficiently different from plants to comprise a separate kingdom.

The higher fungi all follow a characteristic pattern of filamentous growth, with individual filaments or HYPHAE becoming enmeshed into a cottony mass called a MYCELIUM. The mycelium is sometimes organized into elaborate fruiting structures such as puff balls or mushrooms. The cell walls of fungi contain a variety of polysaccharides, but usually also contain CHITIN, the same polymer that hardens the exoskeleton of insects and other arthropods, and/or cellulose, characteristic of the cell walls of higher plants. Lower fungi may lack true cell walls in all phases of their life cycle except for the spore stage.

Protists can be readily classified into phyla on the basis of cell structure. There is certainly little difficulty in distinguishing a flagellate such as *Euglena* from a ciliate such as *Paramecium* or from an *Amoeba* (Chapter 17). But such dramatic variation in cell structure occurs rarely in the fungi; instead the phyla are based on the method and structures associated with sexual reproduction. Three phyla of higher fungi (Phycomycota, Ascomycota, Basidiomycota) can be readily distinguised in this way. But a large number of fungi have no sexual stages; presumably these stages have been lost in evolution. It thus becomes difficult to classify them with any of the three major phyla. Fungi without a taxonomic home are simply dumped into an orphanage known as the FUNGI IMPERFECTI.

[A note on classification: Most botanists prefer the word DIVISION to the word PHYLUM; in this book the term phylum is used throughout. Also, a few authors still follow the older practice of treating the phyla

COLONIAL ALGA Pediastrum boryanum belongs to a family of freshwater green algae called Chlorococcales. Photomicrograph on opposite page, made with the scanning electron microscope, depicts a colony of cells at a magnification of 6100×.

PLASMODIUM OF SLIME MOLD on a decaying log. Slime mold spreads across its substrate by the process of cytoplasmic streaming. Plasmodium engulfs bacteria and other small organisms.

divide mitotically several times before mating, or can function directly as gametes. Two swarmers fuse to form a diploid zygote, which divides by mitosis to form a new plasmodium.

In a sense the slime mold is an all-purpose fungus. Its life cycle contains stages resembling not one but three other major types of organism. First, the plasmodium looks very much like a giant multinucleate amoeba. Second, the formation of a definite fruiting structure containing chitin or cellulose is a fungal characteristic never found in the protozoa. And third, the flagellated swarm cells superficially resemble close relatives of *Euglena*.

The cellular slime molds are a small but interesting group. One member, *Dictyostelium discoideum*, has already been discussed in some detail in Chapter 8. Instead of a plasmodium, the basic vegetative unit of the cellular slime molds is an amoeboid cell called a myxamoeba. Hordes of these cells engulf bacteria and other food particles, reproducing by mitosis and fission. So long as food and moisture are available, the horde of myxamoebas persists indefinitely. When conditions become stringent, the cellular slime molds, like their acellular counterparts, form fruiting structures. The apparently independent myxamoebae aggregate into an irregular mass called a PSEUDOPLASMODIUM. Unlike the true plasmodium of the acellular slime molds, this structure is not simply a giant lump of cytoplasm: the individual myxamoebas retain their cell membranes and therefore their identity. Before forming a fruiting structure, some of the amoeba and their nuclei fuse and undergo meiosis. Like their acellular cousins, the cellular slime molds have stages resembling more than one other phylum. The myxamoebas resemble protozoans, but the fruiting structure is clearly fungal. Moreover, the only other organisms with the same kind of life cycle are the procaryotic myxobacteria (Chapter 18).

of fungi as classes of a single phylum (phylum Fungi). We have placed the names of these classes (for example class Myxomycetes, which is synonymous with the phylum Myxomycota) in parentheses to help the reader make cross references to textbooks that still use the older system.]

Cellular and acellular SLIME MOLDS comprise the two major groups of lower fungi, and belong to the phylum Myxomycota. During most of its life history, an acellular slime mold exists as a wall-less mass of protoplasm which streams over its substrate in a remarkable network of strands called a PLASMODIUM (Figure 1). As the plasmodium spreads over its substrate it engulfs food particles, predominantly bacteria and other small organisms. With enough food and moisture, an acellular slime mold can grow almost indefinitely in its plasmodial stage. Under harsher conditions one of two things can happen: the plasmodium can shrivel into a hardened mass, which rapidly becomes a plasmodium again upon return of favorable conditions; or the plasmodium can transform itself into spore-bearing fruiting structures called SPORANGIA (Figure 2).

The plasmodium is a diploid structure, and during the development of the sporangium its cells reproduce by meiosis. One or more knobs may develop on the end of the stalk, and the outer nuclei wall themselves off and become spores. Eventually the spore-bearing sporangium dries up and the spores are shed. They germinate into wall-less flagellated haploid cells that can either lose their flagellae and

FILAMENTOUS FUNGI

FILAMENTOUS FUNGI (phylum Phycomycota) have characteristics that enable the taxonomist to say, conclusively, "These plants are obviously fungi." They have a cell wall composed either of chitin or cellulose or both. Their long and elaborately branched hyphae may form enormous mycelial mats. Despite these characteristic fungal traits, phycomycetes sometimes lack cross walls, which suggests that they are not truly cellular organisms. Even when cross

SPORANGIA of the slime mold Physarum arise from heaped masses of plasmodium. Nuclei of outlying cells of knobs develop into spores.

walls exist, they contain large perforations through which extensive cytoplasmic streaming occurs. Except for certain reproductive stages, there is no single structural unit that can be called a cell.

The simplest phycomycetes are the chytrids. These organisms, many of which are parasites on aquatic plants, may have no more than one or a few nuclei. Short filaments of cytoplasm digest their way into the substrate, which is either dead organic matter or living plant cells. The chytrids reproduce vegetatively by mitosis, producing numerous flagellated ZOOSPORES. (Many fungi and algae reproduce vegetatively from zoospores, both from haploid and diploid phases of the life cycle.) Sexual reproduction, which has been observed in only a few species, involves conjugation similar to that observed in bacteria and protozoans. One organism produces a slender tube which grows toward another and fuses with it. A nucleus migrates across, and nuclear fusion produces a zygote.

A phycomycete which the reader has certainly seen at one time or another is the black bread mold *Rhizopus stolonifer*. The mycelium creeps over the substrate, growing forward by means of specialized hyphae (Figure 3). Large numbers of sporangiophores are produced, each bearing a single sporangium containing many hundreds of spores. Sexual reproduction is by conjugation, with hyphae of adjacent different strains coming together to fuse and form heavy-walled zygotes.

ASCOMYCETES (phylum Ascomycota) comprise a large and diversified group of fungi distinguished by a unique saclike structure called the ASCUS. Both nuclear fusion and subsequent meiosis take place within individual asci. The resulting ascospores are eventually shed to begin the new gametophyte generation. One member of this phylum, the common bread mold *Neurospora*, is already familiar to the reader as one of the favorite organisms for experimental genetics (Chapter 5).

The ascomycetes can be divided into two broad groups. Those species that enclose their asci within a specialized fruiting structure are collectively called the euascomycetes, while

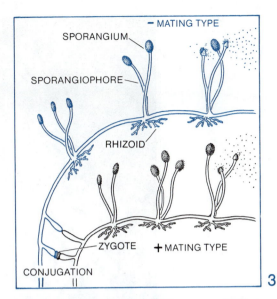

BLACK BREAD MOLD (**Rhizopus stolonifer**) spreads across bread by extending hyphae like the two above. Rhizopus reproduces asexually by shedding spores, or sexually by conjugation.

those that do not are called the hemiascomycetes. Hemiascomycetes are very small in general, and frequently are unicellular. Perhaps the best known are the yeasts, especially baker's or brewer's yeast (*Saccharomyces cerevisiae*). Other yeasts occur naturally on fruits such as figs and, in particular, grapes, and play an important role in the making of wine. Of all the fungi the yeasts are undoubtedly the most important domesticated ones. The yeasts multiply either by simple fission or, in the better known genera, by a process of budding similar to that found in the budding eubacteria. The single cells are haploid.

Among the euascomycetes, the second major group of ascomycete fungi, are several common molds: *Aspergillus*, *Penicillium*, and *Neurospora*, the brown, green, and pink molds, respectively. All three are ubiquitous, frequently occurring on old bread, and are common laboratory contaminants. They reproduce asexually by producing spores.

Penicillium is the organism which produces the antibiotic penicillin, presumably to defend

Photosynthetic plants, the primary producers of organic matter in all food chains, probably outweigh all other organisms by a factor of ten.

4

BRACKET FUNGI grow on tree trunks and dead logs, and produce large fruiting structures like those above. Spore-forming basidia fill pores on one surface of the fruiting structure.

itself from competing bacteria. The two species *Penicillium camemberti* and *Penicillium roqueforti* are the organisms responsible for the characteristic flavor of Camembert and Roquefort cheeses. Not surprisingly, people who are hypersensitive to penicillin frequently react violently if they make the mistake of eating one of these cheeses.

A large number of the euascomycetes are serious parasites on higher plants. The powdery mildews are a group that infects cereal grains, lilacs, and roses, to name but a few host plants. They may also be a serious problem to grape growers, and a great deal of research has been done on ways to control these important agricultural pests.

The euascomycetes also include the BASIDIOMYCETES, some of which produce the most spectacular fruiting structures found anywhere in the fungi. There are puffballs up to two feet in diameter, mushrooms of all kinds (including the poisonous toadstools), and giant bracket fungi often encountered on trees and fallen logs in a damp forest (*Figure 4*). Some basidiomycetes are among the most serious plant pathogens, including the wheat rust (*Puccinia graminis*) and the smut fungi that parasitize cereal grains.

The organization of the elaborate fruiting structure of a higher basidiomycete is illustrated in Figure 5, a diagrammatic cross section of one of the gill mushrooms. On the under side of the cap (or pileus) are found the

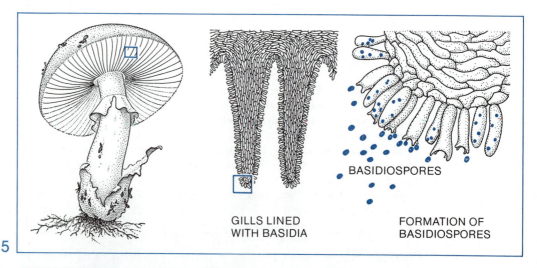

BASIDIOSPORES

GILLS LINED
WITH BASIDIA

FORMATION OF
BASIDIOSPORES

5

GILL MUSHROOMS are basidiomycetes. Gills on underside of fleshy cap are lined with basidia. Tiny spores are dispersed by wind. Small colored square shows area of next section to right.

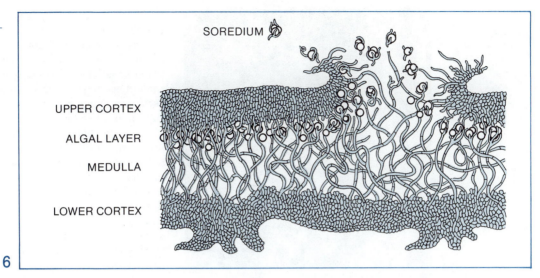

SOREDIUM

UPPER CORTEX

ALGAL LAYER

MEDULLA

LOWER CORTEX

6

CROSS SECTION OF LICHEN THALLUS shows fungal components in color and algal cells in black and white. Upper cortex is protective surface. Algal layer is site of photosynthesis. Medulla apparently serves as food-storage area. Filaments on lower cortex anchor lichen in place. Soredia are specialized reproductive structures consisting of one or a few algal cells surrounded by fungal hyphae. Once detached from the thallus, they are dispersed by air currents.

gills, and on the sides of the gills are enormous number of BASIDIA. The basidia discharge their spores into the air spaces between adjacent gills, and they sift down into air currents for dispersal to start new haploid mycelia.

LICHENS

This unusual group of organisms defies classification. The reason is very simple: a lichen is a meshwork of two radically different organisms, a fungus and an alga. Together they can survive some of the harshest environments on Earth. The fungus is usually an ascomycete, and the alga may be either a bluegreen or a green species.

Lichens grow in all sorts of exposed habitats: tree bark, open soil, or bare rocks. Both the fungal partner and the alga can readily grow alone, and even within the lichen, algal growth may outstrip fungal or vice versa, depending upon environmental conditions. The relationship is an example of mutually beneficial symbiosis. A cross section of a typical lichen is shown in Figure 6. There is a tight upper region of fungal filaments alone, an algal layer, a looser filamentous fungal layer, and finally filaments that attach the whole structure to its substrate. The whole meshwork has properties that enable it to hold water tenaciously. Some nutrients for the algae enter through the fungal hyphae. The meshwork provides a suitably moist environment for algal growth, and the fungi derive fixed carbon from algal photosynthesis. This mutually beneficial association between two different species is an excellent example of symbiosis.

Lichens can reproduce simply by fragmentation of the plant body, which is called the THALLUS, or else by specialized structures called SOREDIA. The soredia consist of one or a few algal cells surrounded by fungal hyphae. They become detached, move in air currents, and on arriving at a favorable location, once again set up the partnership.

THE KINGDOM PLANTAE

The kingdom of organisms known as plants starts with the algae and ends with the seed plants. Photosynthetic plants, the primary producers of organic matter in all food chains, probably outweigh all other organisms by a factor of ten. In general, plants are defined as eucaryotic photosynthetic organisms. The few plants that are not photosynthetic are clearly closely related to other plants that are. We exclude the bluegreen algae as procaryotes that are considered by some taxonomists to be little more than modified bacteria, and we exclude fungi because none of them is photosynthetic.

7

DIATOMS. Despite their splendid diversity of form, all diatoms show either radial or bilateral symmetry. Magnification about 1200×.

ALGAE

Algae probably carry out more than half of all of the photosynthesis occurring on the planet, with higher plants accounting for most of the rest. The overall contribution of the bluegreen algae and photosynthetic bacteria is relatively small. The algae exhibit a remarkable range of growth forms. Some are simply unicellular; others are filaments comprised either of distinct cells or multinuclear structures without cross walls. A few are multicellular and intricately branched or arranged in multicellular, leaflike extensions. In extreme cases the masses are even subdivided into tissues and organs. Almost all these types can sometimes be found within a single phylum — for example, the green algae, phylum Chlorophyta. Life cycles also show extreme variation, but all except the Rhodophyta (red algae) have forms with flagellated motile cells in at least one stage of their life cycle, and some (for example, the Pyrrophyta) are unicellular and motile throughout most of their existence.

The DINOFLAGELLATES (phylum Pyrrophyta), a group of predominantly unicellular algae, are of interest for two reasons. First, the dinoflagellates are probably second in importance only to the diatoms (members of the phylum Chrysophyta, discussed below) as primary photosynthetic producers of organic matter in the oceans. Second, some of the close relatives of the dinoflagellates, although still true members of the Pyrrophyta, superficially resemble amoebas: they lack cellulose walls, possess contractile vacuoles, and occasionally feed on other organisms.

Dinoflagellates are mostly marine, and sometimes reproduce in enormous numbers in warm and somewhat stagnant waters. Certain species produce a potent nerve toxin, and when they reach high numbers (called a bloom) the resulting "red tide" can kill enormous numbers of fish. A particularly severe red tide in the Gulf of Mexico in the summer of 1971 killed tons of fish along the west coast of Florida.

Many dinoflagellates contain a remarkable biochemical system for generating light when physically disturbed. A ship passing through a tropical ocean containing a rich growth of these species produces a bow wave and a wake that glow eerily as millions of dinoflagellates discharge their light system.

DIATOMS and their relatives comprise the phylum Chrysophyta. Most of them are unicellular, although a few filamentous forms are known. All have a predominance of carotenoids in their chloroplasts, giving them a yellow or brownish color.

Architectural magnificence on a microscopic scale is the hallmark of the diatoms. Living in the ocean or in fresh water, these unicelluular organisms produce silicon-impregnated cell walls which show intricate patterns. The cell wall of some members of all classes is constructed in two pieces like the top and bottom of a box of stationery, with the walls of the top overlapping those of the bottom. The taxonomy of the diatoms is entirely based on the wall patterns (Figure 7). Looking down on the top of the "box," one sees one or the other of two basic shapes: diatoms that are radially symmetrical (approximately circular) or those that are bilaterally symmetrical (not circular, but still divisible into right and left halves that are mirror images of each other). Asexual reproduction is by simple cell division, for which the rigid cell wall presents a serious problem: the wall cannot increase in size. The consequences of this problem are illustrated in Figure 8. Both the top and bottom of the "box" become tops of new "boxes" and, as a result, the new cells which started as bottoms must be smaller. If the process continued indefinitely, one cell line would simply vanish. Sexual reproduction largely solves the problem. A given cell forms a pair of gametes, the cell walls are shed, and the gametes fuse.

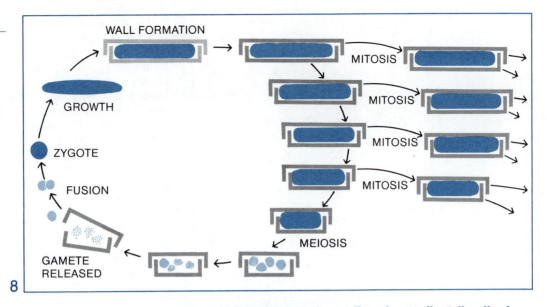

WALL FORMATION

MITOSIS

MITOSIS

MITOSIS

MITOSIS

MEIOSIS

GROWTH

ZYGOTE

FUSION

GAMETE RELEASED

8

DIATOMS REPRODUCE both sexually and asexually. Cell walls, shown edge-on in this schematic diagram, are two-part "boxes." In asexual reproduction (right) the parts separate and each becomes the top of a new box. In the process the offspring within the shell become progressively smaller. Zygotes produced by sexual reproduction (left) grow and lay down new full-size shells.

The zygote increases substantially in size before it lays down a new wall.

Dinoflagellates and diatoms are among the tiniest algae. The BROWN ALGAE (phylum Phaeophyta) are undoubtedly the largest. Some giant kelps, such as *Macrocystis*, may be over 100 feet long. Although they exhibit a variety of life cycles, the brown algae are always multicellular, composed either of branched filaments or leaflike growths called thalli. Commonly called seaweed, these organisms are almost exclusively marine. Some are found floating in the open ocean; the most famous example is *Sargassum*, which forms dense mats of vegetation in the Sargasso Sea in the mid-Atlantic. Most, however, are attached to rocks near the shore. The attached forms all develop a specialized structure called a HOLDFAST, which literally glues them to the rocks. The plant may show extensive differentiation into stemlike stalks and leaflike blades, and many have gas-filled flotation bladders. For biochemical reasons that are only poorly understood, these gas cavities often contain as much as five percent carbon monoxide — enough to be fatal to a canary.

The RED ALGAE (phylum Rhodophyta) include plants that grow near the edges of the sea and also on the bottom. The deepest plants found in the ocean are red algae; very few inhabit fresh water. Most grow attached to some substrate by a holdfast. Almost all of them are multicellular. In a sense, red algae are misnamed. They have the capacity to change the relative amounts of the various photosynthetic pigments depending upon the light conditions where they are growing. Thus the leaflike *Chondrus crispus*, a common northeastern species, may appear bright green when it is growing unshaded in a tide pool, reddish brown when it is growing under other vegetation, and deep red when it is growing at extreme depths. The pigmentation reflects to a remarkable degree the quality of the light which reaches the plants. In deep water, the light penetrating the ocean is mostly in the blue-green part of the spectrum, and the plants accumulate large amounts of phycoerythrin, the pigment which absorbs light of those wavelengths.

Like coral animals, red algae are very important in the formation of tropical reefs. Certain species share with corals the biochemical machinery for depositing calcium carbonate in and around their cell walls. After the death of the cells the calcium carbonate often forms chunks of rocklike material and also acts as a cement that holds the whole reef together.

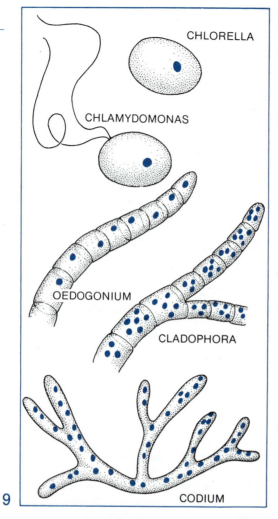

CHLORELLA

CHLAMYDOMONAS

OEDOGONIUM

CLADOPHORA

CODIUM

9

CELLS OF GREEN ALGAE display a wide diversity of form. Chlorella is one of the simplest. Cladophora consists of multinucleate cells. Codium, the largest green alga known, has no transverse cell walls except in its reproductive phase.

The GREEN ALGAE (phylum Chlorophyta) contain the photosynthetic pigments characteristic of the higher plants. Chlorophyll *a* predominates, and a major pigment is chlorophyll *b*, which occurs in none of the other algae. The carotenoids, predominantly β-carotene and certain xanthophylls, are likewise those characteristic of higher plants. The principal photosynthetic storage product is also familiar: long straight chains of glucose (*amylose*) or long branched chains (*amylopectin*) which together comprise starch.

Uniformity of pigmentation and photosynthetic storage product is combined in the green algae with an incredible variety in shape, construction of plant body, and life cycle (*Figure 9*). *Chlorella* is an example of the simplest type: single-celled, immobile, and apparently asexual. *Chlamydomonas* is more complex: unicellular, flagellated, and sexual. *Pediastrum* (*illustrated on the opening page of this chapter*) is colonial but lacks motility. *Oedogonium* is filamentous and sexual, with each cell uninucleate. *Cladophora* resembles *Oedogonium* except that each cell is multinucleate. *Codium* is tubular and coenocytic, forming cross walls only during the development of reproductive structures. *Acetabularia* is a single giant cell with only one nucleus. Finally, *Ulva* is a membranous sheet two cells thick whose unusual appearance justifies its common name of sea lettuce.

Sexual reproduction among green algae runs a long gamut. Some forms reproduce by conjugation; others shed free-swimming flagellated gametes. The plants may be haploid, with meiosis occurring as the first division of the zygote (*Chlamydomonas*), or diploid, with meiosis occurring just before gamete formation (*Bryopsis*). Given the broad range of morphology, cellular construction and life cycles, the reader can perhaps see why biologists are forced to rely on biochemical characteristics when defining the major groups of algae.

THE HIGHER PLANTS

All green multicellular land plants belong to two major phyla: *Bryophyta* (mosses, liverworts, and hornworts) and *Tracheophyta* (vascular plants). Like insects and bacteria, tracheophytes represent one of the triumphs of evolution. They include virtually all of the familiar plants, trees, and shrubs that now dominate the landscape. Probably descended from ancestral green algae, the higher plants first invaded the land in the Silurian period of the Paleozoic era, about 415 million years ago (the 28th day of our geologic calendar).

Most of the characteristics that distinguish

Most of the characteristics that distinguish the higher plants from the algae are adaptations to life on land.

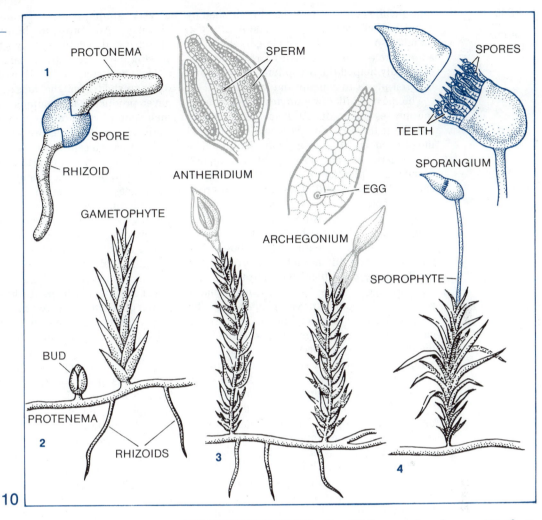

LIFE CYCLE OF MOSSES begins with the sprouting of a spore (A). Shoots of leafy gametophyte plant develop sex organs at their tips, either male antheridia or female archegonia (C). In some species both sex organs occur on the same plant; in other species, on separate plants. After fertilization the egg develops into a sporophyte plant, which is always attached to the tip of the gametophyte. The sporophyte forms a capsule-like sporangium (D), which eventually bursts. The cap disintegrates, and curved "teeth" around the rim fling the spores outwards.

the higher plants from the algae are adaptations to life on land. A woody skeleton provides support against gravity in the absence of the buoyancy of water. No longer bathed, higher plants evolved specialized absorbing tissues, such as roots that project into the soil; specialized conducting tissues to carry minerals and nutrients throughout the plant; and protective coverings to prevent drying out. The reproductive patterns of higher plants also show modifications for life on land, culminating in the invention of the seed.

PHYLUM BRYOPHYTA

Mosses, hornworts, and liverworts are members of an ancient phylum that probably arose in or before the Devonian period. Exactly what the ancestral algae that gave rise to them might have looked like is anyone's guess. The bryophytes have achieved an important evolutionary advance in their sexual cycle. Unlike the algal zygote, which is wholly on its own following fertilization, the bryophyte zygote divides and goes through a distinctive embryo phase while being protected and nourished by a specialized organ called the ARCHEGONIUM, which also shelters the female gametes. This added protection for gametes and zygote undoubtedly helped the bryophytes to be among the first plants to conquer the land.

The most familiar bryophytes are, of course, the mosses (class Musci). It is worth looking closely at the reproductive cycle of a typical moss such as *Mnium* or *Polytrichum*. The spores germinate and develop into a branched filamentous plantlet or PROTONEMA, an organism that looks quite like a branched green alga. Filaments called RHIZOIDS anchor the protonema to the substrate. After a period of growth, cells close to the tips of the photosynthetic branches begin to form buds. The buds eventually differentiate and produce the familiar leafy moss plant with the leaves spirally arranged (*Figure 10*). These leafy plants comprise the gametophyte generation. The gametes develop in special organs, the female gametes in archegonia (mentioned above) and the male gametes in ANTHERIDIA. After fertilization, the zygote of the sporophyte generation develops an absorptive foot, a stalk, and at its tip, a swollen sproangium or capsule.

During the rigidly programmed development of the sporophyte, the archegonial tissue itself is also growing rapidly, and for a time it keeps pace with the rapidly expanding sporophyte. Finally, however, the archegonium loses out and splits apart. The top of the capsule is ultimately shed, exposing the mature spores produced by meiosis. The spores are dispersed when the surrounding air is moist, under the most favorable conditions for subsequent germination. Dispersal is aided by remarkable teeth whose springlike action is triggered by changes in humidity.

Only a few mosses lack this pattern of sporophyte development. A familiar exception is *Sphagnum*. In terms of biomass, it probably outweighs all other mosses put together, flourishing in northern bogs and tundra well into the arctic. The *Sphagnum* sporophyte has

11

LIVERWORTS. Plants of the aquatic liverwort Ricciocarpus float on the surface. Antheridia and archegonia are buried in central grooves.

a very simple capsule containing an air chamber. Air pressure builds up in the chamber and eventually blows the lid off, dispersing the spores with an audible pop.

The LIVERWORTS (class Hepatici) are readily distinguished from the mosses because they never produce a filamentous protonema, and their sporophyte capsules are very simple. The simplest gametophytes are branched plates of cells perhaps a centimeter or so in length (*Figure 11*), producing antheridia and archegonia on their upper surfaces and rhizoids on the lower ones. Spore dissemination in liverworts is sometimes downright peculiar. In some species spores simply are not released until the surrounding capsule wall rots. In others the spores are disseminated by bizarre springlike structures called elaters, which fling the mature spores in all directions.

The HORNWORTS (class Anthocerotae) appear at first glance to be liverworts with very simple gametophytes: flat plates frequently only a few cells thick. But hornworts have two characteristics not found in any other bryophytes. First, the archegonia, instead of being borne on short stalks, are embedded in the gametophytic tissue. Second, of all the bryophytic sporophytes, those of the hornworts come closest to being capable of indefinite growth. In both mosses and liverworts the formation of a capsule at the tip of the sporophyte destroys its potential for further growth. With a plant such as the hornwort *Anthoceros*, however, a region of the stalk below the capsule remains capable of indefinite meristematic activity, continuously producing new spore-bearing tissue above. Sporophytes of hornworts growing in mild and continuously moist conditions have attained lengths of 20 cm (*Figure 12*).

VASCULAR PLANTS

The VASCULAR PLANTS (phylum Tracheophyta) are an extraordinarily large and diverse group, yet they can be said to have been launched by a single evolutionary event. Sometime during the Paleozoic era, probably well before the Silurian period, the sporophyte generation of some long extinct plant produced a wholly new type of cell: the TRACHEID.

Vascular plants are an extraordinarily large and diverse group, yet they were launched by a single evolutionary event: the invention of the tracheid.

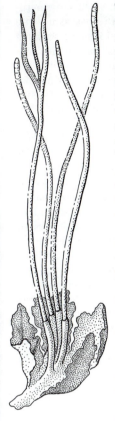

This invention was of crucial importance in the eventual evolution of terrestrial life. As noted in Chapter 10, the tracheid is the principal element of the xylem of all vascular plants except the angiosperms. The evolution of tracheids had two important consequences: first, it provided a pathway for long-distance transport of water and mineral nutrients; second, it provided rigid structural support, something almost completely lacking in algae. Thus a very large plant no longer had to be supported by water, nor did all of its parts have to be in contact with water. The tracheid set the stage for plants to move out of the seas and colonize the land masses of the Earth.

Many contemporary authors (splitters) distribute the vascular plants into several different phyla, basing their divisions on morphology, life cycles and anatomy. We have decided to follow the older practice of including all vascular plants in the phylum Tracheophyta on the grounds that members of every single group possess one peculiar type of cell in the xylem, the SCALARIFORM-PITTED TRACHEID, a tracheid with ladder-like rows of pits on the sides. Even in the primitive angiosperms (flowering plants) one can find xylem composed almost exclusively of scalariform-pitted tracheids.

In studying the vascular plants it is important to learn the details of the basic life cycle. With the exception of those plants which have successfully exploited the seed as a resting stage, namely the gymnosperms and angiosperms, the life cycles of vascular plants are remarkably similar. The life cycle of the fern (Chapter 7) is typical of virtually all seedless plants. Unlike the bryophytes, in which the sporophyte is attached to and completely dependent upon the gametophyte, the tracheophyte sporophyte is the large and obvious plant. The gametophytes are rarely more than a centimeter or two in length and seldom live longer than a few weeks at most. By contrast, the sporophyte of a tree fern may be 15 or 20 meters high and live for years. As in the fungi, algae, and for that matter the bryophytes, the most prominent resting stage in the life cycle is the single-celled spore. Seedless plants require an aqueous environment at one stage of their life cycle because fertilization is accomplished by a motile flagellated sperm.

PSILOPSIDS, the first vascular plants, belong to a now-extinct subphylum (Psilopsida). In the Silurian and Devonian periods of the Paleozoic era, psilopsids dominated the landscape. In 1917, R. Kidston and W. H. Lang first reported some remarkably well preserved fossils of vascular plants embedded in Devonian cherts not far from Rhynie, Scotland (Figure 13). These plants all had a simple vascular system of tracheids and phloem surrounded by a cortex. Many had no central pith. Furthermore, although flattened scales were found on the branches of some, the scales entirely lacked vascular tissue, and thus were not comparable with the true leaves of other vascular plants. Roots were entirely absent; the plants were apparently attached to the soil by horizontal stems that bore rhizoids. These horizontal stems also bore what were evidently aerial branches, with sporangia at their tips.

There is still some debate about whether psilopsids are entirely extinct. In numerous textbooks there are lengthy discussions of two plants, *Psilotum* and *Tmesipteris*, as the only living relics of this early Paleozoic phylum (Figure 14). But David W. Bierhorst has presented powerful arguments based on anatomy, morphology, and development, that these two

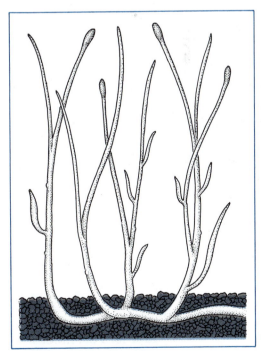

13

RHYNIA. Reconstruction of this extinct psilopsid is based on fossil found near Rhynie, Scotland.

14

PSILOPSID OR FERN? Psilotum nudum may be either a living survivor of a subphylum that became extinct in the Devonian period, or it may be a primitive or a reduced type of fern. Sporangia occur in clusters of three. Plant has no roots, and tiny thornlike scales instead of leaves.

evolution of ferns and the other higher plants lies in the origin and modification of the true LEAF. Up to this point the word leaf has been used rather loosely. In the strictest sense, a leaf is a flattened photosynthetic structure emerging laterally from a stem and possessing true vascular tissue. Although various authors have tried to invent other names for the leaflike structures found in mosses and liverworts, the word leaf has persisted. In our remaining discussion, however, we shall use it in the restricted sense, and hence dismiss the bryophytes and psilopsids as leafless.

This tight definition allows a closer look at true leaves in the Tracheophyta, and it turns out that there are probably two different types, of different origin. The first type of leaf is usually small and only rarely has more than a single vascular strand, at least in plants alive today. Known as a MICROPHYLL, it occurs in two subphyla related to ferns: the Lycopsida (club mosses) and Sphenopsida (horsetails or scouring rushes), of which only a few genera still survive. The principal characteristic of such a leaf is that its vascular strand departs from the vascular system of the stem in such a way that there is scarcely any deformation of the vascular cylinder of the stem. The other type of leaf, called a MEGAPHYLL, is thought to have arisen from the flattening of a branching stem system, with the development of extensive photosynthetic tissue (mesophyll) between the branches (*Figure 15*).

The TRUE FERNS comprise the class Filicinae. In the discussion of hormonal regulation of reproduction (*Chapter 7*), the life cycle of a

genera, which together contain only a handful of species, are very probably either primitive ferns or reduced ferns.

FERNS

The ferns and their relatives (subphylum Pteropsida) comprise one of the major living groups of vascular plants. One key to the

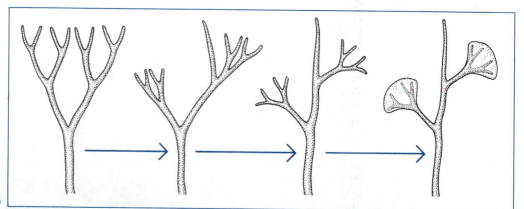

15

PROPOSED ORIGIN OF MEGAPHYLL. Branching stem system (left) became progressively reduced (left center) and flattened (right center). Flat plates of photosynthetic tissue (mesophyll) evolved between small end branches (right). The end branches evolved into the veins of leaves.

typical fern was presented in some detail, and earlier in this chapter it was mentioned as being typical of seedless plants in general. Devonian fossil beds have yielded forms with some characteristics that are psilopsid and some that resemble other subphyla. Thus the genus *Protopteridium* had flattened branch systems with mesophyll, but on the same axes bore terminal sporangia (*Figure 16*). This plant evidently lacked true roots. During late Paleozoic times, the ferns underwent some wild evolutionary experimentation in the structure of leaves and vascular systems. Modern ferns still reflect this extensive evolutionary experimentation, but a true cambium (*see Chapter 6*) is still extremely rare. Even a 20-meter tree fern is constructed of cells which are derived from the shoot apex. Lacking the rigidity of woody plants they do not grow in sites exposed directly to strong winds, but rather in ravines or in the understories of tropical rain forests.

The usual familiar fern is a relatively small inhabitant of shaded moist woodland and swamps. The fern frond is in fact the leaf; moving outward from the base, one must sometimes trace the main axis and several subsets of branches before finding any mesophyll. Some leaves become climbing organs, and may grow

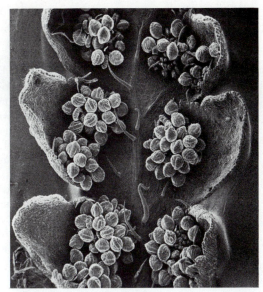

17

SORI are clusters of sporangia on the undersides of fern leaves. Scanning electron micrograph depicts a leaf of the fern Hypolepis tennifolia.

to 30 meters in length. The sporangia are found on the undersurfaces of the leaves — sometimes at the edges, sometimes covering the whole undersurface, and sometimes clustered in groups called SORI (singular, sorus), illustrated in Figure 17.

FOSSIL FERN (Protopteridium) had leaflets and sporangia on the same branches. This genus dates from Devonian period, and lacks true roots.

16

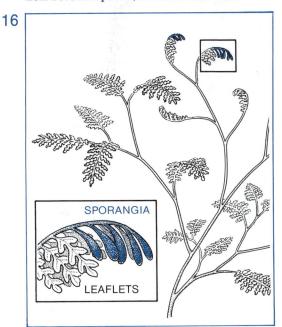

SPORANGIA

LEAFLETS

SEED PLANTS

With the GYMNOSPERMS (Class Gymnospermae) — the pines, firs, cedars, and their relatives — one finally comes to plants that have developed true seeds. In true seed plants the production of male gametophytes as pollen (in fact the whole mechanism of pollination) frees the organisms once and for all from a need for liquid water for fertilization.

Although there are probably fewer than 750 species of living gymnosperms, these plants are second only to the angiosperms in their dominance of the land. The great Douglas fir and cedar forests of the Pacific Northwest, and the massive forests of pine, fir, and spruce that clothe the northern continental regions and upper slopes of mountain ranges, rank among the great vegetation formations of the world. Most of these trees belong to one particular order, the Coniferales — the CONIFERS or cone-bearers (*Figure 18*). Male and female sporophylls — specialized leaves bearing sporangia — are rarely found in the same cones; male and female cones are separate in almost all cases.

The word gymnosperm means naked-seeded. In the conifers the ovules, which develop into seeds upon fertilization, lie on the upper surfaces of the sporophylls. The ovule is a specialized sporangium that produces spores by meiosis. The spores germinate (one per ovule) and grow into mature gametophytes while remaining enclosed within the parent sporophyte tissue. The female gametophyte bears archegonia. Their only protection from the environment is the way that the sporophylls are tightly pressed against one another within the cone. Indeed, some pines have such tightly closed female cones, once fertilization has occurred, that normally only fire suffices to split them open and release the seeds. One important example is the lodgepole pine, which has recolonized vast fire-ravaged areas in the Rocky Mountains.

WHITEBARK PINE (Pinus albicaulis) is a conifer, a member of largest order of living gymnosperms. Trees were photographed at an altitude of 2600 meters in Lassen Volcanic National Park, California.

18

Separate male cones bear male sporangia which produce male spores by meiosis. These spores develop into pollen grains which are eventually shed and carried to the female sporangia by wind. Some species still produce swimming sperm, although the distance to be covered is only a few microns. Others merely release a sperm nucleus close enough to an egg nucleus to fuse without swimming.

All living gymnosperms have an active cambium, and all but a few produce tracheids as the sole type of water-conducting cell found in the xylem. Despite this apparently relatively inefficient design for a water transport system, gymnosperms are among the tallest trees known. The coastal redwoods of California are the record holders; the largest are over 100 meters high. The xylem produced by the gymnosperm cambium is the principal resource of the lumber industry. A single redwood tree is reputed to contain enough lumber for an entire small housing development, a fact that has caused heavy cutting and brought this majestic species close to extinction over large parts of its former range.

The earliest known gymnosperms, dating from the Devonian period of the Paleozoic era, had both psilopsid and fernlike characteristics; but their woody tissue, based on tracheids, resembled that of a gymnosperm.

By Carboniferous times several new lines of gymnosperms had evolved, including one group that possessed fernlike foliage but with characteristic gymnosperm seeds attached to the leaf margins. The first conifers also appeared about the same time. They were not dominant trees, or else they just did not grow where conditions were right for fossilization. In fact, gymnosperms are apparently either at their peak right now, or else they are still an emerging group — despite the extremely long history already behind them.

FLOWERING PLANTS

At the summit of plant evolution stand the FLOWERING PLANTS (class Angiospermae). In earlier chapters, when "plants" were mentioned in discussing processes such as water balance, long distance transport in the xylem and phloem, leaf structure, or hormonal regulation of development, we usually meant this very particular group of plants — the angiosperms. Although a great deal of angiosperm (plant) biology has already been covered, one aspect was deliberately postponed for this chapter: the details of sexual reproduction.

The reason for waiting was that sexual reproduction in angiosperms presents characteristics found nowhere else in the plant kingdom.

The reproductive organ of the angiosperm is the flower. It is essentially a conelike arrangement of modified leaves, some of which bear sporangia. The leaves bearing the male microsporangia are called STAMENS, and those bearing the female megasporangia are called CARPELS. In addition, there are frequently a number of specialized sterile leaves found below the sporophylls; the upper ones are called PETALS and the lower ones SEPALS. The anatomy of an ideal flower (for which there is no exact counterpart in nature) is depicted schematically in Figure 19. From base to apex, the sepals, petals, stamens, and carpels are arranged in whorls attached to a central RECEPTACLE. Towards the tip of each stamen lie four sporangia which collectively comprise the anthers. They produce pollen, the microscopic grains that develop into the male gametophyte and gametes. In contrast to the gymnosperms, angiosperms enclose their female gametophytes within not one but two layers of tissue, forming the OVULE. One or more ovules are then enclosed within a carpel.

The female gametophyte is a remarkably unimpressive structure. One cell within the sporangium divides meiotically to produce four megaspores. Three degenerate, and the remaining spore undergoes mitosis, ultimately producing eight nuclei enclosed by a membrane. This eight-nucleated structure is called the EMBRYO SAC; it constitutes the entire female gametophyte.

The male gametophyte is even less impressive. Meiosis within the four sporangia (fused into two anthers) produces an enormous number of microspores. Before the pollen is shed by the anthers, each microspore normally undergoes one mitotic division. When the pollen arrives at the female STIGMA, a POLLEN TUBE germinates from the grain and penetrates the spongy tissue of the stigma to reach the micropyle. Of the two nuclei present in each pollen grain, one is the TUBE NUCLEUS which lies close to the tip of the grain, and the other is the GENERATIVE NUCLEUS. As the tube digests its way through the tissue of the megasporangium the generative nucleus undergoes a mitotic division, producing two sperm nuclei which are finally released into the cytoplasm of the female gametophyte.

Within the embryo sac there are now ten nuclei: eight female and two male. One of the sperm nuclei next fuses with the egg nucleus producing the diploid zygote. The other sperm nucleus does not degenerate, but fuses with two female nuclei to form a triploid nucleus. While the zygote nucleus begins cell division to form the new sporophyte embryo, this triploid nucleus undergoes rapid mitosis to form a specialized tissue called ENDOSPERM. At this point, the remaining female nuclei simply degenerate.

Shortly after fertilization, the embryo and endosperm begin a program of growth and development (*Figure 20*). As large amounts of nutrient are moved in from other parts of the plant, the endosperm begins accumulating both starch and protein. The embryo becomes differentiated into characteristic root and shoot regions, and the first leaves (COTYLEDONS) appear. A double-layered seed coat derived from the ovule develops, and the carpel eventually becomes the wall of the FRUIT.

This maturing structure contains three different generations in one system. The seed coat comes from the parental sporophyte plant; the endosperm from the female gametophyte; the embryo of course represents the next sporo-

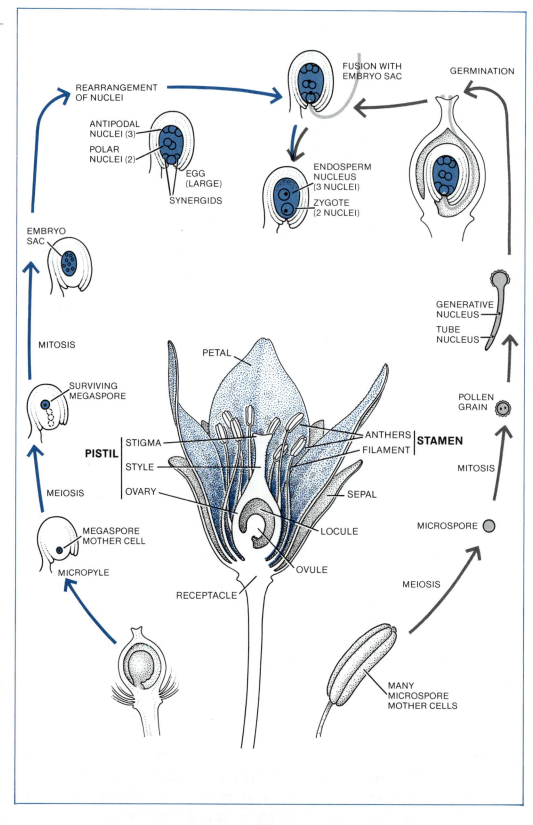

REARRANGEMENT
OF NUCLEI

FUSION WITH
EMBRYO SAC

GERMINATION

ANTIPODAL
NUCLEI (3)

POLAR
NUCLEI (2)

EGG
(LARGE)

SYNERGIDS

ENDOSPERM
NUCLEUS
(3 NUCLEI)

ZYGOTE
(2 NUCLEI)

EMBRYO
SAC

MITOSIS

SURVIVING
MEGASPORE

MEIOSIS

MEGASPORE
MOTHER CELL

MICROPYLE

PETAL

PISTIL

STIGMA

STYLE

OVARY

ANTHERS

FILAMENT

STAMEN

SEPAL

LOCULE

OVULE

RECEPTACLE

GENERATIVE
NUCLEUS

TUBE
NUCLEUS

POLLEN
GRAIN

MITOSIS

MICROSPORE

MEIOSIS

MANY
MICROSPORE
MOTHER CELLS

REPRODUCTIVE CYCLE of flowering plants is illustrated on opposite page. An ideal flower is depicted at center, with female cycle at left and male cycle at right. Each ovule (bottom left) encloses a female sporangium called a micropyle. Meiosis of mother cell produces four megaspores. Three degenerate, and remaining one divides mitotically, producing an egg (embryo sac) with eight nuclei (upper left). Male sporangia are fused into anthers at tip of filament (lower right). Microspore produced by meiosis divides mitotically before becoming a walled pollen grain. When pollen reaches the female stigma, pollen tube grows downward toward ovary and eventually penetrates the embryo sac (top center). Meanwhile the generative nucleus of pollen grain divides mitotically to produce two sperm nuclei, which fuse with nuclei in embryo sac to yield a diploid zygote and a triploid endosperm nucleus.

highly successful. It has produced the dominant land vegetation of the planet.

Something more must be said about the ideal flower which was invented for the above discussion. Because it produces both megasporangia and microsporangia it is said to be PERFECT, meaning that it contains both male and female parts. Many angiosperms produce two types of flower on the same plant, one type male and the other female. Consequently either the carpels or the stamens are nonfunctional or absent in a given flower. In some other species of angiosperms a given plant produces either male or female sporophylls, but never both. In other words, there are truly female plants and truly male plants. In the ideal flower we also illustrated distinct petals and sepals, arranged in distinct whorls, whereas sometimes there is no distinguishing between the two, and they are spirally arranged. In such a case, these appendages are called TEPALS. Sometimes sterile appendages of any sort — petals, sepals, or tepals — are completely absent.

phyte generation. Fruit formation is an ingenious strategy for seed dissemination. Animals eat the fruits and then either spit out the seeds or pass them undigested through their guts.

The strange double fertilization mechanism just described is found in all angiosperms and only in angiosperms. With double fertilization, the angiosperms have finally and conclusively abandoned both the swimming sperm and the archegonium, the ancient device that protected the egg and zygote. Although it is very complicated, double fertilization has been

DEVELOPMENT OF SEED in angiosperms. At left, the ovule within the bud, with the megaspore nucleus in color. Left center, embryo sac within bud of flower. Right center, pollen tube penetrates egg of mature flower through micropyle. Right, fully developed seed within ovary.

20

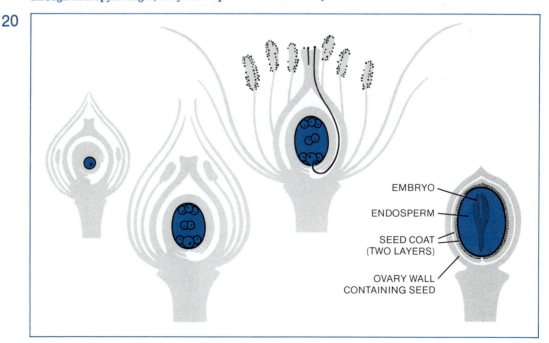

EMBRYO

ENDOSPERM

SEED COAT
(TWO LAYERS)

OVARY WALL
CONTAINING SEED

21

PRIMITIVE FLOWER has many stamens and carpels arranged spirally. Photograph depicts flower of the tulip tree (Liriodendron). Central column consists of many separate carpels, tightly packed.

The most primitive flower, from an evolutionary point of view, has a large number of tepals (or sepals and petals), carpels, and stamens, all spirally arranged (*Figure 21*). Evolutionary change within the angiosperms involved a number of striking changes from this early condition: reduction in the number of each type of organ, differentiation of petals from sepals, stabilization of each type of organ to a fixed number, arrangement in whorls, and finally change in symmetry from radial (as in a lily) to bilateral (as in a sweet pea or orchid) often accompanied by an extensive fusion of parts. The most primitive carpels were clearly modified leaves, appearing as folded but incompletely closed structures — really intermediate between the sporangium-bearing scales of the gymnosperms and the carpels of advanced angiosperms. Primitive stamens were similarly leaflike, little resembling those of the ideal flower illustrated in Figure 19.

Angiosperms are divided into two distinct subclasses, known as the MONOCOTS (subclass Monocotyledonae) and the DICOTS (subclass Dicotyledonae). The names derive respectively from the existence of but a single embryonic cotyledon in monocots, versus a pair of embryonic cotyledons in dicots. There are, however, other major differences between the two groups. Monocots, which include grasses, cattails, lilies, orchids, and palm trees, have leaves with largely parallel veins. They almost never have a cambium, and their vascular bundles are scattered. In the root, bundles of xylem and phloem alternate around the central axis, while in the stem each bundle contains both types of transport tissue. The absence of a cambium means that a palm tree, like a tree fern, is entirely the direct product of the apical meristem.

The dicots include the vast majority of familiar seed plants — most of the herbs, weeds, vines, trees, and shrubs that cover the Earth. The veins of dicot leaves are usually branched or net-like in pattern. The presence of a cambium is common — hence there are many woody species. There is also a characteristic arrangement of vascular tissue in the stem and root: stem vascular tissue is arrayed in a single ring of bundles (with both xylem and phloem) placed around a central pith, and the root system is commonly a single solid star-shaped mass of xylem with phloem between the xylem points. Oaks, willows, violets, sunflowers, and chrysanthemums are examples chosen almost at random from the immense diversity of dicots. Sunflowers and chrysanthemums are representatives of the Compositae, the most specialized dicot family. Their "blossom" is actually a tightly packed mass of flowers on a common receptacle (*Figure 22*). Frequently the outer flowers of the composite blossom are only female, while the inner ones are perfect, having both male and female parts.

There is not space here to discuss the tremendous variation in the structure of flowers, fruits and seeds within the angiosperms, and particularly within the dicots. One can only point out that the taxonomy of the angiosperms is based largely upon these structures, and that there is an enormous technical vocabulary to deal with them.

The origin of the angiosperms is at the present time a paleobotanical mystery. The first clearly angiosperm pollen is found in Cretaceous sediments. By the end of the Mesozoic era, approximately 65 million years ago (half a

22

BLOSSOM OF SUNFLOWER (Helianthus) is actually a large cluster of small flowers. Each of the long outer "petals" is formed by the fusion of five petals of a miniature flower. Center of blossom consists of many tiny "flowerlets," each having functional anthers and pistil.

day on our calendar), the angiosperms were well launched. They underwent an enormous burst of evolution, which is probably still underway, starting in the early Cenozoic era. While various gymnosperms, from the seed ferns to the cycads, have been proposed as possible angiosperm ancestors, one is always left with the curious problem of where and how that one universal and distinctive angiosperm trait, double fertilization, originated in evolution. No living gymnosperm has anything approaching it, and fossil gymnosperms are not likely to be particularly helpful. Direct evidence must come from cytological studies of developing ovules, and unfortunately the fossil gymnosperm ovules don't develop.

READINGS

D.W. Bierhorst, *Morphology of Vascular Plants*, New York, Macmillan, 1971. A detailed and extremely up-to-date treatment of the vascular plants, both living and fossil.

H.C. Bold, *Morphology of Plants*, New York, Harper & Row, 3rd Edition, 1973. An excellent text on the entire plant kingdom.

R.F. Scagel, R.J. Bandoni, G.E. Rouse, W.B. Schofield, J.R. Stein, and T.M.C. Taylor, *An Evolutionary Survey of the Plant Kingdom*, Belmont, Calif., Wadsworth, 1966. An unusual and complete survey of the plant kingdom, by authors who are specialists in each of the various groups.

ORNITHISCHIAN

SAURISCHIAN

PTEROSAUR

PLESIOSAUR

ICHTHYOSA

animals

Anyone looking into the pages of the present handbook will soon find out that the zoology of dreams is far poorer than the zoology of the Maker.

JORGES LUIS BORGES
THE BOOK OF IMAGINARY BEINGS

Unlike plants, animals do not manufacture their own food. They must either search for it or move it toward them, usually by creating currents of water. Much of the anatomy and behavior of animals has been shaped by the requirements of these two methods of feeding.

Sessile animals — those that remain fixed in place for all or part of their life cycles — seem to be more plantlike than animals that move about in search of food. The definition of an animal is somewhat arbitrary, but all animal phyla share certain traits: mobility, well developed sensory and nervous systems, and behavioral responses to outside stimuli. Chapter 17 distinguished between protists and animals on the basis that the former are unicellular and the latter multicellular. That view is a bit oversimplified.

The most important difference between animal-like protists and "true" animals lies in the DIVISION OF LABOR among cells. This change represents a big jump in evolution. Within primeval colonies of single-celled organisms — aggregations similar to those of *Volvox* and other colonial protistans — some cells began to differentiate into specialized types. Figure 1 summarizes the prevailing opinion about the origin of the most primitive animal groups. Most of this evolution took place in Precambrian times, and the few fossils preserved from that time include not only representatives of phyla still living, but also some unidentifiable animals that may have belonged to phyla that became extinct before the Cambrian period (*Figure 2*).

Animals are classified into three main types — acoelomate, pseudocoelomate, and coelomate — according to the anatomy of their soft tissues and organs, and into two other categories — vertebrate and invertebrate — according to the nature of their supporting tissue and skeleton. The reader should familiarize himself with the vocabulary in Box A. The acoelomate animals include sponges, jellyfish, and flatworms. Pseudocoelomates include rotifers and roundworms. The coelomates include all of the so-called higher animals, from invertebrate mollusks and arthropods to vertebrates such as reptiles, birds, and man.

THE ACOELOMATE ANIMALS

The SPONGES (phylum Porifera, from the Latin meaning "pore bearer") are an ancient group of extremely primitive animals. They are almost all marine, with only a few species being found in fresh water. Attached tightly to the sea bottom, the sponge feeds by drawing water through itself and filtering out small organisms and nutrient particles. The first sponges probably evolved from colonial flagellates. They represent an evolutionary dead end, having given rise to no other known phylum. Their gross structure is unique among animals: there is no mouth or true digestive cavity, nor muscles, nor nervous system. For that matter, there are no organs at all in the usual sense of the word.

The body consists of a loose republic of cells built around a unique water canal system. The sponge depicted in Figure 3 is typical of the simpler species. Water currents are set up by the flagellated cells (choanocytes) that line the inside of the body cavity. The water

MESOZOIC REPTILES depicted on opposite page include Triceratops, an herbivore (top); a flying carnivore (left center); Tyrannosaurus, a terrestrial carnivore (right center); aquatic plesiosaur and icthyosaur (bottom). Plesiosaurs had long necks; icthyosaurs were more fishlike.

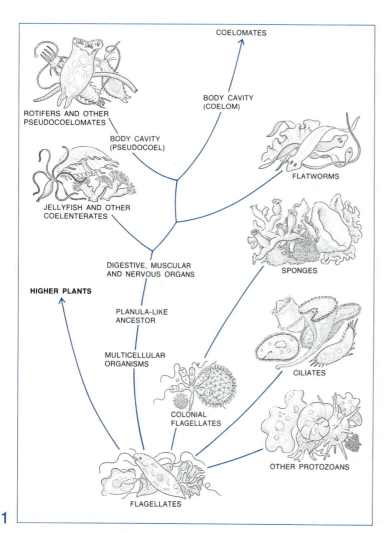

Tests have shown that they are attracted to one another by "aggregation substances," a protein-polysaccharide mixture on the cell surfaces (*Chapter 6*).

JELLYFISH, SEA ANEMONES, CORALS and related forms belong to the phylum Coelenterata. The coelenterates have advanced beyond the sponges by acquiring authentic organs and tissues. They possess nerve nets, tentacles, epithelial cells with muscle fibers, and cells that discharge unique stinging organs called NEMATOCYSTS. There is a true mouth and a blind digestive sac called the GASTROVASCULAR CAVITY. The digestive apparatus, plus the

SIMPLE SPONGES have body cavities lined with flagellated cells that resemble free-living choanoflagellates (bottom). Current created by these cells pulls water into the sponge through pores. Mineral spicules strengthen the wall of the sponge. Arrows indicate flow of water.

1

PHYLOGENETIC TREE OF LOWER ANIMALS summarizes current thinking about their evolutionary relationships. Flagellates, ciliates and other protozoans are classified as protists. Multicellular forms are classified as true animals. Drawings show a few modern members of each group.

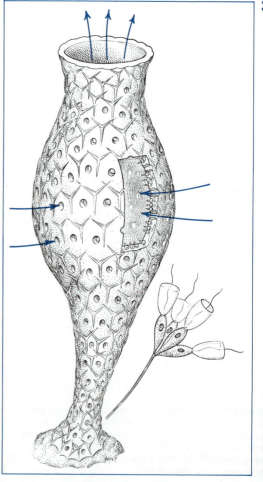

3

flows into the animal by way of minute pores perforating special epidermal cells. It passes up into a chamber of the body and out of a large terminal opening called the osculum. The cells of a sponge are so loosely organized that it is possible to squeeze the entire body of some kinds of sponges through a fine mesh, separating all the cells from one another, without killing the organism. If given the opportunity, the cells reassemble into new sponges.

2

PRECAMBRIAN ANIMAL (Parvancorina) is an offshoot from the evolutionary mainstream, an evolutionary experiment that failed. Fossil of this aquatic animal, which resembles no known living species, is depicted at about its actual size.

row of tentacles that normally surrounds the mouth, enable the animal to capture and swallow a much wider range of food particles than is available to sponges. The nematocysts play an important role in paralyzing and anchoring the prey. A few coelenterates are extraordinarily large. The record is held by the lion's mane jellyfish of the northern oceans (*Cyanea*), whose bell sometimes reaches seven feet in diameter.

Most of the basic coelenterate features are lucidly displayed by the fresh-water animal *Hydra*, familiar from high school biology labs (*Chapter 8*). *Hydra* is considerd a specialized

---

A
CATEGORIES OF ANIMALS

The most primitive animals are ACOELOMATES, *animals whose bodies lack a coelom or any other internal cavity (see diagram). Pseudocoelomates have a false coelom, a kind of body cavity found only in a small number of relatively simple phyla such as rotifers. The* PSEUDOCOEL *is derived directly from the blastocoel of the embryo: the first cavity formed inside the proliferating ball of cells. It provides a liquid-filled space in which many of the body organs float. Coelomate animals have a true* COELOM: *a special kind of body cavity found in the most evolved animal phyla. The coelom is derived from the embryonic mesoderm, and it is lined with a special mesodermal lining called the peritoneum. The internal organs of coelomate animals hang down into the coelom, but do not float free within it; they are slung in pouches of the peritoneum.*

The VERTEBRATE *animals are those in which the nerve chord is enclosed in a backbone composed of bony segments, or vertebrae. The vertebrates are vastly outnumbered by the invertebrates both in number of species and in number of individuals, but because of the extraordinary interest they hold for us (as we are vertebrates ourselves) a basic distinction is usually made between the disciplines of* VERTEBRATE ZOOLOGY *as opposed to* INVERTEBRATE ZOOLOGY. *Zoology courses at the intermediate level are divided this way in the curricula of most college biology departments in the United States.*

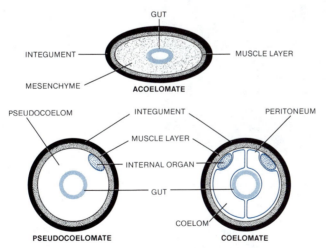

BODY CAVITIES of animals are illustrated in these schematic cross sections. Acoelomates such as sponges, coelenterates and flatworms have no body cavity; spaces between body wall and gut are filled with a jellylike matrix containing cells. Pseudocoelomates such as rotifers and nematodes have true fluid-filled body cavities, but the cavities lack a specialized cellular lining (peritoneum). All higher animals, such as mollusks, insects and vertebrates, have a true coelom lined with a peritoneum, although in some adult arthropods the coelom is almost obliterated, being replaced by a secondarily-formed body cavity called a hemocoel.

5

PLANULA is the free-swimming larva that develops from the fertilized eggs of coelenterates. Larva settles on sea bottom and becomes a polyp. Magnification, about 50×.

6

CORALS are coelenterates that live in warm shallow seas. Polyps tipped with feathery tentacles emerge from their stony skeletons to feed. Polyps of this colony of Galaxea are largely retracted.

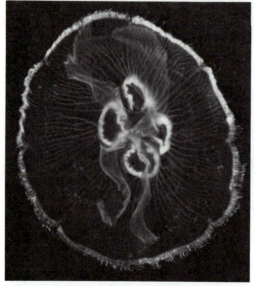

4

AURELIA, a common jellyfish, passes through alternation of generations typical of other coelenterates. At top, three sessile polyps in various stages of development hang from the underside of a rock. Below, the free-swimming medusa has four clearly visible lobes lined with cilia that sweep food into the central mouth. Polyps are reproduced about 3× actual size; medusa, about ½×.

coelenterate. The majority of other coelenterate species, including those considered primitive, pass through an alternation of generations essentially like those of *Aurelia* (Figure 4). The POLYP is the sessile (fixed) stage, in which the body consists of a cylindrical stalk with the end bearing the mouth and tentacles facing upward. The MEDUSA is the alternate, free-swimming stage. Its body is that of the typical jellyfish, shaped like a bell or an umbrella with the mouth and tentacles facing downward.

If all matter except roundworms were to vanish, a ghostly outline of the surface of the Earth could still be seen in the form of the countless nematodes that live in the soil, the water, and in plants and animals.

7

The medusa can be envisioned as a polyp that lost its stalk, turned over, and swam away. Conversely, a polyp can be viewed as a medusa that turned over, grew a stalk, and attached itself. The polyps give rise to medusae by asexual budding, and the medusae generate the polyp state by sexual reproduction. The sexual medusae release reproductive cells into the water. When an egg is fertilized it develops into a free-swimming ciliated larva called a PLANULA (or planula larva) which settles to the bottom and transforms into a polyp. Because the planula possesses such an elementary body plan (Figure 5) it is widely believed to resemble the ancestral multicellular animal that arose from protozoans and gave rise in turn to the coelenterates and higher animals.

The distinction between the polyp and medusa stages must be kept in mind while considering diversity within the coelenterates. The complex "organism" of the Portuguese man-of-war, for example, is actually a large colony of polyps, each specialized to contribute to the locomotion, feeding, or reproduction of the animal. The corals also consist of polyp colonies. In most coral species the individual polyp lays down a skeletal container of calcium carbonate that surrounds and protects its soft body. This skeletal cup is highly regular in form. Coral colonies grow by the budding of individual polyps, with the skeletal cups being added one to the other in a set geometric pattern that varies among species (Figure 6). As a consequence, each coral formation consists of a beautiful but bewildering array of skeletal forms. As the colony grows, old individuals die, leaving their skeletons intact. The living members form a layer on top of the growing coral reef. The reefs, which are often very old, play a major role in the formation of tropical islands, particularly the atolls. They are also the site of the most stable and diverse of all marine ecosystems (Chapter 25).

FLATWORMS (PHYLUM PLATYHELMINTHES)

To most people the term worm conjures up a picture of a long, squirming creature, living in some hidden place without the benefit of eyes, head, or legs. The truth is that many animals possess approximately this body form, and they are very difficult to classify. The simplest worms are the Platyhelminthes. As their Greek name indicates, they are the flat worms (Figure 7). The best known members of the phylum are parasites: the tapeworms (class Cestoda) and flukes (class Trematoda), some of which cause serious diseases in man. The third group, the turbellarians (class Turbellaria) are all free-living; most are marine, but a few live in fresh water or moist habitats on land.

The shape of flatworms has been dictated by their primitive circulatory and excretory systems, which require that each cell be near the surface of the animal (Chapters 9 and 10). Many flatworms are not much more advanced than the planula larva of the coelenterates. In fact, they have probably not evolved very far from the ancient ancestors of all higher animals.

THE PSEUDOCOELOMATE ANIMALS

As noted earlier, the pseudocoelomates include rotifers and roundworms. Many members of the phylum Rotifera, the rotifers or "wheel animalcules" are visible in a drop of pond water at 100× magnification (Figure 8). Most are tiny, not much larger than ciliate protozoans and easily mistaken for them. Each rotifer actually contains between 500 to 1000 cells and highly organized tissues and organs. Rotifers possess a complete gut, one that passes from a mouth at the anterior end to an anus at the posterior end. There is also a body cavity, the pseudocoel defined at the beginning of this chapter, within which the internal organs are suspended. A few rotifer species are marine, but most live in fresh water. A small number, loosely referred to as terrestrial, actually occupy an unusual aquatic habitat. They rest on the surfaces of mosses and lichens in a dry inactive form until a rainfall; then they swim about in the films of water that temporarily cover the plants. Rotifers are remarkably widespread; each species is distributed over most of the Earth, flourishing wherever it can find a suitable habitat.

ROUNDWORMS (PHYLUM NEMATODA)

Ubiquity and abundance are features of this pseudocoelomate phylum. Humans unintentionally eat and drink large numbers of nematodes in their lifetime. A single rotting apple from the ground of an orchard was found to contain 90,000 roundworms belonging to several species. One square meter of mud from the coast of Holland yielded 4,420,000 indi-

8

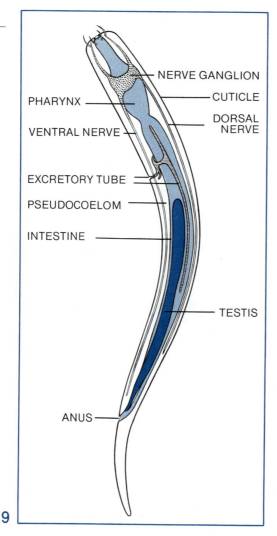

9

PHARYNX

NERVE GANGLION

CUTICLE

DORSAL NERVE

VENTRAL NERVE

EXCRETORY TUBE

PSEUDOCOELOM

INTESTINE

TESTIS

ANUS

NEMATODE WORM. Diagram of male nematode indicates the presence of a gut (light color) with mouth and anus, a nervous system with a ganglion that serves as a rudimentary brain, and a testis (dark color) that opens into the anus.

viduals, and the soil of rich farmland has up to 3 billion to the acre. It has been said that if all matter except roundworms were suddenly to vanish, a ghostly outline of the surface of the earth could still be seen in the form of the countless nematodes that live within the upper layers of the soil, in the bottoms of lakes and streams, in plants, and as parasites in the bodies of most kinds of animals. The bodies of nematodes are about as highly organized as those of rotifers (Figure 9).

THE COELOMATE ANIMALS

The evolution of the coelom made possible the development of the physically largest and most complex organ systems. The mesodermal lining of the coelom, the peritoneum, provides the means to suspend the internal organs in the fluid-filled body cavity while at the same time containing and separating them. The annelid superphylum, together with the echinoderm superphylum discussed later in this chapter, are collectively called the higher animals (see Box B and Figure 10).

SEGMENTED WORMS (PHYLUM ANNELIDA)

Compared with rotifers and flatworms, the earthworm is definitely a higher animal. Its many organ systems are completely formed and it is capable of relatively quick, accurately directed movements during locomotion, feeding, and reproduction. In fact, improved locomotion is the key to the most distinctive aspects of annelid biology. The earthworm is a segmented animal constructed of a number of muscular doughnut-shaped rings separated by thin partitions. The segments can be expanded and contracted in a coordinated fashion to produce waves of bulges moving up and down the length of the body, enabling the animal to push its way rapidly through the soil. The earthworm is an example of the generalization that the groups with the most advanced animal-like traits are those that are required to move around the most in the course of their daily existence. The vast majority of other annelids are marine and aquatic. The major adaptive forms within the Annelida are impressively diverse. Scavengers and soil feeders such as earthworms and their aquatic equivalents comprise most of the class Oligochaeta. The class Polychaeta includes voracious predators that seize and capture animals their own size, and sea-bottom dwellers that live in tubes and trap small organisms with plumelike tentacles. Parasitic forms occur within both the Polychaeta and Oligochaeta, while a third annelid class, the leeches or Hirudinea, includes predators of other invertebrates and temporary external parasites on vertebrates.

MOLLUSKS (PHYLUM MOLLUSCA)

It is hard to imagine animals less alike in outward appearance than a snail, a clam, and a squid. Yet each represents but one class within the phylum Mollusca. Each possesses a FOOT, a muscular organ which serves both in locomotion and as a sturdy underpinning for the viscera. The foot of the squid and other cephalopods is modified into a head equipped with eyes and tentacles. Mollusks also possess a MANTLE, a sheet of specialized tissue that covers most of the viscera like a body wall. The mantle secretes the SHELL, which provides an external armor in most mollusks but has been modified into an internalized backbone in the squids and lost altogether in the octopuses. Within the mantle cavity are found a small number of uniquely constructed, feather-shaped GILLS quite unlike those of fish. A scraping organ, the RADULA, is usually located in the floor of the mouth.

Perhaps the single most notable fact about mollusks is the way the basic body plan has been varied in evolution to permit radically different adaptations to the environment (Figure 11). The CHITONS (class Amphineura) are the most primitive major group. The body is symmetrical, and the internal organs, particularly the digestive and nervous systems, are simply constructed. Development proceeds through a trochophore larva almost indistinguishable from that found in certain primitive annelids. The adult chiton spends most of its life clamped tightly to rock surfaces by means of its large, muscular foot. It is capable of moving slowly by means of rippling waves through the foot. The body of the chiton might, in fact, have been derived from that of some ancestral flatworm. The BIVALVES (class Bivalvia or Pelecypoda) are the familiar clams, oysters, scallops, mussels, and other important edible shellfish, together with a host of similar, lesser known forms. They have two shells, in contrast to the GASTROPODS (class Gastropoda), which have one. Gastropods are relatively primitive mollusks that use their foot primarily for locomotion. The Gastropoda have become the most successful and diverse of the molluscan classes. Among the crawling forms are a bewildering variety of snails, whelks, limpets, slugs, abalones, drills, and the often brilliantly ornamented nudibranchs. Finally, the CEPHALOPODS (class Cephalopoda) are by purely human standards very close to the apex of invertebrate evolution. As exemplified by the squids, octopuses, and chambered nautilus, they possess a well-formed head with a relatively large brain, and are capable of extensive learned behavior. The cephalopod looks at the world with large, bright eyes, superficially similar to our own but radically different in embryonic origins. Its arms and tentacles, another distinctive cephalopod invention, are able to manipulate small objects with impressive skill. The cephalopods never attained the level of intelligence of advanced vetebrates, but of all the invertebrates they came closest to assembling the anatomical structures necessary for such an evolutionary advance.

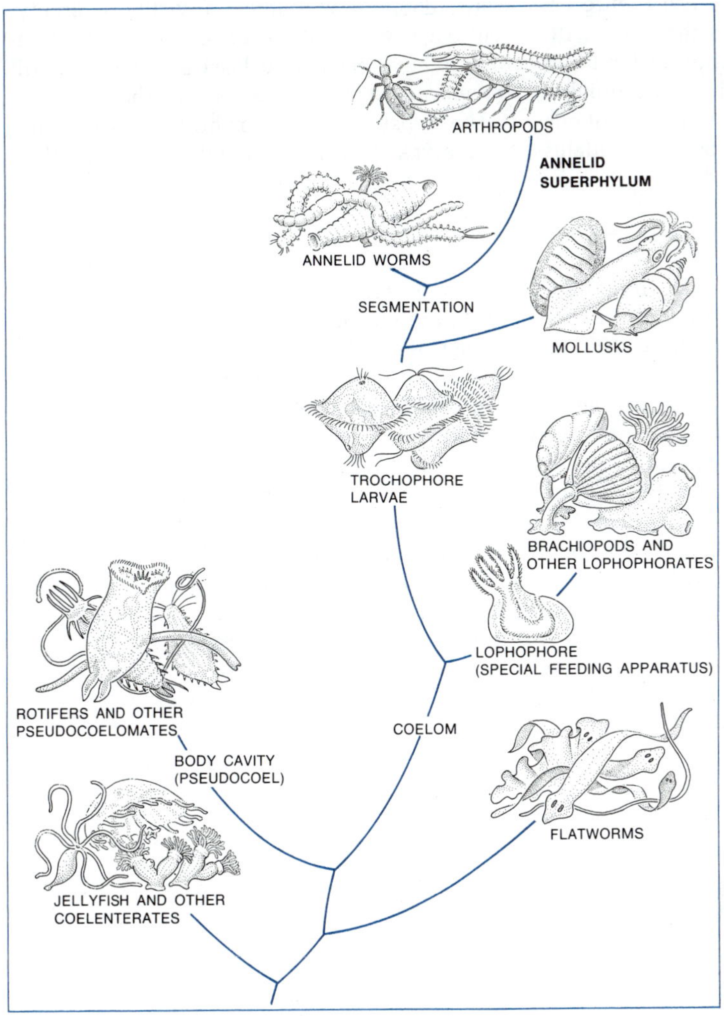

ARTHROPODS

The arthropods are one of the two or three most successful phyla on Earth. The phylum Arthropoda includes trilobites, crustaceans, spiders, centipedes, insects, and related forms. The original arthropods evolved from annelid ancestors in Precambrian times. They acquired a unique armor, the EXOSKELETON or CUTICLE, composed of layers of protein and a tough nitrogen-containing polysaccharide called CHITIN. In some groups of arthropods, espe-

0

ANNELID SUPERPHYLUM, which includes annelid worms, mollusks, and arthropods, is a major addition to the phylogenetic tree depicted in Figure 1. Members of superphylum are linked by common features of development, particularly by the presence of a trochophore larva (Box B).

cially the crustaceans, the exoskeleton is impregnated and toughened with mineral salts such as calcium carbonate and calcium phosphate. The exoskeleton originally afforded protection from predatory animals, a function that it still serves today, but it is also one of the main reasons for the spectacular success of the phylum. Within most of the evolving lines of the Arthropoda, certain appendages were modified from their original walking or swimming functions to aid in eating and reproduction. Others assumed a primarily sensory function. In effect the arthropod exoskeleton is a plastic that nature has molded into an astonishing variety of tools — devices for flying, biting, piercing, cutting, grasping, stinging and even for singing and chirping. The arthropod body plan is depicted in Figure 12. The exoskeleton also provides the body with the support it needs for walking on land and keeps the animal from drying out quickly when it chooses to leave the water. Arthropods were, in short, among the best candidates among the invertebrates to colonize the land. This they accomplished repeatedly, often with extraordinary success. In terms of numbers of species one can safely say that we are now living in the age of arthropods — or more precisely, the age of insects.

Insects originated from centipedelike ancestors at least as far back as the Devonian period, over 350 million years ago (on the 28th day of our evolutionary calendar). This early start gave them an advantage in exploiting the newly formed forests and other primitive forms of land vegetation. By Carboniferous times, a great diversity of types already swarmed over the land, and winged insects had appeared — the first animals to fly. The new environments penetrated by the insects were like a new planet, a virtually empty ecological world. It is not too surprising that insects have become at least as diverse as all the marine invertebrates put together.

The most primitive arthropods were the TRILOBITES (class Trilobita), a group that

B
THE ANNELID SUPERPHYLUM

In Figure 10 a major branch has been added to the expanding phylogenetic tree of the animal phyla. Under the broad and informal grouping labeled annelid superphylum have been placed three strikingly distinct phyla, the annelid worms, the mollusks, and the arthropods. What is the justification for making such an improbable taxonomic cluster? The answer lies in largely hidden details of the early development of these animals, where certain points of correspondence link them with flatworms, rotifers, and others of the lower animal phyla, while separating them from other higher forms of animals. Of equal significance is the possession, in certain marine annelids and mollusks, of the trochophore larva (see photo), a distinctive form propelled by a row of cilia that encircles the middle of the body (the name means "wheel bearer"). Only some of the marine annelids and mollusks go through a trochophore stage during their development, but enough do to suggest that the trochophore is a primitive remnant of the remote ancestor that gave rise to these now very divergent groups. No arthropod possesses this stage, but other evidence strongly suggests that arthropods evolved directly from annelid forebears, perhaps from a line that had already lost the trochophore.

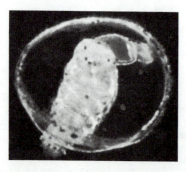

TROCHOPHORE LARVA of the marine annelid Polygordius. Cilated larva has a complete gut and an excretory protonephridium, as well as an embryonic mesoderm. Such larvae are thought to be a primitive survival of the extinct common ancestor of the annelid superphylum.

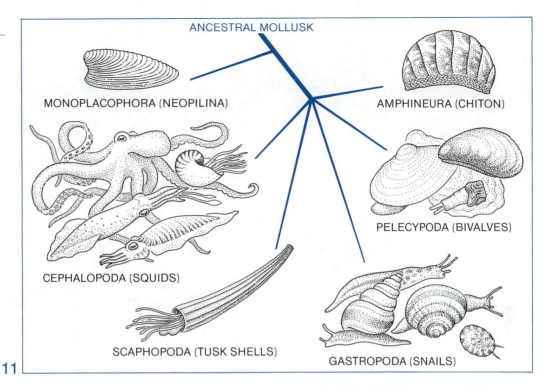

ANCESTRAL MOLLUSK

MONOPLACOPHORA (NEOPILINA)

AMPHINEURA (CHITON)

PELECYPODA (BIVALVES)

CEPHALOPODA (SQUIDS)

SCAPHOPODA (TUSK SHELLS)

GASTROPODA (SNAILS)

11

SIX CLASSES OF MOLLUSK are depicted in this diagram, which suggests the diversity of variations on the basic body plan. Features common to all mollusks include a muscular foot, mantle, shell, gills and radula. Chitons are most primitive mollusks and cephalopods the most advanced.

flourished in the seas of the Cambrian and Ordovician periods but became extinct by the close of the Paleozoic era (*Figure 13*). Trilobites were heavily armored, and the segmentation of their bodies and appendages followed a relatively simple, repetitive plan. Another ancient group is the horseshoe crabs (class Xiphosura). These large animals are common in shallow waters along the east coast of North America and southeast Asia, where they scavenge and prey on mollusks and other bottom-dwelling invertebrates.

The ARACHNIDS (class Arachnida) are relatives of horseshoe crabs that invaded the land and, like the insects, enjoyed an early and lasting success. All have six pairs of appendages, of which the posterior four are legs and the anterior two are modified for feeding. The

most diverse and ecologically important members are the scorpions, harvestmen (daddy longlegs), spiders, mites, and ticks.

The CRUSTACEANS (class Crustacea) are the dominant arthropods of the sea. One group alone, the alga-feeding copepods, are so dense in the plankton that they may well be the most abundant of all groups of animals. Although crustaceans are distinctive in many ways, they can be characterized most simply as arthropods with two pairs of antennae. In addition to the familiar shrimps, sowbugs, sand fleas, lobsters, crayfish, and crabs, all of which belong to phylogenetically advanced groups, there is a vast array of more primitive forms, many bearing a superficial resemblance to shrimps, that are abundant but too small to attract popular notice.

We are now living in the Age of Arthropods, or more precisely, the Age of Insects. The total number of insects on Earth is estimated to be 10^{18} — a billion billion — almost one billion insects for every human being.

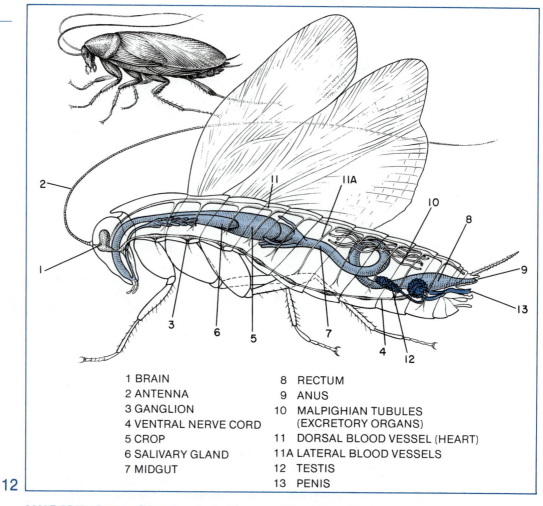

1 BRAIN
2 ANTENNA
3 GANGLION
4 VENTRAL NERVE CORD
5 CROP
6 SALIVARY GLAND
7 MIDGUT
8 RECTUM
9 ANUS
10 MALPIGHIAN TUBULES
    (EXCRETORY ORGANS)
11 DORSAL BLOOD VESSEL (HEART)
11A LATERAL BLOOD VESSELS
12 TESTIS
13 PENIS

12

MALE COCKROACH exhibits the principal features of the arthropod body plan. Armored exoskeleton is segmented and hinged. Various segments have evolved scores of specialized appendages, including those for grasping, chewing, walking, and (in advanced insects) flying. Respiratory system, discussed in detail in Chapter 9, is omitted from this diagram.

The MYRIAPODS are an assemblage of four classes of terrestrial animals: the centipedes (hundred legs), the millipedes (thousand legs), the symphylans, and the pauropods. The centipedes, predators of insects and other small animals, and the millipedes, scavengers and plant eaters, are the most abundant and diverse of the four groups. The symphylans, which resemble small white centipedes and are relatively abundant in the soil, are most notable as the probable ancestors of the insects.

INSECTS

The INSECTS (class Insecta), the little conquerors of the land and air, differ from the arthropods in the possession of three basic body parts (head, thorax, abdomen), a single pair of antennae on the head, and three pairs of legs originating from the thorax. Respiration is accomplished by means of air sacs and channels. The latter are called TRACHEAE (singular, trachea); they branch and extend from external openings inward to tissues throughout the body (*Chapter 9*). The higher insects are distinguished by the power of flight. The adults, but not the immature stages, possess two pairs of stiff, membranous wings attached to the thorax.

The number of described insect species is rapidly approaching one million. One authoritative estimate based on statistical analysis places the actual number of species alive on Earth at approximately three million. The number of individuals, incidentally, is believed to be in the neighborhood of 10^{18}, or a billion billion — roughly one billion insects for every human being. Entomologists divide the known living species into about 26 orders. One can make some immediate sense from this bewildering variety by sorting the species into four major adaptive levels, or EVOLUTIONARY GRADES. At the first (lowest) level are the primitive wingless insects, including springtails (order Collembola) and silverfish (Thysanura), small insects that mostly live hidden in the soil, beneath rocks, or in the crevices of tree trunks. Some, particularly the thysanurans, are not far removed from the ancestral myriapods.

Winged insects belong to the three higher grades. The second grade consists of insects with wings that cannot be folded against the body. They are often excellent flyers but need a great deal of open space in which to maneuver when they land. These insects include dragonflies (order Odonata) and mayflies (Ephemeroptera). The third grade includes insects that can fold their wings over their backs when not in use, a trick that bestows considerable versatility on the animals that can do it. They can fly from place to place, then upon landing tuck their wings away and crawl into tight places just like wingless insects. Many orders have this ability, including Orthoptera (grasshoppers, crickets, cockroaches and walking sticks), Isoptera (termites), Dermaptera (earwigs) and Homoptera (aphids, cicadas and leafhoppers). Some third-grade orders such as Anoplura (lice) have lost their wings during evolution and have become secondarily flightless.

The fourth (highest) grade includes those insects that not only can fold their wings, but also undergo COMPLETE METAMORPHOSIS during development. In other words, the newly hatched insects are not merely nymphs, which look like miniature adults, but are wormlike larvae that change first into a pupa, often encased in a cocoon, before emerging as an adult. The adaptive advantage of complete metamorphosis lies in the extreme specialization of the life stages. The larvae are totally adapted to feeding and growing, while the adults are specialized for reproduction and dispersal. Often the adult has a different diet, so that larvae and adults can exploit more of the envi-

13

FOSSIL TRILOBITE belongs to a class of arthropod that flourished in the seas of the Cambrian and Ordovician periods. Traces of the shell are visible at the top of this fossil impression, which is reproduced about actual size.

ronment without competing with one another. The fourth grade includes several of the most diverse and successful of all insect orders: Coleoptera (beetles), Lepidoptera (butterflies and moths), Diptera (flies), Siphonaptera (fleas) and Hymenoptera (bees, wasps and ants).

ECHINODERMS

As Figure 14 indicates, the rest of the higher animals, from starfish to man, belong to a diverse group known as the Echinoderm Superphylum. The characteristics of this superphylum are summarized briefly in Box C.

The echinoderms proper (phylum Echinodermata) include sea stars, sea urchins, sea cucumbers, sand dollars, and related forms. The echinoderms are truly different animals, so distinctive in body plan from other higher animals that if they were not already so familiar as sea animals they might seem to have originated from some other world. Their single

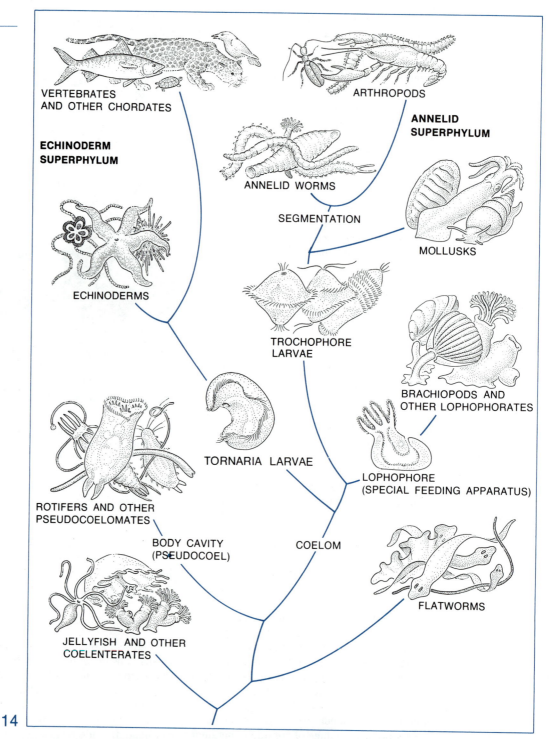

VERTEBRATES
AND OTHER CHORDATES

ARTHROPODS

ECHINODERM
SUPERPHYLUM

ANNELID
SUPERPHYLUM

ANNELID WORMS

SEGMENTATION

MOLLUSKS

ECHINODERMS

TROCHOPHORE
LARVAE

BRACHIOPODS AND
OTHER LOPHOPHORATES

TORNARIA LARVAE

LOPHOPHORE
(SPECIAL FEEDING APPARATUS)

ROTIFERS AND OTHER
PSEUDOCOELOMATES

BODY CAVITY
(PSEUDOCOEL)

COELOM

FLATWORMS

JELLYFISH AND OTHER
COELENTERATES

14

ECHINODERM SUPERPHYLUM completes the phylogenetic tree depicted in Figure 10. Similarities in embryonic development, and between the larvae of echinoderms and those of certain primitive chordates, provide strong evidence that the two groups arose from a common ancestor.

C

**THE
ECHINODERM
SUPERPHYLUM**

At first glance a starfish and a bird seem completely unrelated. Yet the primitive members of the Echinodermata and Chordata, the respective phyla of starfish and birds, are very similar in their early stages of development. Many species in the echinoderm superphylum share a characteristic form of larva, in this case the tornaria larva, easily recognized by the winding, longitudinal bands of cilia by means of which it swims through the open sea. The echinoderms and chordates are also distinguished from other animals by the earliest stages of embryonic development. These embryonic characteristics indicate that ancestors of the echinoderms and chordates split off from the rest of the coelomate animals at a very early stage in their evolution.

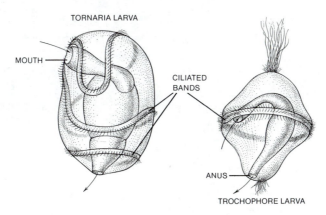

TORNARIA LARVA of echinoderms and primitive chordates is compared with the trochophore larva of annelids and mollusks in diagram above.

most characteristic feature is the WATER-VASCULAR SYSTEM, an array of canals and tubelike appendages that serve several functions simultaneously, principally locomotion and the capture of food. The structure of this system can be roughly grasped by examining a diagram such as that of the starfish shown in Figure 15. Although the total functioning of the system has never been fully understood, the mechanical operation of the basic unit, the TUBE FOOT, is clear enough. Each tube foot is a little adhesive organ that creates suction by hydraulic expansion and contraction. The tube feet serve starfish and other echinoderms in a variety of other ways.

They are used in locomotion and as organs of touch. Because of their thin walls, they also function as the equivalent of lungs or gills during respiration.

Echinoderms are distinctively armored. They possess an internal skeleton of calcareous plates that are either articulated, permitting flexibility, or else fused to form a rigid skeletal box. Warty or spiny projections extend outward as extra protective devices — the word echinoderm in fact means spiny skin. The echinoderms are a very ancient and diverse group. In addition to the starfish (class Asteroidea) there are four major living groups: brittle stars (class Ophiuroidea), sea urchins

Where did the vertebrates come from? Two theories have been advanced but neither can be tested, because the primordial vertebrates have not yet been found in the fossil record.

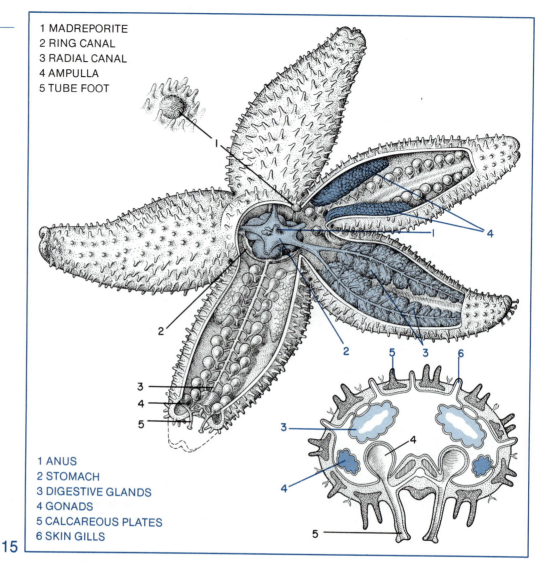

1 MADREPORITE
2 RING CANAL
3 RADIAL CANAL
4 AMPULLA
5 TUBE FOOT

1 ANUS
2 STOMACH
3 DIGESTIVE GLANDS
4 GONADS
5 CALCAREOUS PLATES
6 SKIN GILLS

15

CUT-AWAY VIEW OF STARFISH reveals the water-vascular system, digestive glands (light color), and sex organs (dark color). Function of water-vascular system is explained in text.

and sand dollars (class Echinoidea), sea cucumbers (class Holothuroidea) and sea lilies (class Crinoidea).

CHORDATES (PHYLUM CHORDATA)

Most of the chordates in the sea, and all on the land, are vertebrates. The remaining species, the INVERTEBRATE CHORDATES, share certain similarities in their embryonic development and display, at some point of their life cycle, the three diagnostic characteristics of the phylum Chordata: 1, a dorsal, hollow nerve cord; 2, clefts in the wall of the throat region, usually referred to as gill slits, which function to circulate water during feeding and respiration; and 3, a notochord, a unique stiffening rod located along the back. The body plan of one species, amphioxus, is depicted in Figure 16. Amphioxus is a small fishlike animal that burrows in the sand and feeds by filtering food particles from the water.

VERTEBRATES

The VERTEBRATES (subphylum Vertebrata) rank with the arthropods as dominant animals of sea, land, and air. The fishes in particular are the principal large-bodied carnivores and scavengers of the sea and fresh water, while amphibians, reptiles, and many of the birds and mammals occupy the same position on the land. Other kinds of birds, and a large array of mammals, from mice and rabbits to kangaroos, antelopes, and elephants, are among the most important plant feeders. Although low in numbers of individual organisms, the vertebrates rival the major invertebrate phyla in diversity and ecological significance. But, of course, the single fact of overriding consequence is that the vertebrates produced us, and thus deserve our closest attention.

A vertebrate is an animal with a series of segmented bones, vertebrae, which surround the notochord and nerve cord (Figure 17). In the evolutionarily more advanced groups the notochord is present only during embryonic stages and is supplanted in its protective role by the vertebrae during most of the life of the organism. The vertebrate organism is further characterized by a complicated, closed circulatory system in which the blood, containing hemoglobin-filled red blood cells, is pumped by a chambered heart through dense systems of capillaries that penetrate deeply into all of the active tissues of the body. This better-designed circulatory system makes it possible for animals to grow large in size while remaining physically very active. A second trait associated with increased size and activity is the better-developed nervous system. The brain of even the most primitive vertebrates is superior to that of the invertebrates. In the birds and mammals it has advanced to a wholly new level of complexity and organization. A third development is the improved sensory apparatus, including large eyes capable of sharp image perception, ears that serve in primitive forms as organs of equilibrium and in higher forms for both equilibrium and hearing, and, in the more primitive, aquatic forms, a lateral-line system that provides a sixth sense capable of detecting slight changes in water pressure and currents.

Where did the vertebrates come from? Two theories have been advanced but neither can be tested because the primordial vertebrates have not yet been found in the fossil record. They are, however, closely approached by the jawless fishes, or AGNATHS (class Agnatha), a group that dates back at least to Ordovician times. A diversity of agnaths, many of them possessing strangely shaped armored bodies, flourished over a period of tens of millions of years, then disappeared almost completely as more advanced kinds of fishes became abundant. Only a few relict forms — the lampreys, hagfishes, and slime eels — survived beyond the Devonian period down to the present. All of these creatures are degenerate predators, making their living either by sucking blood from living fish or eating the flesh of dead or dying fish. Their eel-like bodies have completely lost the bones that characterized the early agnaths.

In the Devonian period, generally referred to as the Age of Fishes, an immense variety of more advanced types came to throng the seas and fresh water in company with the agnaths (Figure 18). Among the most primitive were the archaic placoderms (class Placodermi). These jawed fishes attained a distinctly higher evolutionary grade than the agnaths. Many also developed elaborate sets of swimming fins and sleek body forms that must have improved their maneuverability in the open water. A few attained huge size and were probably the top-level predators of the world at that time. The placoderms, however, were a short-lived group. All but a very few disappeared by the close of the Devonian period, and none survived to the end of the Paleozoic.

During the Devonian two other major groups began careers that were to be crowned with success down to the present time. The first were the sharks, skates, and chimaeras — the chondrichthyans (class Chondrichthyes). Their bodies possess a single type of stabiliz-

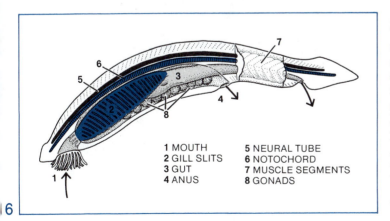

1 MOUTH	5 NEURAL TUBE
2 GILL SLITS	6 NOTOCHORD
3 GUT	7 MUSCLE SEGMENTS
4 ANUS	8 GONADS

AMPHIOXUS is an invertebrate chordate. About four or five centimeters long and shaped more or less like a fish, it spends much of its time buried in sand beneath coastal waters, feeding by sifting microorganisms from the water passing through its mouth and gills.

6

ing system which they share with all other higher fishes: a pair of pectoral fins just behind the gill slits, and a pair of pelvic fins just in front of the anal region. The sharks and their relatives "chose" speed and mobility over the more sluggish, turtle-like existence of most earlier fishes.

The "real" fish, in the minds of most people, are the BONY FISHES or osteichthyans (class Osteichthyes). They are indeed the most abundant and diverse of the fish classes in both fresh and salt water. Their variety is so great that no concise definition of their class is possible. As their name implies they have a more extensive internal bony skeleton than all other fish groups. But they are basically different because they originated in fresh water. The numerous marine groups that we know today are secondary invaders of the oceans. The

primitive bony fish possessed lungs, which supplemented the gills in respiration. Perhaps they used these extra organs to overcome occasional oxygen shortages encountered in the fresh-water habitats. The great majority of later bony fish converted the lungs into swim bladders, which now serve as organs that help to keep the fish suspended in water.

The invention of lungs by the early bony fishes set the stage for the invasion of land by their descendants. The osteichthyans which accomplished this feat were the crossopterygians, or LOBE-FIN FISHES, close allies of the lungfishes. This group flourished from Devonian into Mesozoic times.

The first land vertebrates, the AMPHIBIANS (class Amphibia), arose from crossopterygians in Devonian times. This epochal event (Figure 19) did not require a new design of the res-

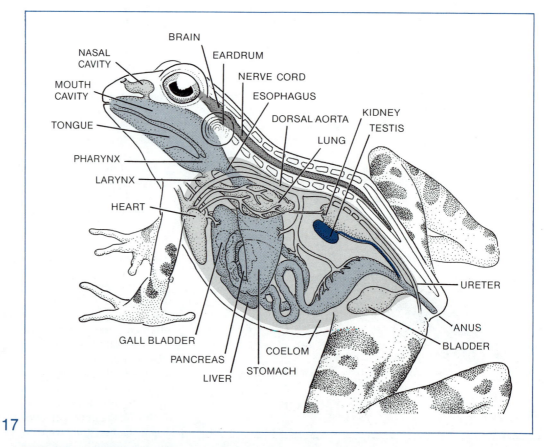

17

THE FROG exemplifies the characteristics of a typical vertebrate: segmented vertebrae surrounding the nerve cord; closed circulatory system; and a well-developed nervous system with advanced sensory organs. In advanced vertebrates the notochord is present only in embryos.

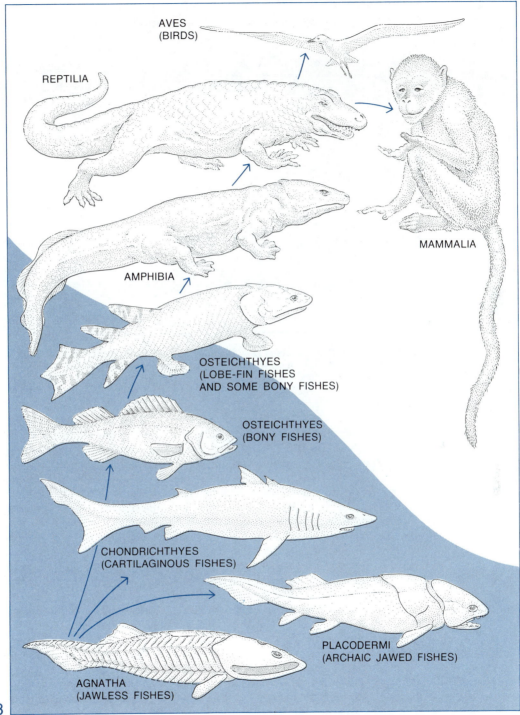

AVES
(BIRDS)

REPTILIA

MAMMALIA

AMPHIBIA

OSTEICHTHYES
(LOBE-FIN FISHES
AND SOME BONY FISHES)

OSTEICHTHYES
(BONY FISHES)

CHONDRICHTHYES
(CARTILAGINOUS FISHES)

PLACODERMI
(ARCHAIC JAWED FISHES)

AGNATHA
(JAWLESS FISHES)

18

EVOLUTION OF VERTEBRATES. The first vertebrates are thought to have resembled the jawless fishes (agnaths) that later virtually disappeared as more advanced fishes became abundant. Lobe-fin fishes gave rise to amphibians, which marked the beginning of the transition to land. Reptiles were first true land animals, and they in turn gave rise to both birds and mammals.

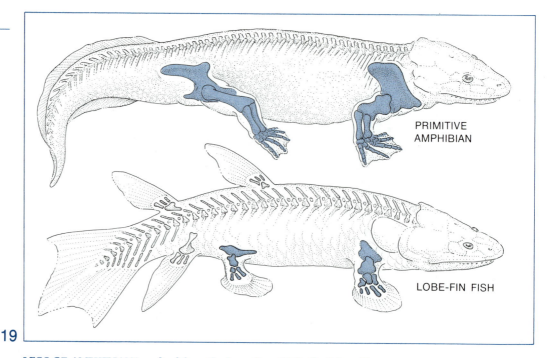

19

LEGS OF AMPHIBIANS evolved from the bony fins of lobe-fin fishes. The fishes were probably able to crawl from stream to stream or from pond to pond on their stubby fins.

piratory system, for the ability to breathe air had been achieved earlier when the ancestral bony fishes evolved lungs at the beginning of their history. Instead, the crucial step was the evolution of the stubby fins of the crossopterygians into the walking legs of the amphibians — legs whose basic design has since been bequeathed all the way on down to us in their basic design. Quite likely the Devonian crossopterygians were able to crawl from one pond or stream to another by pulling themselves along on their fins. The earliest amphibians, which remained very fishlike in most of their body structure, merely perfected this particular locomotory ability.

Most of the modern amphibians remain chained to the water for at least part of their life cycle. The bulk of the anurans (frogs and toads) and salamanders spend part or all of their adult lives on the land, typically in a moist habitat, but return to fresh water to lay their eggs. The primitive structure of the egg requires this reversion to the ancestral home. The eggs are small, surrounded by delicate membranes, and contain only limited supplies of yolk. Usually they give rise to an aquatic larva, such as the tadpole of the frogs and

toads, which pursues a fishlike existence for some period of time before transforming into the terrestrial adult form.

To understand the vulnerability of the amphibians is to understand a great deal about the advance in evolutionary grade that brought about the origin of the REPTILES (Class Reptilia). Primitive reptiles arose from primitive amphibians in the early Pennsylvanian period at the latest, some 310 million or more years ago. Reptiles became the first true fully liberated land animals. Those that live mostly in water, such as the majority of turtles, entered this habitat later as a secondary adaptation. The liberation of the reptile life cycle from the water was achieved by the AMNIOTIC EGG (*Chapter 17*). This type of egg is familiar to everyone. It is preserved in hens and other birds which are the direct descendants of reptiles. It has a leathery, calcium-impregnated shell that resists evaporation of the precious fluids inside. Within the shell and surrounding the embryo are several membranes, including the amnion from which the egg gets its name, that give added protection from drying and assist the embryo in excretion and respiration. Finally, the embryo is supplied with large quantities of yolk that permit it to attain a relatively advanced state of development before it must break the shell and face the outside world. Such an egg need not be laid in the water. It can be deposited on the land, even in a dry place, a circumstance that permits the adult reptile to move for unlimited distances away from bodies of fresh water.

Reptiles are advanced beyond the amphibians in other ways. Fertilization is achieved within the body of the female by copulation, permitting mating on land. The ventricle of the heart is partially divided into chambers that separate the freshly oxygenated blood from unoxygenated blood and permit it to be pumped to needy tissues more efficiently. Respiration is improved by a bellows-like movement of the ribs. Even the brain is more advanced. In the reptiles one finds the first small cerebral hemispheres, which in the mammals later became the centers of the higher mental faculties. The total result of all these changes was the transformation of the reptile into a more alert and adaptable animal.

During the Mesozoic era, often called the Age of Reptiles, the reptiles went through an extraordinary diversification and became the dominant large animals of the land (see drawing on opening page of this chapter). During this time two orders, the Ornithischia and Saurischia, popularly called the dinosaurs, were the prevalent large reptiles. But many other forms flourished, from the familiar and relatively humble turtles, snakes, and lizards to extreme aquatic and marine groups such as the plesiosaurs, ichthyosaurs, and mosasaurs. By the end of the Mesozoic, most of the great assembly disappeared, to be replaced in the Cenozoic by an equally impressive group of mammals.

Well into the Mesozoic, at least three separate lines of reptiles achieved the capacity for sustained flight by a flapping movement of wings. Two of the groups, collectively called pterosaurs (winged reptiles), still retained a basically reptilian anatomy. They became extinct by the close of the Mesozoic era. The third group advanced so far as to deserve the distinction of being placed in a class of their own. These are of course the BIRDS (class Aves), which survived the demise of the ruling reptiles to become one of the dominant vertebrate groups of the Cenozoic era. Zoologists sometimes lightly refer to birds as "glorified reptiles" and an important truth is embodied in the jest. Virtually all of the important differences between the birds and their reptilian ancestors have arisen as adaptations to flight. The single most characteristic feature is their feathers, which are highly modified versions of the old reptilian scales. The body skeleton was extensively modified to serve the requirements of flight. One of the most conspicuous changes was in the shape of the sternum, which has been transformed into a large vertical keel for the attachment of the breast muscles. These muscles (the familiar white or breast meat of chicken and other fowl) pull the wings downward during flight. Flying also requires a constant high temperature, so birds have become warm-blooded. It requires a highly efficient circulation, which has been achieved by the complete division of the ventricle of the heart into two chambers. One chamber pumps "used" blood to the lungs; the other receives the freshly oxygenated blood from the lungs and pumps it to the rest of the body. The brain is much larger than that of reptiles. The change is not in the cerebral cortex, the principal seat of intelligence in mammals, but rather in the centers of sight and muscular coordination.

MAMMALS

In a real sense the highest evolutionary grade of all animals is the MAMMAL (class Mammalia). About 65 million years ago — only 11 hours ago on our calendar — the Mesozoic era, the Age of the Reptiles, gave way to the Cenozoic era, the Age of Mammals. The change was not simply that mammals arrived on the scene to displace an obsolete set of reptiles. On the contrary, mammals appeared in the early part of the Mesozoic and coexisted as lesser partners of the ruling reptiles for tens of millions of years. When the dinosaurs and other dominant reptile groups disappeared at

During the Mesozoic era reptiles became the dominant large animals of the land. By the end of the Mesozoic most of the great assembly had disappeared, to be replaced in the Cenozoic by an equally impressive group of mammals.

the close of the Mesozoic, the mammals increased dramatically in numbers and diversity. Why the reptiles declined is not really known; it has often been suggested that the mammals gained enough momentum to assist in the process. Whether or not they played the role of exterminators, they inherited that part of the ecological world controlled earlier by reptiles.

Humans should not hesitate to designate themselves, and other mammals, as superior animals, physically and mentally (*Figure 20*). Virtually everything that is distinctively mammalian is geared to an increase in speed, in alertness, in intelligence, and in protection given the young. As the name implies (from *mammae*, or milk glands), only the mammals suckle their young with nutritive fluid. Except in the most primitive forms, the embryos are not deposited in shelled eggs but are nurtured within the body of the mother until they reach

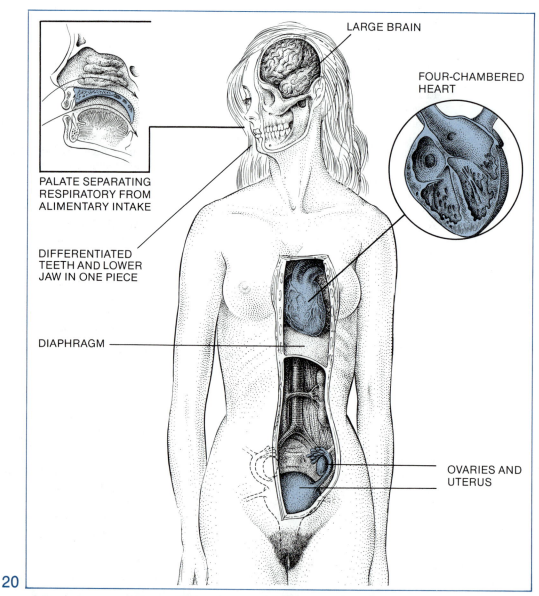

LARGE BRAIN

FOUR-CHAMBERED HEART

PALATE SEPARATING RESPIRATORY FROM ALIMENTARY INTAKE

DIFFERENTIATED TEETH AND LOWER JAW IN ONE PIECE

DIAPHRAGM

OVARIES AND UTERUS

20

WOMAN AS A MAMMAL. In addition to mammary glands, several other anatomical traits are characteristic of the class Mammalia. Some of them are identified above; others are discussed in the text. Uterus is somewhat enlarged, as it would appear early in pregnancy. Features illustrated here are typical of placental mammals; marsupials have an external pouch for carrying young.

an advanced stage of development. The jaw is strengthened, and teeth are differentiated into types variously specialized for cutting, chewing, and grinding. A muscle wall, the DIA-PHRAGM, completely separates the chest cavity from the abdominal organs and increases the depth and efficiency of breathing. The heart, like that of birds, is improved by a complete division of the ventricle into two chambers. The legs are swung beneath the body, permitting a clean back-and-forth motion of the legs and greater speed than is possible with the sprawling posture of the reptile. The mammal is warm-blooded; its temperature is kept high and relatively constant, allowing it to be active in a wide range of temperatures and climates. The brain is greatly enlarged, chiefly in the cerebral hemispheres. The key to mammalian success is internal activity guided by intelligence. Almost every living mammal is capable of more learned behavior than any living amphibian or reptile. Human evolution, the apogee of this development, is described in detail in Chapter 27.

READINGS

R.D. BARNES, *Invertebrate Zoology*, 3rd Edition, Philadelphia, Saunders, 1974. One of the best general textbooks, covering all groups from the Protozoa to the invertebrate chordates.

W.D. RUSSELL-HUNTER, *A Biology of Lower Invertebrates*, New York, Macmillan, 1968. A somewhat more advanced treatment of functional anatomy and evolution. It gives a scholar's view of the current "state of the art" of invertebrate zoology.

W.D. RUSSELL-HUNTER, *A Biology of Higher Invertebrates*, New York, Macmillan, 1969. The companion book to the above.

F.M. BAYER AND H.B. OWRE, *The Free-Living Lower Invertebrates*, New York, Macmillian, 1968. Beautifully written and illustrated, this is an authoritative and esthetic introduction to the acoelomate animals.

A.S. ROMER, *The Vertebrate Body*, 4th Edition, Philadelphia, W. B. Saunders Company, 1970. One of the best and most enjoyable accounts of vertebrate zoology, with a strong emphasis on evolution.

PART FOUR THE STRATEGY OF EVOLUTION

Evolution is the ultimate existential game, a game most species are doomed eventually to lose. Defeat — extinction — is final. Victory brings only the meager privilege of staying for another round, a round in which nature is both the opponent and the umpire, selecting the winners by rules that change unpredictably. In one generation a species may have to survive prolonged drought; in the next unexpected cold, followed perhaps by a food shortage or a new disease.

The species alive today, from *E. coli* to man, comprise a small group of temporary winners. Shaped by the interaction of life and Earth for roughly four billion years, modern organisms are formidable competitors, genetically adapted to survive in harsh and variable environments. At times they seem to follow a collective strategy for survival. Strictly speaking, however, living organisms do not consciously evolve "in order to" meet a particular environmental challenge. They evolve because certain combinations of genes possessed by some organisms enable those organisms to meet the test of the environment better than others, and hence these organisms are able to survive and reproduce when others die. In retrospect one can observe certain trends in evolution — patterns of play common to all living things — and for verbal convenience they can be called a strategy. In this sense, then, one can say that life has staked its survival on a strategy of expansion and diversity. Each species tends to multiply and occupy new territory, expanding aggressively until halted by natural conditions: a food shortage, a geographic barrier, predators, and so on. Diversity enables life to hedge its bets in the contest with nature. The more kinds of organisms that exist, the smaller the chance that all life might be destroyed by any one environmental disaster. And an assortment of organisms can exploit the resources of an environment more fully than can a single species.

Diversity is a result of genetic variability. The most successful players of the game of evolution somehow strike a balance between adaptiveness (the physiological capacity to cope with their present environment) and adaptability (the capacity to produce offspring with new combinations of genes, combinations that may be better suited to future environments).

Mutation and sexual reproduction are the mechanisms that create genetic variability, mutation by creating new genes and sexual reproduction by rearranging genes into new combinations. Imagine that two individuals in a species both possess *Aa* genotypes for a single locus. If they could only reproduce asexually their offspring would all be *Aa*. But if they could mate, their offspring would be a mixture of *AA, Aa* and *aa*. Collectively the sexual offspring could adjust to environmental conditions that favored any of the three genotypes. As a group, they are more adaptable than the otherwise equivalent asexual offspring.

The more variable a species — that is, the more heterozygous combinations it contains — the faster it can adjust to new conditions in the environment through evolution. But the greater its variability, the less perfect the adaptation to any particular environment at a given moment. The genotype *AA* may be superior in one environment and *aa* in another. Adaptiveness and adaptability are antagonistic requirements, and each species must maintain a degree of genetic variability that represents a compromise between the two.

Evolution is not a game of individuals but of groups of organisms. The basic unit of evolutionary studies is the population — a group of individuals of the same species who exchange genes by interbreeding. For this reason the study of evolution is to a large extent a branch of population biology.

Individual mortality is part of the price that a species pays for long-term adaptability. Each individual is doomed to perish regardless of whether or not its species survives. If individual organisms developed the potential to live forever, and ceased reproduction to avoid the risk of being replaced by their offspring, there would be no evolution. A species of immortals would, in effect, be staking its existence on an all-or-nothing solution to the problems of survival. The odds against success would be overwhelming. Living organisms must continually invent new solutions to the challenge of the environment. Nature selects the best solutions and in the process chooses the players for the next round of evolution.

Chapter 21, *The Process of Evolution*, introduces the basic theories of evolution. The remaining chapters in this section provide a glimpse of the game in action, beginning with the multiplication of species (*Chapter 22*). The factors that govern the distribution of animals and plants are described in the chapter on biogeography (*Chapter 23*) and in the following two chapters, which treat populations and ecosystems. The section ends with two chapters on the emergence of superorganisms — animals that can multiply their capabilities by uniting in societies of highly specialized individuals. The path of social evolution has led to two triumphs of organization, one attained by insects (*Chapter 26*) and the other by man (*Chapter 27*).

21

the process of evolution

There is grandeur in this view of life, with its several powers, having been originally breathed into a few forms or into one; and that, whilst this planet has gone cycling on according to the fixed law of gravity, from so simple a beginning endless forms most beautiful and most wonderful have been, and are being, evolved.

CHARLES DARWIN

Evolution — is it a fact or a theory? The question is important because it can shed light on the scientific distinction between fact and theory. The process of evolution is a fact. Biologists have measured its progress at the level of the gene. They have created new species in the laboratory and in the experimental garden. They have collected much fossil evidence, in many cases so complete that it cannot be rationally explained by any nonevolutionary hypothesis. But exactly *how* evolution occurs has been the subject of rival theories. Of the two leading theories of the nineteenth century, Lamarckism and Darwinism, the Darwinian explanation has been accepted to the point where it is now almost universally used to account for evolutionary phenomena. The modern version of Darwinism fits so well into genetics, paleontology, systematics, and other branches of biology, that it must be regarded as one of the most reliable explanatory systems in all of science.

LAMARCKISM

To understand Darwinism fully, one needs to look back briefly in history to see how it contrasts with its chief rival. In 1809 Jean Baptiste de Lamarck, the great French systematist, proposed a very ingenious theory of evolution. He suggested that organisms evolve into new kinds of organisms because they want to. They try to do something new, or they try to improve on what their ancestors were doing before them, and the effort results in a change in the size of the particular organ or capacity that is exercised. If this change is passed on in some degree to the offspring of the organism that made the effort, the result is evolution. Lamarck cited some arresting examples from nature that he believed could be explained by his theory. Giraffes have long necks, he argued, because for many generations they have stretched to reach the succulent branches and leaves high up in trees. As the members of each generation succeeded in lengthening their necks, they passed some or all of this acquisition to their offspring, who were then in a position to add to it still further on their own. The stork, to take a second of Lamarck's hypothetical examples, has a long neck because it has attempted over many generations to catch fish in ever deeper water. To this Lamarck added an extra bit of speculation of the kind that often got him into trouble with his contemporaries: the long legs of the stork can be explained by the fact that the bird does not like to get its belly wet.

Lamarckism, also known as the theory of evolution by the inheritance of acquired characteristics, actually has some good features as theories go: it is simple, clear, provocative, and easy to test. But the tests showed that it is wrong. It is clear from a modern understanding of genetics that DNA directs the formation of the phenotype — the length of the giraffe's neck, for example — but does not accept instruction back from the phenotype. Of necessity something else directs the modification of DNA.

H.M.S. BEAGLE, depicted on opposite page, was the ship that carried young Charles Darwin on round-the-world voyage that led to his theory of evolution. Ship was painted in 1841 by Owen Stanley as she lay at anchor in Sydney harbor.

DARWINISM

The directing force was described independently in 1858 by Charles Robert Darwin and Alfred Russell Wallace in the *Journal of the Linnean Society of London*. In 1859 Darwin published the book *On the Origin of Species by Means of Natural Selection*. This classic work became the means by which the theory gained support and, in the process, came to be called Darwinism.

The theory of Darwin and Wallace can be expressed as follows: Every population of individual organisms contains genetic variability. Some of the hereditary traits permit individuals to survive and reproduce better than others. As a consequence, these superior traits become more prevalent in later generations. In other words, evolution has occurred. As long as there is genetic variability in the population evolution can continue.

Table I presents the approximate form of the original 1858 argument. What is the fundamental difference between Lamarckism and Darwinism? Lamarckism is a primitive conception based on the idea that change is transmitted in a directed manner from individual to individual. Darwinism, in contrast, is based on the concept of the population as the unit of evolutionary change. In the Darwinist view it is the hereditary content of the population as a whole, and not just that of the single individual, which evolves (*Figure 1*).

Natural selection, as envisioned by Darwinism, occurs whenever some genotypes produce more representatives in the next generation than others. This greater fitness can come from a superior resistance to parasites and predators, or to cold, heat, drought, and other extremes of weather. It can result from a

greater capacity to invade habitats not previously occupied by the species. It can also result, of course, from the capacity to breed faster. Any or all of these qualities in combination can serve as the components of fitness. With each passing generation the superior genotypes will become more numerous relative to the less fit ones.

To complete this process of evolution one must imagine, like Darwin and Wallace, that new genotypes appear spontaneously within a population. They are then tested against the other genotypes by natural selection. Some will prove superior and increase in number. Others will prove inferior and disappear. Under these conditions evolution can continue indefinitely. Eventually each population will replace a large part of its original genetic material. The result is the creation of a very different kind of organism.

THE MODERN CONCEPT

The Darwin-Wallace theory is still regarded as essentially correct by biologists. In the language of modern genetics evolution can be broadly defined as a change in the heredity of a population. Population genetics permits an even more precise definition: EVOLUTION IS ANY CHANGE IN GENE FREQUENCY IN A POPULATION. Consider two alleles occupying the same chromosome locus within a given population. For convenience one allele can be labeled A and the other a. Each organism in the population will therefore consist of one of the following three diploid combinations: AA, Aa, or aa. Suppose further that in a certain population 60 per cent of all the alleles at the locus were A and 40 per cent a. The GENE FREQUENCIES are therefore 60 and 40 per cent respectively; in decimal notation, 0.60 and 0.40. In the usual symbols of genetics, the first frequency is denoted p and the second q. Then $p + q = 0.60 + 0.40 = 1.00$. Suppose that a researcher followed the frequencies p and q over three successive generations and found that they changed from one generation to the next as follows: $0.60 + 0.40 = 1.00$; to $0.59 + 0.41 = 1.00$; to $0.57 + 0.43 = 1.00$. In this case p, the frequency of A, steadily decreased while q, the frequency of a, increased by the same amount. This is evolution at the simplest possible level.

In such an elementary case it is easy to imagine a reversal in the trend, as in the following example: $0.57 + 0.43 = 1.00$; to $0.59 + 0.41 = 1.00$; to $0.60 + 0.40 = 1.00$. In practice, however, evolution of the whole population in

DARWIN-WALLACE THEORY of natural selection is a logical system of postulates and deductions. Brief synopsis in this table illustrates the conceptual framework of the theory.

I

POSTULATE	DEDUCTION
EACH POPULATION OF PLANTS OR ANIMALS TENDS TO GROW GEOMETRICALLY—THE MORE INDIVIDUALS THAT EXIST, THE FASTER THEIR NUMBER INCREASES. BUT THE SPACE AND FOOD THEY HAVE AVAILABLE TO LIVE ON INCREASE SLOWLY OR NOT AT ALL.	THERE IS A CONTINUING STRUGGLE FOR EXISTENCE AMONG THE MEMBERS OF THE GROWING POPULATION.
HEREDITARY DIFFERENCES EXIST AMONG MEMBERS OF THE POPULATION THAT AFFECT THEIR ABILITY TO SURVIVE AND TO REPRODUCE.	THE RESULT IS A CONTINUING PROCESS OF THE SURVIVAL OF THE FITTEST (NATURAL SELECTION).
NEW HEREDITARY VARIATION CONTINUES TO APPEAR IN THE POPULATION INDEPENDENTLY OF THE SELECTION PROCESS.	ORGANIC EVOLUTION OCCURS.

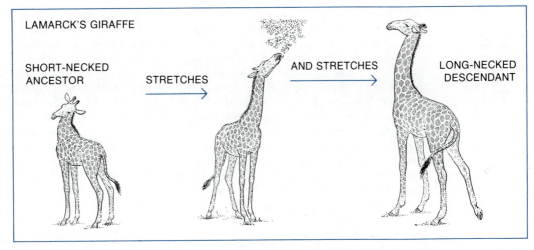

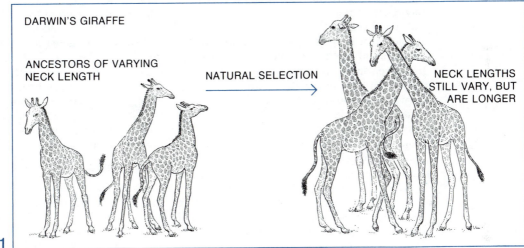

RIVAL EXPLANATIONS OF EVOLUTION were advanced by Lamarck and Darwin. According to Lamarck, the long neck of the giraffe arose as a result of generations of stretching to reach food. According to Darwin, giraffes with longer necks arose as a result of natural selection.

all of its traits is seldom reversed. The reason is simple. The heredity of most species of organisms is based not on a single locus but on thousands or tens of thousands of loci, many of them represented in the population by more than just one or two alleles. It is extremely unlikely that most or all of this huge assemblage of allele systems would undergo complicated changes in relative frequency, then reverse themselves and return, all together, back to the starting point.

Even without the aid of this particular argument, biologists long ago recognized that EVOLUTION SELDOM IF EVER REVERSES ITSELF TO ANY SIGNIFICANT EXTENT. This generalization is called DOLLO'S LAW, after Louis Dollo, a Belgian paleontologist who first derived it from his studies of fossils. Actually, Dollo's Law is

*Evolution seldom if ever reverses itself. As species evolve —
and all species do evolve, almost all the time — they always
diverge from one another.*

not a true law like the laws of physics and chemistry, but rather a strong inference based on the fossil record and some persuasive theoretical deductions from population genetics. The law has some far-reaching implications. One is that once a major anatomical structure has been lost or transformed into another structure, it can never be regained in its original form. The hind legs of seals, for example, which evolved into flippers in the remote past, can never return to a walking form identical with that of land-dwelling mammals. Similarly the eyes of blind cavefish, having been genetically lost, cannot be regained. No mammal can ever evolve back into the ancestral reptile, no reptile can evolve into the ancestral amphibian, and so on down the phylogenetic tree.

A corollary of Dollo's Law states that as species evolve — and all species do evolve, almost all of the time — they always diverge from one another. All bird species have always diverged from one another and, as a group, from all mammal species. Occasionally there is some convergence in life habits or in body form. Bats, for example, are mammals that acquired birdlike wings and the power of flight. But the resemblance is strictly superficial. Modern species of bats differ genetically from birds more than their flightless ancestors.

The problems of the science of population genetics can be reduced to two fundamental questions stemming from the original Darwin-Wallace model of evolution. First, what is the origin of the basic units of genetic variation at the levels of the gene and chromosome? Second, what are the causes of changes in their frequencies in populations? Mutations and grosser alterations in the structure of chromosomes create the raw materials of evolution. By themselves they do not cause more than trivial changes in gene frequencies, except in the few cases where their rates are abnormally high. The major thrust of evolution, consisting of changes in the gene frequencies of entire populations, is caused by a complex array of other causes. Of these by far the most important is natural selection.

Even without the benefit of modern knowledge the early Darwinian notion of random genetic fluctuation turned out to be an adequate approximation of the effects of mutations and chromosome aberrations. The development of evolutionary biology since about 1920 is often referred to as the MODERN SYNTHESIS, or NEO-DARWINISM, by which is meant that Mendelian genetics has been fused with the theory of natural selection, creating the basic discipline of population genetics. Population genetics has in turn been applied with great success to the reshaping of such subjects as systematics, speciation theory and biogeography, which will be explored in some detail in subsequent chapters.

THE HARDY-WEINBERG LAW

The keystone of modern evolutionary theory is the Hardy-Weinberg Law, named after the British mathematician and the German biologist who developed it independently in 1908. The law states that THE PROCESSES OF SEXUAL REPRODUCTION DO NOT OF THEMSELVES CHANGE THE FREQUENCIES OF GENES (for example, of A and a) or of the diploid genotypes

---

A

DAUGHTERS OF THE REVOLUTION?

The members of a women's patriotic organization proudly trace their descent from ancestors who aided the cause of independence in the American Revolution. Let us examine the genotype of a woman descended from one of the Minutemen. Half of her genes came from each of her two parents, one fourth (on the average) from each of her four grandparents, one eighth from each of her eight great-grandparents, and so on back through time. She is separated from her Minuteman ancestor by approximately seven generations. She had no fewer than $2^7 = 128$ ancestors in that generation. She can be proud of her Minuteman ancestor, but he contributed only about 1/128th of her genes — the equivalent of less than one half of one of the 46 chromosomes in each of her cells. Where did the rest of the Minuteman go? Provided he founded a lineage of average size, his genes have been dispersed to a very large array of people now alive. If he and his descendants followed the American pattern of large families, this number is likely to be even more than 128 persons (Figure 2).

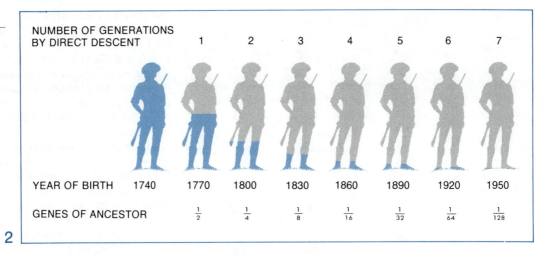

NUMBER OF GENERATIONS BY DIRECT DESCENT	1	2	3	4	5	6	7
YEAR OF BIRTH 1740	1770	1800	1830	1860	1890	1920	1950
GENES OF ANCESTOR	$\frac{1}{2}$	$\frac{1}{4}$	$\frac{1}{8}$	$\frac{1}{16}$	$\frac{1}{32}$	$\frac{1}{64}$	$\frac{1}{128}$

2

DISPERSION OF GENOTYPE in successive generations is a result of meiosis and sexual recombination. Living man at far right, seven generations removed from ancestor who lived during American Revolution (here depicted as Minuteman), would carry only 1/128 of ancestor's genes.

(*AA*, *Aa*, and *aa*). In other words, evolution cannot be caused by sexual reproduction. At first this conclusion may seem obvious. But no normal cellular event is more complex or disruptive than gametogenesis, the first step of sexual reproduction. In diploid organisms a crucial event is the first meiotic division, during which the homologous chromosomes come together in synapsis, exchange segments, and then separate into different daughter cells. Repeated over many generations, this process increasingly fragments and disperses the original genotype. From the example in Box A one can see why population geneticists like to think of evolution as occurring in a GENE POOL (all of the genes of an entire population) rather than down separate lineages of organisms. When evolution occurs over more than a few generations (it often occurs over thousands) it is almost impossible to think in terms of individual organisms and of the fate of individual genotypes. Actually the individual genotype has no integrity over long stretches of time. Within several generations the "chromosome chopper" of meiosis has begun to dissolve it into the gene pool.

The Hardy-Weinberg law allows one to predict gene frequencies from genotype frequencies and vice versa. For the simplest possible case of two alleles *A* and *a* on the same locus, in each generation the frequencies of these two alleles, *p* and *q*, will remain constant. Moreover the frequencies of the genotypes will remain constant; they will be related to the allele frequencies as follows:

IN WORDS

Frequency of *AA* + Frequency of *Aa* + Frequency of *aa* = 1

IN SYMBOLS

$p^2_{(AA)} + 2pq_{(Aa)} + q^2_{(aa)} = 1$

To understand why this equation must be so, first examine the idealized life cycle of a diploid organism shown in Figure 3. Since the basic Mendelian laws of heredity are followed, the set of interbreeding individuals is called a MENDELIAN POPULATION. It is necessary to add that breeding must occur at random within the population. Up to this point we have been referring loosely to the concept of a population. Now we must be more specific and say that THE POPULATION IS A GROUP WITHIN WHICH ANY ADULT IS JUST AS LIKELY TO SELECT ANY MEMBER OF THE OTHER SEX AS A MATE, REGARDLESS OF LOCATION. Such a randomly mating population is called a DEME. It is the basic unit of evolution. A deme can consist of all the mice infesting a small house, for example, or all the robins nesting in a woodlot. The geographic limits of demes and the number of organisms comprising them differ widely among different kinds of plants and animals.

In a deme the gametes are mixed at random. As illustrated in Figure 3, each gamete carries either *A* or *a*. To see how the Hardy-Weinberg formulation is derived from a simple knowledge of *p* and *q* (the frequencies of the two

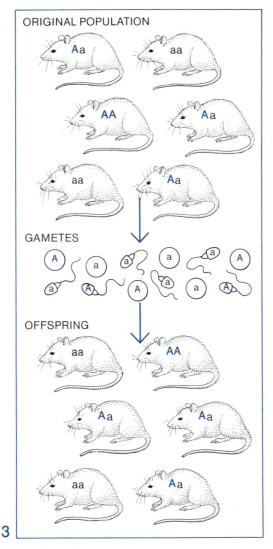

ORIGINAL POPULATION

Aa   aa

AA   Aa

aa   Aa

GAMETES

A   a   a   a   a   A
a   A   A   a   a   A

OFFSPRING

aa   AA

Aa   Aa

aa   Aa

3

HARDY-WEINBERG LAW states that frequencies of two alleles remain constant from generation to generation in a Mendelian population. Diagram shows two generations of mice. Despite meiosis and sexual recombination, frequencies of alleles A and a remain unchanged in the second generation.

fore also 0.5.) Six out of ten random selections of a sperm or egg will bring up an A allele. And since $p + q = 1$, the probability of drawing an a allele is $1 - 0.6 = 0.4$. Now the probability of joining two A-bearing gametes together at fertilization is (according to the theorem just cited) $p \times p = p^2 = 0.36$. Therefore 0.36 or 36 percent of the offspring in the next generation should have AA genotypes. Also, the probability of bringing two a-bearing gametes together at fertilization is $q \times q = q^2 = 0.16$. Table II shows that there are two ways of producing a heterozygote: an A sperm matched with an a egg, the probability of which is $p \times q$; and an a sperm matched with an A egg, the probability of which is also $p \times q$. Consequently the probability of obtaining a heterozygote from both ways is $2 pq$.

From the Hardy-Weinberg formula,

$$p^2_{(Aa)} + 2pq_{(Aa)} + q^2_{(aa)} = 1,$$

it is easy to show that the gene frequencies p and q remain constant each generation. Notice that in the sum on the left side of the formula a total of $p^2 + pq$ consists of A alleles. The fraction that this comprises of all the alleles is

$$\frac{p^2 + pq}{p^2 + 2pq + q^2} = p^2 + pq = p^2 + p(1 - p) = p.$$

By similar reasoning it can be shown that the frequency of a in the next generation will be q. Thus the original gene frequencies are preserved.

A fundamental trait of Mendelian demes is that no matter what the proportions of the diploid genotypes at the beginning, they will come to fit the Hardy-Weinberg distribution in the next generation and remain true to it every generation afterward. For example, suppose one started a new population by mixing 6000 AA individuals with 4000 aa individuals. No Aa individuals at all are included in this first generation. In the mixture $p = 0.6$ and $q = 0.4$. Applying the Hardy-Weinberg formula one can predict that in the next (that is, the second) generation approximately 36 percent of the individuals will be AA, 48 percent Aa, and 16 percent aa. This ratio will remain constant in all later generations.

The Hardy-Weinberg formula is the basic theorem of evolutionary genetics. In the study of real populations it is used to test hypotheses of random mating and of evolutionary change. If a local population is really a deme — if its members are truly breeding randomly — and if it is undergoing no evolutionary change,

alleles) one needs only to recall an elementary theorem of mathematics: the probability of the union of independent events is equal to the product of their separate probabilities. In the Hardy-Weinberg formulation the probabilities are identical to the gene frequencies. If the frequency (p) of A genes is 0.6, to take an arbitrary example, then the probability of any given sperm or egg bearing an A allele as opposed to an a allele is also 0.6. (Similarly, 0.5 of tossed coins come up heads, and the probability of a given coin coming up heads is there-the diploid frequencies will fit the Hardy-Weinberg predictions. If the diploid frequencies deviate significantly from the expected Hardy-Weinberg values, one must look for evidence of either nonrandom mating among the genotypes (such as a preference of individuals for mates of their own genotypes) or a change in frequencies due to evolution occurring from one generation to the next. In some cases both happen. At this point the reader should solve some problems using the Hardy-Weinberg formula to make sure he really understands it (see Box B).

SEXUAL REPRODUCTION

The genetic effect of normal sexual reproduction is to create new diversity in the diploid stage at each generation — without altering the frequencies of the genes involved. Sex is immensely effective in this, its primary role. Among asexual organisms (in which no recombination occurs) the only way for a population to increase its variability is through new mutations or immigration of new types from the outside. Among sexual organisms, on the other hand, new combinations of genes can be assembled on the same chromosome through crossing over at each gametogenesis (see Chapter 5). If the organisms are diploid, even more new combinations of chromosomes can be created at each fertilization. The collaboration of crossover, which generates new combinations of genes on the same chromosome, and

HARDY-WEINBERG FORMULA can be derived by calculating the probability of each possible combination of two alleles in sexual reproduction. Table shows worked-out examples in which p = 0.6 and q = 0.4. Sum of probabilities is always 1.

independent assortment, which changes the combinations of chromosomes, produces enormous genetic diversity. Consider the simplest possible case: an allele system in which two alleles on one locus produce just three diploid genotypes, AA, Aa, and aa. Addition of a second locus with two alleles, B and b, makes possible nine genotypes: $AABB$, $AABb$, $AAbb$, $AaBB$, $AaBb$, $Aabb$, $aaBB$, $aaBb$, $aabb$. The number of possible genotypes goes up swiftly with the addition of more such loci. In fact, the number in a population containing n of the loci is 3^n. This remains true where each locus contains only two alleles each. The number is usually much higher in actual cases. In general, if the number of genotypes possible at the first locus is m_1, and the number possible at the second locus is m_2, and so on, the total number of genotypes that can be put together from all of the n genotypes is simply the product of all these numbers; it is $m_1 \times m_2 \times m_3 \times \cdots \times m_n$. In most Mendelian populations, both the number of loci and the number of alleles per locus are large. For example, in both *Drosophila melanogaster* and *Homo sapiens* it is estimated that there are at least 10,000 loci, many of them carrying numerous alleles. The total number of conceivable diploid genotypes in such species is astronomical; in fact in these two examples the number is greater than that of all the atoms in the visible universe! Changing its heredity like a kaleidoscope, the sexual species exposes a new array of genotypes to the environment at each generation, while holding the frequencies of its alleles nearly constant. As a result sexually reproducing organisms enjoy an adaptability in the face of a changing environment far beyond the range of asexual species. This adaptability seems to be the reason why sexuality is nearly universal in nature. It has been relinquished only in a few groups of organisms where peculiarities in the form of adaptation — such as the need to reproduce rapidly by means of budding — set asexual reproduction at a special premium.

AGENTS OF EVOLUTION

Sexual reproduction creates diversity but does not force changes in gene frequencies. In other words, the population "uses" sex to increase the diversity it needs to evolve, but sexual reproduction alone does not cause the population to evolve. The essential changes in gene frequencies are caused by four evolutionary agents: MUTATION PRESSURE, GENETIC DRIFT, GENE FLOW, and NATURAL SELECTION.

Gene flow and selection are the most important in controlling MICROEVOLUTION — simple changes in gene frequencies among local populations. Selection alone is the guiding agent in the larger progressive changes of evolution, such as the origin of new structures and new major taxonomic groups. Such creative changes are based on hundreds or thousands of separate microevolutionary events, which are assembled and focused by the selective pressures exerted by the environment.

MUTATION PRESSURE

Mutations are of two basic kinds: POINT MUTATIONS and CHROMOSOME ABERRATIONS. Chapter 4 noted that point mutations are molecular events involving the substitution of one or more "wrong" nucleotide pairs in the DNA molecule. Chromosome aberrations are major structural changes that can be observed under the microscope. They involve hundreds or thousands of nucleotide pairs. Chromosome aberrations are often regarded by students as rather difficult to master, but there is an easy way to comprehend them in a few minutes. Think of a chromosome as a stick of putty and imagine all of the ways one can modify it without twisting or pulling it out of shape. The list, if complete, will form a catalog of chromosome aberrations. The putty stick, for example, can be broken up to create a larger number of smaller sticks (increase in chromosome number). It can be joined to another putty stick (fusion, leading to reduction in chromosome number). One can take a piece out of the stick (deletion); double an existing piece (duplication); remove a piece, flip it over, and reinsert

---

B

PROBLEMS IN POPULATION GENETICS

Problem. Twenty-five percent of the individuals of a population of cavefishes are found to be albino, a trait known to be controlled by a single recessive gene. Estimate the frequency of the gene in the population as well as the frequency of its heterozygotes.

Answer. Let us arbitrarily label the albino allele a, and the "normal," non-albino allele as A. It was stated that $q^2_{(aa)} = 0.25$. Then q = 0.5, the frequency of a in the population. And from this p = 1 − 0.5 = 0.5. The frequency of the heterozygotes is then estimated to be 2 pq = 0.50.

Problem. A species of flowering plants contains two alleles, a_1 and a_2, which control the color of the seed coats. In a breeding experiment a new population is started by mixing 5000 homozygotes of the allele a_1 with 95,000 homozygotes of the allele a_2. Predict the genotype frequencies in the next generation and in each generation thereafter.

Answer. From the data given, a = 0.05 and p = 0.95. It can be predicted that in the next generation, and in each succeeding generation, the frequency of a_1a_1 individuals will be $q^2 = (0.05)^2 = 0.0025$; the frequency of a_2a_2 individuals will be $p^2 = (0.95)^2 = 0.9025$; and the frequency of a_1a_2 individuals will be 2 pq = 0.0950.

Problem. The alleles at one locus in a population of moths is known to consist initially of 70% M and 30% m, where m controls a color trait recessive to that controlled by M. After a severe frost, and before the moths have a chance to breed again, the genotype frequencies are found to be 55% MM, 44% Mm, and 1% mm. Interpret the result.

Answer. If p = 0.70 and q = 0.30 as given, we should expect the following genotype frequencies: $p^2_{(MM)} = 0.49$, $2pq_{(Mm)} = 0.42$, and $q^2_{(mm)} = 0.09$. But after the frost the population contains far fewer mm individuals than predicted by the Hardy-Weinberg formula, somewhat more Mm individuals, and even more MM individuals. The most reasonable hypothesis is that a higher proportion of individuals possessing the M gene survive than those possessing the m gene. If the explanation is true, the moth population can be said to have undergone a small amount of evolution.

it (inversion); or transfer a piece to another stick (translocation). It is also possible simply to duplicate the entire stick. If the full set of chromosomes is duplicated — if the chromosome number per cell is doubled, tripled, or more — the result is called polyploidy. In their genetic consequences chromosome aberrations can be thought of in much the same way as point mutations. They have similar effects on the phenotype, in that their presence commonly alters such traits as size, color, fecundity, and behavior. With the exception of polyploidy, all segregate and recombine according to the same Mendelian laws.

Mutations, both point mutations and chromosome aberrations, create all of the genetic raw material on which evolution is based. But can they also *direct* evolution? Can mutations alone progressively change the frequencies of alleles or chromosome types (for example, a set of four such alleles or types: a_1, a_2, a_3, and a_4) that can mutate one into the other and back again:

$$a_1 \rightleftarrows a_2 \rightleftarrows a_3 \rightleftarrows a_4$$

The answer is a qualified yes. Mutations are usually reversible, as suggested by the arrows that lead backward as well as forward. Suppose that the forward rates were very much higher than the backward rates. If no other evolutionary agent (such as natural selection) intervened to reverse the process, a population composed originally of a_1 alleles might well end up composed of a_4 alleles. Such a change would be evolution by MUTATION PRESSURE.

However, the possibility that much evolution occurs by mutation pressure in any particular case is remote. The reason is that mutation rates are too low. Rates ranging as high as one in a thousand per generation are seldom encountered; one in a million is a much more typical number. The low rates are sufficient to create the variability needed to permit evolution to occur, but they are not great enough to drive evolution along by mutation pressure. If a mutation is occurring at the rate of one organism per million, this means a change in the gene frequency of only one in a million (say,

0.264318 to 0.264319), a negligible alteration that will almost certainly be offset by other more powerful agents such as natural selection and gene flow.

GENETIC DRIFT

Genetic drift is the alteration of gene frequencies through chance alone. It is most effective in very small populations: those containing fewer than 100 individuals. Suppose an investigator crossed $Aa \times Aa$ individuals of *Drosophila*, and selected four individuals from among the offspring to start a new population. If the offspring were chosen at random, would he expect them to yield the exact 1:2:1 Mendelian ratio? When only four individuals are involved, such a ratio would occur only part of the time. On other occasions, chance alone would dictate that the ratios would be one AA and three Aa, or two AA and two aa, or all AA, or all aa, and so on. To take a close analogy, when two coins are flipped they sometimes come up as two heads, sometimes as two tails, and sometimes as one head and one tail. If a population is small enough to have its gene frequencies altered by chance alone in this fashion, the result is evolution by genetic drift. It is even possible for an allele (or chromosome type) to be lost entirely from a population in one short step.

Genetic drift, like mutation pressure, is probably of minor significance in most populations, most of the time. But the process is likely to become important when the number of breeding individuals in a population drops to a few tens or hundreds. Then there is a good chance of some alleles becoming lost while others become simultaneously fixed. To use our familiar two-allele case again, the population might come to consist entirely of AA or aa organisms rather than of a mixture of organisms bearing AA, Aa, and aa. Geneticists have found that such a reduction in genetic variability not only lowers the capacity of a population to adapt to changes in the environment, but it also tends to reduce the overall fitness of the population as well. The reason is that certain of the alleles produce impaired

Natural selection is the sole agent that guides the larger progressive changes of evolution, such as the origin of new structures and major taxonomic groups.

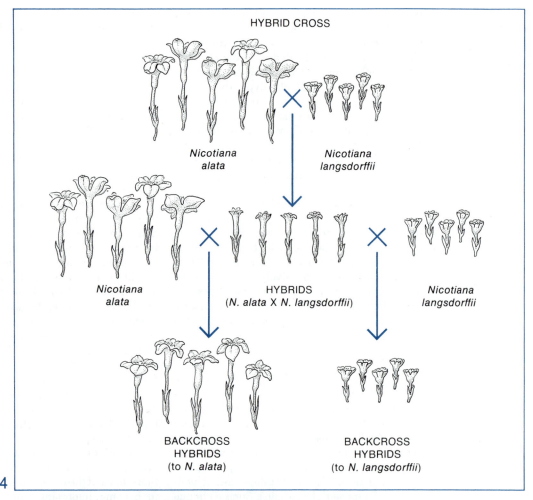

HYBRID CROSS

Nicotiana alata

Nicotiana langsdorffii

Nicotiana alata

HYBRIDS
(*N. alata* X *N. langsdorffii*)

Nicotiana langsdorffii

BACKCROSS
HYBRIDS
(to *N. alata*)

BACKCROSS
HYBRIDS
(to *N. langsdorffii*)

4

INTROGRESSIVE HYBRIDIZATION of two species of tobacco. Interbreeding between two species (top) produces hybrids with intermediate characteristics. When these hybrids are then backcrossed to one or both of the parent species (middle) the offspring are highly variable (bottom).

fertility or survival ability when in the homozygous condition. If species become too small in population size, they may reach the point where the accidental fixation of less adaptive genes speeds their decline still more.

Another way in which genetic drift can operate is through the FOUNDER EFFECT. New populations in new places are usually started by a very small number of pioneering individuals lucky enough to reach a place where they can survive and reproduce. Because they contain only a small fraction of the alleles found in the source population, they must be different from the source population at the very beginning. The new populations will also differ among themselves because of chance differences in the genes of their founders.

Geneticists have often used the founder effect in an attempt to explain the small, seemingly meaningless differences between populations of the same species.

GENE FLOW

Most demes are only partially isolated from other demes of the same species. Some organisms are still able to migrate from one to the other. If the migrants can survive and breed in their new home, they add to the gene pool of the deme. This injection of new genes by migration is called GENE FLOW. In many cases the immigration rate is relatively low, adding perhaps no more than one in a million individuals per generation. At this level it can add new genetic material but not contribute significantly to the alteration of gene frequencies. In other cases the rate of immigration is high: as many as one or more immigrants for every hundred individuals in the host population. Gene flow of this magnitude can serve both as a source of new genetic variability and as a prime mover of evolution, on a par with natural selection.

Of equal importance to the amount of gene flow is the degree of genetic difference that separates the immigrants from their host population. If the two are genetically identical, then of course no evolution occurs, no matter how intense the gene flow. But if they are very different, a small amount of immigration can result in a great deal of evolution, especially if some of the immigrant genes prove to be adaptively superior in the new environment. The extreme case of such exchange occurs during HYBRIDIZATION: the interbreeding between populations that are different enough to be ranked as species. If the hybrid offspring are able to breed back with one or both of the parent populations, the process is called IN-TROGRESSIVE HYBRIDIZATION. It results in the production of radically new genetic combinations within the parent species. Most of these products are less well adapted to the local environment and destined for early extinction in the process of natural selection. But some, acting like daring new mutations, may cause the parent populations to take new directions in evolution. An example of introgressive hybridization is given in Figure 4.

NATURAL SELECTION

Natural selection is the only agent of evolution that specifically adapts populations to their immediate environment. Its result is always the same: some genotypes gain in the population at the expense of others. The selective force can act on the variability of a population in one or another of several radically different patterns, producing the array of effects illustrated in Figure 5. It is important to understand the consequences of each of these patterns separately.

STABILIZING SELECTION involves a disproportionate elimination of the extremes. This process occurs repeatedly in all populations.

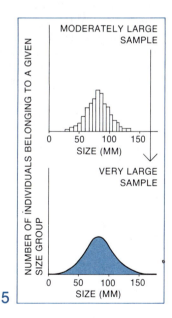

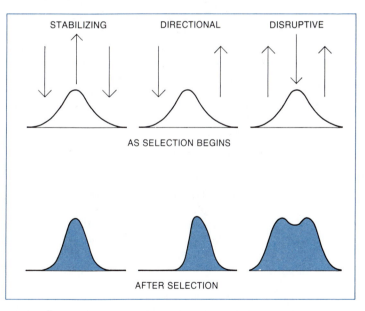

EFFECTS OF NATURAL SELECTION on a population are plotted on two sets of curves. Graphs illustrate the impact of the three principal types of selection — stabilizing, directional and disruptive — on the size of an imaginary organism. In this particular species, most individuals are roughly from 65 to 85 millimeters long, as indicated by the 'normal' distribution curves at left. Height of curve corresponds to the number of individuals having a given size. In the set of curves at right, arrows pointing upward indicate phenotypes (in this case, sizes) favored by natural selection; downward arrows, sizes disfavored by selection. Selection alters the range of sizes, as shown by curves at bottom right.

5

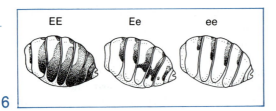

6

ABDOMEN OF FRUIT FLY illustrates the effects of balanced polymorphism. Differences in color shown here are due to a pair of alleles (E and e) on the same locus. Heterozygote (Ee) has superior fitness to both homozygotes, so neither allele eliminates the other by natural selection.

In all but the smallest and most isolated populations the variability is increased each generation by mutation and gene flow. Stabilizing selection then "pulls in the skirts" of population variation by reducing or eliminating all but the genotypes best suited to the local environment. A special form of stabilizing selection makes possible the phenomenon of BALANCED POLYMORPHISM. This is the coexistence of two or more phenotypes not connected by intermediate phenotypes. Sexual differences, and differences between developmental stages of the kind that occur in the life cycle of a single individual, are excluded from this definition of balanced polymorphism. One example is the blood type of human beings. Each condition of the A, B and O series, as well as the Rh positive and Rh negative factors and other less familiar series, are controlled by single alleles. An example of balanced polymorphism involving *Drosophila* is given in Figure 6. In this case we follow the practice of the original investigator of labeling the alleles *E* and *e*, although for our purposes they could be equally well labeled *A* and *a*. The heterozygote (*Ee*) is superior in fitness to either of the two homozygotes (*EE* and *ee*). Since the superior heterozygote contains both of the alleles, *E* and *e*, neither one can eliminate the other in evolution. Much of the natural genetic variability of populations is based on the coexistence of such multiple alleles. Heterozygote superiority is perhaps the most important mechanism for perpetuating the genetic variability created by mutation.

DISRUPTIVE SELECTION is a rarer phenomenon, or at least one less well documented at the present time. It can create balanced polymorphism by breaking up the population into two or more distinct types, and therefore it can serve as an alternative to heterozygote superiority. In extreme cases populations subjected to disruptive selection might even divide into two or more species, although such a process has not been demonstrated outside the laboratory.

MICROEVOLUTION

Two case histories illustrate the rich diversity of forms that microevolution can take. In each example natural selection, particularly the directional forms of natural selection, is the principal agent. This reflects the general rule that when microevolutionary change is closely analyzed natural selection is almost always involved, and is usually of overriding importance.

EYE COLOR IN DROSOPHILA

Certain mutations in the fruit fly *Drosophila* alter the normal dark red color of the adult's eyes. The new shades have been descriptively labeled by geneticists as raspberry, garnet, white, and so forth. These eye-color mutants are quite rare in natural populations, which leads to the belief that they are selectively inferior to the "normal" dark red allele. This supposition is easily confirmed in laboratory experiments. When eye-color mutants are mixed with normal alleles in confined populations, they are almost invariably replaced by the normal alleles in a few generations. They lose out even when they are given an initial

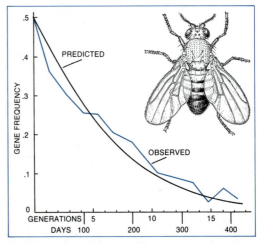

7

RESULTS OF BREEDING EXPERIMENT with the fruit fly. Curves indicate predicted and observed declines in the frequency of the mutant gene for raspberry eye color. In competition with flies of normal eye color, mutant individuals almost died out in 17 generations (about 400 days).

numerical advantage. One such experiment involving the raspberry gene is illustrated in Figure 7. The gene declines because males possessing the raspberry eye color are only about 50 percent as successful in mating as their competitors with normal eye color. The predicted curve in Figure 7 consists of a projection based on this simple observation. The fit of the observed curve to the predicted curve is very close, offering one more bit of evidence to confirm classical population genetics at the elementary level. The case also reveals how rapidly microevolution proceeds when the genes are being tested by a moderate degree of directional natural selection.

THE SICKLE-CELL TRAIT

The case of sickle-cell anemia in human beings is the classic textbook example of balanced genetic polymorphism. Its cause has been traced all the way down to the molecular level, and the reason for the high frequency of the gene in human populations has apparently been explained in full by classical population theory.

Sickle cell anemia is a severe hereditary disease that afflicts a large percentage of the native populations of Africa. It is caused by an abnormal form of hemoglobin, called hemoglobin S, whose molecules have a peculiar tendency to link up into long rods. The rods are sufficiently rigid to distort the red blood

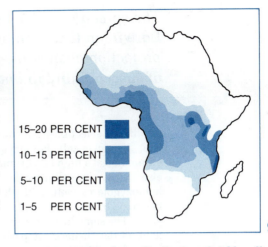

15–20 PER CENT

10–15 PER CENT

5–10 PER CENT

1–5 PER CENT

9

MAP OF AFRICA shows distribution of sickle-cell gene. The range roughly coincides with that of the malaria-causing protozoan Plasmodium falciparum. Darker color indicates higher frequencies.

cells from their normal disc shape to a twisted, sickle-like form (*Figure 8*). These cells tend to clog the smaller blood vessels. The automatic response of the body is then to destroy them, resulting in a shortage of red blood cells: anemia. The sickle-cell trait is controlled by a single gene. This gene produces hemoglobin S by directing the substitution of a single incorrect amino acid in the hemoglobin molecule (*Chapter 2*). Persons who are heterozygous for the gene, carrying one normal and one sickle-cell gene, manufacture both normal hemoglobin and hemoglobin S. They show signs of anemia only under conditions of stress, such as breathing at high altitudes, when an unusual demand is placed on the hemoglobin. Homozygous carriers, on the other hand, are much more severely afflicted; most die in childhood. The remainder suffer from chronic anemia with periodic crises when blood is temporarily cut off from various body organs.

Although sickle-cell anemia is virtually lethal in the homozygous state, the gene is abundant in many parts of Africa (*Figure 9*). As many as 40 per cent of the members of some of the tribes carry the trait, mostly in the heterozygous condition. Among black Americans approximately nine per cent are heterozygous for the gene, and about 0.25 per cent are homozygous. How can a gene imposing such lowered fitness on its homozygotes rise to these levels? Whenever population geneticists encounter a situation of this kind, they immediately consider the possibility of

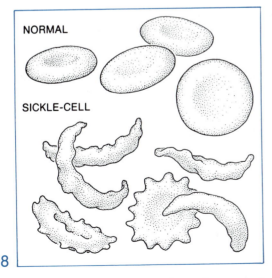

NORMAL

SICKLE-CELL

8

RED BLOOD CELLS of humans change from normal disc shape (top) to sickle shape (below) when blood of a person who is homozygous for the sickle-cell gene is deoxygenated during respiration.

The spread of sickle-cell anemia is a case history of Darwinian theory in action, and it involves an ironic twist: protection against one deadly disease was bought at the price of susceptibility to another.

stabilizing selection by balanced polymorphism. If sickle-cell homozygotes are being eliminated at a high rate by anemia, something else must be eliminating the "normal" homozygote as well, with the result that the heterozygote is more fit than both homozygotes. The something else appears to be malignant tertian malaria, a particularly virulent form of this blood disease caused by the protozoan *Plasmodium falciparum*. Not only do the ranges of the sickle-cell gene and *Plasmodium falciparum* roughly coincide, but field studies have indicated that persons carrying hemoglobin S are more resistant to the parasite than those lacking it.

This example of Darwinian theory contains an ironic twist: protection against one deadly disease has been bought at the price of another. Nothing could illustrate more clearly that it is the population, not the individual, that constitutes the ultimate unit of evolution. Individuals do not become hemoglobin heterozygotes by any kind of personal design or physiological adaptation, nor are they able to confer their valuable heterozygosity to particular offspring by personal choice. By sheer chance some of their children will be heterozygotes and others homozygotes. Among the homozygotes chance will also decree who will be susceptible to malaria and who will die from anemia. In a sense the anemics can be regarded as losers in the evolutionary game — the population's mindless response to the selective pressure exerted by the malarial parasites.

A similar effect may be involved in schizophrenia. Abundant evidence suggests that this very common mental disease has a strong hereditary component. According to one hypothesis it is controlled by a single dominant gene with low penetrance — meaning that only a small minority of the carriers develop the disease to any perceptible degree. The high frequency of schizophrenia in populations around the world, despite the relatively high mortality and low reproductive replacement of human beings afflicted by it, suggests the operation of a hidden selection pressure favoring its survival in less overt ways. This adaptive factor could be the greater ability of schizophrenics to withstand shock from wounds and operations, an hypothesis that is still the subject of controversy and continued research. In considering the deeper meaning of such an evolutionary trade-off it is useless to ask whether complete sanity is preferable to resistance to wound shock, just as it is useless to ask whether it is better to resist malaria than to avoid anemia. In permitting the operation of balanced genetic polymorphism, nature does not ask such questions.

MACROEVOLUTION

Each of the examples of microevolution mentioned above involves shifts in the frequencies of small numbers of genes. In these and hundreds of other cases, biologists have been able to watch the beginnings of evolutionary change in many kinds of plants and animals, and they have used these opportunities to test the assumptions of population genetics that form the foundations of modern evolutionary theory. But can more extensive evolutionary change — macroevolution — be explained as an outcome of such microevolutionary shifts? For example, did birds really arise from reptiles by an accumulation of gene substitutions like the one for raspberry eye-color in *Drosophila*?

The answer is that it is entirely plausible, and no one has come up with a better explanation consistent with the known biological facts. One must keep in mind the enormous difference in time scale between the observed cases of microevolution and macroevolution. For birds to evolve from reptiles, for example, required many millions of years. As paleontologists explore the fossil record with increasing care, transitions are being documented between increasing numbers of species, genera, and higher taxonomic groups. The reading from these fossil archives suggests that macroevolution is a very gradual process. Its slow rate suggests that it is based upon hundreds or thousands of gene substitutions no different in kind from the ones examined in these case histories.

READINGS

P.R. EHRLICH, R.W. HOLM, AND P.H. RAVEN, eds., *Papers on Evolution*, Boston, Little, Brown, 1969. Many of the basic articles from modern evolutionary biology are reprinted here, with commentaries by the editors. This is a useful adjunct to the two textbooks suggested below.

G.L. STEBBINS, *Processes of Organic Evolution*, 2nd Edition, Englewood Cliffs, N.J., Prentice-Hall, 1971. Perhaps the best short textbook on general evolutionary biology and a good book to read next. Well-balanced presentation of population genetics and speciation theory, with both plant and animal examples.

E.O. WILSON AND W.H. BOSSERT, *A Primer of Population Biology*, Sunderland, Mass., Sinauer Associates, 1971. This brief self-teaching textbook is designed to form the bridge between the elementary textbook and more advanced treatments of population genetics, evolutionary theory, and ecology.

the multiplication of species

Virtually all of the classification of plants and animals is based on the concept of species — the next level of organization above the individual organism. After assigning a specimen to a particular species, the taxonomist can construct a simple hierarchy. He subdivides species into geographical sets of populations, called subspecies, or clumps them into sets of similar species called genera (singular: genus). The genera are in turn clumped to form still higher taxonomic categories (see Chapter 17).

The species is the basic unit in almost all modern ecological and evolutionary studies. In describing the process by which populations and communities of organisms interact and evolve, the biologist resorts almost unconsciously to the species as the lowest operational unit.

It is curious that although Charles Darwin entitled his book On The Origin of Species, he did not actually explain it. Darwin identified natural selection as the chief mechanism of evolution, revealing how a given species can change its genetic constitution through time. But he failed to show how one species can split into two or more. The reason for this omission is quite simple. Darwin was preoccupied with proving that species are flexible arbitrary units that natural selection alters relentlessly through time. He said, in opposition to the prevailing opinion of the day, that "no clear line of demarcation has as yet been drawn between species and subspecies — that is, the forms which in the opinion of some naturalists come very near to, but do not quite arrive at the rank of species; or, again, between sub-species and well-marked varieties, or between lesser varieties and individual differences. These differences blend into each other in an insensible series; and a series impresses the mind with the idea of an actual passage." In other words, Darwin believed that species could not exist as discrete units.

THE CONCEPT OF SPECIES

Today biologists take an intermediate position in assessing the problem. They agree that species are indeed changeable and encompass a great deal of variation, as Darwin was anxious to point out. But biologists also recognize that sharp discontinuities exist between populations. The gaps are based on the absence of free interbreeding among the populations that live in the same place. Since reproductive isolation is a natural criterion, not one based on some arbitrary degree of difference selected by the taxonomist, it is referred to as the biological species concept. The concept defines the species as a population or series of populations within which free gene flow occurs under natural conditions. This means that all of the normal, physiologically competent individuals in a given place at a given time are capable of breeding with all of the other individuals of the opposite sex belonging to the same species, or at least that they are capable of being linked genetically to them through chains of other breeding individuals. By

COURTSHIP FLIGHT of male sulfur butterfly as it would look to an insect. Photograph on opposite page, made with ultraviolet light, shows male (right) approaching female, flashing the bright upper surface of his wings. The flash area is conspicuous to the insect (whose eyes are sensitive to ultraviolet) but not to humans. Wing-flashing is one of the mechanisms that enable sulfur butterflies to recognize members of their species, and thus prevents interbreeding with other species.

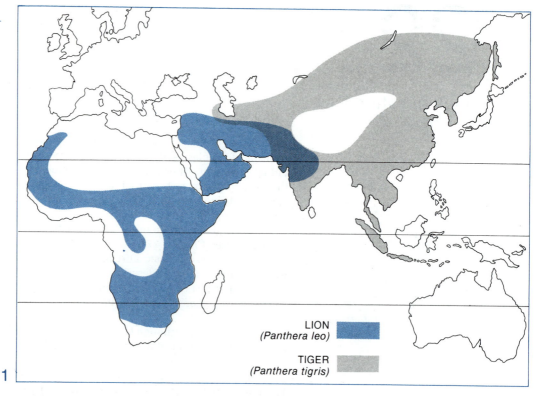

1

RANGES OF LION AND TIGER were once far more extensive than they are today. Until the nineteenth century the two species overlapped throughout much of India (darkest shading).

definition the members of a species do not breed freely with members of other species.

In establishing the limits of a species it is not enough merely to prove that genes of two or more populations can be exchanged under experimental conditions. The populations must be demonstrated to interbreed fully in the free state. A familiar case with some surprising implications involves lions and tigers. The two species are genetically closely related, despite their obvious differences in outward appearance. They are frequently crossed in zoos to produce hybrids, called "tiglons" (tiger as father) and "ligers" (lion as father). But this accomplishment does not prove them to belong to the same species. The ability to hybridize in a suitable experimental environment is necessary to fulfill the definition, but it is not enough. The important question is whether the two forms cross freely in the wild. Lions and tigers coexisted over most of India until the 1800's, when lions began to be reduced by hunting (*Figure 1*). Now they are nearly extinct there, limited to a few hundred individuals in the Gir Forest in the state of Gujurat. No doubt

lions and tigers were fully isolated reproductively during their coexistence, for no tiglons or ligers have ever been found in India. Suppose that lions and tigers had been shown to be wholly intersterile under experimental conditions. This could have reasonably been interpreted to mean that they are distinct species, because the condition could be assumed to hold in nature also. But the opposite evidence means nothing, since many other genetic devices in addition to mere intersterility might (and obviously do) isolate them in nature. In fact the lion and the tiger differ strongly in their behavior and the habitats they prefer. The lion is more social, roaming in small groups called prides, and it prefers open country. The tiger is solitary, and is found more frequently in forested regions. These differences between the two forms could be great enough to account fully for their failure to hybridize.

The biological species concept has proven useful because it accounts for many of the discontinuities in phenotypic variation. Its validity can be further substantiated by comparing species concepts among different human cultures. In 1927 Ernst Mayr, an evolutionist who has specialized in the study of species, led an expedition to collect and identify the birds of the remote Arfak Mountains of northwestern New Guinea. He was able to recognize 138 species using the biological criterion. The native Arfak hunters among whom he lived distinguished and assigned special names to 137 of the 138 forms which he had recognized as species and labeled with technical names derived from Greek and Latin. Biologists do not in fact always perform as well as aboriginal peoples. Six tribes of Amerindians living in the rain forests of the Amazon-Orinoco basin use a total of about 5,000 names to distinguish more than half of the tree species that grow within their hunting grounds. The tribes depend on the trees for their livelihood and also refer to them constantly in their folklore and religion. Each of several individual tribes knows over 1,000 species — an impressive accomplishment even for a professional botanist. Not only do the Indian classifications usually match the genetic concepts of the botanists, but in some cases the Indians have called the attention of the botanists to obscure species previously overlooked by professionals.

SPECIES AND PHENOTYPE

When a biologist distinguishes a species on the basis of the modern biological concept, he does more than simply point out the existence of a state of reproductive isolation. The rank of species normally has two other basic attributes:

1. To be reproductively isolated, the species has probably enjoyed a relatively long evolutionary history during which it adapted to the environment in its own unique way.

2. Consequently, the species differs from others in many phenotypic characters in addition to the ones first noticed by the taxonomist. These characters involve morphology, physiology, biochemistry, and (in the case of animals) behavior. Moreover the species occupies a distinctive geographic range of its own.

The power of the species concept is most effectively illustrated in examples of the so-called sibling species, which are difficult to tell apart by ordinary means. Sometimes taxonomists stumble upon the existence of groups of sibling species by discovering gaps in the apparently continuous variation among relatively obscure characteristics. Upon detailed analysis, taxonomists find that the gaps are maintained by reproductive isolation between biological species. Later other previously unsuspected phenotypic gaps are noticed. Among the important examples of sibling species unveiled in this manner are the malaria mosquitoes of Europe. Before 1934 malaria in Europe was thought to be carried by a single species, called *Anopheles maculipennis*. Entomologists were puzzled, however, by the abundance of the mosquito in some parts of Europe where malaria was absent. They noticed that in a few places the mosquito ignored man while continuing to feed on domestic animals. In some localities it bred in brackish water and elsewhere in fresh water. All of this variation had no particular meaning until finally it was shown to be correlated with constant differences in the egg color and shape of the egg masses. Further analysis revealed that most of the variation, including the ability to transmit malaria versus the lack of the ability, can be used to separate genuine biological species. Some species in the *Anopheles maculipennis* complex can transmit malaria and some cannot; the matter is as simple as that. This understanding has helped greatly in the control of malaria in Europe.

WEAKNESSES OF THE CONCEPT

The definition of the species as a closed gene pool is the best devised thus far, but in comparison with other scientific concepts it has some glaring weaknesses. The definition looks very good in cases like the Arfak Mountain bird fauna and the *Anopheles* mosquitoes of Europe. In general, where the biotas (fauna and flora together) are considered at just one place and over a period of time too brief to permit evolution, the species concept is strong. We must stress the importance of one place and one time — the smaller the place and shorter the time the easier the concept is to use. For this reason it is often said that the ideal species to which the definition can be applied is the nondimensional species, a population observed over too small an area and through too short a time for these dimensions to play a significant role. For it is the dimensions of space and time, when their effects on variation are fully considered, that cause the greatest difficulty in the biological species concept. These difficulties can be summarized as follows:

1. Populations that do not occur in the same places cannot be evaluated precisely, because the investigator does not know whether they

would interbreed freely with each other if they came together under natural conditions.

2. Populations that do not occur together in time cannot be evaluated for the same reason. Figure 2 reveals the gradual evolution of one species of snail, as shown by fossil shells from successive Pliocene strata. The snails at the beginning and at the end of the progression would almost certainly be considered distinct species solely on the strength of the anatomical differences that separate them. But if this proposition is accepted one is forced to ask: which of the connecting populations should be separated as additional species? The answer is that the biological species concept is quite meaningless in this case. It is necessary to turn to a different, quite arbitrary species definition based on the amount of difference in the phenotypic characters preserved in the fossils. This criterion is admittedly selected more for convenience than for its future theoretical value.

3. Populations that are either completely asexual or self-fertilizing also lie outside the domain of the biological species concept, because the concept is based wholly on the exchange of genes during sexual reproduction. Where sexual reproduction is lacking, arbitrary species definitions must be based on the degree of difference among the phenotypes.

4. Populations that exchange genes with other populations to an intermediate degree are common in some groups, especially plants. Such populations cannot "decide" whether to interbreed freely with each other, in which case they would all form a single species, or whether to maintain strict reproductive isolation, in which case they could easily be classified as distinct species. These popula-

tions are often called semispecies, and pose unusually difficult problems for the taxonomist who must attempt to separate and classify them.

REPRODUCTIVE BARRIERS

How do two species coexist in the same place and time and yet not interbreed? Their reproductive isolation is a secondary result of the genetic differences between them. The differences are generally called intrinsic isolating mechanisms, in order to contrast them with such environmental (extrinsic) barriers as mountain ranges, rivers and straits, which separate some populations physically. Intrinsic isolating mechanisms comprise all the genetic differences between species which (regardless of the origin and the primary adaptive significance of the differences) reduce the chance of interbreeding. In other words, they are the sum of everything that can go wrong in reproduction between two populations.

For example, if one species has evolved so that it is active only during the day, and the other only during the night, this difference in adaptation isolates the two species reproductively. It is an intrinsic isolating mechanism. Biologists who study speciation (the process of species multiplication) have found it convenient to classify the isolating mechanisms into two kinds: those that operate to keep the species apart before mating occurs (premating isolating mechanisms) and those that prevent hybrid offspring from developing or breeding (postmating isolating mechanisms). A more complete classification is given in Table I.

Intrinsic barriers tend to act in sequence. For example, if there are no ecological barriers, or if they exist but are insufficient, temporal dif-

2

FOSSIL SNAIL SHELLS from successive rock strata of the Pliocene epoch demonstrate how a macroevolutionary change can occur in a series of microevolutionary steps. Snails depicted on these two pages were fossilized in fresh-water deposits in Yugoslavia sometime between three and ten million years ago. Specimens at far left and far right scarcely seem to belong to same species.

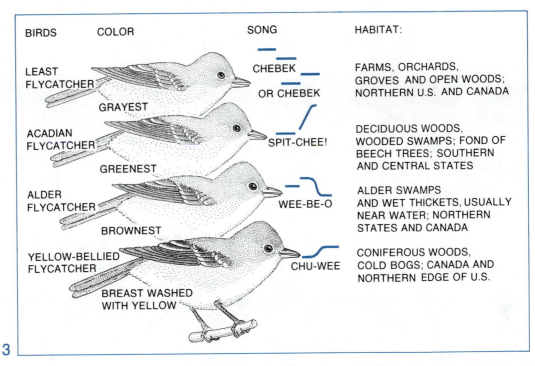

BIRDS	COLOR	SONG	HABITAT:

LEAST FLYCATCHER — GRAYEST — CHEBEK / OR CHEBEK — FARMS, ORCHARDS, GROVES AND OPEN WOODS; NORTHERN U.S. AND CANADA

ACADIAN FLYCATCHER — GREENEST — SPIT-CHEE! — DECIDUOUS WOODS, WOODED SWAMPS; FOND OF BEECH TREES; SOUTHERN AND CENTRAL STATES

ALDER FLYCATCHER — BROWNEST — WEE-BE-O — ALDER SWAMPS AND WET THICKETS, USUALLY NEAR WATER; NORTHERN STATES AND CANADA

YELLOW-BELLIED FLYCATCHER — BREAST WASHED WITH YELLOW — CHU-WEE — CONIFEROUS WOODS, COLD BOGS; CANADA AND NORTHERN EDGE OF U.S.

3

ISOLATING MECHANISMS OF BIRDS include song, habitat and color of plumage. These characteristics, which enable ornithologists in the field to distinguish between species, also serve to keep different species from interbreeding. Flycatchers shown are natives of eastern U.S.

INTRINSIC ISOLATING MECHANISMS keep two species separate by preventing them from mating or by preventing hybrid offspring from developing or breeding.

I

PREMATING ISOLATING MECHANISMS	
1.	ECOLOGICAL: THE SPECIES OCCUPY DIFFERENT HABITATS OR AT LEAST BREED IN DIFFERENT HABITATS. THESE DIFFERENCES, LIKE ALL OF THE OTHER INTRINSIC MECHANISMS LISTED BELOW, HAVE AN HEREDITARY BASIS.
2.	TEMPORAL: THE SPECIES BREED AT DIFFERENT SEASONS OR AT DIFFERENT TIMES OF THE DAY.
3.	BEHAVIORAL: IN THE CASE OF ANIMALS, THE COURTSHIP BEHAVIOR DIFFERS.
4.	MECHANICAL: MORPHOLOGICAL OR PHYSIOLOGICAL DIFFERENCES PREVENT NORMAL MATING.
POSTMATING ISOLATING MECHANISMS	
5.	MECHANICAL: THE SPERM ARE UNABLE TO REACH OR TO FERTILIZE THE EGGS.
6.	MORTALITY: THE HYBRID DIES AT SOME STAGE PRIOR TO MATURITY.
7.	STERILITY: THE HYBRID MATURES BUT IS STERILE TO SOME DEGREE.
8.	FITNESS: THE HYBRID IS FERTILE BUT THE F_2 HYBRID GENERATION HAS LOWER FITNESS.

ferences might suffice as barriers. If these also are lacking, certain behavioral differences might keep the species apart; and so on down the list. Sexual reproduction consists of an elaborate series of exceedingly delicate steps. Anything that can go wrong at any place in the sequence qualifies as an intrinsic isolating mechanism. Species, in the course of their evolution, obey Dollo's Law (*see Chapter 21*) and always continue to diverge from each other in genetic composition. As a result, they will tend always to reinforce the intrinsic isolating mechanisms that separate them, because the greater their genetic difference, the more things there are that can go wrong if they attempt to interbreed.

A great many familiar facts of natural history can be usefully interpreted in terms of this particular aspect of speciation. For example, three species of birds called flycatchers are found in eastern North America. They are nearly identical in size, plumage, and other outward characteristics. In other words, they are sibling species. In Roger Tory Peterson's *A*

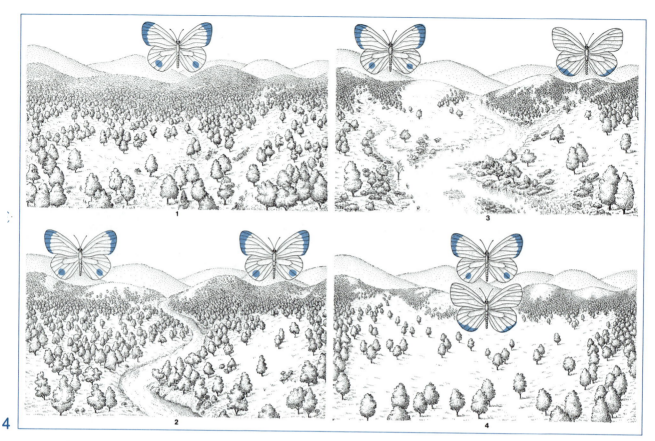

4

SIMPLE FISSION is one biogeographic process that leads to the formation of new species. Drawing 1 at upper left shows a single species of butterfly living in its habitat. In drawing 2 (lower left) the intrusion of a natural barrier — in this case, a river valley — divides the species into two isolated populations. Eventually they evolve into two distinct species (3). If the barrier then disappears — for example, if the river dries up or changes its course — the two new daughter species can coexist in the same habitat without interbreeding, as shown in the drawing at lower right.

Field Guide to the Birds, birdwatchers are advised to take note of differences in habitat preference and song when attempting to identify flycatchers to the species level (*Figure 3*). These nonanatomical characteristics are indeed strong. In fact, they are the way the species are kept apart. The segregation of habitats, combined with radically different songs used during territorial defense and courtship, are probably sufficient to prevent free interbreeding of the species. As might be expected, many of the diagnostic characteristics by which one classifies species are the intrinsic isolating mechanisms by which the organisms themselves maintain the separateness of the species.

HOW SPECIES MULTIPLY

How do separate species arise? There are two basically different mechanisms: GEOGRAPHIC SPECIATION involves the intervention of external barriers that split populations, allowing them to diverge to species level. SYMPATRIC SPECIATION involves internal factors that subdivide populations on the spot, without the aid of external barriers.

In geographic speciation, a single population (or series of populations) is first divided

New species are created by two different mechanisms. In geographic speciation, populations are split by barriers like rivers and mountains; in sympatric speciation, by internal factors such as polyploidy and disruptive selection.

by an extrinsic barrier — a river, a mountain range, an arm of a sea. The isolated populations then diverge from each other in evolution because of the inevitable differences of their environments. Since all populations evolve when given enough time, all extrinsically isolated populations must eventually diverge. By this process alone the populations can acquire enough differences (intrinsic isolating mechanisms) to reduce gene flow to the point where they evolve into a separate species.

Figures 4 and 5 show the two most common ways in which geographic speciation occurs. They represent, in a deliberately oversimplified form, the process of speciation by fission and by multiple invasion. Virtually all geographically separated populations live in different environments. This statement may seem extreme at first. But recall that the physical environment does shift from point to point on the globe, often dramatically and over relatively short distances. The total climate and chemistry of the soil change in going inland from a coast, or up from the coastal plain to the mountains, or out across the sea from a continent to an island. Even sharper changes occur in the vegetation and in the composition and relative abundance of the animal species living on the vegetation. Geographically isolated populations are probably always subjected to different selective pressures. Consequently when they evolve they will diverge from one another. When the populations have diverged enough so that effective intrinsic isolating mechanisms have come into existence, they

must be ranked as separate species even though they have not yet come into contact. The coming together of the newly formed daughter species is the final step of speciation, but it is not the crucial one.

When populations come together after a period of geographic isolation, one of three things can happen:

1. If the populations have attained the level of full species and are very different from each other, they may coexist side by side with no significant interaction.

2. If the populations have failed to attain the level of species, they will interbreed to some extent. They may fuse completely into a single species, or else maintain an intermediate level of gene flow, producing what is known as a semispecies.

3. The populations may have attained the level of species but interfere with one another in such a way as to cause one or both of them to diverge still further in evolution. This peculiar form of post-contact evolution is called CHARACTER DISPLACEMENT. Interference can occur in two ways: first, the two newly formed species may still be so similar in food habits, habitat preference and other ecological traits that they compete with each other. This means that there is less food, nesting space, or other necessities of life for both species. It will therefore be advantageous for the two species to become less alike; genotypes that differ from those of the competing species will be favored in evolution. As a result, the two species will diverge in evolution more rapidly where their

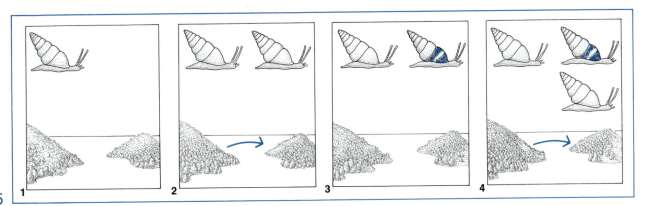

5

MULTIPLE INVASION also leads to the formation of new species. Drawing 1 (far left) depicts a single population of snails on one side on an unchanging geographic barrier. Somehow a few individuals manage to cross the barrier (2), starting a new and largely isolated population. Eventually the two populations evolve into separate species (3). If a second colonization follows, the two daughter species will overlap without interbreeding (4).

6

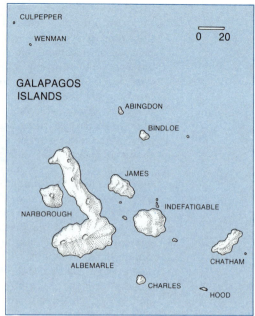

GALAPAGOS ISLANDS lie on the equator, 600 miles west of Ecuador. Small colored square in map above shows area of detailed map at right. Here Darwin found 14 species of finch, all descended from a single species.

populations come into contact. The same thing can happen when the populations coexist as semispecies. The presence of semispecies means that interbreeding is not complete.

Attempts to interbreed between the two newly formed species produce fewer offspring than usual. Either the potential mate fails to respond to the mating attempt, or else it results in the production of hybrid offspring which are less successful at developing and breeding on their own. The genotypes of either species that avoid such attempts to interbreed will be favored by natural selection. The two semispecies will then diverge from each other and eventually become full species.

DARWIN'S FINCHES

Figures 6 and 7 illustrate a classic example of geographic speciation by multiple invasion. The finches in the drawing belong to a group of species found only on the Galapagos Islands. When Darwin visited this archipelago on the equator, 600 miles off the coast of Ecuador, the differences he observed among the populations of plants and animals in going from island to island convinced him that "species are not immutable." This realization came first from a rather casual study of the mockingbirds found on several of the Galapagos Islands. It was confirmed later by examinations of other kinds of animals, including the giant tortoises and remarkably diverse finches, that are found only on the Galapagos. Approximately a century later David Lack, a British ornithologist, undertook a more detailed study of Darwin's finches, and it was he who discovered the example presented here.

The Galapagos Islands are a series of more or less rounded volcanic peaks that project above the deep waters of the Pacific. All of the native species of plants and animals originated from a few lucky stragglers that were accidentally transported over the ocean. Most of the birds came from the coast of South America, but a very few others came from Polynesian islands far to the west. Because the Galapagos Islands are separated from each other by open ocean, speciation has occurred chiefly by means of multiple invasions from one island to the other. All of the 14 living species of finches, comprising an entire subfamily of their own, are believed to have evolved from a single species that reached the islands sometime in the last several million years. The ancestral species and its descendants proceeded to divide many times through the process of multiple invasion. As newly formed species accumulated, they faced the problem of interference through competition and hybridization, and character displacement may have occurred frequently. The example shown in Figure 7 is in an early and unusually well marked stage. The three seed-eating species undoubtedly descended from a common ancestor. The larger the bill, the harder the seed that can be cracked open and eaten. Thus the three species probably avoid competition by feeding on different kinds of seeds. The shape of the bill may also serve during courtship as a premating isolating mechanism. The depth of the bill is important in preventing interference between the species, and this is the character that has undergone displacement.

Character displacement has a significance that goes well beyond its role in speeding up evolution after the completion of speciation. It is also the means by which species are packed into ecological communities. Evolution by displacement is rather like the deformation that occurs when a full suitcase is closed. The ability to displace increases the number of species that can be fitted onto an island or other ecological community, just as the ability of clothing to be folded and squeezed increases the number of items that can be forced into a suitcase. The question of how many species can coexist in one place is called the species packing problem. Studies of character displacement have provided an important means of partially solving this fundamental problem. It has been found, for example, that where displacement has occurred in a single characteristic, such as beak depth in the species of Geospiza, the species usually evolve apart until they differ by approximately 30 to 50 percent. This is exemplified in the two species, fuliginosa and fortis, that show the most clear-cut pattern of displacement. Obviously, if there is a minimal degree of difference that species must attain in order to coexist, there is a corresponding maximal number that can be packed into any one place. For example, the number of species of Geospiza ranges from three on the largest islands to one on the smallest. The exact mathematical relation between the minimal character difference and the maximum number of species that can coexist has not yet been solved, but it is a problem on which theoretical biologists are now actively working.

SYMPATRIC SPECIATION

Sympatric speciation can take one of two very different forms: POLYPLOIDY or DISRUPTIVE SELECTION. When there is an accidental doubling in the number of chromosomes in some of the offspring, the resulting polyploid individuals will usually not be able to interbreed with the normal population that gave rise to them. If they are able to interbreed among themselves, the polyploids constitute an instantaneously created new species. In disruptive selection some new selective force removes intermediate individuals from a pre-

7

VARIATION IN BEAK DEPTH of Galapagos finches is a classic example of character displacement. Drawings at top compare beaks of three species. Thicker beak enables a finch to feed upon seeds too hard for competing species to crack. Bar graphs show variation in beak depth among populations on several islands. Vertical scale shows percentage of birds having beaks of the depth indicated on horizontal scale. On each of two islands inhabited by a single species (two bottom rows) finches have beaks of intermediate depth. On four islands (top row) fortis is squeezed by two other species, displacing the beak depths of all three. On Charles and Chatham islands (second row) fortis and fuliginosa compete with each other. Beak size of fortis is displaced toward the right of the graph — that is, toward the larger beak size of magnirostris, a species that has become extinct on Charles island.

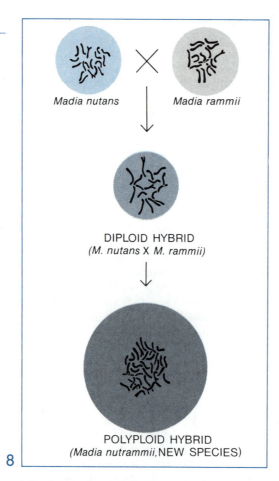

8

DIPLOID HYBRID
(M. nutans X M. rammii)

POLYPLOID HYBRID
(Madia nutrammii, NEW SPECIES)

NEW POLYPLOID SPECIES can be created from a diploid hybrid in one generation. Diagram shows how a new species of California tarweed arose spontaneously from two diploid species growing in an experimental garden.

a new species in only one or two generations. The increased number of chromosomes interferes with the process of meiosis, and makes breeding between a diploid species and its polyploid offspring difficult or impossible. A little reflection on the mechanics of meiosis will make this clear. Consider (to take an actual case, from the plant genus *Galax*) a species with a diploid chromosome number of 12. This means that in each cell there are two copies of six different chromosomes (the haploid number). At meiosis the two chromosomes of each kind must pair with each other to complete the process of gamete formation. Each gamete will end up containing one of each kind of chromosome, a total of six. Now suppose that due to an accident during meiosis some new plants are produced with exactly twice that many chromosomes, four of each kind or 24 in all. These polyploid plants can produce gametes. The polyploid gametes will contain 12 chromosomes (two of each kind), instead of just six, and they can combine to form fertile offspring with the polyploid number of 24. They can also combine with haploid gametes (one of each kind) to form hybrids with a total of 18 chromosomes (6 + 12 = 18). However, these 18-chromosome hybrids cannot produce normal gametes. The reason is that when meiosis begins and the time comes for the chromosomes to pair, there will be three of each of the six kinds, two from the polyploid parent and one from the diploid parent. Two chromosomes of a kind can pair easily, but three run into severe mechanical difficulties, causing uneven distribution of chromosomes in the daughter cells. Meiosis is often interrupted at this point, rendering the hybrid partially or wholly sterile. If diploids can breed freely with other diploids, and polyploids with other polyploids, but hybrids of the two are sterile, the diploids and polyploids comprise two different species. A new polyploid species has been created in one step.

Another and even more important way of creating polyploid species is by multiplying the number of chromosomes of diploid hybrids. Two distinct plant species occasionally produce sterile hybrid offspring because the chromosomes from the two parent species are too different to pair successfully. But if the hybrid doubles its chromosome number it has no difficulty at meiosis. There are two of each kind of chromosome from each parent instead of just one, and these two pair in the usual

viously freely interbreeding population. This splits the population as effectively as an external barrier.

Geographic speciation appears to be the more frequent process in both plants and animals. Polyploidy is responsible for the origin of approximately half of the living species of higher plants, but it is a relatively insignificant process in animal speciation. Disruptive selection has not been demonstrated with complete certainty in nature. It remains a potentially important mechanism for the rapid multiplication of species.

SPECIATION BY POLYPLOIDY

Polyploidy, the exact multiplication of the number of chromosomes in the cell, can create

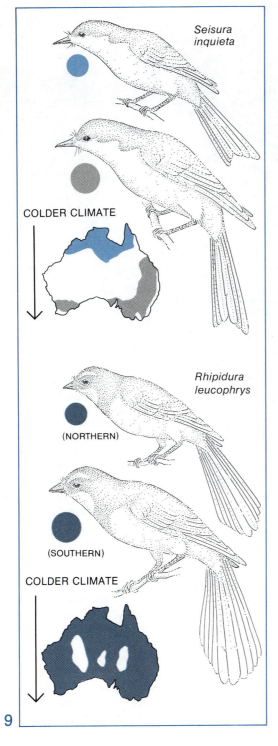

CLINE in size of two Australian birds. Larger birds are found in colder climates. Average size of southern specimens of Rhipidura is 11 percent larger than northern ones. In Seisura, which is divided into three populations, difference is 22 percent.

fashion to permit the completion of meiosis. The hybrid polyploid is a new species, isolated from its diploid parents for the same reason that a non-hybrid polyploid is isolated. A typical example of speciation by hybrid polyploidy is given in Figure 8.

RACES AND SUBSPECIES

A population that differs significantly from other populations belonging to the same species is referred to as a geographic race or subspecies. We can restate the generalization just made about geographic variation by saying that subspecies show every conceivable degree of differentiation from other subspecies. At one extreme are the populations that fall along a CLINE — a simple gradient in the geographic variation of a given character. In other words, a particular character varies gradually from one end of the species' range to the other. The closer together the populations of the species are throughout the range, the more even is the clinal variation from population to population. Figure 9 shows two examples of clines that exemplify this generalization. They also illustrate another feature of clines, namely that the trend in the variation can often be correlated with climatic differences that occur within the range of the species. The two Australian bird species illustrate Bergmann's Rule, which states that the larger subspecies of warm-blooded vertebrates occur in colder climates. The colder climate in this case is of course the southern, more temperate portion of Australia. The adaptive significance of Bergmann's Rule appears to be the greater ability of larger animals to conserve heat, because they have a lower ratio of body surface exposed relative to the total body weight. The same explanation holds for Allen's Rule: that in warm-blooded animal species the subspecies living in colder climates tend to have shorter protruding body parts, such as bills, tails, and ears. A third such climatic rule (out of a great many others that could be cited) is Gloger's Rule, that subspecies living in warm and humid areas are usually more darkly pigmented than those living in cool and dry areas. These rules, which are named after the nineteenth century zoologists who first proposed them, are not laws in any strict sense but at most only statistical generalizations based on the study of a great many species.

Another important generalization about

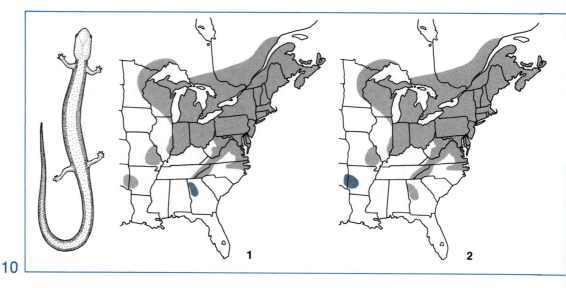

10

GEOGRAPHIC VARIATION in four traits of the North American salamander Plethodon cinereus. Maps 1, 2 and 3 show ranges of subspecies that differ in stripes and coloration. Map 4 shows range of salamanders with a particular number of vertebrae. Distribution pattern varies from trait to trait, a common occurrence that makes classification into subspecies difficult and often misleading.

geographic variation is that it can occur in virtually every kind of characteristic — from skin color and head shape to body temperature, urine composition, and inherited patterns of behavior. Any trait subject to genetic variation (and hence capable of evolutionary change) can show geographic variation. Species often differ drastically from one another in the particular traits by which they vary. One species, for example, may show great geographic variation in skin color and none in head shape, while a second otherwise similar species varies significantly in head shape but not in skin color.

Despite the seeming clarity of the patterns in the geographic variation of many species, and the convenience of labeling their geographic divisions as subspecies, the subspecies category has some inherent weaknesses that make it very difficult to use in analysis and classification. The first difficulty comes from the definition of the population. Few species are actually broken up into ideal populations: isolated or semi-isolated groups of randomly breeding individuals. Most species are struc-

tured in more complex, ambiguous patterns. Take for example that famous animal species, the African gorilla. Approximately 10,000 individuals of *Gorilla gorilla* exist in the eastern part of the range of the species. All are confined to a central portion of the continent. They appear to be grouped into about 60 populations. Each population occupies ten to 100 square miles of mountain country. In the center of the range lies a large area in which gorillas appear to be scarce but evenly distributed. The exact limits of these populations are unknown because the rate at which gorillas move from one area to another to breed is unknown. In the language of population genetics, the rate of gene flow is uncertain. Without that crucial fact, very little more can be concluded about the population structure of gorillas. *Gorilla gorilla* is not at all unusual in this respect. In fact it is much better known at the present time than the vast majority of animal and plant species.

The second major difficulty of the subspecies concept is the tendency of various characteristics to show differing patterns of

Difficulty in distinguishing subspecies lies at the heart of the problem of defining race in the human species.

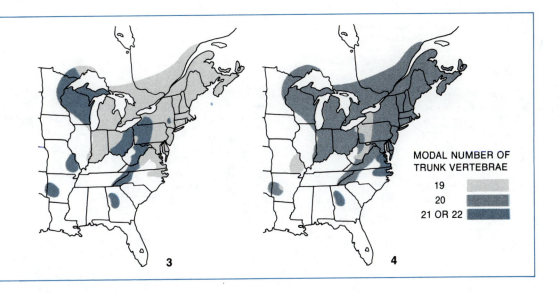

MODAL NUMBER OF
TRUNK VERTEBRAE

19
20
21 OR 22

3   4

geographic variation within the same species. The red-backed salamander *Plethodon cinereus (Figure 10)* is quite typical in this regard. If a taxonomist defines the subspecies of this very abundant American vertebrate on the basis of any one of the four morphological characters chosen, he will have no trouble. To define the subspecies with two of the characters is difficult but possible. When all four characters are used as criteria, however, the task becomes so difficult that one wonders if subspecies are worth the effort. In many species tens or even hundreds of characters show a comparable degree of variation. In fact, the more characters that are added to the analysis, the more subspecies turn up, and the less practical the classification.

Difficulty in distinguishing subspecies lies at the heart of the problem of defining race in the human species. Skin color alone follows one pattern of geographic variation, nose shape follows another, blood types still another, and so on down an unending list.

Anthropologists who attempt to classify subspecies, or distinct races, or whatever they choose to call them, run into predictable difficulties. Within the past 30 years various anthropologists, using their own favorite characters as the criteria, have produced widely varying estimates of the number of geographic races in the entire human species. One recognized five races, another six, another 30, and still another 37 plus about 30 "subraces." Of course one cannot deny that mankind is a geographically variable species. On the contrary, he shows a respectable amount of such variation when compared with other animal and plant species. But he is quite typical in the discordant nature of his patterns. It is necessary to realize that subspecies are not a realistic unit of classification in most cases. As a consequence, more and more biologists have begun to analyze geographic variation in terms of the individual character patterns instead of attempting to distinguish populations on the basis of geographical ranges.

READINGS

V. GRANT, *Plant Speciation*, New York, Columbia University Press, 1971. A work complementary to the Mayr text, cited below, dealing with plants instead of animals.

E. MAYR, *Populations, Species, and Evolution*, Cambridge, Mass., Belknap Press of Harvard University Press, 1970. An abridged and very readable version of the definitive work on modern speciation theory as it applies to animals. It is, however, more of a scholarly review than a textbook.

G.L. STEBBINS, *Processes of Organic Evolution*, 2nd Edition, Englewood Cliffs, N.J., Prentice-Hall, 1971. This intermediate-level text gives the best short review of speciation.

23

biogeography

To do science is to search for repeated patterns, not simply to accumulate facts; and to do the science of geographical ecology is to search for patterns of plant and animal life that can be put on a map.

ROBERT H. MACARTHUR,
GEOGRAPHICAL ECOLOGY

Why did the lion, which once ranged from Africa to India, and the tiger, which ranged from Persia to Siberia, not expand their ranges to cover all of Africa and Asia, or even the entire earth? On a larger scale, why are the rats and mice (family Muridae), common frogs (family Ranidae), and pond turtles and their relatives (family Emydidae) now dominant groups that are spreading over almost the entire Earth; while at the same time other groups, such as the rhinoceros (family Rhinocerotidae) and tapirs (family Tapiridae), have given up their wide ranges and are retreating to odd corners of the tropics? This kind of question is the concern of biogeography.

The biogeographer must collect piecemeal data concerning the distribution of species provided by taxonomic studies and somehow explain this information in terms of the principles of evolutionary theory, ecology, geology, and climatology. Biogeography has two disparate major branches: historical biogeography, in which the biologist attempts to account for the histories of the species of plants and animals, from the moment of their origin to the time they propagate other species or become extinct; and ecological biogeography, in which the biologist deals with the adaptations the species have evolved and the physical position that they occupy in local habitats. In the past, biologists have often pursued these two subdisciplines separately, as though they were different subjects. Actually they are really two aspects of the same phenomenon and must be joined into a single science.

ADAPTIVE RADIATION AND CONVERGENCE

Once a population has acquired intrinsic isolating mechanisms, thus becoming a full fledged species, it is destined to diverge forever in its genetic composition from all other species. But this does not mean that it will evolve independently of other species. The direction in which a species evolves is determined by the opportunities open to it, and these in turn are set by the other species of animals and plants with which it shares its environment. Obviously, an insect species cannot become specialized for eating pine seeds if no pine trees grow within its range. Less obviously, the specialization it chooses depends in part on the kinds of other specialists already present. If a species is given a choice between pine seeds and fir seeds, where only the pine seeds are being utilized by other insect species, the species will be more likely to evolve toward specialization on fir seeds. In this case competition is the force that moves it in one direction as opposed to another. Each animal species survives by eating other species — plant and/or animal — and competes with other species for the food. And each species is eaten by another. Thus each is tightly constrained within an ecological community, and there is a definite limit on the adaptive pathways it can follow in evolving.

TYPICAL AFRICAN ANIMALS in photograph on opposite page include plains zebras, Grant's gazelles, yellow baboons and a single bustard (bird at lower left). Photo was made at a watering place near Lake Amboseli in Kenya.

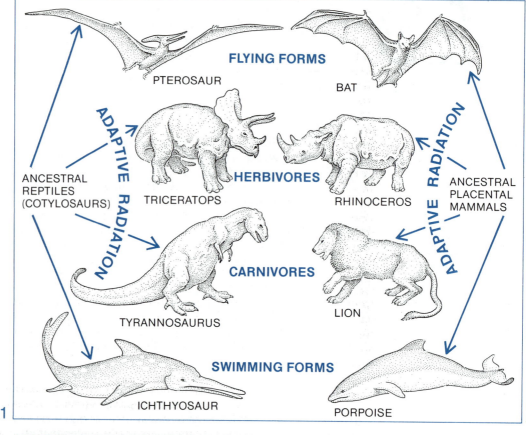

ADAPTIVE RADIATION of reptiles during the Mesozoic Era produced an assortment of animals that are more or less convergent with forms produced by adaptive radiation of modern mammals.

Again and again, on different continents and islands and in different geological times, evolution has created ecological communities that resemble one another in their broadest features. The origins of the species that make up these separated communities are often drastically different, but the final products are usually remarkably similar. The proliferation of species into different niches of a community is called ADAPTIVE RADIATION. The similarity acquired by species that come to occupy approximately the same ecological position but in different communities is called CONVERGENCE. Figure 1, an example of two adaptive radiations that occurred at quite different geological times, indicates the complementary relationship between adaptive radiation and convergence.

WAVES OF EVOLUTION

A common characteristic of evolution in general is the dynastic succession of groups, each of which flowered into adaptive radiation as it supplanted the group preceding it. According to conservative estimates, over 99 percent of past evolutionary lines have become extinct. The vast majority of these lines were products of adaptive radiation that gave way to succeeding radiations. In some cases, as in the amphibian-reptile-mammal lines, the ances-

Pioneers from a single ancestral species can invade a new continent or a group of islands and diversify into new species that spread out and fill many niches.

tors of the new radiations were products of the previous ones (*Figure 2*). But in many other cases entire groups were supplanted by lines phylogenetically remote from their own, sometimes invaders from other continents.

Adaptive radiation not only occurs repeatedly through time, it also occurs simultaneously on different, isolated parts of the earth. Perhaps the most dramatic example of simultaneous radiation is the origin of the modern mammalian fauna of Australia. The continent of Australia has been a giant island, isolated from Asia by the ocean straits of Indonesia, since before the beginning of the Cenozoic Era approximately 70 million years ago. Very early in this long spell of isolation, or even before it began, Australia was colonized by primitive egg-laying mammals called monotremes. These strange little animals make up only a minute fraction of the present Australian fauna; the echidnas and the duckbill platypus are the only surviving representatives. The early colonists also included the marsupials. These are mammals somewhat more advanced evolutionarily than mono-

tremes, and they are distinguished by their peculiar means of reproduction. The young are born while still very small and undeveloped; they then crawl into a pouch (the marsupium) located on the mother's belly and attach themselves firmly to the mother's teats to complete their development. The marsupials have been an outstanding success in Australia. Today they comprise about two-thirds of the native mammal species of the continent. During the past 70 million years the marsupials have undergone an adaptive radiation of the first magnitude, filling most of the niches open to land-dwelling mammals (*Figure 3*). In other lands equivalent radiation was achieved by placental mammals. The degree of convergence that has occurred during the radiations of the two groups is astonishing. In some instances, for example the flying squirrel (placental) *versus* the sugar glider (marsupial), the woodchuck (placental) *versus* the wombat (marsupial), and the placental *versus* the marsupial versions of the wolf and the mole, the external resemblances are so close that special knowledge is needed to place a given species

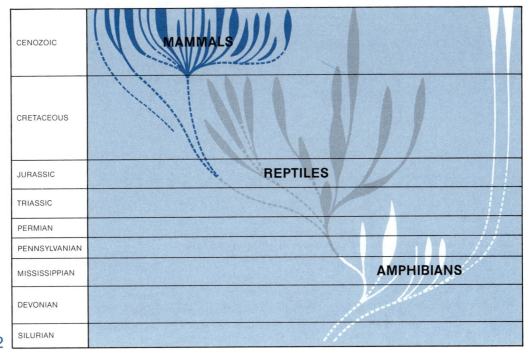

2

DYNASTIC SUCCESSION of amphibians, mammals and reptiles is depicted in this highly schematic diagram. Each group held a dominant position for a time and then was replaced by another group. Width of branches indicates abundance of species. Dashed lines indicate unknown ancestral forms.

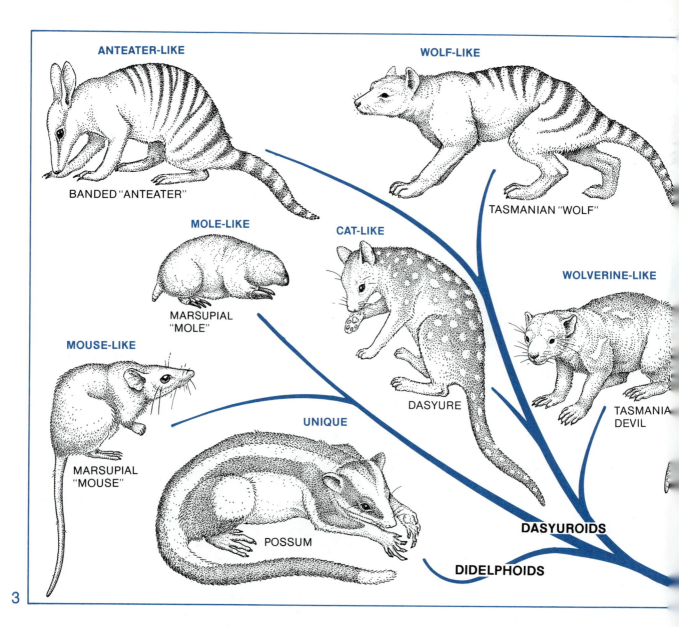

ANTEATER-LIKE

BANDED "ANTEATER"

WOLF-LIKE

TASMANIAN "WOLF"

MOLE-LIKE

MARSUPIAL "MOLE"

CAT-LIKE

DASYURE

WOLVERINE-LIKE

TASMANIA
DEVIL

MOUSE-LIKE

MARSUPIAL "MOUSE"

UNIQUE

POSSUM

DASYUROIDS

DIDELPHOIDS

3

ADAPTIVE RADIATION OF MARSUPIALS in Australia parallels that of placental mammals on other continents. Diagram shows only a few evolutionary lines. Some lines, such as those leading to kangaroo and flying phalanger, contain many species. Unique species with no convergent equivalents have developed among both placentals and marsupials; kangaroo and possum are marsupial examples.

in the proper group. In addition, unique forms have been produced on both sides. For example, the kangaroos are the large herbivores (plant-eaters) of Australia. Some species live in trees and browse on the vegetation, while others are specialized for grazing grass and low herbiage like sheep or cattle. The body form of the kangaroos is unique. While filling

the niches for large herbivores in Australia, these animals have developed their own peculiar means of locomotion and the anatomy that goes with it. The placental mammals of the remainder of the world also include unique specialists, from the horses and other hooved herbivores to the aerial bats and exclusively aquatic and marine forms such as manatees, porpoises, and whales.

Adaptive radiation is produced by speciation. It is possible for a single ancestral species, originating from a few pioneering organisms, to invade a new continent or group of islands and to generate species that spread out and fill multiple niches in the radiative pat-

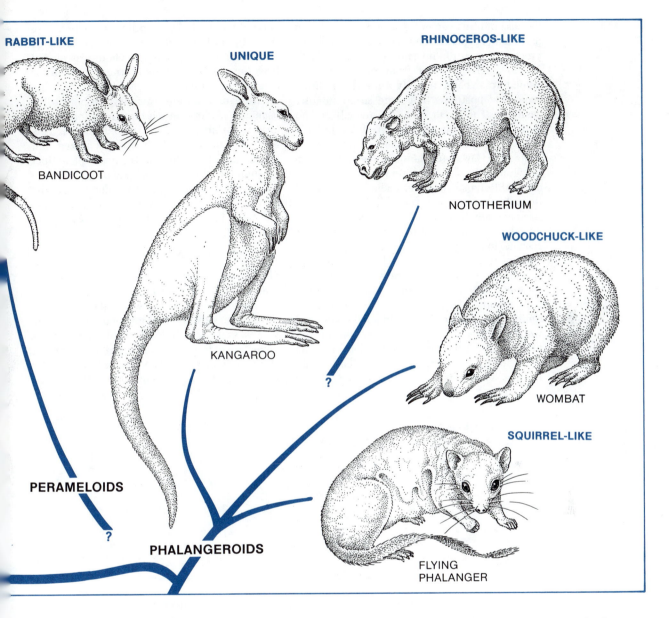

RABBIT-LIKE

BANDICOOT

UNIQUE

KANGAROO

RHINOCEROS-LIKE

NOTOTHERIUM

WOODCHUCK-LIKE

WOMBAT

SQUIRREL-LIKE

FLYING PHALANGER

PERAMELOIDS

PHALANGEROIDS

?

tern. How such an event can occur is illustrated by a case history of adaptive radiation and convergence on a smaller scale.

One of the most striking examples is illustrated in Figures 4 and 5. The Hawaiian honeycreepers, comprising a distinct family of their own (Drepanididae), are believed to have originated from a single species of small goldfinch-like birds that colonized Hawaii from the temperate zone of Asia or North America. When the ancestral drepanidids landed some million years ago, they found a rich new environment devoid of most of the kinds of bird characteristic of the faunas in the rest of the world. Hawaii contained no finches,

parrots, woodpeckers, or hummingbirds. Even today the drepanidids share the Hawaiian islands with only about 17 other native land and fresh-water bird species, such as the native crow and the Nene, or Hawaiian Goose. The reason Hawaii had such a poor fauna even earlier is of course its isolation from other land masses. Only a very few plant and animal species have succeeded in reaching and colonizing Hawaii. Those that do make the landfall are presented with unusual opportunities for the exploitation of unfilled niches. Speciation occurs easily among the expanding populations, because the water gaps separating the islands are powerful barriers to gene flow. In the case of birds generally, and drepanidids in particular, speciation has occurred through the process of multiple invasion (*see Chapter 22*). As the species diverged from each other, they tended to expand quickly into the unfilled ecological positions that were so abundantly available. The result was an unusually broad radiation with the production of a relatively small number of species. When Europeans first settled the Hawaiian Islands in the early 1800s, they found the bird fauna relatively intact. The original Polynesian settlers who had preceded them by some 2,000 years had not yet destroyed enough of the native habitats, or introduced enough harmful animal and plant species, to damage the natural living environment of Hawaii. Among the drepanidids were species that physically resembled finches, warblers, and parrots, and possessed similar food habits. There was a "woodpecker", *Hemignathus wilsoni*, that chiseled open dead wood with its stiff lower bill and picked out insects with its sharp, curved upper bill. The familiar true woodpeckers of the family Picidae accomplish the same goal in other parts of the world by chopping into wood with the entire bill and retrieving the insects inside with a long, sticky tongue. There was also at least one unique adaptive form: *Pseudonestor xenophrys*, which made its living by tearing open dead wood and trees with a parrot-like beak and feeding on the large longhorn beetle larvae that are abundant in such places. This bird, like many of the most specialized and interesting of the drepanidids, became extinct throughout almost all of its original range when its environment was altered by the European settlers of Hawaii — a possibly irreplaceable loss to evolutionary biology and to human esthetics.

BIOGEOGRAPHIC REGIONS

The world is not a continuous sphere of ecologically uniform terrain. Much of the land is divided into islands, while the six continents are separated from one another by oceans, seas, and straits. The continents themselves are extremely diverse, being comprised of broad deserts, forests, grasslands, and rivers and lakes. The consequence of this elementary

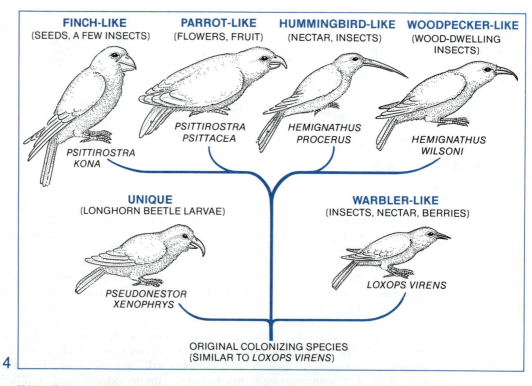

4

HAWAIIAN HONEYCREEPERS probably originated from a single species of finch-like bird that colonized the islands and later radiated by means of multiple invasion. This drawing shows six of the more extreme adaptive forms. In the early 19th century 22 species inhabited the islands.

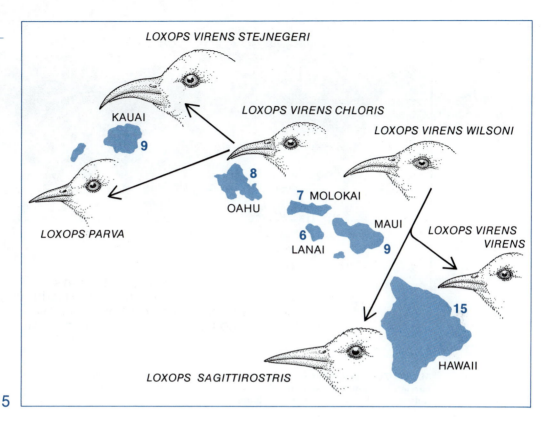

5

MAP OF HAWAIIAN ISLANDS shows number of honeycreeper species originally present on each island. Species and subspecies of the genus Loxops originated in isolated populations on different islands. In the course of their divergence they developed bills of different shapes.

generalization is that from the point of view of dispersing organisms the world is a great and complicated system of Australias and Hawaiis. Wherever enduring barriers exist, the floras and faunas separated by them tend to evolve in their own directions, to produce their own adaptive radiations, and thus to create self-contained communities. As a result, the Earth can be divided into biogeographic regions, each of which contains a distinctive assemblage of plant and animal species. Figure 6 is a map of the classic zoogeographic regions of the world, which are the most distinctive areas based on the animal species alone. A similar (but not identical) map could be drawn for the phytogeographic regions, based on the plants.

The mammals are a typical animal group. Australia and the immediately adjacent islands, comprising the Australian Region, are distinguished by kangaroos, bandicoots, the koala, the wombat, and other products of the marsupial radiation, as well as by the echidnas

and the platypus. Africa south of the Sahara Desert, the Ethiopian Region, has the African elephant, a large array of distinctively African antelopes, two species of rhinoceros, the chimpanzee, the gorilla, and was the home of early man (Chapter 27). Tropical Asia, the Oriental Region, is distinguished by the Indian elephant, the Malay tapir, the water buffalo, and the tiger. The Neotropical Region, comprised of tropical Mexico and Central and South America, has the llama, the spider monkey, marmosets, and the great anteater. The Nearctic Region (North America) is distinguished by the pronghorn antelope and grizzly bear, among others, while the Palaearctic Region (Europe, North Africa, and temperate-zone Asia) has the hedgehog, the wild ass, and the chamois. These lists of diagnostic species could be enormously lengthened with the names of additional species that represent virtually every other major group of plant and animal. Each species comprises, to a greater or lesser extent, one

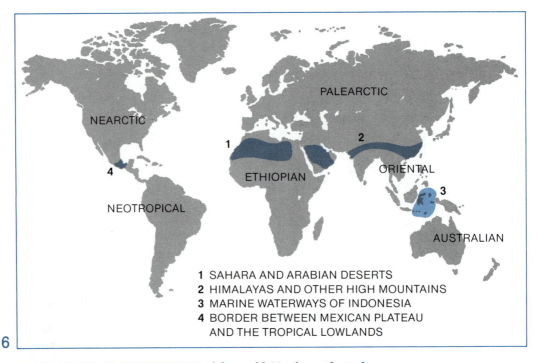

1 SAHARA AND ARABIAN DESERTS
2 HIMALAYAS AND OTHER HIGH MOUNTAINS
3 MARINE WATERWAYS OF INDONESIA
4 BORDER BETWEEN MEXICAN PLATEAU
AND THE TROPICAL LOWLANDS

MAJOR ZOOGEOGRAPHIC REGIONS of the world. Numbers refer to the barriers that separate them. Barriers are shown as broad colored bands to indicate gradual change from one fauna to the next.

product of an adaptive radiation that was hemmed in by the great physical barriers dividing the continents. Similar but less distinctive biogeographic regions can be distinguished in the life of the sea. Here the barriers consist primarily of the continents.

It must be added that the biogeographic regions serve only for a first, crude description of the collective distribution of life. The regions are not separated from one another by sharp boundaries. Most of the diagnostic species occur only within a small part of the region they represent, and no two have exactly equal ranges. As a result, biogeographers often attempt to subdivide the regions into biotic provinces and other, lesser divisions. Many of the older textbooks on biogeography and evolution are filled with this kind of information. But the effort is seldom useful, because the limits of each such subdivision depend entirely on the set of species chosen to define it. In other words, the limits change as the lists of diagnostic species change. There exists a close analogy between this shifting quality of biogeographic units and the difficulties that accompany the delimitation of geographic subspecies (see Chapter 22). The geographic

limits of the subspecies usually depend on the choice of the characteristics used to define them. The variable characteristics, like the species themselves within the biogeographic region, tend to be independent in their patterns of distribution. Moreover, not all of the species obey the limits of the biogeographic region. Many are found on both sides of the lines drawn by biogeographers to separate the regions. This is the reason that the "line" is more accurately represented as a band or zone within which the representatives of a region become increasingly prevalent as one travels toward the region. The only precise line that can be drawn is the one that represents the shift from a majority of species characteristic of one region to a majority characteristic of the other.

Not all groups of organisms conform to the classical zoogeographic regions in their distribution. The 19th century biologist Alfred Russell Wallace proposed that a north-to-south line through the Indonesian archipelago separated the Oriental flora and fauna from species of Australian affinities. But insects and higher plants, for example, do not conform to Wallace's line at all. The transition from mostly

Oriental to mostly Australian species occurs farther east, between New Guinea and the continent of Australia itself. Even the tropical mountain forests of northeastern Australia contain large numbers of species of Oriental origin.

THE BALANCE OF SPECIES

The "balance of nature" is not just a figure of speech. It can be measured precisely in the numbers of species of plants and animals that occupy a given island or other piece of the Earth's surface over a specified period of time. These numbers tend to remain constant for long intervals. As new species colonize the area old species become extinct, in approximately a one to one relationship. The equilibrium number can be rather quickly moved up or down by changing the rate at which species are added to the system, or by altering the environment of the area.

One of the most reliable indicators of the existence of a balance in the number of species is the area-species curve. One example is given in Figure 7. The relationship between area and the number of species is most clearly expressed when the faunas and floras of different islands are compared with one another. Only species that are members of some well-defined taxonomic group can be reliably used for this purpose. For example, one can include all of the land and fresh-water birds, as shown in Figure 7, or all of the birds belonging to some lesser subdivision of the birds such as the family Drepanididae, or the snails, or the higher plants, and so forth. Very roughly, the number of species belonging to such a taxonomically uniform group increases according to the formula, $S = CA^z$, where S is the number of species present, A is the area of the island, C is the value of S when $A = 1$ (its value is not important for our purposes), and z is a number that varies from one taxonomic group to another and from place to place. In the great majority of cases z falls somewhere between 0.25 and 0.35. In the case of the birds of the

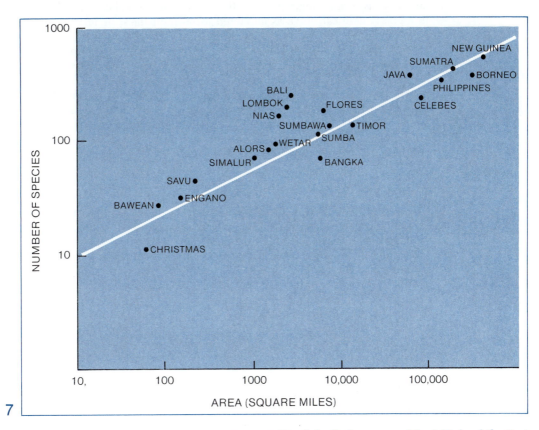

AREA-SPECIES CURVE of the fresh-water and land birds of the East Indies. Straight-line curve indicates that the numbers of species on individual islands are approaching theoretical limit.

East Indies, for example, z is 0.28. Another way of expressing this formula is to say that when the area is increased tenfold, the number of species approximately doubles. An island with an area of 100 square kilometers supports about twice as many species as one with an area of ten square kilometers. A great advantage of the area-species formula is that it allows one to predict the number of species present on a given island before the island is even explored. The formula is useful in dealing with any isolated biologic community. It can be applied to lakes, ponds, and streams — islands of water in a sea of land. It also applies to habitat islands — patches of habitat surrounded by other kinds of habitats. To the animals preferring spruce forests, for example, a patch of spruce forest in the middle of grassland is an island. A single tree standing alone in a grassy field can serve as an island to insects and other small creatures that are strongly dependent on the tree.

Several factors can cause a group of species on an island (or habitat island) to deviate from the area-species curve. One is failure to attain equilibrium. If the island is young, or has suffered a recent catastrophe that eliminated much of the life existing on it, the species may still be in the act of colonizing it. This means that the rate at which new species are arriving is greater than the rate at which they are becoming extinct. The number of species present on the island is thus increasing toward its equilibrium level on the area-species curve. A second cause of deviation is the distance effect: the number of species present on an island is smaller than expected because the island is remote from the sources of immigrant species. Because fewer colonists are arriving, a smaller number of species are present. This continues to be true even when the number reaches equilibrium. Also, fewer genera and higher taxonomic groups will be represented on the island. In extreme cases, islands lack such groups as fresh-water fishes, snakes, ants, and even conventional trees. Such faunas and floras are referred to as DISHARMONIC. Ireland has a disharmonic flora and fauna. For example there are no snakes, a basic life form found throughout Europe and most of the rest of the world. The snakes were not banished by St. Patrick, as folklore tells it. If they ever lived in Ireland in the first place, they were eliminated by glaciers during the ice ages of the Pleistocene, and they have not succeeded in recolonizing this relatively isolated European island since that time.

READINGS

S. CARLQUIST, *Island Biology*, New York, Columbia University Press, 1974. A superb general book on biogeography that goes far beyond its main topic of islands to treat continents as well. It analyzes most of the basic topics of biogeography in considerable depth, yet in clear, simple language and with a wealth of illustrations. This book is strongly recommended as the next book to read on biogeography after you finish this chapter.

P.J. DARLINGTON, *Zoogeography: The Geographical Distribution of Animals*, New York, John Wiley & Sons, 1957. The classic in its field, written in clear language with many examples.

R.H. MACARTHUR, *Geographical Ecology: Patterns in the Distribution of Species*, New York, Harper & Row, 1972. For students with a good mathematical background, this is the best introduction to general quantitative theories of biogeography.

NOTES

the growth and interaction of populations

Ecological explosions differ from some of the rest by not making such a loud noise and in taking longer to happen. That is to say, they may develop slowly and they may die down slowly; but they can be very impressive in their effects, and many people have been ruined by them, or died, or forced to emigrate.

CHARLES ELTON,
THE ECOLOGY OF INVASIONS

Imagine that one could select a single bacterium at random from the surface of this book, and could somehow endow it and all of its descendants with the power to grow and to reproduce without any restriction. In a month this bacterial colony would weigh more than the visible universe, and would be expanding outward at the speed of light. Similarly, a single pair of Atlantic cod and their descendants reproducing unhindered would in six years fill the Atlantic Ocean with their packed bodies. After a few more decades they, too, would weigh as much as the visible universe and be expanding in volume at the speed of light.

Man is one of the most slowly maturing and breeding of all organisms. But if the existing human population could somehow achieve the impossible feat of continuing to increase at its present rate, which is less than its maximum potential, it would come to weigh as much as the entire earth in 2000 more years. Four thousand years later it would weigh as much as the visible universe — and, of course, be expanding at the speed of light.

All populations have this potential for explosive growth. The reason is quite simple. Since all, or nearly all, mature individuals in the population can produce offspring, the rate at which a whole unrestricted population increases is exactly proportional to the number of individuals in the population. Therefore the larger the population, the faster it grows. And the faster it grows, the sooner it becomes still larger, and so on upward at an accelerating pace. In other words, the size of a population continuously accelerates when restraints are removed. This form of increase is called EXPONENTIAL GROWTH, and it is expressed mathematically in the following way:

$$\text{Rate of increase in number of individuals} = \left(\text{Average birth rate} - \text{Average death rate}\right) \times \text{Number of individuals}$$

The difference between the average birth rate and average death rate, the term enclosed in parentheses in the above equation, is called the INTRINSIC RATE OF INCREASE. To see this relationship more clearly, consider the exponential growth of mankind. At the present time, as everyone knows who has heard of the population crisis, the number of human beings is increasing at an accelerating rate. In South America, to take one example, the total population is now approximately 200 million. The birth rate, according to one authoritative estimate, is 42 children born for every 1000 persons every year, or 4.2 percent. The death rate is 19 persons out of every 1000 persons every year, or 1.9 percent, which less than balances the birth rate. The birth rate minus the death rate is the number of new persons added to every 1000 every year; this figure is $42 - 19 = 23$, or 2.3 percent. Translating the percentages into decimals we have a birth rate of 0.042 per person per year, a death rate of 0.019 per person per year, and an intrinsic rate of

SCHOOL OF BAITFISH (Stolephorus purpureus) in the underwater photograph on opposite page scatters before the attack of a predatory member of the tuna family (Euthynnus affinus).

increase of 0.023 per persons per year. In South America therefore, the population is increasing at approximately the following rate:

Rate of
increase in = Intrinsic × Number of
number of   rate of increase   individuals
individuals

$$= (0.042 - 0.019) \times 200,000,000$$
$$= 0.023 \times 200,000,000$$
$$= 4,600,000 \text{ per year}$$

When South America acquires a population of 500 million (this will occur around the year 2010 if the present trend continues), its growth rate will be up to 11,500,000 additional persons per year. At one billion (projected time: 2037) it will be adding 23,000,000 per year. Both estimates are made simply by multiplying the expected population size in that year times the present intrinsic rate of increase (0.023). These figures are only approximate, as they are based on data collected around 1960. If living conditions change and the proportions of individuals belonging to different age groups shift with respect to one another, the rate of increase also changes. For this reason we can use the expression intrinsic rate in only a loose sense when describing the particular case of South America.

The intrinsic rate of increase also varies enormously from species to species. In the laboratory rat, for example, it is 5.4 (540 per cent) per year. In most bacteria it is between 10 and 100 (1000 to 10,000 per cent) per day. The intrinsic rate of increase also varies among populations belonging to the same species, depending on the physiological condition of the organisms and the particular environment in which they exist. In extreme cases the intrinsic rate drops to zero. This is merely a formal way of saying that if organisms are living in excessively harsh circumstances, they cannot reproduce.

Of course, no population can maintain exponential growth indefinitely. At any given time most populations have a steady size, or else their numbers fluctuate up and down around a constant average value. In other words, the actual rate of increase in the number of individuals is zero. For this reason, ecologists often refer to the normal long-range condition of populations — including those of man — as one of zero population growth. Once in a while a population is temporarily reduced by catastrophe to a very low level. Or like man, it finds a new way to exploit the environment and thus enters temporarily unfilled areas. Then its numbers begin to increase rapidly. At first the growth conforms approximately to the exponential type expressed by the equation. But this condition is short-lived — no population can weigh as much as the visible universe! What happens is that the growth curve soon bends over to assume an S shape, as shown in the right-hand side of Figure 1. The number of individuals increases to the point where their environment can support no more. At this population size, called the CARRYING CAPACITY of the environment (for that particular population at that particular time), the death rate equals the birth rate. The rate of increase in the number of individuals is then zero. Temporary deviations from zero will probably occur in the future, causing the population to grow for a short time, or to decline, but the average value over long periods of time will be zero. As a result the population will fluctuate up and down around the carrying capacity of the environment for most of its existence. For most of its history, to take a familiar example, the human species had an actual rate of increase at or close to zero. Only within the last several thousand years have circumstances permitted an exponential increase. These circumstances are certainly temporary.

Why does expansion up to the carrying capacity of the environment follow a logistic (S-shaped) form? The reason is that as numbers increase, the birth rate slows down, or the death rate increases, or both occur together. Eventually the death rate comes to equal the birth rate, and the carrying capacity of the environment is reached. This more realistic model of logistic growth must be substituted for the exponential growth model given earlier. In fact, ecologists spend a great deal of

Ecologists often refer to the normal long-range condition of populations — including those of man — as one of zero population growth.

their time working out sophisticated variations of logistic growth, in attempts to make ever more precise predictions of the effects of various kinds of environmental change on the fate of populations.

SURVIVAL AND OPTIMAL YIELD

There are two fundamental ideas in ecology to be gained from a study of the logistic growth curve and the equation on which it is based. The first is that the carrying capacity of the environment is independent of the intrinsic rate of increase. How fast a population can grow has no bearing on the size it can finally attain, and vice versa. Slow-breeding elephants are capable of filling up the entire world (given a little extra time), but the fastest breeding population of bacteria will soon become extinct if it is limited to a drying pool of water. Ecologists have used this principle to show that when attempting to control a pest species, such as rats infesting dumps or seagulls endangering aircraft around landing strips, it is much more efficient to reduce the breeding sites or food supply of the population than it is to try to exterminate the animals themselves. Efficient garbage removal gets rid of rats better than rat traps. Taking away the resources of the species reduces the carrying capacity of its environment, a permanent alteration, but killing off part of the population is a temporary measure which the animals swiftly counteract through their awesome power of exponential increase. The same principle is important in a different way in the planning of conservation policies. When trying to protect a rare animal species, the most important step is to provide it with a preserve, an area containing its favorite habitat, to ensure that it will have a sufficiently high upper population limit. Once this is accomplished, it is possible to remove part of the population by hunting or capturing specimens for exhibition without endangering the species or even significantly reducing its numbers.

The second basic point about population growth is that the greatest amount of increase does not occur either when the population is just beginning to grow or when the population has reached the carrying capacity of the environment. It occurs at one particular point in between. Look again at the idealized logistic curve of Figure 1. The steepest rise of the S-shaped curve is in the center of the S, exactly halfway on the vertical axis between the baseline and the carrying capacity of the environment. At this point the greatest number of new individuals is added to the population in a given amount of time. This maximum rate of population growth is referred to as the OPTIMAL YIELD. Because not all growth curves fit the ideal logistic form, the "optimal yield problem" can seldom be solved so simply, and ecologists must resort to more sophisticated techniques. But almost without exception the optimal yield point occupies an intermediate position on the growth curve. Suppose a trapper were interested in getting the maximum number of foxes from a forest, or a fisherman in trying to extract the maximum number of fish from a pond, lake, or ocean fishing bank. The ideal economic strategy for either to follow would be to "crop" the population down to the point of optimal yield. If left near the maximum that the environment can support, or if exploited to the point where the organism becomes scarce, the yield will decline. One of the most important activities of applied ecology, therefore, is the solution of optimal yield problems. One theoretical but realistic example is given in Figure 2. This case, incidentally, illustrates a third principle: that the heavier the exploitation, the more the population will come to consist of younger and smaller individuals, which are also usually the least valuable commercially.

Already many of the great fishing areas of the world have had their populations driven below the point of optimal yield by excessive harvesting. The Georges Bank off the coast of New England, source of the cod, halibut, and

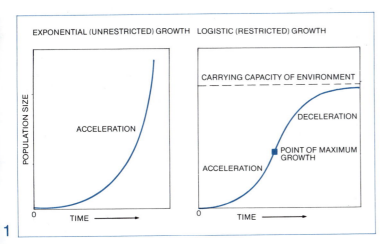

POPULATION GROWTH CURVES. Beginning with a small number of individuals, a population can increase in either exponential (left) or logistic fashion (right).

1

other prime food fishes, is one example of acute concern to the United States. As the fish populations are reduced, more expensive and sophisticated ships and trawling techniques are required to produce the same amount of yield, and prices rise accordingly. Another example of excessive harvesting, one of truly tragic proportions, is provided by the whaling industry. The blue whale, the largest creature that ever lived on the land or the sea, has now been hunted to the point of near extinction. The same fate has overtaken the humpback whale, while others among the great whales, notably the sperm whale and the sei whale, are being decimated and their existence will eventually be threatened. The senselessness of this destruction is deepened by the fact that the whaling industry has carried its activities to the point where whaling is now only marginally profitable and is in danger of wiping itself out.

AGE AND THE PROBABILITY OF DEATH

The intrinsic rate of increase of a population depends on two statistical qualities: the ages of the organisms comprising the population, and the reproductive performance of the organisms

FISH POPULATIONS at point of optimal yield (right) can be harvested with much less effort than populations reduced by overfishing (left), and the proportion of large fish caught is higher.

in each age group. Obviously, a population consisting entirely of individuals too old to reproduce has a zero rate of increase and is doomed to early extinction. A population made up mostly of immature individuals has a low rate of increase but is destined to improve it soon, and it probably has a secure future.

The age distribution of a population is the proportion of individuals in each age group. For example, the population of the United States in 1950 could be analyzed as follows: 29.0 per cent were 0–19 years old, 45.7 per cent were 20–49 years old, 16.2 per cent were 50–64 years old, and 9.1 per cent were 65 years old or older. When a population is allowed to exist in a constant environment for several generations or more, so that its birth and death rates become constant, its age distribution becomes stable — that is, the distribution is maintained without change from that time forward. The stable age distribution, like the intrinsic rate of population increase, differs greatly from species to species and depends to some extent upon the environment of the population. It further resembles the intrinsic rate of increase in being an ideal condition seldom maintained for long in nature.

One of the elements that determines the form of the age distribution is the survivorship curve. Examples of the three basic shapes this curve can take are shown in Figure 3. The oyster is an example of a species in which vast numbers of young stages are produced, and the majority quickly die. Only a tiny fraction succeed in attaching themselves to a rock or to some other support, the necessary step for completing the life cycle. The survival rate among oysters reaching this point is much higher. The hydra exemplifies species with constant mortality rates. An individual is just as likely to die when it is one year old as when it is one day old. Man and fruit flies, in contrast to the oyster and hydra, are species with a definite period of senescence (old age). Provided a good environment, most individuals live to a certain age in reasonably sound health. In man the age of senescence is approximately the biblical "three score and ten" (70) years. Then the diseases and infirmities of old age begin to set in, and death becomes increasingly probable with each passing year.

No human being has been reliably certified to have lived beyond 120 years. At this point it is worth asking a peculiar sort of question as part of our review of ecology. Why do senescence and death occur at all? To the evolutionary biologist no question having to do with any regular process of life, or its termination, is out of order. Each process has an evolutionary history, and is susceptible to explanation by means of the theory of natural selection. The explanation for the origin of senescence and scheduled death most generally accepted among biologists is the "broken test tube" theory of Peter Medawar, the distinguished British physiologist. Consider, Medawar said, a row of test tubes that are in constant use in a laboratory. They are sure to be broken at a regular rate just through everyday accidents. After an equally predictable period of such exposure, most will have perished and been replaced. A survivorship curve can be constructed for the test tubes, although they are inanimate objects, just like that for certain organisms — for example hydras and sparrows. If organisms were exactly like test tubes and did not age, they still would die at a steady rate from accidental causes. For most individuals an accidental death consists of being eaten by predators, killed by disease, or starving. If by age x (in primitive man x might have been 70 years) the great majority have been eliminated, natural selection will favor individuals that are vigorous and able to breed at earlier ages. Genes that cause senility and increase the probability of death after age x would not be penalized — what difference would they make if nearly everyone has already been eliminated by accidents? If these same genes also added vigor and reproductive ability to individuals younger than x, they would be favored through natural selection. Senility is therefore the price paid by individuals who manage to live beyond the usual time allotted by accidental death. They succumb to "play now, pay later" genes. Senility becomes a common condition only when populations are sufficiently well fed and protected to reduce mortality by predation, disease, and other external causes. Senility used to be rather rare in human populations. In the earliest days of civilization, the average life expectancy was only about 15

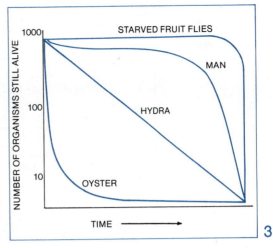

3

SURVIVORSHIP CURVES for four species of animal. Starting with equal populations of 1000 individuals, each species follows a characteristically shaped curve. Time scale is adjusted to cover same distance for each species.

years. Mortality was very high in infancy and childhood, and only a few individuals reached what would be today regarded as middle age.

REGULATION OF POPULATION GROWTH

In logistic growth, what slows the population down and finally brings it to a halt? Almost anything can. The factors of population control are exceedingly diverse, and differ from species to species.

To be eaten, to starve, to fall victim to disease — these and all other kinds of ill fortune met by some members of the population will increase the overall death rate of the population, or depress its birth rate, or both. But not every one of these effects helps to hold the population at the carrying capacity of the environment. To regulate population growth, a process must increase in its intensity as the population grows larger. When it does this, the process is known as a DENSITY DEPENDENT EFFECT (some textbooks use the less precise term

Senility becomes common only in populations sufficiently fed and protected to reduce mortality from disease, predation and other external causes. It is the price paid by individuals who survive beyond the usual lifespan alloted by accidental death. They succumb to "play now, pay later" genes.

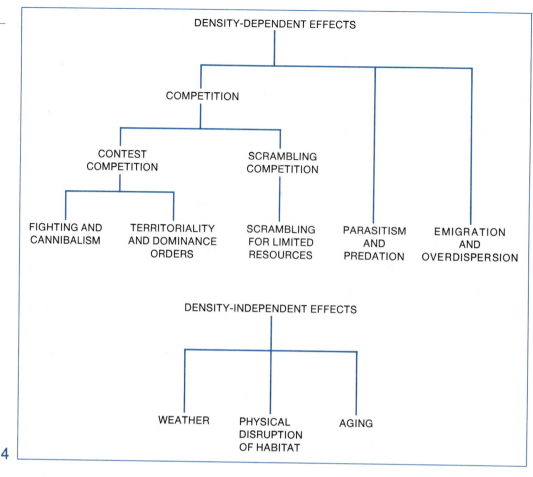

4

FACTORS THAT LIMIT LOGISTIC GROWTH include both density depen-dent and density independent effects. Density dependent effects tend to stabilize population size. Independent effects also affect size of population, but cannot hold it constant. Diseases are included under parasitism.

density dependent factor). Figure 4 presents the principal categories of density dependent effects. As a population of organisms grows more dense, its predators have an easier time finding prey. Infectious disease is more easily transmitted (epidemics among humans occur most often in the crowded cities). Competition increases, food and space grow scarcer. Mortality goes up and the average birth rate declines. Emigration may also increase as the organisms become more restive, and the popu-lation growth rate declines still further. In the end, one or more of these density dependent effects drives the mortality high enough and the birth rate low enough that the two come to be equal, and there is no longer any net growth in the population size. The population has at-tained zero population growth.

In contrast to the governing role played by

density dependent effects, DENSITY INDEPEN-DENT EFFECTS operate regardless of population size. A flash flood in the desert destroys a cer-tain fraction of the populations it touches, whether they are sparse or dense at the time. The same kind of random destruction is caused by many other adverse events that occur periodically in the physical environ-ment, from landslides and volcanic eruptions to unfavorable changes in the weather. The physiological process of aging is also largely independent of the population density. Den-sity independent effects can push population growth upward or downward, but they cannot hold the population size at a constant level.

COMPETITION

Competition, as ordinary experience teaches us, occurs when there is not enough of something desirable to go around. The ecologist defines competition as the active demand of two or more organisms for a common vital resource. An animal that aggressively challenges another over a piece of food is obviously competing. So is a plant that absorbs phosphates through its root system at the expense of its neighbor, or cuts off its neighbor from sunlight by shading it with its leaves.

The techniques of competition are extraordinarily diverse. As indicated in Figure 4, some ecologists make a rough distinction between contest competition, in which the organisms confront one another and actively contend for the resource, and scramble competition, in which the winning organism is the one that is able to get to the resource first or proves more efficient at using it up. Extreme contest competition involves direct aggression. When barnacles of the species *Balanus balanoides* invade rock surfaces occupied by a second barnacle species, *Chthamalus stellatus*, they eliminate these competitors by physically removing them. In one population studied in Scotland, ten percent of the individuals in a colony of *Chthamalus* were overgrown by the shells of the *Balanus* within a month, and another three percent were undercut and lifted off during the same period. A few others were crushed from the side by the expanding shells of the *Balanus*. By the end of the second month 20 percent of the *Chthamalus* had been eliminated, and eventually all disappeared. Individuals of *Balanus* also destroy one another, but at a slower rate than they destroy members of the competitor species.

Ant colonies are notoriously aggressive toward each other, and colony warfare both within and between species has been witnessed by many entomologists. The colonies often establish their territorial limits in this fashion, and sometimes they fight to the end.

Contest competition does not always take the form of open combat. Threats and mutual avoidance of animals serve the same purpose in many species. And in the scramble forms of competition we encounter an array of complex and subtle phenomena that are often uncovered only by close study. Flies provide some of the purest examples of scramble competition. The individuals most likely to survive are often simply those able to eat and digest the available food the fastest.

What keeps species from completely elim-

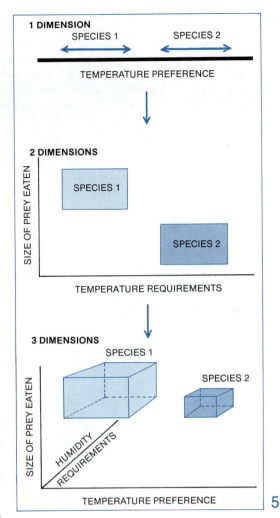

ECOLOGICAL NICHE can be represented by a one-, two- or three-dimensional graph. Niche may consist of more than three variables, and variables may differ from the three plotted here.

inating their less successful competitors? Why haven't the *Chthamalus* barnacles, for example, become extinct? If two species are too similar to each other in their requirements, one of them does indeed become extinct. This generalization is usually referred to as Gause's principle, or the PRINCIPLE OF COMPETITIVE EXCLUSION. It can be stated in broad but concise terms as follows: No two species that are ecologically identical can long coexist. The idea of ecological identity and difference can be more easily grasped by considering the concept of the niche. Figure 5 presents a simplified version of the way ecologists view the niches of species. Each species, or more precisely each local population, has a temperature range in which it can successfully live and reproduce itself. It also has an array of kinds of food on which it can subsist. In the case of plants, each population has a certain list of required nutrients it must obtain from the soil in addition to the basic requirement of radiant solar energy. And the population can succeed only within a particular range of humidity.

Figure 5 shows that up to three dimensions of a niche can be conveniently represented by a simple graph. But of course in nature there are many more than three dimensions. One must add, for example, the time of day or year in which the species is active, the major habitat in which it lives, the place in the habitat where it lives, and so on. These additional components cannot be added to this particular graph, but it should be clear that the biologist analyzes them, component by com-

ponent, to characterize much or all of the total niche of a population.

The principle of competitive exclusion, in other words, states that unless the niches of two species differ, they cannot coexist. This implies that if the two species are so genetically similar that their niches are the same, they cannot occupy the same geographical range. Conversely, two species can be genetically quite different but find themselves forced to live in the same place and do the same things in a particular environment. In this case too, one species will replace the other. The pair of species can coexist only if new pieces are added to the environment, so that one part favors one species and the other part favors the second species. An example of this aspect of the principle of competitive exclusion is provided in Figure 6.

The concept of the niche provides only a

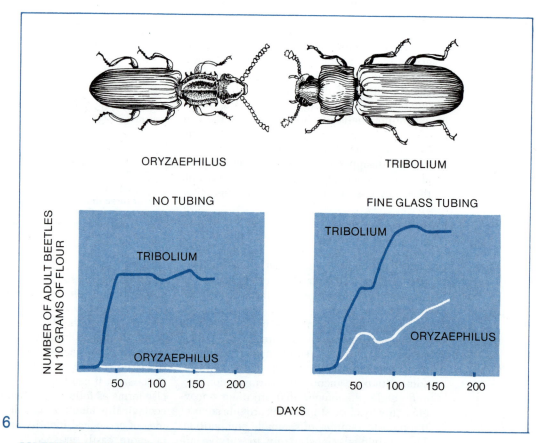

COMPETITIVE EXCLUSION in two species of flour beetle. In pure flour Tribolium defeats Oryzaephilus (left), but when fine glass tubing is added to the flour — creating tiny shelters to which Oryzaephilus can escape — the two species can coexist, as shown by graph at right.

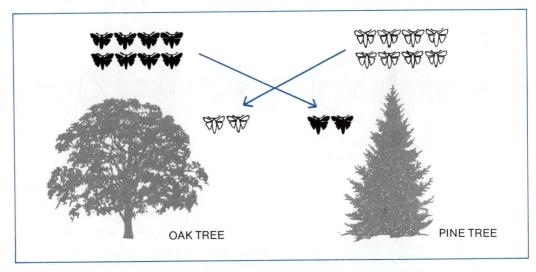

OAK TREE

PINE TREE

COMPETING SPECIES COEXIST by changing their niches through evolution. In diagram above, black moth has specialized for life on oak tree, and white moth for life on pine. Each species can attempt to invade trees of the other, but carrying capacity of "home" trees is too small to support enough invaders to compete successfully on the rival's home trees.

crude understanding of competitive exclusion. There is more to the story than just the tolerance limits of the species. The flour beetle example in Figure 6 illustrates that there is also a connection between competition and logistic population growth. In the experiment the two species are permitted to increase upward from low population levels. Each tends to grow according to the S-shaped logistic curve, finally reaching the carrying capacity of the environment. Competitive exclusion occurs if one species produces enough individuals to prevent the population of the other from increasing. By adding a new piece to the environment, in this case the glass tubing that favors the *Oryzaephilus*, the losing species can sometimes be taken beyond the reach of its competitor. Another way of accomplishing the same thing is for the losing species to change its niche through evolution. The environment remains the same, but the losing species has now been able to take itself far enough from its competitor to coexist. This is one form of the evolutionary process of character displacement, described in Chapter 22. Figure 7 presents an imaginary example to illustrate the relationship between logistic growth, specialization into different niches, and competitive exclusion. Here the dimension of the niche that has been diversified is the place where the moths live and breed. One could just as easily

have made it the season in which they live and breed, or the part of the forest they favor, or many other ecological properties, utilized singly or in combination.

EMIGRATION, MIGRATION, AND POPULATION CYCLES

One of the most effective devices operating in the control of population size is emigration: the simple departure of individuals from the mother population. Emigration of people to the New World, and later to Australia and Africa, was of course a principal means by which Europe eased its human population problem for several centuries. When emigration is density dependent, it can serve as one of the most efficient and finely tuned regulators of population size. The behavior of animals prone to such emigration is typically different from that of other members of the population. They are more restless and they tend to travel long distances in a single direction. The trip is evidently designed not to move them from one particular spot to another, but only to transport them away from the crowded mother population. In short, emigration is in many cases not merely an accidental departure from the range of the population but a distinctive behavior pattern designed to disperse the population. The individual organism that emigrates is gambling that somewhere it will find a habitat

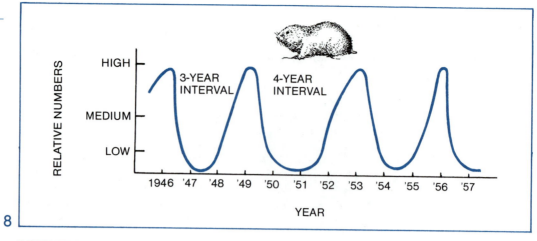

8

POPULATION CYCLE of the brown lemmings near Barrow, Alaska. Number of individuals oscillates widely through a cycle that lasts three or four years. Reason for cycle is unknown.

more favorable to life than the crowded one left behind. The gamble pays off frequently enough to make emigration a favored behavioral trait in natural selection.

One can find many examples of density dependent movement. Some aphid populations, for example, begin to produce more winged individuals as the populations become crowded. These migratory forms then fly away to start new populations. Lemmings march outward in search of new land to occupy; when they enter water, it is to swim to the other side — not to commit suicide. Emigration, whether density dependent or not, always has the same two results: lessening of the population pressure in the home area, and an increase in the likelihood of starting new populations elsewhere. In the long term it benefits both the individuals who remain behind and the pioneers who find new places to live.

Emigration should not be confused with migration. As practiced by birds, migration is the seasonal, back-and-forth flight between two favorable living areas. The migratory bird species come to the north temperate zones in the spring, when the exceptional bursts of vegetation and insect life make the rearing of their young especially favorable. They remain there through the summer and fall until both the weather and the food supply begin to deteriorate. Then they migrate southward to warmer climates to spend the winter. The long lives and great flying power of individual birds have made possible the evolution of this very special kind of oriented movement.

Another peculiar phenomenon associated with the regulation of population growth is the POPULATION CYCLE. As exemplified by the case of the lemming (*Figure 8*), the population cycle is a more or less regular oscillation in population size. It differs only in degree from the irregular fluctuation in numbers displayed by most animal and plant populations. Biologists have always been fascinated by the great sweep and regularity of some of these cycles, which are sometimes accompanied, as in the lemming, by mass emigrations during the times of greatest population density. No single explanation has proved satisfactory in all cases. According to one hypothesis concerning the brown lemming cycle, the population "crash" occurs when the population becomes so dense that it consumes all of the arctic vegetation on which it depends for a living. This may occur at some places (for example, Barrow, Alaska) but not at others. According to current theory, other populations fluctuate cyclically either because of density dependent hormone changes or because of short-term genetic changes that occur as the populations increase in size. This is one of the many fields of ecology which is in an early and very active stage of investigation.

PREDATORS AND PREY

Predation is the act of consuming another organism. Predators are either herbivores (devourers of plants), carnivores (devourers of other animals), or omnivores (devourers of both). Predators and prey can control each other's population size. When the prey become too numerous, they are cropped back in a density dependent manner by the predators. When the predators become too numerous, they crop the prey down to a low level, which causes them to run out of food and to suffer a population decline of their own.

A simple and instructive example of the balance between predator and prey is that of the wolves and moose of Isle Royale. Isle Royale is a 210-square-mile island located in Lake Superior near the Canadian shore. It is kept in its primitive condition by the U.S. National Park Service. Early in this century moose colonized Isle Royale, probably by walking over the 15-mile stretch of ice from Canada during the winter. In the absence of timber wolves and other predators, the moose population increased rapidly. By the mid 1930's the herd numbered between 1,000 and 3,000 animals. At this point the moose population far exceeded the carrying capacity of the island for moose, and the low vegetation on which they depend for existence was soon consumed. A population crash ensued, reducing the herd to well below the carrying capacity. As the vegetation grew back, the herd expanded rapidly again — and crashed again in the late 1940's. In 1949 timber wolves crossed the ice from Canada to Isle Royale. Their appearance had a marked and beneficial effect on the Isle Royale environment. The wolves reduced the number of moose to between 600 and 1,000, somewhat below the carrying capacity. The browse vegetation has returned in abundance, and the moose now have plenty to eat. Their numbers are controlled by predation rather than by starvation. The timber wolf population has remained steady at between 20 and 25.

What controls the number of timber wolves? Why don't they just keep eating moose until none of these prey are left, then suffer a population crash of their own? The answer is very simple. The wolves catch all of the moose they possibly can, but it is very hard work to trap and kill a moose. The wolves travel an average of 15 to 20 miles a day during the winter. All this effort yields a "crop" of about one moose every three days. That is enough to provide each of the wolves with an average of 10 to 13 pounds of meat per day. Apparently the wolves simply cannot increase the yield beyond this point, and their number has consequently stabilized. The moose, by unwillingly supplying the wolves with one of their members about every three days, have

stabilized their own population. In short, the predator-prey system is in balance. As a curious side effect, the moose herd is kept in good physical condition, since the wolves catch mostly the very young, the old, and the sickly ones. And, finally, because the moose population is not permitted to increase to excessive levels, the vegetation on which they feed remains healthy.

The Isle Royale example is one of the best documented cases of a balanced predator-prey system. It is generally true that herbivorous animals separated from their predators tend to increase to excessive levels and to strip the landscape of their food supply, often with disastrous effects on the environment. Insect pests are usually nothing more than species that have been introduced into new countries or new environments without their predators. Unshackled from these density dependent controls, the populations increase explosively. They can then destroy the crops or the shade trees, or whatever plants they are adapted to feed upon. Entomologists are sometimes able to solve the problem by means of biological control, which is usually the introduction of predators or disease organisms that specialize on the herbivore back in its native country.

When the European rabbit increased to excessive numbers following its introduction into Australia, it was brought under partial control by the deliberate addition of the virus that causes myxomatosis, a usually fatal disease of rabbits. Plants, too, often become serious pests when introduced into new countries without their natural herbivores. An especially dramatic example is that of the prickly pear (*Opuntia*), a cactus introduced into Australia as an ornamental plant sometime prior to 1839. Some of the plants escaped from cultivation and spread rapidly over 60 million acres in Queensland and the warmer parts of New South Wales. Much of this land was solidly covered by groves of the cactus. Biologists then found a moth, *Cactoblastis cactorum*, that is an effective natural enemy of such plants in South America. When freed in Australia, *Cactoblastis* multiplied and spread rapidly. It destroyed the cactus over much of its range by literally eating it up. Today both the cactus and the *Cactoblastis* coexist at relatively low densities. In other words, the herbivore-plant system has attained the same kind of balance as the animal predator-prey systems described earlier.

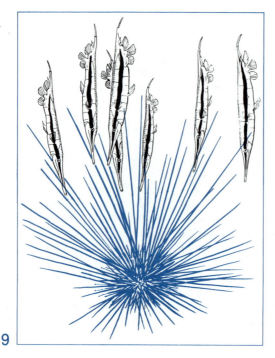

9

SYMBIOSIS between tropical fish Aeoliscus strigatus and a sea urchin is an example of commensalism. Fish is protected by sheltering spines without affecting sea urchin one way or the other.

SYMBIOSIS

Translated literally from the Greek, symbiosis means life together. Biologists define symbiosis as the association of two species in a prolonged and intimate ecological relationship, ordinarily involving frequent or permanent bodily contact. Symbiosis can take one or the other of the following three basic forms: in PARASITISM one species benefits at the expense of the other; in COMMENSALISM one species benefits while the other species neither benefits nor is harmed; in MUTUALISM both species benefit from the relationship.

Because of the complexity of the subject, the study of symbioses is virtually a science unto itself. Symbioses of one kind or another occur in all the major groups of organisms, from protists to mammals. They are extraordinarily diverse in kind, and the more advanced types employ bizarre adaptations that completely transform the life cycles and even the anatomy of the participants. As one might expect, the life cycles of symbionts are the most complex found in nature.

The most familiar parasites are the pathogens, the organisms that cause disease in their hosts. The reader is probably familiar with a long list of parasites that live in human beings. Into this group fall the many kinds of viruses (measles, German measles, mumps, smallpox), rickettsiae (typhus, Rocky Mountain spotted fever), *Corynebacterium diptheriae* (a bacterium causing diphtheria), *Pasteurella pestis* (a bacterium causing the plague), *Vibrio cholerae* (a bacterium causing cholera), several species of *Plasmodium* (protozoans causing malaria), *Ancylostoma duodenale* (hookworm), *Diphyllobothrium latum* (tapeworm), and so on. A large part of the practice of medicine is simply applied parasitology. Few people realize that for every one of the parasitic species that cause serious disease in man and other organisms, there are many others that give their hosts little or no trouble. The deadliest of the pathogens are usually the ones

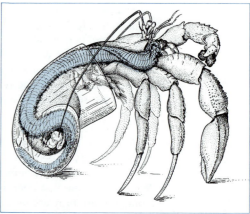

10

NEAR-PARASITIC SYMBIOSIS is involved in the association between a hermit crab and the polychaete worm Nereidepas fucata. Drawing depicts a crab forced to live in an artificial glass shell instead of usual mollusk shell. Worm (color) has moved in with the crab and steals some of its food.

The deadliest pathogens are usually those most poorly adapted to the host. The ideally adapted parasite is one that can flourish without reducing its host's ability to grow and reproduce.

11

LICHENS are symbionts that colonize harsh environments. Those depicted here, about actual size, were growing on an outcropping of rock in the Connecticut countryside.

that are the most poorly adapted to the host species. In the words of Lewis Thomas, "Disease usually results from inconclusive negotiations for symbiosis, an overstepping of the line by one side or the other, a biologic misinterpretation of borders." Frequently, the organism that succumbs is a secondary host that has picked up the disease from the primary

host species by accident. The ideally adapted parasite is one that can flourish without reducing its host's ability to grow and reproduce. As Thomas puts it, "The man who catches a meningococcus is in considerably less danger for his life, even without chemotherapy, than meningococci with the bad luck to catch a man."

Perhaps the most familiar examples of organisms that live in commensalism, or neutral symbiosis, are the bryophytes, mosses, bromeliads, orchids, and other plants that grow on the trunks and branches of trees. They flourish at no visible expense to the trees because they occupy the surface of what is in effect protective tissue, which is not thwarted in its primary role by the relatively lightweight symbionts. Commensalism is widespread throughout the animal kingdom, and is especially common among marine invertebrates. The host organisms are typically slow-moving or sessile, housed in structures such as shells or burrows, which can be readily shared by the smaller commensal species. One excellent example involving a fish and an echinoderm is shown in Figure 9. In a second example involving two invertebrates (*Figure 10*), we see the fine line that separates commensalism from parasitism. The polychaete worm lives by taking some of the food collected by the hermit crab. In this sense it is mildly parasitic, although it does not harm its host by a direct

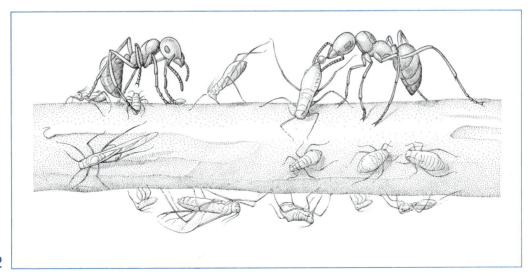

12

MUTUALISM between ants and aphids. When stroked by ant's antennae, aphid excretes sugar-rich "honeydew" (top center). In return, ants protect aphids from parasitic wasps and other enemies.

attack in the manner usually employed by parasites. In any given case it is very difficult to say that a species is a pure commensal, because this is equivalent to concluding that it has no effect on the host organisms whatever.

Mutualism, like parasitism and commensalism, occurs widely throughout most of the principal plant and animal groups and includes an astonishing diversity of physiological and behavioral adaptations. Some of the most advanced, and ecologically most important, examples occur among the plants. Nitrogen-fixing bacteria of the genus *Rhizobium* live in special nodules in the roots of legumes. In exchange for protection and a constant environment, the bacteria provide the legumes with substantial amounts of nitrates which aid in their growth (*Chapter 8*). Lichens are the ultimate symbionts belonging to this category. The reader will recall from Chapter 19 that they are actually compound organisms consisting of highly modified fungi that harbor green and bluegreen algae among their

hyphae. This improbable combination has proven especially efficient at occupying habitats that are inhospitable to most other plants: rock surfaces, the bark of trees, and bare hard ground (*Figure 11*). The lichens, including the so-called reindeer moss, are among the dominant plants of the treeless arctic tundra, and they are common among the pioneer organisms that colonize newly exposed rock and soil.

A radically different form of mutualism is illustrated in Figure 12. Many kinds of ants depend partly or wholly upon aphids and scale insects for their food supply. They "milk" these inoffensive little creatures by stroking them with their fore legs and antennae. The "cattle" respond by excreting droplets of honeydew, which is simply partly digested plant sap that has passed all the way through their guts. In return for this sugar-rich food, the ants protect their charges from parasitic wasps, predatory beetles and other natural enemies.

READINGS

J.H. Connell, D.B. Mertz, and W.W. Murdoch, eds., *Readings in Ecology and Ecological Genetics*, New York, Harper & Row, 1970. This collection of research articles published in the field of ecology in recent years is recommended to the student who wishes to examine some of the source materials on his own.

S.M. Henry, Ed., *Symbiosis, Volumes I, II.* New York, Academic Press, 1966, 1967. The most thorough and authoritative review of all aspects of symbiosis. It is unfortunately somewhat detailed and advanced for the beginning student.

W.W. Murdoch, ed., *Environment: Resources, Pollution and Society*, 2nd Edition, Sunderland, Massachusetts, Sinauer Associates, 1975. A superior collection of articles on many aspects of applied ecology.

E.P. Odum, *Fundamentals of Ecology*, 3rd Edition, Philadelphia, W.B. Saunders Company, 1971. A very good general introductory textbook of ecology, clearly written and rich in ideas and examples from population ecology.

R.W. Poole, *Introduction to Quantitative Ecology*, New York, McGraw-Hill, 1974. For the student with basic college training in mathematics, this is an excellent textbook on the quantitative aspects of ecological theory.

R.E. Ricklefs, *Ecology*, Newton, Massachusetts, Chiron Press, 1973. The most comprehensive and clearly written recent textbook that combines ecology and evolution.

NOTES

25

ecosystems

It is interesting to contemplate an entangled bank, clothed with many plants of many kinds, with birds singing on the bushes, with various insects flitting about, and with worms crawling through the damp earth, and to reflect that these elaborately constructed forms, so different from each other, and dependent on each other in so complex a manner, have all been produced by laws acting around us.

CHARLES DARWIN,
ON THE ORIGIN OF SPECIES

Every organism on Earth is a member of an ecosystem, a unit that consists of the other organisms that affect it (the biotic environment) plus the nonliving matter and radiant energy that make up its physical environment. Because these numerous elements act on one another in a reciprocal manner, the ecosystem is an almost endlessly ramifying and complicated unit. In this respect it accurately reflects the state of ecology, the science that studies it. Ecology in fact differs from the other disciplines of biology in that it does not treat the organism simply as an isolated unit, with a narrowly defined input of radiant energy, nutrients and stimuli. Ecology attempts to account for the effects the organism has on the remainder of the ecosystem, and vice versa. The movements of the organism, its discharge of waste products, and ultimately its contribution of material through the death and decay of its body, all change the environment to some degree.

It is easy to think of a pond or an island in the middle of the ocean as a single ecosystem. But it is often equally useful to deal with far less discrete units: a patch of pine forest, for example, or the grassy border of a highway. The limits drawn around ecosystems are always arbitrary and are selected only for convenience, because ecosystems are never entirely closed. All patches of nature are linked to the surrounding environment. Even a pond receives its water from precipitation plus the drainage from nearby land. Much of the organic matter in a pond is leached from the surrounding soil or is conducted to it in the bodies of immigrating organisms.

LAND BIOMES

The most superficial classification of ecosystems is into biomes and biome types. The biomes are the great communities of species that occupy the major patches of the environment, assemblages that have a strikingly different physical appearance from one another. The grassland of the western United States is one biome, and the nearby desert is another. A grassland and a desert also exist in southern South America, but the species that comprise these biomes are almost wholly different from their counterparts in the United States. Biomes that resemble one another in physical appearance but differ in species composition comprise a worldwide biome-type. All of the grasslands of the world are said to comprise the grassland biome-type.

Ecologists often differ in their classifications of the major patches of the environment. The biomes, like the subspecies and biogeographic regions described in earlier chapters, are strictly defined by the elements put into them. The borders of the biome shift as one adds or subtracts the species that are used to define them. The great boreal forests, for example, are usually designated as one of the biomes of

TROPICAL CORAL REEF is the most complex marine environment and contains the richest community of organisms. Photograph on opposite page depicts reef covering a vertical underwater cliff off Sharmesh Sheikh in the Red Sea.

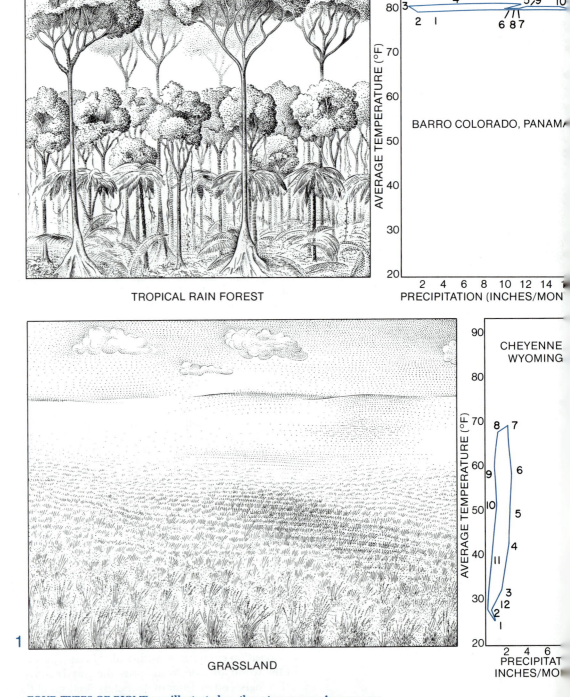

TROPICAL RAIN FOREST

BARRO COLORADO, PANAMA

GRASSLAND

CHEYENNE
WYOMING

1

FOUR TYPES OF BIOME are illustrated on these two pages. Accompanying climographs indicate year-round changes in temperature and precipitation. Numbers on climographs refer to months, with January as month number 1, February number 2, and so on. Place names on climographs indicate stations where readings were taken. Drawings depict generalized biome types, not specific places.

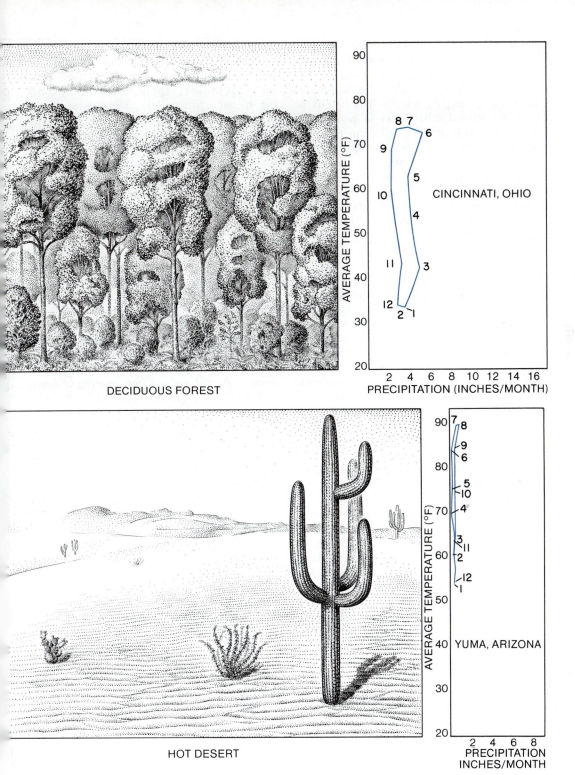

DECIDUOUS FOREST

CINCINNATI, OHIO

AVERAGE TEMPERATURE (°F)

PRECIPITATION (INCHES/MONTH)

HOT DESERT

YUMA, ARIZONA

AVERAGE TEMPERATURE (°F)

PRECIPITATION
INCHES/MONTH

North America. They consist of dense stands of coniferous (cone-bearing) trees such as spruce, fir, and larch. The boreal forest biome is sometimes referred to lightly as the "spruce-moose biome" by ecologists, who include one of the principal animal inhabitants in naming the biome. The ecologists are well aware that parts of the coniferous forest lack the spruce, other parts lack the moose, while still other parts have neither spruce nor moose. The point is that the biomes seldom exist as sharply defined patches. They have broad borders, and the species that comprise them have weakly correlated geographic ranges.

Despite this basic limitation, the biome concept is very useful for pointing out two important generalizations about life on this planet. The first is that the physical environment is all-important in determining the gross appearance of the organisms that exist in an ecosystem, especially the height, the profile, and the arrangement of the vegetation. The second generalization is a corollary of the first: given a particular physical environment in different parts of the world, the species of plants and animals will adapt to it by acquiring much the same outward body forms, regardless of their phylogenetic origins. This last statement should seem vaguely familiar. The classification of biome-types is in fact simply another way of describing convergent evolution (see Chapter 23).

Figure 1 depicts four of the biome types of the world. A simplified view of each of the plant formations is given, together with a characteristic "climograph" that documents the year-round changes in temperature and precipitation. Because temperature and precipitation are among the most crucial physical factors in determining the structure of the vegetation, in a large percentage of cases it is possible to predict the biome type at a given locality from the information provided in such climographs alone. Of course other physical factors are important too. The structure and chemistry of the soil can be equally vital. Certain trace elements, for example, are essential for the full development of plants. These include boron, chlorine, cobalt, copper, iron, manganese, molybdenum, sodium, vanadium, and zinc. When one or more of these substances is in short supply, land which according to the temperature-precipitation climograph should carry a forest will instead be covered with scrubby vegetation or be grassland bearing a few scattered trees. A single factor such as temperature, humidity, or the concentration of trace elements can determine the presence or absence of species and thus the character of the entire biome. This generalization is sometimes expressed as the LAW OF THE MINIMUM: the factor that is most deficient is the one that determines the presence and absence of species.

One of the most distinctive biome types is the tundra (from the Finnish *tunturi*: a treeless plain), the cold treeless land that encircles the arctic and elsewhere occupies the highest mountain tops above the treeline. The vegetation of the tundra resembles grassland but is actually made up of a mixture of lichens, mosses, grasses, sedges, low-growing willows and shrubs. A permanent layer of frozen soil, the permafrost, lies from a few inches to a few feet beneath the surface. It prevents the roots of trees and other deep-growing plants from becoming established, and it slows the drainage of surface water. As a result the flat portions of the tundra are dotted with shallow lakes and bogs, and the soil between them is exceptionally wet.

At the opposite extreme is the tropical rain forest, the biome type known to the popular imagination as the teeming jungle. A lowland tropical rain forest in prime condition is actually a glorious sight. The higher trees reach 100 feet, with a few soaring to 120 feet or more. Beneath the highest canopy are many lower layers of trees. Vines and palm trees are abundant, and dense clusters of orchids and other commensalistic plants cover many of the trunks and branches. The tight multiple canopies of the trees allow little sunlight to reach the forest floor. As a result few shrubs and herbaceous plants grow there; it is relatively easy for a man to walk through a rain

A single square mile of the richest tropical rain forests contains hundreds of species of trees, hundreds more of birds, reptiles, amphibians, insects, and dozens of species of mammals.

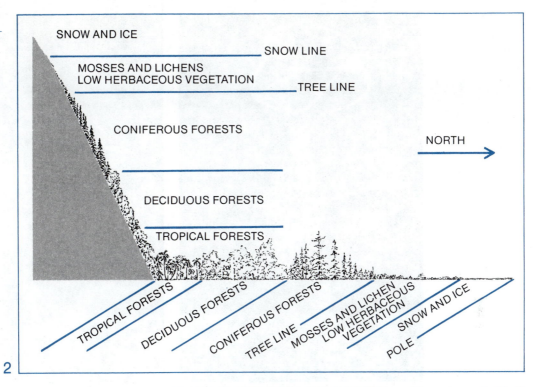

SNOW AND ICE

SNOW LINE

MOSSES AND LICHENS
LOW HERBACEOUS VEGETATION

TREE LINE

CONIFEROUS FORESTS

NORTH

DECIDUOUS FORESTS

TROPICAL FORESTS

TROPICAL FORESTS

DECIDUOUS FORESTS

CONIFEROUS FORESTS

TREE LINE

MOSSES AND LICHEN
LOW HERBACEOUS
VEGETATION

SNOW AND ICE

POLE

2

EFFECT OF ELEVATION AND LATITUDE on biome types. In North America mean temperature falls as one travels north or climbs upward from sea level. On mountainsides a rise of about 200 feet in elevation corresponds to a northward shift of one degree of latitude at sea level.

forest. Fallen leaves and dead wood decompose so rapidly that humus is thin, and even missing in spots on the forest floor. The diversity of life is the greatest found anywhere on Earth, either on land or in the sea. A single square mile of the richest forests contains hundreds of species of trees and hundreds more of birds, reptiles, amphibians, butterflies, ants and dozens of species of mammals. The environment is divided up to an astonishing degree by specialists among the plants and animals. For example there is an entire flora of lichens, mosses, and other small plants that grow as commensals on the leaves of the forest trees. In this microvegetation is hidden a little fauna of insects, mites, nematode worms, and other small invertebrates.

Such diversification is probably a result of the great age and stability of the moist tropical regions of the world. It also reflects the fact that for most of the geological past the tropical rain forest covered a much greater part of the world than it does today. Sixty million years ago such forests grew as far north as the southern United States and the British isles. The

tropical rain forests, and perhaps also the rich adjacent savannas and thorn scrubs, have served as the headquarters from which dominant groups of vertebrates have repeatedly arisen and from which they have then spread to other biome types around the world (see *Chapter 23*).

When climbing a high mountain, whether on foot or by automobile, one can observe the changes in biome types over relatively short distances upward. The Sierra of California, the high Rockies, and the White Mountains of New Hampshire are especially good examples. As Figure 2 suggests, a rise of about 200 feet in elevation is the rough equivalent of traveling 1° latitude northward at sea level. This apparent telescoping of space is due, of course, to the rapid decrease in temperature with altitude, usually (but not always) accompanied by an increase in precipitation and humidity.

THE SEA

Oceans and seas cover 70.8 percent of the Earth's surface, but except for the shallow water at their edges they are far less complex

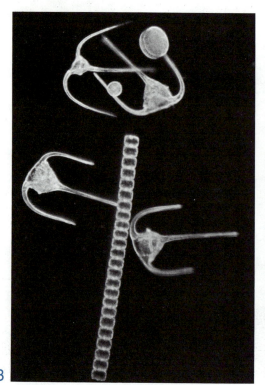

3

PLANKTON consists of microscopic plants such as diatoms and dinoflagellates (phytoplankton, left) and tiny animals such as copepods (zooplankton, right). Photos are not reproduced to same scale.

in structure and productivity than the land. Instead of forests and grasslands, the green plants of the open sea consist mostly of phytoplankton, the microscopic diatoms and other plant cells that drift in the lighted layers of the water (*Figure 3*). Consequently, as one looks down into open ocean water signs of plant life are rare. It is nevertheless there in abundance. The phytoplankton are eaten by a great array of invertebrate animals, ranging in size from protozoans to large medusae. These are in turn consumed by other, larger invertebrates and vertebrates. Like the phytoplankton, some of the invertebrates also drift with the water currents and are therefore collectively called zooplankton. The drifting plants and animals together are simply called PLANKTON. Other animals, the NEKTON, swim actively in search of prey. From the point of view of the biogeographer an ocean or sea can be thought of as a large basin, with shallow water lapping the rim and sunlight penetrating for a limited distance into the upper layers (*Figure 4*).

Because of the vastness and uniformity of this basin, and the relatively rapid mixing of the organisms within it, ecologists do not attempt to divide the oceans into biome types like those on land. Instead, they distinguish the waters of the shallow continental shelves (the NERITIC PROVINCE) from those of the main part of the basin (the OCEANIC PROVINCE). Next, they distinguish the bottom of the ocean at all depths (the BENTHIC DIVISION) from the open water above the bottom (the PELAGIC DIVISION). Finally, they distinguish the upper, lighted zone of the water from the deep, lightless zone — often referred to as the ABYSSAL ZONE. The exact boundary between the lighted and abyssal zones at any given locality depends upon the intensity of the sunlight, which in turn depends on the latitude and upon the turbidity of the water. Usually no light penetrates below 600 meters; beyond this depth the phytoplankton cannot grow. The animal life of the deep sea depends instead on the rain of dead and dying organisms that drift down from the lighted layers above.

The physical environment of the ocean is most diverse in the shallow water along its margin, where there is the greatest variation in temperature, salinity, light intensity and water turbulence. And here is the maximum luxuriance of both plant and animal life. The shoreward portion of this marginal strip, a subdivision of the neritic province, is often referred to as the LITTORAL ZONE. It extends from the uppermost line of tidal wave action out into the water to the depths at which the water is no longer thoroughly stirred by tides and waves. Here, in contrast to the deeper water, the plant life includes many larger forms of multicellular algae (the "seaweeds") and a few higher plants such as the ubiquitous eel grass (*Zostera*). But most plant growth is out of sight, in the form of microscopic algae.

The most diverse — and most interesting — of all marine formations is the coral reef (*Figure 5*). There are many parallels between the coral reef and the tropical rain forest. The reefs are limited to shallow tropical waters less than 60 meters deep. They are built up principally from the lime skeletons of millions of coral polyps. At any given moment the living generation of polyps forms a thin, growing layer on the tops of the massed skeletons of their ancestors. Some other organisms, such as lime-secreting species of plants (especially algae), single-celled foraminiferans, bryozoans, mollusks, and serpulid worms, add to the material and in a few localities are among the principal contributors. The individual reefs are relatively stable and often ancient — thousands or millions of years old. They are also complex in form, like tropical rain forests. The reason is that many kinds of coral add

their own distinctive skeletal forms — the staghorn corals, the organ corals, the brain corals, the fire corals, and many others. These supporting structures present other kinds of animal with a diversified landscape within which they can specialize and radiate. In the reefs of Port Galera, in the Philippine Islands, for example, are found 111 species of corals, 70 of chaetopod worms, ten of sipunculid worms, and between 200 and 250 species of crustaceans, as well as thousands of other species of brittle stars, crinoids, holothurians, mollusks and representatives of virtually every animal phylum.

SUCCESSION

The tropical rain forest and the coral reef do not spring full-blown om the ground or sea floor. They take possession of a patch of land or sea bottom by a long process of succession, in which empty space is first filled by a simple community of pioneer species, then gradually by more complex and bulky aggregations. The final community, or at least the most stable and longest-lived one, is referred to as the CLIMAX COMMUNITY. It is the end of the succession, and usually can be identified as belonging to one of the biome types or marine communities described above. An example of a succession is presented in Figure 6. Here the empty space is a new beach created by the waves of Lake Michigan. The first plants to take hold in the bare sand are grasses. As these bind the sand together and add humus, the next group of plants is able to take root. The transition flora, chiefly pine, oak and cottonwood, add still more humus. Finally, after a few hundreds or thousands of years the climax forest, comprised chiefly of beech and maple, takes over. There are many other ways to create new space and to start a new succession. Destruction of existing ecosystems by fire, severe wind storms, flooding, and landslides are among the most frequent. Since these are the recurrent catastrophes of nature, they ensure that no spot on Earth supports a climax community forever. At any given time, part of the natural environment is always at or near a climax condition, while other parts are passing through successions. This dynamic equilibrium on a grand scale ensures that species specialized for successional stages always have habitats available to them, and thus they do not go extinct.

Successions also occur within the lesser ecosystems. When a tree falls, its wood creates a space for decomposer species, and as it de-

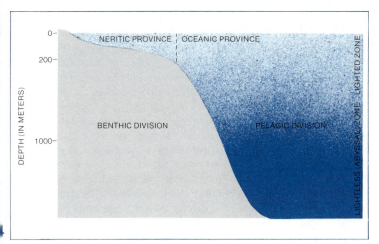

OCEANIC ECOSYSTEM is classified not into biomes but into provinces, divisions and zones. Special local communities such as coral reefs and intertidal zones are not included in diagram.

5

CORAL-DWELLING FISH show a wide range of specialization in the positions they occupy while resting and feeding on the reef. Drawing depicts two kinds of branching coral found in the Marshall Islands of Micronesia, together with a few of the fish that inhabit them. Corals with long branches, such as the one at right, usually grow on parts of reef that are sheltered from rough waves.

cays and crumbles, these species succeed one another in a highly regular pattern. When a rock or piece of metal or glass is dropped into the sea, its surface is colonized by a regular progression of bacteria, algae, coelenterates, bryozoans, polychaete worms, tunicates, barnacles, and other invertebrates. One can observe this particular succession merely by placing some empty bottles in a tidal pool or some other quiet, convenient body of marine water and checking them periodically over a period of a few weeks or months.

Why do successions occur? Why don't the first colonists simply take the space and hold it against subsequent intruders? Two forces oppose such pre-emption. In some instances the species alter the environment in a way to make it more favorable for other species than for themselves. For example, the first insects to attack a dead tree are specialized for boring into hard wood. As they crumble the wood, in collaboration with fungi and bacteria also specialized for this early stage of decay, the wood becomes less favorable to them and their offspring and more favorable to the insect species that prefer trees in an advanced state of decay. The species specialized for each stage of succession eat themselves literally out of house and home, leaving behind ruins that provide an excellent habitat for the next set of species. In a parallel way, plants sometimes foreclose future reproduction by the process of their own growth. In Australia, for example, the *Eucalyptus* trees of the open, sunny savan-

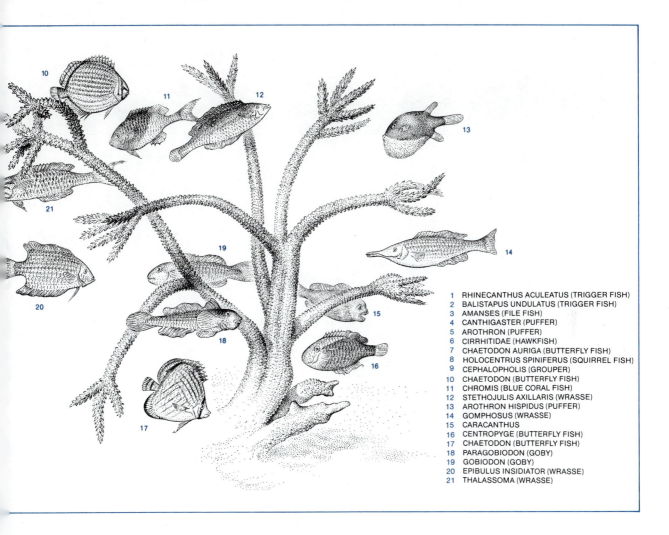

1	RHINECANTHUS ACULEATUS (TRIGGER FISH)
2	BALISTAPUS UNDULATUS (TRIGGER FISH)
3	AMANSES (FILE FISH)
4	CANTHIGASTER (PUFFER)
5	AROTHRON (PUFFER)
6	CIRRHITIDAE (HAWKFISH)
7	CHAETODON AURIGA (BUTTERFLY FISH)
8	HOLOCENTRUS SPINIFERUS (SQUIRREL FISH)
9	CEPHALOPHOLIS (GROUPER)
10	CHAETODON (BUTTERFLY FISH)
11	CHROMIS (BLUE CORAL FISH)
12	STETHOJULIS AXILLARIS (WRASSE)
13	AROTHRON HISPIDUS (PUFFER)
14	GOMPHOSUS (WRASSE)
15	CARACANTHUS
16	CENTROPYGE (BUTTERFLY FISH)
17	CHAETODON (BUTTERFLY FISH)
18	PARAGOBIODON (GOBY)
19	GOBIODON (GOBY)
20	EPIBULUS INSIDIATOR (WRASSE)
21	THALASSOMA (WRASSE)

nas provide shade in which young trees from the nearby rain forests can sprout and grow. When these intruders reach their full size, they cast too much shade for the *Eucalyptus* to reproduce themselves. The *Eucalyptus* finally die out as the rain forest takes over the land.

Not all successional stages prepare the way for their own decline and fall. In the eastern United States, mixed forests of pine and oak modify the soil in a way that makes it more favorable for the growth of their own seedlings than for those of their competitors. Succession in such cases occurs simply because slower-growing trees rise to dominance at a later time, and alteration of the environment therefore may have nothing to do with the replacement. Ecologists recognize two broad classes of species involved in this automatic kind of succession. Opportunistic species, sometimes referred to more colorfully as fugitive species, are able to disperse widely, and they grow and

breed rapidly. Many annual weeds belong to this category. The opportunistic species fill the pioneer and early successional positions. In the course of their evolution they have adopted the strategy of finding and utilizing empty space before other species pre-empt it. They are prepared to reproduce quickly and get out (send out their seeds or other dispersal stages) before other species take over. Stable species, on the other hand, specialize in competitive superiority. Forest trees of most kinds obviously belong to this category. They grow and disperse more slowly, but in their encounters with opportunistic species they are able to win the ground and hold it for a longer time.

FOOD WEBS AND ENERGY FLOW

As noted in Chapter 3, almost all of the energy utilized by living organisms comes from the sun. Even the fossil fuels — coal, oil, and natural gas — upon which so much of the

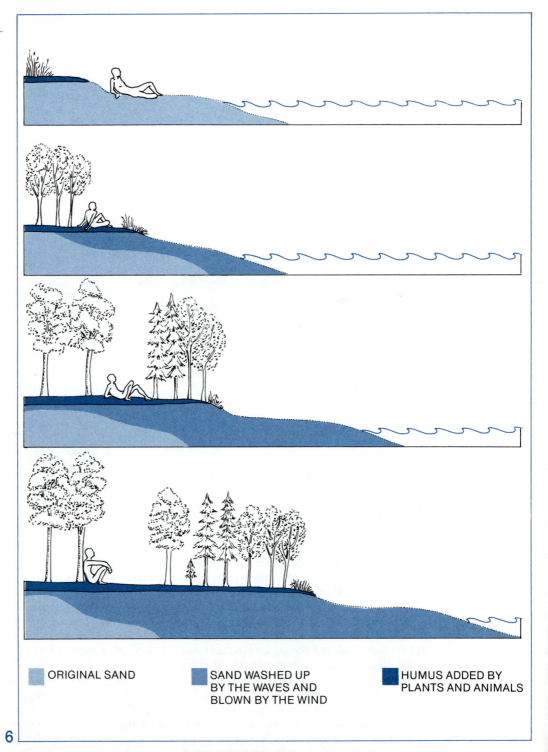

ORIGINAL SAND

SAND WASHED UP
BY THE WAVES AND
BLOWN BY THE WIND

HUMUS ADDED BY
PLANTS AND ANIMALS

6

SUCCESSION OF PLANTS has altered the beach dunes on the Indiana shore of Lake Michigan. First plants to invade sand are grasses. The final climax vegetation belongs to the deciduous forest biome.

economy of modern civilization is based, represents reserves of captured solar energy that were locked up by organisms millions of years ago. The amount of life on the Earth has always been strictly limited by the available solar energy. Each year this planet receives 5×10^{24} calories of energy from the Sun, of which 4×10^{20} are captured in photosynthesis. This primary solar energy cannot be increased. But energy is lost at each step as it is used by organisms. In time all of it is lost as heat, the form of energy that cannot be recovered.

The Earth is thus an open system with respect to energy. Energy passes quickly through the living organisms that cover the Earth; they cannot hold on to it. The best they can do is pass it on in ever diminishing quantities to their offspring and to the organisms that consume their bodies. To understand fully this basic principle of ecological energetics, one must understand the nature of the food chain. A food chain is most commonly a sequence of prey and predator species; one species is eaten by another, which is eaten in turn by a third, and so on. Examples of typical food chains from terrestrial and marine environments are shown in Figures 7 and 8. Ecosystems are vastly more complex than these deliberately simplified diagrams suggest. They contain a great many more chains, and most of the chains are tied together by cross-connectives, as exemplified by the three chains leading to the Great Horned Owl. For this reason the complete network of chains representing an entire ecosystem is often called the food web.

Each species forms a step, or link as it usually is called, in one or more food chains. The position located on the chain is referred to as the trophic level. Thus the green plants, which are the producers for the entire community, comprise the first trophic level. The second trophic level is formed by the herbivores, which are the consumers of the green plants, the third trophic level by the carnivores, which eat the herbivores, the fourth trophic level by the secondary carnivores, which eat the carni-

vores, and so on. In almost all ecosystems there exist top carnivores — one or more large, specialized animal species that browse on the animals in the lower trophic levels but are not ordinarily consumed by predators themselves. The larger whales enjoy this status, as do lions, wolves and man, the most gluttonous of all the top carnivores. In addition to the producer-to-carnivore chains there are parasite chains, in which small animals feed on their larger animal hosts, and decomposer chains, in which bacteria, fungi and a huge diversity of animal species feed on the dead bodies of organisms from all trophic levels.

As energy flows through the various food chains it is being constantly divided into three channels (Figure 9). Some of it goes into PRODUCTION, which is the creation of new tissue by growth and reproduction, as well as the manufacture of energy-rich storage products in the form of fats and carbohydrates. Some of the energy is lost from the ecosystem by EXPORT, the emigration of organisms coupled with the passive transport of dead organic material out of the ecosystem by the actions of wind and water. The rest of the energy is lost permanently to the ecosystem — and to all other ecosystems — by RESPIRATION. The leakage due to respiration is very high. In fact, only a small fraction of the energy is transferred successfully from one trophic level to the next. Ecologists put the figure at about ten percent. The exact measurement used to make this important generalization is ECOLOGICAL EFFICIENCY, the ratio of production available to one trophic level to the rate of production available to the next trophic level, on which it depends for its food.

Consider the following very simple ecosystem: a field of clover, the mice that eat the clover, and the cats that eat the mice. According to the "ten percent rule" of ecological efficiency, one would expect that for every 1,000 calories of new clover grown, about 100 calories of mice and 10 calories of cats would be added. Notice that the calories present in

No animal species preys on tigers — not because tigers are so formidable, but because tigers, being rather scarce, produce too few calories to make it worthwhile.

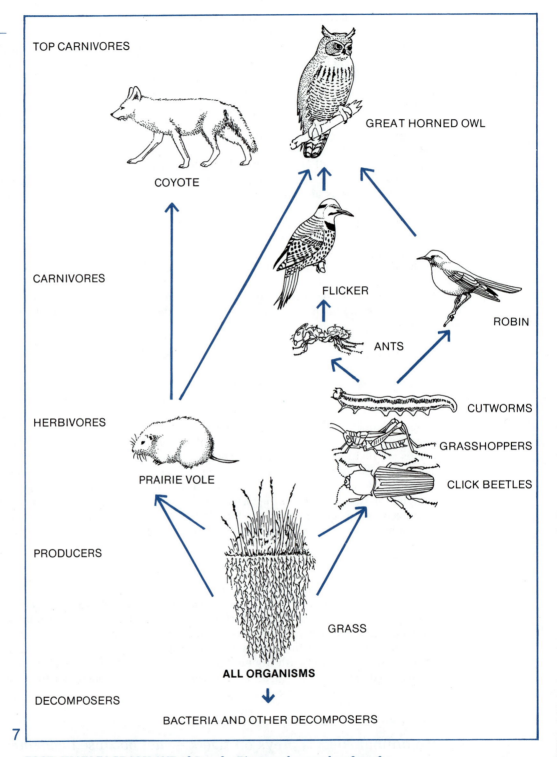

TOP CARNIVORES

COYOTE

GREAT HORNED OWL

CARNIVORES

FLICKER

ANTS

ROBIN

HERBIVORES

CUTWORMS

GRASSHOPPERS

CLICK BEETLES

PRAIRIE VOLE

PRODUCERS

GRASS

ALL ORGANISMS

DECOMPOSERS

BACTERIA AND OTHER DECOMPOSERS

7

FOOD CHAIN IN GRASSLAND of Canada. Diagram shows only a few of the chains and branches in this biome. Chains vary from one to five links. Trophic levels are indicated at far left.

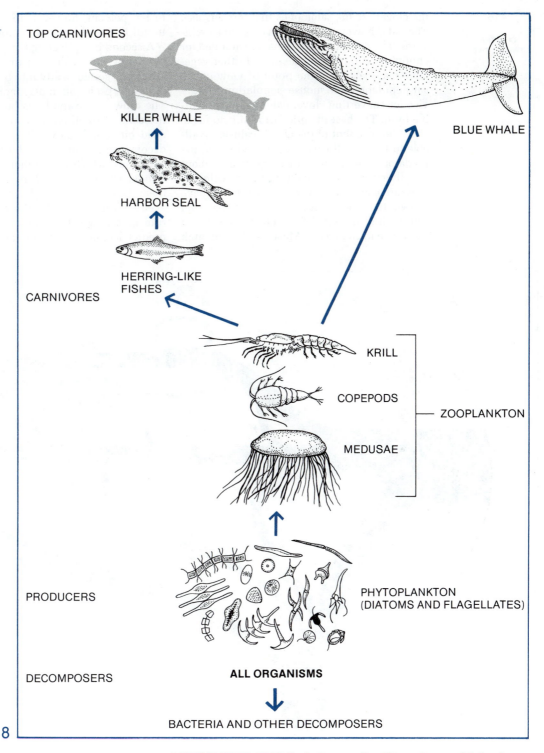

TOP CARNIVORES

KILLER WHALE

BLUE WHALE

HARBOR SEAL

CARNIVORES

HERRING-LIKE
FISHES

KRILL

COPEPODS

ZOOPLANKTON

MEDUSAE

PRODUCERS

PHYTOPLANKTON
(DIATOMS AND FLAGELLATES)

DECOMPOSERS

ALL ORGANISMS

BACTERIA AND OTHER DECOMPOSERS

8

**MARINE FOOD CHAINS. As in preceding illustration, trophic levels are
indicated at far left. Most of the organic matter that feeds marine organisms is synthesized by phytoplankton in the sunlit surface waters.**

the clover are the same ones that the cats use. The cats, however, are carnivores, not herbivores. The mice have teeth that are adapted for chopping and grinding seeds and other vegetable material. From the point of view of the cat population, the mouse population is a device for converting clover calories into an edible form. The best the mice can do is to make about ten per cent of the clover calories available to the cats. If cats were preyed upon, the predator would find these animals about equally efficient at converting mouse calories. Measurements in diverse ecosystems have shown that the ecological efficiencies from link to link in actual food chains vary from about five to 25 percent. Most are close enough to ten percent, however, to make this figure useful for rough approximations.

A second complicating factor in the analysis of energy flow is the fact that species cannot always be sorted neatly into trophic levels. Individual species often play more than one role. The crow, for example, is both a predator of insects and small animals and a scavenger of dead birds and mammals — and therefore a decomposer. Some other birds feed on fruit and seeds, which makes them herbivores, and also on a wide variety of insects, which makes them first and second level carnivores. Yet even with these qualifications, the ten percent rule of ecological efficiency can be used to predict accurately a major consequence in the

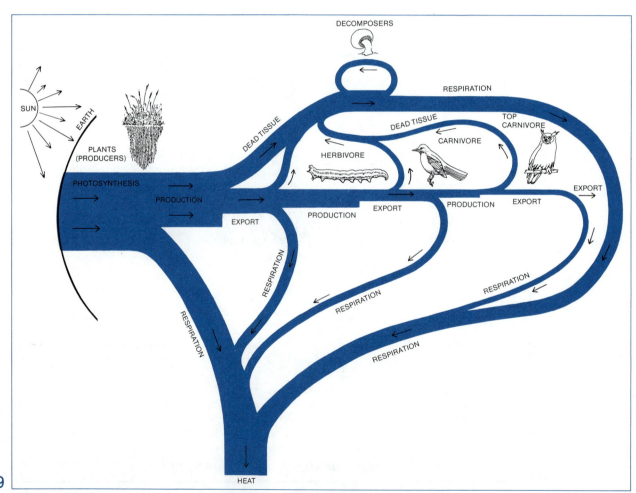

FLOW OF ENERGY through the ecosystem is divided into three channels, as shown by this flow chart. Relative amounts of energy channeled into production, export and respiration are not drawn to scale. Only about ten percent of the energy is transferred from one level to the next. Addition of new energy by importation of organisms and dead organic matter from outside the system is omitted.

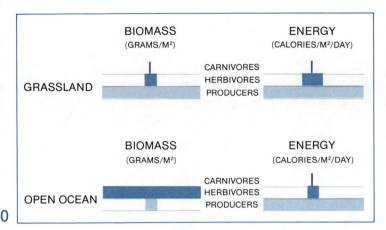

BIOMASS AND ENERGY PYRAMIDS for two very different ecosystems. The form of energy pyramids is similar from one system to another, but form of biomass pyramids varies considerably.

organization of ecosystems: food chains seldom have more than four or five links. The reason is that roughly 90 percent reduction in productivity with each step results in only $\frac{1}{10} \times \frac{1}{10} \times \frac{1}{10} \times \frac{1}{10} = \frac{1}{10,000}$ as much food at a trophic level four links removed from the green plants. In fact, the top carnivore that is producing only one-ten-thousandth as many calories as the plants on which it ultimately depends must be both sparsely distributed and far-ranging in its movements. Wolves must travel as much as 20 miles a day to find enough energy (*Chapter 24*). The territories of individual tigers and other great cats often cover hundreds of square miles. Such organisms are simply too sparse to support predators on their own. No animal species preys on tigers — not because tigers are so formidable, but because they produce too few calories to make it worthwhile.

The reader should note that there is only a loose relationship between the amount of living material present in an ecosystem, usually called the BIOMASS, and its rate of production. It is quite possible for a grassland with a relatively low biomass to outproduce a forest with a biomass hundreds or thousands of times greater. All that is required is that the forest have a lower production of calories per gram. In general, early successional stages have a lower biomass and a higher production than the climax stages that replace them.

This quality of communities can be compared with a similar quality in single populations: the optimal yield of populations, the reader will recall, occurs at an intermediate stage of population density and not at the maximum level permitted by the environment. The loose connection between the biomass and production is portrayed with particular vividness when the same trophic levels in different kinds of communities are compared. Figure 10 presents examples of the kind of diagram used by ecologists to summarize the quantitative properties of whole communities. The biomass pyramid shows the total weight of organisms belonging to the different trophic levels that can be found at any instant in a circumscribed area, in this case one square meter. The energy pyramid shows the rate of production per unit time in the different levels. As one might expect from the ten percent rule, energy pyramids are very similar from one ecosystem to the next, but biomass pyramids vary greatly. The reason is that plants, the producer organisms, exhibit such extreme variability in their ability to photosynthesize. The algae of the ocean can greatly outproduce most land plants, including grasses, on a per gram basis. Consequently they are able to support a proportionately much larger biomass of herbivores. Note that the production of ocean herbivores is still only about ten percent that of algae. Their biomass is greater, however, because the algae are growing and being eaten at an exceptionally high rate.

CYCLING OF MATERIALS

Although the world is an open system with respect to energy, it is a closed system with respect to materials. Water, carbon, nitrogen, phosphorus, and the other elementary building materials of living tissue are cycled perpetually through organisms to the environment and back again through organisms. The carbon and nitrogen atoms of which we are constructed at this moment are the same atoms that composed dinosaurs, insects, and trees in the Mesozoic Era. Most of the atoms are built into complex molecules by the organisms, only to be released in the form of small molecules during the process of death and decomposition. For a period of time that varies from a few seconds to millions of years — according to the kind of substance involved and the laws of chance — the small molecules circulate in the environment. Eventually most are recaptured by organisms and incorporated into large molecules to begin the cycle again. Because the chemical elements are being passed back and forth between biological organisms and their geological (physical) envi-

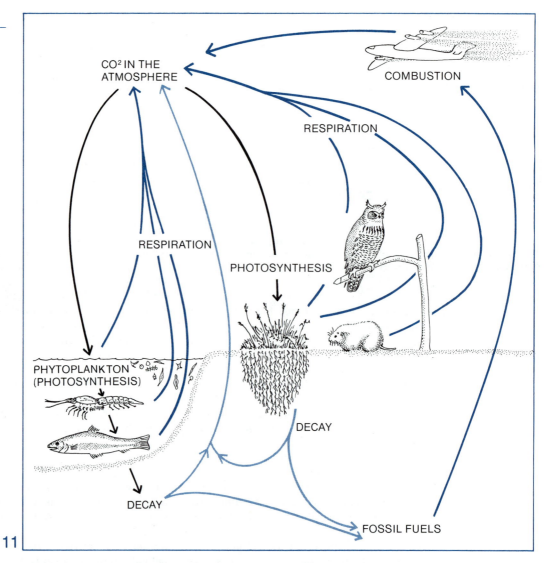

THE CARBON CYCLE. Carbon is removed from the CO_2 pool in the atmosphere by the process of photosynthesis. Animals obtain carbon by eating plants or other animals that feed on plants. Animal respiration, decay of organic matter, and combustion of fossil fuels return CO_2 to atmosphere.

ronment, the material cycles are usually referred to as BIOGEOCHEMICAL CYCLES.

Between 30 and 40 of the chemical elements participate in such cycles. Some are obscure trace elements that are required only in tiny amounts. Iodine, for example, is essential for the endocrine physiology of vertebrates. Molybdenum is essential for nitrogen-fixing organisms; when it is deficient, as it is in the soils of certain parts of Asia and Australia, vegetation is sparse and scrubby. Other substances, such as the radioactive form of strontium (Sr^{90}), have been put into circulation in

abnormally large quantities by the activities of man.

Each of the biogeochemical cycles has its own distinctive pathway and rate of flow. Those followed by carbon and nitrogen are displayed in Figures 11 and 12. Several general qualities of cycles are exemplified in these two cases. Circulation occurs chiefly when the carrier molecules are in the gaseous phase and become part of the atmosphere. There are two very different routes which a given atom of carbon or nitrogen can follow. One leads through organisms and is relatively quick. The

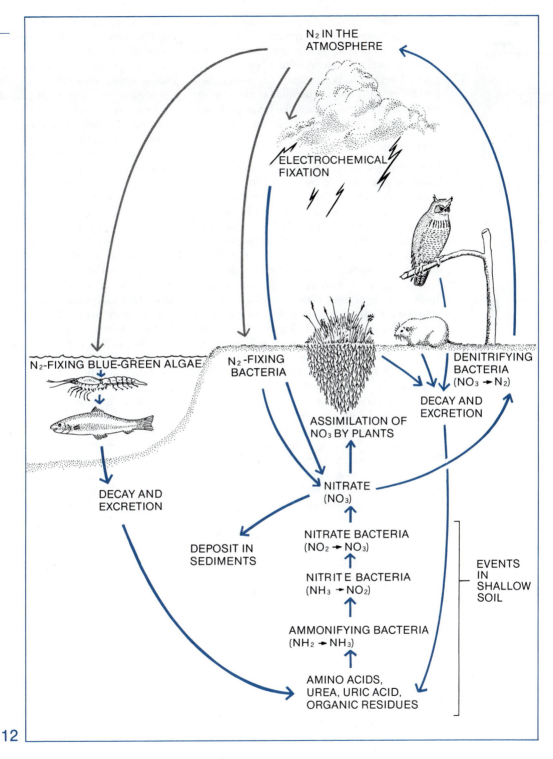

N₂ IN THE
ATMOSPHERE

ELECTROCHEMICAL
FIXATION

N₂-FIXING BLUE-GREEN ALGAE

N₂-FIXING
BACTERIA

DENITRIFYING
BACTERIA
($NO_3 \rightarrow N_2$)

DECAY AND
EXCRETION

ASSIMILATION OF
NO_3 BY PLANTS

DECAY AND
EXCRETION

NITRATE
(NO_3)

DEPOSIT IN
SEDIMENTS

NITRATE BACTERIA
($NO_2 \rightarrow NO_3$)

EVENTS
IN
SHALLOW
SOIL

NITRITE BACTERIA
($NH_3 \rightarrow NO_2$)

AMMONIFYING BACTERIA
($NH_2 \rightarrow NH_3$)

AMINO ACIDS,
UREA, URIC ACID,
ORGANIC RESIDUES

12

THE NITROGEN CYCLE. Bacteria and bluegreen algae convert atmospheric nitrogen to a form that other organisms can use. Lightning has the
same effect. Plants absorb nitrates through their roots. Animals obtain
nitrogen by eating plants or other animals. Dead organic matter and waste
products of animals return nitrogen to the soil. Denitrifying bacteria recycle it back into the atmosphere.

other leads through decay, entombment in geological deposits, and liberation after very long periods of time by exposure and weathering or through the intervention of man. A third general principle is that man has added new pathways to the biogeochemical cycles. He has also altered to some degree the lengths of each of the cycles, with far-reaching consequences.

POLLUTION

The word pollution has many meanings and implications for different people. The ecologist, with special insight, defines it as the misplacement of resources by alteration of the biogeochemical cycles. Only very recently has man become aware of his own profound influence on these cycles. Previously he thought the world to be a bottomless sink into which waste materials could be poured forever. They would disappear, he believed, into the uncharted depths of the soil and of the sea. When they returned, if ever, they would come back to us transformed into the pure, "natural" products that he first exploited. Now he knows better. The evidence shows that mankind is an inordinately demanding passenger on this spaceship Earth. He has drastically altered the direction and rate in which certain materials are entering the biogeochemical cycles. Many of these substances are now changing the complexion of entire ecosystems as large as forests, estuaries, rivers, lakes, and bays; and some are beginning to act as outright poisons.

No matter what is done to the biogeochemical cycles, they will still be cycles. Eventually the rates of input and output of each of the substances, for example nitrogen into nitrates and nitrates back into nitrogen, will strike a balance or at least begin to approach a balance. Why, then, should anyone worry about disturbing the system? The answer is that both the disturbances and the new balances that follow them are unhealthy for man and most species of plants and animals. They are unhealthy for the reason that man belongs to ecosystems that evolved for millions of years in the old, unpolluted environment.

Pollution alters the distribution of materials in two ways. It changes the rate of flow of the basic materials, and it introduces some of the materials in the form of new, man-made molecules with which decomposer species cannot cope. One of the most pervasive consequences of the first type of pollution is EUTROPHICATION, the addition of excess nutrient materials to water. Man, the great and wasteful consumer, injects excess phosphates, nitrates, ammonia and diverse forms of organic residues into the lakes, streams and estuaries. Most are passed in sewage and industrial waste. Other substantial amounts come from the drainage of farmlands, where manure and synthetic fertilizers are spread on the ground during the winter and flushed into streams during the spring thaws and rains. Considerable additional quantities of nitrates are added from the combustion of coal and other fossil fuels. Although these materials are mostly distasteful excretory products for us, they are vital nutrients to aquatic plants. The plants are thus provided with a eutrophic (richly nourished) environment. Algae multiply until they form "blooms" that turn the water green and coat the surface with a scum resembling green paint. As eutrophication proceeds, the original dominant elements of the freshwater phytoplankton, the desmids, are replaced by diatoms, which in turn give way to flagellates and other green algae. Finally, the bluegreen algae take over. As the masses of algae and other aquatic vegetation decay, the water is depleted of oxygen. Anaerobic bacteria become the prevalent decomposers. These organisms cannot break organic molecules all the way down to CO_2. Instead, the end products of their respiration are such substances as hydrogen sulfide and propionic and butyric acids. The water becomes offensive in smell and taste to human beings. At the height of its eutrophication Lake Monona, in Wisconsin, was typical when it was compared by one observer to "a foul and neglected pig sty." Finally, as the oxygen is used up much of the original fauna declines, to be replaced by midges, certain types of "trash fish" and other resistant forms of animal life.

Man, while thoughtlessly flushing excessive nutrients into the waters all around him, has

The greatest destroyer of ecosystems in history, except for man himself, is the humble and winsome European rabbit.

also been adding abnormal quantities of metals and other elements that were previously scarce in ecosystems. Lead and mercury are two familiar substances that have been increased to hazardous levels in many places. Perhaps the most insidious of all additions is strontium 90, a radioactive isotope that originates in the fallout from atomic tests and also as a by-product of fission in nuclear reactors. When strontium 90 is absorbed by plants and then passed along the food chains, it becomes more and more concentrated in the bodies of organisms. This enrichment is especially pronounced in vertebrates, where strontium tends to replace calcium in the construction of bones. There it lodges for the lifetime of the organism, steadily bombarding the surrounding tissues with ionizing radiation.

Of equal significance to ecosystems are the new organic molecules produced and discarded by man. So long as these substances are biodegradable (capable of being broken down to simpler components by organisms in the ecosystem) they pose no greater threat than the natural products (such as phosphates and nitrates) that contribute to pollution. But when substances cannot be easily degraded by organisms, they create a wholly new kind of problem. Fifteen years ago artificial detergents of the kind used daily in households were accumulating at an alarming rate in the freshwater systems around the cities and towns of the United States. Although not poisonous at low to moderate levels, they became a nuisance because of the large masses of foam they generated in streams, filtration plants, and other places where the water is stirred vigorously. These early detergents built up for the simple reason that the organic portions of their molecules had been synthesized in a form which microorganisms found extremely difficult to metabolize. When manufacturers deliberately switched to the production of biodegradable forms of detergents, that particular problem disappeared.

Unfortunately, no easy way out exists in the case of the hard pesticides; the chemicals which are effective in killing insects and other pests are also difficult for organisms to digest. The roll call of these persistent poisons is a familiar list of household words: aldrin, chlordane, dieldrin, DDT, heptachlor, lindane, and others. DDT is by far the most important in its impact on the environment. One of the first of its class of insecticides, it was introduced with

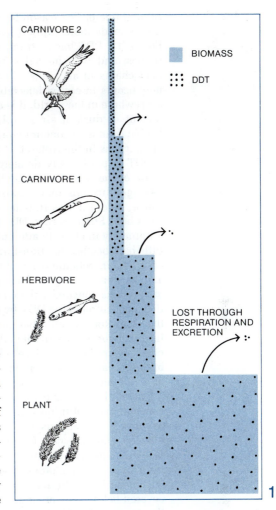

DDT IS CONCENTRATED by the ecosystem. At each level of the food chain, more than half the biomass is lost through excretion, respiration and decay. But most of the DDT is retained, and as a result becomes highly concentrated in carnivores.

spectacular success during the early years of World War II. By eliminating insect vectors such as mosquitoes, flies and lice, DDT permitted the control of malaria, typhus and other diseases over large parts of the world. It is credited with having saved over half a billion human lives. It also saved agriculture from uncounted millions of dollars damage from a variety of plant-eating insect pests. DDT in fact has been one of the great benefactors of mankind, and it justly won a Nobel Prize for its discoverer, Paul Mueller of Switzerland. But because it also degrades slowly and is relatively toxic, it has now turned into an ecologi-

13

cal nightmare. The reason for this transformation is made more explicit by the diagram in Figure 13. DDT has a strong affinity for fatty tissues, and it is moved efficiently through food chains in all kinds of ecosystems. DDT now occurs in easily detectable levels almost everywhere in the world. It is a principal component of dust in air samples from remote Pacific islands. It contaminates the fatty tissue of penguins in the Antarctic.

DDT is particularly dangerous to birds because of the way it is concentrated during its passage through food chains. In its early career, before it was indicted by Rachel Carson's book *Silent Spring*, DDT was used indiscriminately in cities in attempts to control the little bark beetles that transmit Dutch elm disease. The insecticide reached the soil of yards and gardens in high concentrations. When earthworms consumed the contaminated soil, they concentrated the DDT by a factor of more than ten. Robins and other songbirds that ate the earthworms concentrated the DDT again — often above their own level of tolerance. The bird populations were temporarily reduced by 30 to 90 percent.

There is irony — and eventual hope — in the conception of pollution as the misplacement of resources. If man has witlessly altered the biogeochemical cycles to his own disadvantage in the past, he can, with appropriate planning and effort, change them back to his advantage in the future. Nutrients can be recovered before they enter streams to continue the process of eutrophication. Or they can be diverted to places where eutrophication is desirable, for example commercial fish ponds and bays and estuaries where high productivity of crustaceans, mollusks, and food fish is desired. Poisonous minerals, such as lead and mercury can be filtered out more effectively and perhaps even to economic advantage. Some substances such as copper, must be recovered and recycled more effectively because they are now being lost to the sea in forms that are not economically retrievable by any current technology. The organic molecules synthesized by industry for the manufacture of plastics, pesticides, detergents, and other disposable materials must always be quickly biodegradable. At the present time the industrial nations of the world are creating trash and liquid pollutants faster than these materials can be metabolized and recycled by microorganisms. This trend must somehow be reversed, unless we wish to see the biosphere of the Earth gradually covered over with a "detritosphere" of refuse.

THE FRAGILE ENVIRONMENT

Ecologists are fond of quoting the following epigram about the environment: "You cannot do just one thing." You cannot exterminate one species, or dam one river, or cut down one forest, without altering other parts of the environment — often drastically. Consider, for example, the lesson taught by the Aswan Dam. This great structure was built across the Nile River by the Egyptians with the help of the Soviet Union. The idea was to replace the unchecked floods that have covered the lower Nile Valley since before recorded history, with a continuous release of the stored water to permit year-round artificial irrigation. Like most such schemes the program worked — at least at the beginning. But unforeseen side effects are now appearing that threaten the value of the whole enterprise. First, the flood plains of the lower Nile have been deprived of the annual load of fertile silt that was deposited on them each year in the past. Farmers will now have to import artificial fertilizers to maintain the desired level of agricultural production. Second, the silt is being deposited behind the dam, which therefore may in time have to be abandoned altogether. Third, the employment of controlled irrigation without periodic flooding and flushing of the soil increases the level of salinity, with adverse effects on agriculture. Fourth, the loss of the nutrients that were previously carried into the eastern Mediterranean Sea by the flooding Nile has drastically reduced the sardine population. The annual catch has dropped from 18,000 tons a year to 500 tons. Fifth, the reduction of flow of fresh water from the Nile has increased the salinity in the eastern Mediterranean, causing an increase of fish species introduced from the Red Sea (via the Suez Canal) at the expense of native species. Sixth, there has been an increase in the incidence of the dreaded disease schistosomiasis. The pathogen, a blood fluke, depends on snails to complete its life cycle, and freshwater snails are much more abundant in irrigation canals than in natural river systems.

The reader may be surprised that the greatest destroyer of ecosystems in history, next to man himself, is the humble and winsome European rabbit, *Oryctolagus cuniculus*. The European settlers of Australia and New Zealand deliberately introduced rabbits into these countries during the 1800's, in the belief that they would provide a ready supply of meat that could be harvested from the wild countryside. The rabbits exceeded this expectation to a tragic degree. Protected from predators and harsh weather by the labyrinthine retreats of their warrens, they developed abnormally dense populations that stripped vegetation and loosened topsoil over wide areas. Both farming and sheep herding were made impracticable in the hardest hit areas, and many farmers were forced to abandon their land — or to try to make a living by farming rabbits! The native marsupials of Australia have been diminished and pushed back over part of their native ranges by the habitat destruction caused by this single alien species.

READINGS

J.H. CONNELL, D.B. MERTZ, AND W.W. MURœDOCH, eds., *Readings in Ecology and Ecological Genetics*, New York, Harper & Row, 1970. See comments at end of Chapter 24.

D.W. EHRENFELD, *Biological Conservation*, New York, Holt, Rinehart & Winston, 1970. An excellent little book for the student who wants to pass directly into readings on applied aspects of ecology.

P.R. EHRLICH AND A. EHRLICH, *Population, Resources, Environment*, 2nd Edition, San Francisco, W.H. Freeman, 1972. A definitive and excitingly written source book on population and environmental problems.

E.P. ODUM, *Fundamentals of Ecology*, 3rd Edition, Philadelphia, W.B. Saunders Company, 1971. See comments at end of Chapter 24.

R.W. POOLE, *Introduction to Quantitative Ecology*, New York, McGraw-Hill, 1974. An excellent introduction to the quantitative analysis of ecosystems.

R.H. WHITTAKER, *Communities and Ecosystems*, 2nd Edition, New York, Macmillan, 1975. A good, brief textbook that stresses ecosystems and cites many examples from the recent literature.

26

sociobiology

A solitary ant, afield, cannot be considered to have much of anything on his mind; indeed, with only a few neurons strung together by fibers, he can't be imagined to have a mind at all, much less a thought . . . . It is only when you watch the dense mass of thousands of ants, crowded together around the Hill, blackening the ground, that you begin to see the whole beast, and now you observe it thinking, planning, calculating. It is an intelligence, a kind of live computer, with crawling bits for its wits.

LEWIS THOMAS,
THE LIVES OF A CELL

Social behavior enables a species to transcend the limits of individual organisms. The members of societies can specialize to perform different functions, then form groups and integrate their skills. The complexity and efficiency of the group can exceed that of the individual. These two basic qualities, specialization of individuals and integration of groups, are the yardsticks for measuring animal societies. When integration is sufficiently elaborated and tightened, and combined with high levels of individual intelligence, it can lead to the next great step in the evolution of behavior: the invention of culture, which consists of the transmission of learned information from one generation to another.

ELEMENTARY SOCIETIES

A society is a group of individuals that belong to the same species and are organized in a cooperative manner. Reciprocal communication leading to cooperative behavior is the essential ingredient. We can reasonably exclude the simplest AGGREGATIONS, such as masses of ladybird beetles or rattlesnakes that come together to hibernate, because they do not respond to outside stimuli as an organized group. A pair of mating animals could be construed as a society only if one stretches the definition to absurd limits. On the other hand it is useful to regard combinations of adults and offspring as elementary societies providing they are truly bound together by reciprocal communication. There is in fact a special reason for calling them societies: parental care has often served as the evolutionary stepping stone to the most complex forms of society. Some degree of parental care is displayed by a great variety of animal groups, including crustaceans, spiders, centipedes, roaches, beetles, bees, wasps, fishes, lizards, snakes, crocodilians, birds and mammals.

Another form of very elementary society is the MOTION GROUP; the most familiar examples are swarms of locusts, schools of fish, flocks of birds and herds of mammals. The members of these assemblages communicate with one another just enough to stay together as they move from place to place in search of food, water, or resting places. There are certain advantages to membership in such a group. Perhaps the most general and obvious benefit is the superior protection it provides agains predators. One can witness how this works by quietly approaching a flock of pigeons or other birds that are feeding on the ground. When one comes too close, his presence will probably be detected first by the individuals at the edge of the flock, who will then either move away toward the center of the flock, or take flight directly. Both responses will eventually alert the entire group. The important

LIVING BRIDGE OF ARMY ANTS (Eciton burchelli) spans two logs in the photograph on opposite page. A symbiotic silverfish can be seen crossing the bridge at center. Eciton is a native of the forests of South and Central America.

point is that many of the members will be made aware of an intruder's presence at a greater distance than if they had been feeding alone. This is why a predator has less chance of catching a particular bird if it is a member of a flock. Many species of social birds have special signals such as sharp alarm cries or conspicuous flicking of the wings that serve to galvanize the flocks more quickly into alertness.

Once a predator launches an attack, membership in a group gives added protection to the individual. Figure 1 shows how flocks of the common European starling thwart the efforts of a predator by presenting a united front that makes assault dangerous. The peregrine falcon, one of their principal natural enemies, catches birds in flight by swooping down from above at speeds in excess of 100 miles per hour. This tactic renders the hawk very vulnerable to injury by collision. It must catch a bird alone and strike it with its talons first. The starlings make this difficult by clumping together in a tight formation. The falcon can hit any one bird easily enough, but it is also likely to crash into others with its head and wings as it plummets down through the flock. The falcon obtains a meal only by making passes to one side of the flock until it is able to catch a member that has lost contact with the others through inferior maneuvering. Numerous other examples of such group defense exist in birds, fishes, and mammals. Adult musk oxen,

when confronted by a pack of wolves, form a tight ring with their heads facing outward and the young protected in the center. On the open tundra, where these large animals live, the circle tactic works very well against wolves, but it leaves the oxen helpless when hunted by men with guns. As a result the species has been exterminated over most of its original range during the past 100 years.

Motion groups are believed to serve other functions besides defense against predators. In some species they may serve as a means of discovering food more efficiently, of coordinating reproductive activity and increasing the likelihood of mating and even, in the case of fish schools, of improving the ease with which the members can move through the water. Most of these ideas, however, remain to be tested by critical research.

HIGHER SOCIAL BEHAVIOR

Parental behavior, aggregations and motion groups are only short first steps in the evolution of social behavior. How do species achieve more complex, highly integrated forms of organization? Several elements must be added to the behavior of individuals to make societies possible. The first of these is leadership: action by one or a few individuals that controls the activities of other members and orients the group as it moves from place to place. The existence of a leader is not as important for social behavior as one might at first imagine. Fish schools, for example, are led by the individuals who simply happen to be on the foreward fringe at the moment. When the school changes direction a new set of temporary leaders takes over. Reindeer herds are led from one grazing area to another by a vanguard of individuals, mostly females, who seem to fall into their role simply because they are the most timid and restless members. They are the first to finish eating, the first to rest, and the first to start roving again. In their travels they are followed rather mindlessly by their more contented associates. Wolves, elephants, and some primates such as rhesus macaques, baboons and gorillas, have a leader in the more human sense of the word: a dominant individual, almost invariably a male, who leads the other members from place to place, watches over their activities, occasionally intrudes to settle their disputes, and guards them against enemies. But such a role has emerged in only a tiny minority of all the social animal species.

1

FLIGHT PATTERN of a group of European starlings changes during attack by peregrine falcon. Starlings usually fly in scattered pattern (left), but assemble in tight formation when attacked (right).

A second important element in the evolution of social behavior is the territory. Biologists define a territory simply as any defended area, or to be somewhat more technical, an area occupied more or less exclusively by an animal or a group of animals by means of repulsion through aggressive defense or advertisement. Such a place must be distinguished from the home range, which is all of the area that is regularly patrolled for food. Sometimes the home range and the territory are the same; the animals defend all of the ground they patrol. But more typically the animals defend only a limited space around their nest, while sharing parts of their home range with their neighbors. Individuals of some species occupy territories only during the breeding season, and they become either solitary wanderers or join social groups at other times.

Territorial behavior varies among different species from the prompt, aggressive exclusion of intruders all the way to the use of chemical signposts unaccompanied by threats or attacks. The employment of aggressive behavior has been well documented in many different kinds of animals. Dragonflies of many species, for example, patrol the ponds or lake borders in which their eggs are laid and drive out intruders belonging to the same species by darting flying attacks. A more common and somewhat less direct way to defend a territory consists of repetitive vocal signaling. A large part of the bird song heard in spring consists of advertisement by males warning intruders away. Additional examples include some of the monotonous songs of crickets and frogs. Such vocalizing is not directed at individual intruders but is broadcast as a "to whom it may concern" message. There is only one restriction: normally only members of the same species pay attention.

A pattern of behavior closely akin to territoriality is individual distance, the distance an animal strives to keep between itself and another animal. It is maintained while the animal is both stationary and on the move, and is a kind of floating territory. The constant operation of the effect produces the astonishingly regular spacing patterns often seen in roosting birds and swimming schools of fish. Individual distance is sometimes maintained by overt aggression. For example, a brooding hen begins to show agitation when another hen comes within 20 feet. If the approach is narrowed to under ten feet, the hen lowers her wings and prepares to attack.

DOMINANCE

The dominance hierarchy is a set of aggressive relationships that develops within a group of animals living together. It can be conveniently viewed as the equivalent of territorial behavior in which the participating animals all coexist within one territory. The result is that one of the group members acts as the territory holder and comes to dominate the others. When food is limited, it eats while the others wait. It takes precedence in the selection of mates and resting places. This "alpha" individual makes frequent use of threats to maintain its position, and usually resorts to violent attack if the threats prove inadequate. Below the dominant individual there may exist a second, "beta" animal that controls the rest of the membership in the same way, and beneath it a third-ranking individual, and so on. Because biologists used flocks of chickens in many of the early studies of dominance hierarchies, the hierarchies are sometimes referred to loosely as peck orders. Of course there are many other ways in the animal kingdom of establishing status besides the pecking, wing-flapping and spur-gouging of the sort practiced by barnyard fowl. Dominance hierarchies are very widespread among social animals of many very different kinds. They are known, for example, in social wasps, bumblebees, hermit crabs, fish, lizards, birds and mammals. The relationships display one or the other of several geometric patterns. They may consist of simple despotism, in which a single individual is in complete charge and all the subordinate members are of equivalent rank. They may take the form of a straight chain, in which the alpha individual dominates all other members, a second-ranking individual dominates all but the alpha, a third-ranking individual all but the alpha and beta, and so on down the line. Finally, they can include closed loops, in which one member dominates a second, who dominates a third, who in turn dominates the first (Figure 2).

Hierarchies are formed in the course of the initial encounters among the group members by means of threats and fighting, but after the issue has been settled, each member gives way to its superiors with a minimum of hostile exchanges. The life of the group may eventually become so peaceable as to hide the existence of such ranking from the observer — until some minor crisis happens to force a confrontation. Troops of the yellow baboon (*Papio*

2

PECK ORDERS in chickens. Dominance hierarchy takes three forms. Alpha chicken may dominate all others (left); a second chicken may dominate all but the alpha bird (center); or birds may form a closed loop (right) in which one chicken dominates another, who dominates a third, who dominates the first.

cynocephalus), for example, often go for hours without enough hostile exchanges to reveal the form of the hierarchy. Then in a moment of tension — a quarrel over an item of food is sufficient — the ranking is displayed with startling clarity. In its more pacific state the hierarchy is often supported by status signs. The identity of the leading male in a wolf pack is unmistakable from the way he holds his head, ears and tail, and the confident face-forward manner in which he approaches other members of his group. In the great majority of encounters he is able to control his subordinates without any overt display of hostility. Fighting leading to injury or death is very rare.

Control and hostility — these are the harsh ingredients of social organization considered up to now. It is easy to understand how such selfish behavior can be fashioned by straightforward Darwinian natural selection. Numerous experiments with a variety of social animals have confirmed that it pays to be on top. The dominant animals are healthier, they live longer, and they reproduce more frequently. But selfishness is the least essential part of the

social repertory. There is a paradox in the rest of it. Why should animals subordinate themselves to their dominant group-mates in such a resigned way? How can any form of submissive behavior be evolved through natural selection? One explanation that appears to be emerging is that it is better to be a subordinate member of a group than no member at all. Consider European rabbits in their warrens. Each group of eight to ten individuals is organized into a dominance hierarchy. The top-ranking males and females have access to the central burrows, and as a result they enjoy longer lives and produce more offspring. The lower ranking animals are not so fortunate, but at least their membership in the group gives them some chance of surviving and reproducing. If they were forced to wander on their own outside the territorial limits of the established groups, they would soon be caught by a predator or die of starvation.

In short, half a loaf is better than no loaf at all for the social animal. But this is still not the whole story. In some animal societies one can find examples of altruism. In biology, as in everyday life, altruism is defined as self-destructive behavior for the benefit of others. An altruistic animal is one that reduces its own fitness, its chances of survival or of producing offspring, to increase the fitness of another animal. Altruism is expressed in many forms. A mother bird stays on her nest to resist a predator's attack on her newly hatched young, and risks her own life in the process. A worker bee flies out of the hive to sting an intruder. Because the sting is lined with barbs that anchor it in the enemy's flesh, it remains in place and part of the viscera is pulled out with it when the bee flies off. The result is invariably fatal to the bee. A male baboon falls back to challenge a leopard stalking its troop. The troop escapes — perhaps at the cost of the life of the male. How can such acts be explained as the outcome of natural selection? The answer in each case is that the altruistic act preserves genes in other individuals that are shared with the altruistic individual by common descent. Notice that the baboon does not risk its life in behalf of the leopard, or even baboons in other troops. It does so for individuals that are probably related to itself and therefore more likely to share the genes, including the ones that caused the altruistic behavior. Kinship, together with the altruism it makes possible, has been one of the essential ingredients in the evolution of the most complex forms of social behavior.

TWO KINDS OF SOCIETY

Billions of years of evolution have seen two kinds of organisms, the vertebrates and the insects, climb separate pathways to the highest pinnacles of social behavior. Biologists are fortunate to have both of them for close comparative study. The social insects differ so much from the vertebrates in phylogenetic origin, brain structure, and elemental modes of communication, that for all practical purposes they might as well have originated on some other planet. Perhaps they are the closest we shall ever see to a truly independent form of social being.

The most complex vertebrate societies, and particularly those of birds and mammals, are distinguished from insect societies by a single trait so overriding in its consequences that the other characteristics seem to flow from it. This trait is personal recognition among the members of the group. As a rule each adult animal knows and bears some particular relationship to every other member. Status is extremely important. Where dominance hierarchies exist, elaborate signaling is employed to implement them, as in the case of wolves. Parent-offspring relationships are specific to individuals, tightly binding and of relatively long duration. Within primate societies, personal groupings often ascend to the level of cliques and, in the case of the chimpanzee, even to the level of child care by "uncles" and "aunts" — adults who form temporary alliances with parents by catering to their offspring. Group members spend large amounts of time and energy in establishing and maintaining these many-sided personal bonds. Associated with the personal form of communication is a prolonged period of socialization in the young. By constant experience, much of it acquired in the form of play, the young animal leaves its personal relationship to its parents and establishes an early status among its age-peers. Some division of labor exists: there are weakly defined specialists in defense, leadership and foraging. But it is not based upon physiologically different castes. Each member of the group is basically the same, and pursues its own interest as best it can.

In contrast, the higher insect societies, particularly those of the ants, wasps, bees and termites, are for the most part impersonal. Although limited dominance hierarchies are found in a few species, the sheer size of the colonies and the short life of the members makes it inefficient, if not impossible, to establish individual bonds. The average adult honeybee, for example, lives for only about one month in the midst of a rapidly changing hive population of up to 80,000. An army ant worker has only a few weeks or months to become acquainted with a million or more nestmates. Of course it cannot do this, nor does

How can any form of submissive behavior evolve through natural selection? One explanation holds that it is better to be a submissive member of a group than no member at all.

it even try. Even the relationship of the workers to the queen of the insect colony is impersonal. She is recognized by her pheromones. When little pieces of cork are impregnated with these substances, the pieces are treated for a while as if they were queens. Where songbirds have individual calls that are identified by their territorial neighbors, and mammals use individual scent marks, members of an insect colony employ signals that are for the most part uniform throughout the species. The single known exception is the colony odor, which is used to distinguish nestmates from all members of alien colonies. Socialization is minimal in the insects, and play is apparently absent.

Yet the insect colony, as a unit, can equal or exceed the accomplishments of the nonhuman vertebrate society. The nests of some termites, for example, are more complex in design than those of any mammal or bird. The raids of army ants are more intricate and coordinated than those of a wolf pack. The insects achieve their results with rigid divisions of labor among castes. Colony members are specialized to perform in reproduction (the queens), in defense (the soldiers), and in ordinary labor (the workers). The workers often specialize further as nurses, nest-builders, or foragers. When all of these specialists combine their efforts in the correct way, the colony functions as a superorganism capable of matching the feats of vertebrates. Had vertebrates followed the same route in evolution, the results might have been spectacular. The individual bird or mammal, possessing a brain vastly larger than that of an insect, might have been shaped into a much better specialist. The vertebrate society then could have evolved into an unimaginably more complex, efficient unit — at the price, of course, of independent action on the part of its members. Vertebrates have instead remained chained to the cycle of individual reproduction. This forever enhances freedom on the part of the individual at the expense of efficiency on the part of the society.

INSECT SOCIETIES

The typical insect society is a family. The central figure is the queen, the mother of the colony. Her daughters, and in termites her sons as well, make up the worker force. Their sex can be identified, but they are smaller in size than the queen and they do not normally engage in reproduction. In extreme cases their gonads are completely degenerate. The worker's role is to protect the queen and to rear its brothers and sisters. They accomplish these tasks with almost total altruism. When the colony has reached a certain size and is in good health, some of the offspring develop into fully functional males and virgin queens. These individuals mate, travel to new locations, and become the parents of new colonies.

Advanced social organization based upon an altruistic worker caste of reproductive neuters has originated at least 12 times in the insects: at least twice in the wasps, once or twice in the ants, at least eight times in the bees, and once in the termites. The wasps, bees, and ants all belong to the single order Hymenoptera, while the termites comprise the entire order Isoptera. The tendency of the Hymenoptera toward altruistic social evolution is a striking curiosity. It can be at least partially explained in the following way: Sex is determined in the Hymenoptera by haplodiploidy, which means simply that females come from fertilized eggs and males from unfertilized eggs. Males are therefore genetically haploid, and are also homozygous. Sisters are genetically very close to each other, because each one receives identical genetical material from the father. Sharing more genes than is usually the case between sister animals, they are more prone to behave altruistically to each other. The reason is that to favor such an exceptionally close relative is to favor — in large part — one's own genes.

ANT ORGANIZATION

Figure 3 shows part of a colony of a relatively primitive ant species. Many of the basic fea-

Had vertebrates followed the same evolutionary path as insects, the results might have been spectacular. Possessing brains vastly more powerful than those of insects, birds or mammals might have evolved an unimaginably more complex, more efficient society — at the price of the individual freedom of its members.

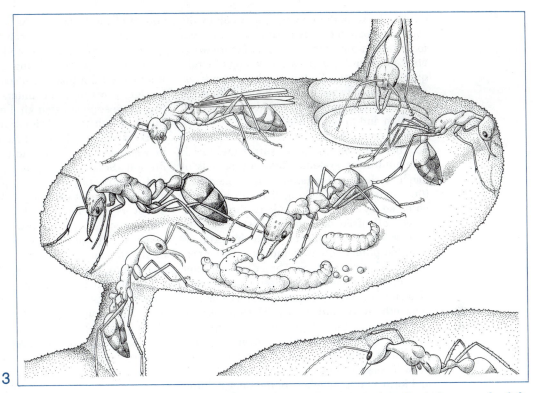

3

COLONY OF BULLDOG ANTS (Myrmecia gulosa) shows much of the social behavior characteristic of Hymenoptera in general. Queen stands at left of chamber with a winged male behind her. Workers to the right attend larvae and pupae of queen. Worker at far right doubles up its body as it lays an egg. In a moment it will feed egg to one of the larvae, as another worker is doing at bottom center. Cocoons containing pupae — also offspring of the queen — are piled at the back of the chamber.

tures of social life in ants, and social Hymenoptera as a whole, are recorded in this drawing. The queen and worker castes are seen to be anatomically distinct from each other. The queen is a normal female, rather closely resembling the females of the solitary wasps from which the ants evolved. In addition to an ordinary and fully functional reproductive system, each mother queen of the kind shown in the drawing is equipped with two pairs of wings when she first emerges from the pupa. She uses them to fly from the parent nest and to participate in a nuptial flight with other young queens and males. After mating she descends to the ground, breaks off her own wings at the base and sets out to start a colony of her own. This she accomplishes by digging a short burrow in the ground and laying her first small cluster of eggs. When the larvae hatch, they are fed by nutrients metabolized from the queen's wing muscles, together with insects the queen

is able to capture in the vicinity of her little nest. All of the members of the first brood are destined to mature into workers. As shown in the drawing, workers are females but they are smaller than the queen. They never possess wings or wing muscles, and their thoraces are proportionately more slender. Those of a few species, such as the Myrmecia, lay eggs, but in most cases feed them promptly to the larvae or another adult member of the colony. Soon after the first group of adult workers emerges from the pupal stage, they take over all of the labor from the queen, from nest building to foraging for food and the care of their immature nestmates. The queen then becomes an egg-laying machine. Once this point in colony development is reached, the population grows rapidly. When the colony reaches a mature size, which varies according to species, it begins to produce virgin queens and males. These reproductive individuals then depart on nuptial flights of their own to a new colony life cycle. The parent colony meanwhile continues to live and grow by the addition of more workers. Each year it produces a new crop of virgin queens and males. So long as the queen lives, which in some species is as long as ten or 15 years, the colony she founded continues to function and to reproduce.

Members of the ant colony coordinate their activities by means of a surprisingly rich communication system. There are signals of touch, including the moment-by-moment jostling, tapping, and licking that can be easily observed by watching a colony through the glass wall of an observation nest. Some ant species also communicate in part by stridulation, the production of sound by rubbing certain body parts together. The effect is a faint squeak, like the sound produced when a piece of metal is scratched lightly against glass or some other very hard surface. The ants use the signal as a call for help, when they are buried in a cave-in or under attack by an enemy. Most communication, however, is by means of pheromones, chemical signals passed back and forth among members of the same colony. As a rule the pheromones are produced as secretions of glands that open to the outside of the body, and transmitted to nestmates in liquid or gaseous form. When tasted or smelled, each pheromone induces its own particular response. The vocabulary of ants thus consists of a set of chemical substances released (one by one or in combinations) to direct the behavior of other colony members.

Figure 4 shows ant workers engaged in three of the principal forms of chemical communication. Alarm substances are released when a worker is attacked or when it encounters an intruder in the vicinity of the ant nest. Often the pheromones are sprayed out with poisonous secretions used in defense. In ants of the genus *Formica*, as illustrated in the drawing, the worker discharges a stream of formic acid mixed with undecane. The acid serves as a poison to repel invaders, while the undecane serves as a pheromone to alert other worker ants. In still other species, the poison and the pheromone are the same. Because the two functions are so intimately connected among the ants, the use of the secretions is sometimes referred to as an alarm-defense system. At low concentrations the alarm pheromones are typically simple attractants that arouse the curiosity of workers and draw them closer to the

source of the discharge. As the workers enter zones of ever higher concentration, they become increasingly excited and may commence a true alarm frenzy, during which they rush frantically from place to place and attack any foreign object they encounter.

A more sophisticated kind of pheromone is the trail substance. Odor trails are used by workers to guide their nestmates to food and to new homes. The species shown in Figure 4 is a fire ant (*Solenopsis*), a member of a species that manufactures the trail pheromone in a gland attached to the sting. When a worker discovers a piece of food too large for it to move, such as a large dead insect, it heads homeward dragging its extruded sting behind it. The trail substance is dispersed through the sting onto the ground. As the liquid trace evaporates, it forms a long, narrow cloud of attractive vapor through which other workers move in search of the food. An identical form of trail communication is used when worker ants discover a superior nest site and attempt to recruit other members of the colony to it.

A still more elaborate form of communication is TROPHALLAXIS, the feeding of liquid substances by one worker to another. The most common and basic form of trophallaxis, also illustrated in Figure 4, involves the regurgitation of stored liquid food from the gut. Trophallaxis plays at least two major roles in the organization of ants and other social insects. First, it homogenizes the stomach contents of the colony members. This process was demonstrated in experiments in which radioactive syrup was fed to single workers who were then allowed to return to their nests. At regular intervals afterward, other workers were removed from the nest and their stomach contents examined for radioactivity. It was found that regurgitation rapidly diffuses liquid food throughout the colony. In some species, every worker of the colony received at least some of the material within 24 hours. Rapid trophallaxis ensures that each member of the colony has roughly the same stomach content as every other member. By this means it is made continuously aware of the nutritional condition of the colony as a whole. When the worker acts for itself out of hunger or thirst, it acts for the colony as a whole. Trophallaxis is a form of altruistic behavior that stands in sharp contrast to the selfish maneuvering for status and resources that characterizes the primate society. It also serves a secondary role in the transfer of some of the pheromones, which must be tasted or even eaten in order to produce an effect.

HONEYBEE ORGANIZATION

Figure 5 illustrates the life of another highly advanced form of social insect, the honeybee. Each hive of bees contains from about 20,000 to 80,000 workers, all of which are females. There is a single mother queen, who exercises a remarkable kind of dominance over her daughters. From her mandibular glands she secretes a continuous supply of a pheromone called queen substance. The workers eagerly lick this material up along with other attractive pheromones secreted by the queen's body and pass it among themselves by regurgitation. The queen substance has two inhibitory effects: it prevents the workers from rearing other queens (a psychological effect) and it inhibits the ovaries of the workers from producing eggs (a physiological effect). If the mother queen dies or is deliberately removed by the beekeeper, the resulting shortage of queen substance is felt within hours by the workers. They commence building special enlarged cells, inside which new queens are reared. A few days later some of the workers

also undergo an enlargement of their ovaries and develop the capacity to lay eggs. The queen substance, then, is a subtle but powerful device which the mother queen uses to prevent the creation of rivals for her supreme reproductive position.

The life cycle of the honeybee is basically similar to that of ants, but with one outstanding difference. When the queen successfully completes her nuptial flight by mating with one or more drones, she does not attempt to start a new colony on her own. Instead, she returns to the parent hive and becomes the new queen there. But what, you may ask, happened to the old queen? The answer is that she left before the new queen emerged from the pupa. She was accompanied by a flying swarm of thousands of workers, who comprised a large fraction of the colony population. Together they searched for a new hive or suitable nesting site. Thus the honeybee colony multiplies in a curious way that is the reverse of the usual procedure in animals generally: it is the parent, not the offspring, who leaves the safety of the established territory and assumes the role of pioneer.

Honeybees are famous for the use of the waggle dance in the recruitment of nestmates to newly discovered food sources and nest sites. The dance is a symbolic representation of the flight from the hive to the target locality and as such is one of the most sophisticated forms of communication found in the animal kingdom (see Chapter 15). Honeybees, like ants, rely primarily on chemical communication to organize the remainder of their social life. In addition to the queen substance there are alarm substances, several kinds of attractants used to recruit nestmates, a peculiar footprint pheromone that is used as a kind of trail substance, and a hive odor by which nestmates are distinguished from strangers. All of these classes of pheromones have close parallels in ant colonies.

TERMITE ORGANIZATION

Figure 6 shows a section of a termite colony. The most remarkable fact about these insects is that they are so similar in the details of their social organization to the ants, despite having an utterly different phylogenetic origin. Termites can be thought of as social cockroaches. They originated evolutionarily from primitive cockroaches no later than the middle of the Cretaceous Era, over 100 million years ago.

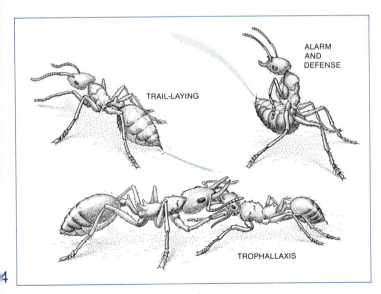

TRAIL-LAYING

ALARM AND DEFENSE

TROPHALLAXIS

THREE MAJOR TYPES OF SIGNAL are included in the chemical repertory of various ant species. At top left a fire ant worker lays an odor trail from its extruded sting. As the pheromone evaporates its vapor guides other workers. At top right a worker of Formica polyctena sprays a mixture of formic acid and undecane at an enemy. The acid repels the intruder, and the alarm pheromone alerts other ants in the nest. In bottom drawing a Daceton worker regurgitates food to a larger nestmate.

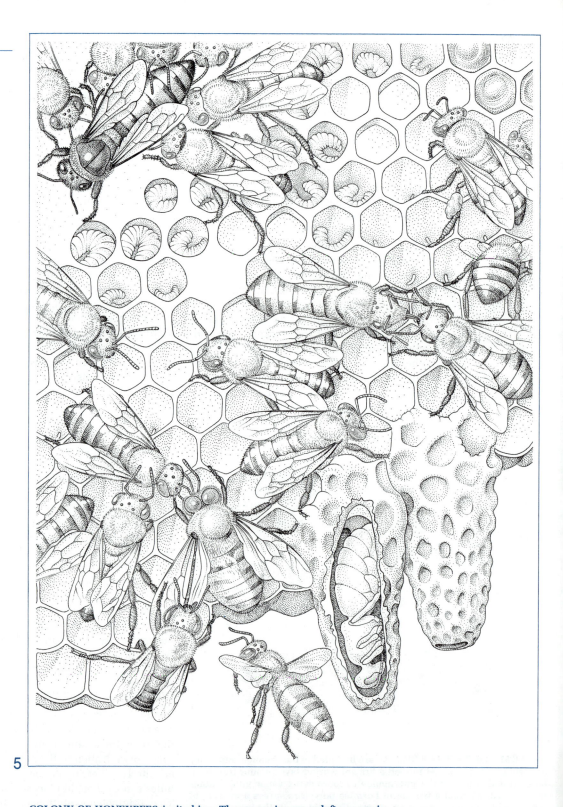

COLONY OF HONEYBEES in its hive. The queen in upper left corner is surrounded by attendants. Two bees at right center exchange pollen and nectar. Many of the open cells contain eggs or larvae. Cells at lower right contain pupae of future queens. Male (drone) is visible to left of pupae.

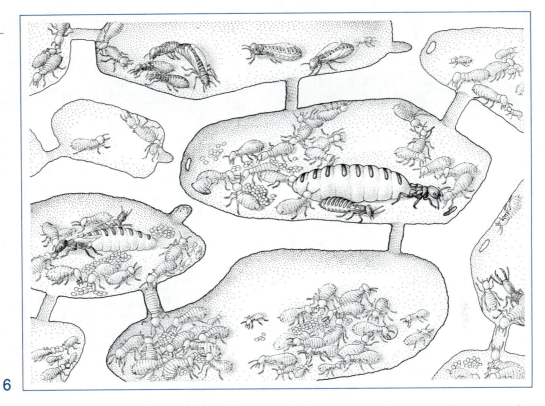

6

COLONY OF TERMITES (Amitermes hastatus) consists of interconnecting chambers. In the central chamber the primary queen, her abdomen swollen with eggs, rests with the primary king. Secondary queen occupies chamber at left, while immature queens and males occupy upper chamber. In chamber at lower right corner a soldier (distinguishable by its large mandibles) is accompanied by a nymph that is developing into a soldier. Elsewhere workers attend piles of eggs and larvae wander about.

Like the ants, they have become dominant land animals and have produced a wide range of adaptive types. They differ basically from ants, however, in their food habits. From the beginning the termites have been eaters of dead wood and other dead vegetable material. Their great achievement in evolution, next to the invention of society itself, was to enter into symbiosis with protozoans and bacteria capable of digesting cellulose. The termites chew up the vegetable fibers and pass them down their intestine, where the cellulose is broken down by hordes of the microscopic symbionts. In exchange for part of the nutrients the protozoans and bacteria are given protection and a constant supply of new raw materials. The combination is an extremely efficient one, enabling termites to become the chief animal decomposers of wood over much of the Earth. The most primitive living termite species, *Mastotermes darwiniensis* of Australia, is also

one of the most formidable insect pests in the world. Its workers attack poles, fences, wooden buildings, wharves, bridges, oil-soaked wood, living trees, crop plants, wool, horn, ivory, vegetables, paper, hay, leather, rubber, sugar, flour, human and animal excrement, billiard balls, and the plastic linings of electric cables. Entire ranches have been reduced to dust in only one or two years — house, fences and all.

The members of termite colonies communicate primarily by pheromones. The basic signals of the termite chemical language and those of the ants and other social Hymenoptera are an example of evolutionary convergence (*Chapter 23*). The termites secrete queen substances that inhibit the development of rival queens within the colony. They also have "king substances" that prevent the appearance of new reproductive males. Alarm and trail pheromones and distinctive colony odors are also well developed. Various forms of trophallaxis have been evolved, including both the regurgitation of liquid food and the licking of attractive substances from the body surface. The symbiotic protozoans carried by some species are passed from one termite to another by the unique method of anal trophallaxis, in which one termite extrudes an anal droplet containing the symbionts and feeds it to a nestmate.

PRIMATE SOCIETIES

Among the higher primates the vertebrate form of society reached its zenith. Excluding man, with his unique language and his capacity for cultural transmission, one can regard the primate society as a logical extension of the personalized association found in birds and lower mammals. The difference is one of degree and not of kind. An efficient way, therefore, to gain a view of vertebrate societies as a whole is to examine one particular primate species in depth.

One of the best-studied primates is the rhesus monkey *(Macaca mulatta)*, or rhesus macaque as it is sometimes called, the species of monkey most frequently used in laboratory research. The rhesus society is based to a large degree upon aggression and status. Each adult male occupies a position in a rigid dominance hierarchy that is maintained, virtually moment by moment, through an extraordinarily diverse set of vocal signals, body postures, and facial expressions. The adult females, and to a much lesser extent the juveniles, are also organized into loose hierarchies. In hostile exchanges, which break out frequently, the vocabulary of signals is used to communicate intensity of feeling and, with it, the probability that the signaling animal is about to attack or submit. An individual that is merely expressing a superior attitude toward another may do nothing more than look in the direction of the rival for a brief time. Greater hostility is expressed by a sustained stare. If a man walks into a room containing caged (and therefore unhappy) rhesus monkeys, some will probably hold their bodies rigidly while giving him what appears

to be a prolonged, curious stare. The monkey is not expressing curiosity; it is attempting to convey its feeling of hostility, to put fear into the observer. As aggression mounts among competing monkeys, the next strongest signal is opening the mouth wide. To a human observer the monkey looks astonished, but it is simply angrier. If the opponent does not yield, the monkey may bob its head up and down while continuing to hold its mouth open. A still stronger signal is to lunge forward while slapping the hands on the ground.

Hostile encounters are not random events. They occur principally between individuals of nearly equal rank in the dominance hierarchy or when circumstances cause a momentary conflict of interest between two monkeys of unequal rank. Most of the time order is maintained by a recognition of status among the monkeys. As with wolves, an individual displays its rank by its posture. High-ranking males are instantly recognizable by their manner. With their heads held high and their tails pointed upward with a little crook in the end, they stroll along slowly and deliberately, regarding each monkey in their path with a seemingly indifferent expression. A low-ranking monkey shows the opposite behavior. When faced by a leading male it moves discreetly out of his line of vision. Its head is kept low, its gaze averted, its tail lowered. Its movements are nervous and furtive (*Figure 7*).

There can be no question as to the advantages won by dominant males: they have first access to food and mates. Subordinate males are also able to mate, but less frequently and only with females not claimed at the moment by their superiors. A female that becomes the sexual partner of a dominant male temporarily acquires his social position and thus gains a greater measure of security during at least part of her pregnancy.

The rhesus society is governed by a communication system that far exceeds in complexity those controlling the simple schools, flocks, herds, and other elementary societies mentioned earlier. When the full range of sexual and parent-offspring interactions are

WALK OF DOMINANT MALE

WALK OF LOW-RANKING MALE

7

POSTURE of rhesus monkey indicates its status in the dominance hierarchy. High-ranking male walks with head held high and tail hooked upward, in contrast to slouch of low-ranking male.

The study of primate societies is a relatively new subject. The bulk of existing knowledge comes from studies conducted in the field in Africa, Asia, and tropical America only during the past 20 years. Many of the species of monkeys, baboons, and anthropoid apes have been demonstrated to live in societies as complexly organized as those of the rhesus monkey. However, there is a surprising amount of variation among the species in some of the basic details. Territoriality and dominance behavior are particularly subject to variation. The rhesus monkey and its fellow macaque species, together with the baboons, occupy one end of the scale in aggressive behavior. Some anthropoid apes, particularly the chimpanzee and the gorilla, are nearer the opposite, more benign end of the scale. Bands of gorillas, for example, roam broadly overlapping home ranges but do not interact aggressively when they meet. Each group is typically under the very tolerant control of a single older male. Fights occasionally occur between members of the same band, but they are less frequent and less violent than fights in the rhesus troops. As yet there is no adequate evolutionary explanation of these differences in the primate societies, which promise to shed light on the origins of human social behavior.

READINGS

H. Kummer, *Primate Societies: Group Techniques of Ecological Adaptation*, Chicago, Aldine Publishing Co., 1971. A short but insightful and thought-provoking book on social evolution in monkeys and baboons.

J. Van Lawick-Goodall, *In the Shadow of Man*, Boston, Mass., Houghton Mifflin, 1971. A beautifully illustrated account of the behavior of wild troops of gorillas, the most manlike of the anthropoid apes.

E.O. Wilson, *The Insect Societies*, Cambridge, Mass., Belknap Press of Harvard University Press, 1971. A comprehensive review of the social insects and other arthropods.

E.O. Wilson, *Sociobiology: The New Synthesis*, Cambridge, Mass., Belknap Press of Harvard University Press, 1975. A study of the evolution of social behavior in organisms from bacteria to man.

added to the codes of territory and dominance, the known repertory of communicative acts known to be utilized by a rhesus society comes to 56. Add to this 27 combinations of these same signals which when presented simultaneously convey still different meanings, and the number of distinct signals totals 83. The individual rhesus monkey not only utilizes most of these signals in the course of its life, but it directs them selectively toward specific group members, each of which has a particular significance for its rank within the society and the scope of its daily activities. *Macaca mulatta* is thus a species totally committed to a social existence in the vertebrate mode. A large part of its forebrain is devoted to this part of its life. The forebrain is so large, in fact, because of the requirements placed on it by the exceptionally complex social existence of the species.

human evolution

If I were forced to call the human species apes, I should at least show that they are the culture-apes, the cumulating-tradition apes, the ideal conception of all apes. That man is naked is irrelevant. He might just as well be furry.

KONRAD LORENZ

Fifty million years ago Africa was an island continent. Together with Arabia it was cut off from Europe to the north and India and Asia to the east by the great Tethys Sea, a body of shallow tropical water that stretched between the Atlantic and Indian Oceans (*Chapter 24*). During its isolation Africa, like all great islands, evolved a distinctive fauna and flora. Many groups of mammals in particular gave rise to spectacular adaptive radiations, parallel and simultaneous with the developments on the island continents of Australia and South America (*Chapter 23*). Elephants and their kin originated in Africa, together with a great assemblage of less familiar types — elephant shrews, hyracoids, tenrecs, arsinotheres, barytheres, moeritheres, and others. The Tethys Sea did not form a total barrier, and occasional groups were able to cross from Europe or Asia at various times and to colonize the island continent. During the Oligocene epoch, primitive forerunners of the modern pigs, buffalos, antelopes, and lions invaded Africa and began secondary radiations that rivaled, and in some instances replaced, the older ones. Africa, in short, became a major center of mammalian evolution.

One of the earliest imports was the primates (order Primates). The exact place of origin of this key group has not been established by fossil discoveries. The most primitive primates were similar to the living tree shrew illustrated in Figure 1. They descended some time in the Cretaceous period from small animals that probably resembled living shrews belonging to the primitive mammalian order Insectivora. By Eocene time at the latest, primitive primates had established themselves in Africa. By the end of the Miocene, a few tens of millions of years later, they had completed a remarkable adaptive radiation that produced lemurs and similar primitive forms, baboons and other kinds of Old World monkeys, anthropoid apes, and man himself. Meanwhile, in South and Central America, an independent radiation had produced the New World monkeys, including the woolly monkeys, squirrel monkeys, spider monkeys, howlers, marmosets, tamarins, and others.

Simply to define a primate is to understand a great deal of the evolution that made the origin of man possible. The primates are a group of mammals that are fundamentally adapted to life in trees by means of improved balance, leg and arm flexibility, and vision. The primitive tree shrews, lemurs, monkeys, and apes were apparently all tree-dwelling. They retained generalized limb structures, with five digits terminating each limb. In early stages of evolution the limbs acquired the ability to rotate and to move in many directions, so that the body could be moved or supported while the animal was in any of a wide variety of positions in the trees. The fingers and toes became capable of grasping the trunks and branches. The claws changed into flat nails, and the inner surfaces of the hands and feet were sculpted into sensitive, hairless pads. Balance and motor control were correspondingly improved. Surrounded as they were by the open spaces of

BUSHMAN MOTHER in the photograph on opposite page prepares the skin of a freshly killed animal. Hunter-gatherers such as the African bushmen are modern survivals of an early stage in the evolution of human society.

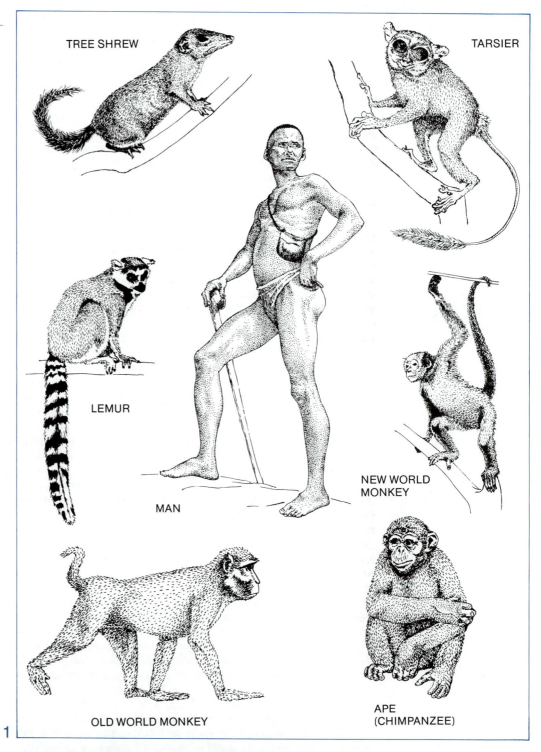

TREE SHREW

TARSIER

LEMUR

MAN

NEW WORLD
MONKEY

OLD WORLD MONKEY

APE
(CHIMPANZEE)

1

PRIMATES are mammals adapted to life in trees. Illustration depicts living representatives of the seven major groups. The earliest primates probably resembled the tree shrew at upper left.

their habitat, and depending completely on the judgment of distances simply to move about, the primates' sense of smell was deemphasized while their vision was improved. The eyes were enlarged and moved forward in the head, providing overlapping images and instant three-dimensional perception.

During the African radiation some primates were able to enlarge their size by one or the other of two very different means. The first was brachiation — movement by grasping branches over the head and swinging along beneath them. This is the only means by which a large animal can travel easily through the tree tops. The true apes — the chimpanzee, gorilla, gibbon, and orangutan — became highly modified for brachiation. They have long, powerful arms and shoulders and relatively heavy, deep chests. Their hind legs are short and their feet shaped for grasping. As a consequence locomotion on the ground is usually on all fours, with most of the weight being supported on the knuckles of the feet and hands. It is important to realize that these several traits, which in the popular mind are the essence of the ape's nature, are not primitive characteristics. They are advanced specializations of large animals for life in trees. Man never passed this way in evolution. Although his ancestors in the early and middle Cenozoic could be called primitive apes, he did not descend in a direct line from any creature closely resembling one of the modern apes.

The second way large size could be achieved by primates was simply to abandon an arboreal existence and return to the ground, where bigness is no impediment to locomotion. In Africa two principal lines of primates made this transition independently — the baboons and man (*Figure 2*). The baboons have regained a close approach to the quadrupedal (four-footed) locomotion of their ancient ground-dwelling ancestors. They run swiftly on all fours over the open land of the savanna. At night they often climb trees to sleep more safely. The ancestors of man, the man-apes as they are usually called, also left the forests to hunt and sleep in the savannas of Africa. In so doing, they adopted a form of locomotion destined to have enormous secondary consequences in their subsequent evolution. This is the bipedal (two-footed) stride, in which the body is held in a fully erect position and moved exclusively by the stiff swinging motions of the hind legs beneath it. The bipedal stride is unique in the animal kingdom. It freed the hands and gave the man-apes the potential for carrying weapons and tools.

FROM MAN-APES TO MEN

Figure 3 depicts the phylogeny of the family Hominidae — man and his immediate ancestors — which has been deduced from fossil material discovered in the past one hundred years throughout Europe, Asia, and Africa. In this book there is not enough space to describe the story of the arduous search for those remains, which has provided the piece-by-piece solution of the puzzle of man's origin. Suffice it to say that it has been, and continues to be, one of the most exciting chapters in the history of science. This chapter will focus on several of the key biological generalizations that have emerged.

The first generalization is that man's erect posture and bipedal locomotion, the qualities which at a distance make an individual look human, appeared full blown long before the great enlargement of the brain that truly characterizes our species. Two million years ago the man-ape *Australopithecus africanus* had a human posture combined with a brain not much larger than that of a gorilla. These man-apes were only moderately impressive. They used stones as weapons and possibly also as crude cutting instruments. They may also have wielded sticks and bones as additional weapons — but chimpanzees do no less. They strip small branches of their leaves and use them to fish out termites from their burrows. Chimpanzees also crumple leaves in their hands and use them to sponge up water from

Two million years ago the man-ape Australopithecus africanus had a human posture combined with a brain only slightly larger than a gorilla's.

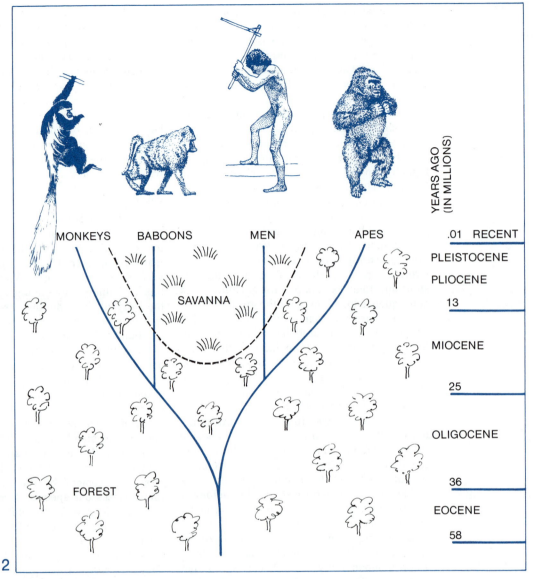

**ADAPTIVE RADIATION produced the higher primate groups of Africa.
Baboons and man have adapted to life on the savanna, the tree-studded
grassland adjacent to the tropical forests.**

narrow recesses. The great leap forward in human evolution occurred during the period from about 1,500,000 to 500,000 years ago, when *Australopithecus* gave rise to *Homo erectus*. This species, which spread from Africa through large parts of Europe and Asia, was the first true man. He had a cranial capacity intermediate between those of *Australopithecus* and modern man, and used fire and tools as advanced as stone axes.

What caused *Australopithecus* the man-ape to cross the threshold to *Homo* the man? The following proposition, while difficult to prove with finality, is supported by a growing amount of circumstantial evidence. Early *Australopithecus africanus* was a modest hunter, subsisting at least in part on birds, reptiles, and rodents and other small mammals. Probably he also consumed fruits and other vegetable material such as tubers. Later, he acquired the capacity to hunt larger mammals, such as antelope, adding a rich new source of protein to his diet. How this change was accomplished is of great importance for reasons having nothing to do with nutrition; it appears likely that at about this time man began to hunt in groups.

Pack-hunting is a rare phenomenon in nature. It occurs, for example, in wolves and some other wild dogs and in killer whales. It requires not only cooperation but a considerable amount of communication and division of labor. The further these qualities can be refined, the more successful the hunt. Early man evolved a complex social structure in which the hunting groups were primarily or exclusively male. Women and children probably remained secluded in caves and bivouacs while the hunters were abroad. An elaborate language, flexible yet precise, was developed to coordinate the actions of the hunting groups and the complicated social relations necessitated by the division of labor between the sexes at home. The brain size grew, providing the neurobiological basis of the evolving language. The larger brain also facilitated a dramatic increase in other dimensions of intelligence. A fortunate final result was the ability to invent true culture, the handing of knowledge and traditions from one generation to the next.

In previous chapters evolution was de-

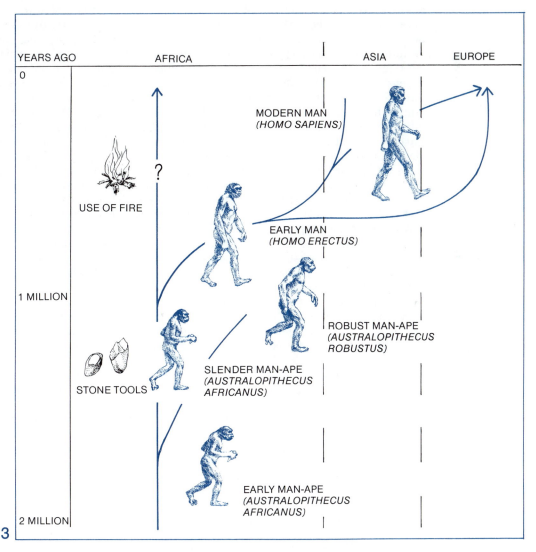

MAN AND HIS ANCESTORS comprise the family Hominidae. Breakthrough in human evolution occurred between 1.5 and 0.5 million years ago when Australopithecus gave rise to Homo erectus.

scribed as opportunistic. Each important adaptive radiation produces some types that are "thrown away" — doomed to early extinction; others survive and give rise to the next evolutionary step. Human evolution was no exception. As *Australopithecus africanus* progressed toward *Homo*, it lived side by side with a second man-ape species, *Australopithecus robustus*. The massive jawbones and teeth of *A. robustus* were best suited for grinding roots and seeds, and suggest that it became more vegetarian at the same time *A. africanus* was turning increasingly to the pursuit of animals. This remarkable creature became extinct a little less than a million years ago without having contributed to the ancestry of *Homo*. A similar radiation of short-lived lines occurred at the racial level during approximately the last 100,000 years of the evolution of modern man (*Homo sapiens*). The most famous example is Neanderthal Man (*Homo sapiens neanderthalensis*), who had a full-sized brain but heavy skull bones and brutish features dominated by a thick brow ridge. The Neanderthal race lived from about 100,000 to 35,000 years ago around the Mediterranean and northward for some distance into Europe. Other relatively primitive races that have since passed into extinction were Solo Man of Java and Rhodesian Man of Africa.

Where did the man-apes originate in the first place? This question is the equivalent of asking when did the ancestors of the man-apes begin to diverge from those of the chimpanzee, gorilla, and other modern apes. Few fossils have been discovered that shed light on this crucial episode. The best candidate for the primal ancestor is *Ramapithecus punjabicus*, known only from a few teeth and jaw fragments collected at two localities in India and Kenya. The specimens, which date from the transition between Miocene and Pliocene times approximately 14 million years ago, are more manlike than apelike in structure. *Ramapithecus* has accordingly been placed in the Hominidae. It is very tentatively judged to be an ancestor of man in a direct line. But it is also so similar to the primitive apes that it lies near the point at which the Hominidae first branched off as a separate group.

ADAPTATIONS OF MODERN MAN

The evolution of the man-apes and early man has shaped the biology of the modern human species. All of the uniquely human traits are adaptations to environments in which man evolved in the millions of years before history. These adaptations include profound modifications of primate anatomy, physiology and behavior.

The form of the skeleton is adapted to fully erect posture and bipedal locomotion (*Figure 4*). The spine is curved in a way that distributes the weight of the trunk more evenly down its length. The chest has been flattened, a geometrical change that moves the center of body weight back toward the spine. The pelvis is broadened to provide adequate attachment for the powerful striding muscles of the upper leg, and realigned to convert it into a basinlike support for the viscera. The tail has disappeared and its shortened vertebrae now curve down and in to form the coccyx, which makes up part of the floor of the pelvis. This transformation of the tail vertebrae is unique among the vertebrates. The occipital condyles, those bony points by which the skull pivots on the neck vertebrae, have rotated far beneath the lower surface of the skull until the weight of the head is nearly (but not exactly) balanced upon them. The leg is straight, aligned in such a way that it transmits the entire weight of the body through the knee joint. The feet are specialized to accommodate the bipedal stride. They have been narrowed and lengthened, and the toes have been shortened to become little more than flexible extensions of the feet. The result is that the entire weight of the body can be placed on one foot at a time and passed from the heel to the ball and toes as the foot rocks forward during the walking motion. The opposable thumb of the hand, a primate trait, attains its maximum development in man. It is

The reproductive biology of man has undergone extensive evolutionary changes — changes that favor the formation of intimate, long-lasting partnerships between men and women.

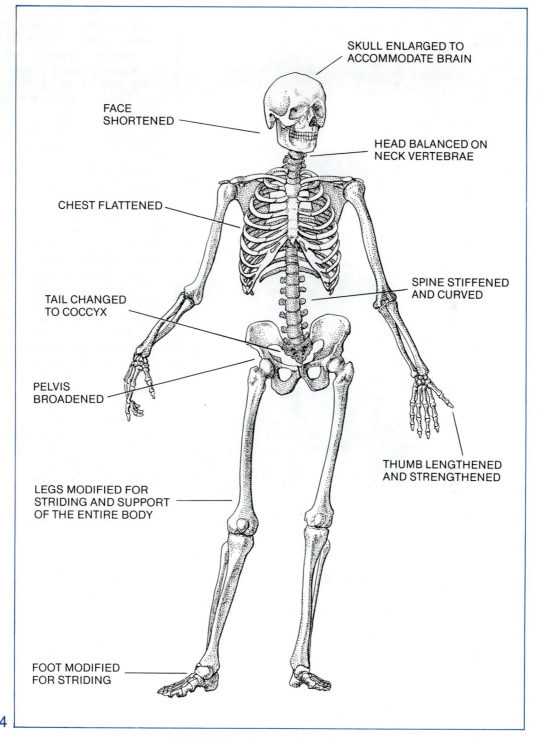

SKULL ENLARGED TO
ACCOMMODATE BRAIN

FACE
SHORTENED

HEAD BALANCED ON
NECK VERTEBRAE

CHEST FLATTENED

SPINE STIFFENED
AND CURVED

TAIL CHANGED
TO COCCYX

PELVIS
BROADENED

THUMB LENGTHENED
AND STRENGTHENED

LEGS MODIFIED FOR
STRIDING AND SUPPORT
OF THE ENTIRE BODY

FOOT MODIFIED
FOR STRIDING

4

HUMAN SKELETON reveals modifications for upright posture and bipedal stride. Straight leg, curved spine, flattened chest and positioning of head all serve to align weight of body along a vertical axis. Broad pelvis provides attachment for powerful striding muscles, and together with curved coccyx, supports viscera. Shortened toes facilitate heel-and-toe walking.

FISH

REPTILE

RABBIT

MAN

5

EVOLUTION OF THE BRAIN. Extraordinary increase in size of the human brain relative to that of other vertebrates is due almost entirely to growth of cortex. Older parts of the brain that control coordination, reflexes and instinctive behavior came to lie beneath the cortex.

relatively very long and powerfully muscled, providing the hand as a whole with a strong precise grip. The face has retreated back into the skull, presumably to assist the shift of gravity of the head in the direction of the occipital condyles. As the brain grew in size, the forehead expanded, completing the evolution of the main structure of the human face.

Man lost most of his hair. The reason why he became a "naked ape," is still a matter of con-

jecture. Perhaps the most plausible explanation is that nakedness was originally a device to assist in the cooling of the body during the strenuous pursuit of prey in the hot sunlight of the African plains. It is correlated with man's heavy reliance on sweating to reduce body heat. The human body is uniquely adapted for producing large quantities of sweat. It contains between two and five million sweat glands, a number far in excess of that found in any other primate.

The reproductive biology of man has undergone extensive changes. The estrus cycle of the female has been altered in two ways that affect sexual behavior. Menstruation, the loss of blood which occurs at the end of the cycle whenever fertilization fails to occur, has been intensified. The females of many primates experience slight bleeding at this time, but only in women is there a heavy loss of blood from the sloughing uterine wall (*Chapter 12*). The estrus itself, the period of "heat" during which ovulation occurs and the female actively seeks copulation, has been virtually erased. Instead of the intensive bouts of copulation during estrus encountered in other primates, there is more or less continuous sexual activity throughout the human cycle (*Chapter 7*). And in place of the sexual signals of estrus, such as changes in skin color surrounding the female sexual organs or release of distinctive odors, copulation is initiated through extended foreplay between the partners. Physical attraction is based on conspicuous anatomical features including the breasts of women and the pubic hair of both sexes. These changes favor the formation of intimate, long-lasting partnerships between men and women. Human sexual evolution is believed by many zoologists and anthropologists to have occurred as part of the unique division of labor between the sexes that arose when man turned to hunting in groups. According to this view, continuous relationships were needed to support individual women and their young children because they were sequestered in the camps and not permitted to join the chase. The period of childhood lengthened to permit the more elaborate learning procedures required to join the societies. Life beyond the reproductive period — middle and old age — was greatly extended, probably because of the adaptive advantage of ensuring the care of the last offspring.

The brain was enormously increased by the growth of the cortex, which contains the centers for memory and complex computation

During the million or so years required for the transition from Australopithecus to Homo the brain doubled in size. The expansion of the cortex was associated with the most important development of all: the human form of language.

(*Figure 5*). During the approximately one million years required for the transition from *Australopithecus* to *Homo* the brain doubled in weight and intelligence increased to a degree unprecedented in all earlier animal evolution.

The expansion of the cortex was associated with the most important development of all: the development of the human form of language. Chapter 15 noted that animal communication consists of the exchange of a limited repertory of signals which are hereditarily fixed and relatively invariable within species. Almost all of these signals pertain to the immediate present. They may consist of a threat, a plea for food, or an invitation to "follow me" — in other words single statements that are simple, urgent, and direct. Human language, by contrast, has been liberated in time and space. It can refer not only to immediate circumstances but also to past and future times and to distant places. The basis of this change, like the one that transformed human sexuality, has been speculated to lie in the necessities imposed by the group hunting and division of labor adopted by early human societies. The increase in information was made possible by several new qualities in the vocal exchanges. Words became arbitrary in meaning — different sounds could be applied to different objects (and later, to different ideas) according to the desires of the speakers. True syntax was added — the meaning of words changed according to the other words with which they were spoken and the order in which all appeared. These two traits, arbitrary symbolism and syntax, made possible a productive language, one in which the number of messages can be multiplied without limit simply by adding newly coined words or joining old ones in new combinations. Man is the only species that succeeded in creating a truly productive language, and it is the scaffolding of human culture and civilization.

READINGS

B. CAMPBELL, *Human Evolution: An Introduction to Man's Adaptations*, Chicago, Aldine Publishing Co., 1966. Easily the best written and most authoritative account of the adaptive basis of human evolution. This is the ideal book to read after the present chapter.

F. C. HOWELL et al., *Early Man*, New York, Life Nature Library, Time Inc., 1965. Recommended for its superb illustrations.

W. HOWELLS, *Mankind in the Making: The Story of Human Evolution*, Revised Edition, New York, Doubleday, 1967. This book covers much the same ground as Campbell's but lays more emphasis on the modern races of man. It also provides a fascinating account of the discovery of human fossils.

D. MORRIS, *The Naked Ape*, New York, McGraw-Hill, 1969. A popular and entertainingly written version of the subjects covered by Campbell's work. This is the book to suggest to friends who do not have the amount of biological background equivalent to an elementary college course.

PART FIVE ALTERNATE FUTURES

The Romans named the first month of the year after Janus, a god with two faces who could look backward in time as well as forward. Man is the first animal to have the ability to look backward, in the sense that he can study the process of evolution that produced him. Even more important, he has the ability to look forward, aware that he now has the power to shape not only his own evolution but also that of other species. Man will decide — either consciously or by default — the next steps in the future of life on this planet.

It is a grave responsibility. In the words of biologist Clifford Grobstein, "Man has long assumed purpose in the universe — a purpose that he has personalized in a Divine Being. Man is now challenged to provide purpose himself, since science is placing in his hand increasing power to control events and hence to implement purpose."

Is our species equal to the task? Man's performance to date raises some doubt about his competence as overlord of Earth. He has squandered its resources as if there were no tomorrow, and has driven one species after another to extinction. He has just begun to realize that overconfidence in his mastery and an arrogant disregard of his place in nature jeopardize his own chances for survival. One biologist observed that it is not at all certain that superior intelligence confers a long-term survival advantage on the species that possesses it.

Restoring the equilibrium of the Earth is a worldwide problem. Mankind as a whole will be required to act with a degree of cooperation and altruism heretofore more characteristic of insect than of human societies. Human nature changes slowly if at all, and so do human institutions. Can the reaction time of our culture be speeded up enough to cope with the series of crises that are now rushing toward us?

Prophecy is a risky game, but the task can be simplified by projecting present trends into the future, preferably the near future. This process leads us to predict that most of the ecological crises discussed in the next two chapters will be resolved by the end of the 21st Century, either by man's reason or by nature's indifference. We have attempted to make some educated guesses about the outcome of a few of these crises, particularly the most urgent one: the population explosion. But our guesses are colored by our individual, personal and largely nonscientific judgments of the nature of man. We found ourselves in disagreement on this subject, and our views of man's prospects spread across a spectrum ranging from tempered optimism to gloom.

The optimists among us believe that our species possesses the flexibility and generosity, or at least the enlightened self-interest, to find a reasonable solution to our predicament; and that although the future may hold some difficult readjustments for man, catastrophe is neither inevitable nor probable. The view that the biological sciences, intelligently applied, will lead man to a new harmony with nature is reflected in Chapter 28.

The pessimists argue that there is no precedent in human history for the required degree of global altruism or even global cooperation; that it may already be too late for whole nations to reverse the disastrous momentum of rising population, soaring expectations, and commitment to economic growth — in short, that man's cultural reaction time is too slow to solve so many urgent problems so quickly. The view that man is about to enter one of the Dark Ages of human history is the theme of Chapter 29. The reader can decide for himself which of our alternate futures he prefers to believe; and as a member of the next generation, he can help to shape it.

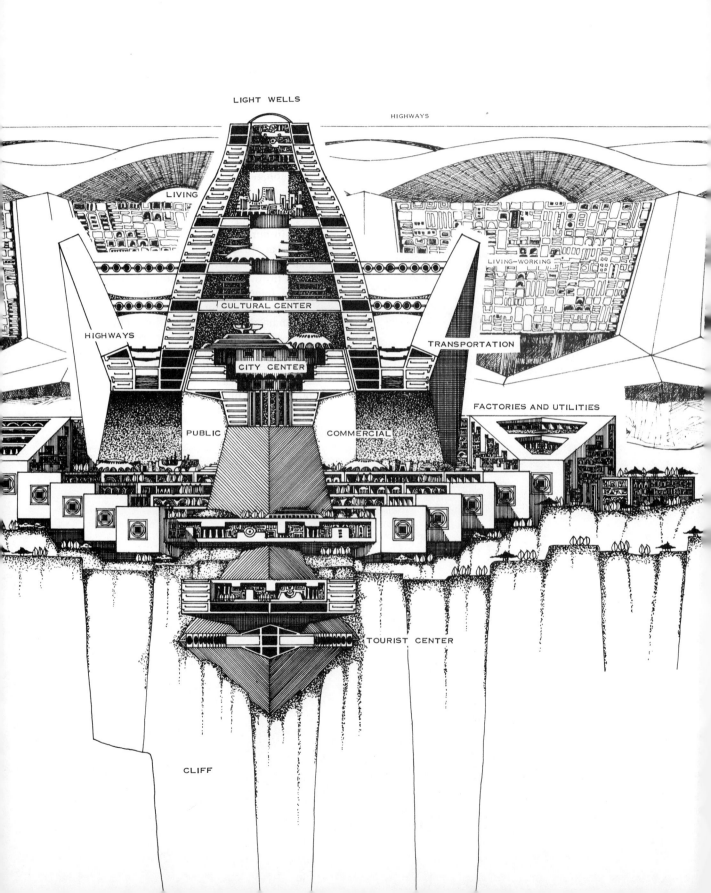

28

toward a
stable ecosystem

*Have you descended to the springs of the sea
or walked in the unfathomable deep?
Have the gates of death been revealed to you?
Have you ever seen the doorkeepers
of the place of darkness?
Have you comprehended the vast expanse of the world?
Come, tell me all of this, if you know.*

THE BOOK OF JOB,
THE NEW ENGLISH BIBLE

Devoid of science and impoverished of philosophy, the human species for the greater part of its existence has occupied a fragile and subordinate position in nature. Near the beginning of recorded history, according to the Bible, God could select Job as a representative of humanity and mock him for his ignorance.

The rise of modern science and technology has utterly changed man's position. He has personally descended to the deepest trenches of the sea and can orbit the Earth at just beneath escape velocity. He controls nuclear power and transmits messages through space at the speed of light. He understands the chemistry of the gene and has taken the first steps toward the artificial synthesis of life in the laboratory. He has walked on the moon, an achievement that would surely have strained even Job's capacity for dreaming.

But man has not freed himself. Like the mythical Greek hero Prometheus, who tricked Zeus and stole fire from the gods, man's success brings with it the risk of cruel punishment. The control of atomic power is accompanied by the danger of atomic war. Man's eminence as a biological species has temporarily abolished the old population controls and launched him into the population explosion, an unchecked growth that can eventually consume the resources of the world and cancel all the benefits of science and technology. It has been argued that even if man escapes atomic war he will achieve essentially the same result — devastation of the planet — simply by continuing to reproduce at the same rate for another hundred years. For this reason the population factor must weigh heavily in all attempts at prophesy.

THE POPULATION EXPLOSION

The alarming facts concerning human population growth can be quickly comprehended by studying the growth curve reproduced in Figure 1. The number of humans remained almost constant for hundreds of thousands of years, then began to climb sharply. About three billion are alive today. In only about 30 more years, at the current growth rate, the world population will double again, to about 6 or 7 billion. The biological basis of the problem is disclosed by the dip in the curve seen in the 14th Century. This temporary decline, in 1348–1350 to be exact, was caused by the Black Death (the Germans called it the Great Dying) which destroyed a quarter of the population of Europe. The disease responsible was plague, caused by the bacterium *Pasteurella pestis*, which is transmitted by fleas from rats to man. People who take for granted the hygienic comforts of western civilization find it hard to appreciate the threat imposed by such infectious diseases in earlier times. Typhus, malaria, syphilis, cholera, yellow fever and a long list of other deadly diseases collaborated with famine

MEGASTRUCTURE on opposite page is envisioned by architect Paolo Soleri as a component of a Utopian city of the future. A cluster of these enormous multilevel structures, taller than the Empire State Building and much more massive, would house all the activities of a city. By eliminating suburbs and other one-level forms of urban sprawl, Soleri's high-density city would leave surrounding countryside unspoiled.

481

and war to make early death highly probable. This was the reason there was no population problem throughout most of man's history. Man's enemies were the density-dependent controls in whose presence he had first evolved. A high natural birth rate was required to maintain the population at its constant low level. The population biology of mankind, in short, was typical of an animal species. With the advent of agriculture and later of more careful hygiene and modern medical practice, the death rate was drastically reduced in all the societies that could make use of these technologies. But the birth rate remained the same. As a result, the population of most countries grew at a net annual rate of between one and three percent.

Mankind has already passed the point where it places a severe strain on the resources of the globe. Our species already consumes more food than all other land vertebrates combined. We have stripped or altered most of the forests and grasslands, and harvested many of the richest fishing banks of the sea below their point of optimal yield. Although human populations are expanding exponentially, the size of the Earth remains finite. In the time of Christ the density was approximately one person per square kilometer. Today it is approximately $20/km^2$ and in less than 30 years it will exceed $40/km^2$. In the year 2600, if the present growth is somehow miraculously preserved, there will be just enough land for each person to stand shoulder to shoulder. The food energy available can be increased by improvements in agricultural technology, but in the long run it too will be a finite resource. All of the algae and other plant growth of the oceans together produce somewhere from 5.0 to 13.5×10^{16} calories per year, and all the plants of the land produce between 5.0 and 7.2×10^{16} calories per year. Thus a maximum total of 1 to 2×10^{17} calories are annually made available to animals over the entire earth in the form of plant food. A single human being requires an average of 2,200 calories per day or, 8×10^5 calories per year, and the entire human race requires 8×10^{15} calories per year. In short, mankind already preempts at least one percent of all the energy available, and will continue to do so even if every person were to go on to a strictly vegetarian diet. It is already too late for the whole world to eat steaks because, as mentioned in Chapter 25, cows and other herbivores are very inefficient devices for gathering and packaging energy. In a few more decades of unabated population growth, it may be too late for the whole world even to eat grass.

The population crisis is the gravest problem that confronts mankind. In the case of atomic war, it is necessary to perform the drastic act of starting the war in order to suffer its terrible effects. But in the case of the population crisis, nothing more is required than for mankind simply to continue mindlessly doing what it is already doing. Fortunately an awareness of the problem has begun to grow among educated people, and there are signs from India to the United States that political leaders are willing to try to do something about it. It is a cardinal principle that every nation must undergo a demographic shift (*Figure 2*). In the most primitive human societies, such as the high-

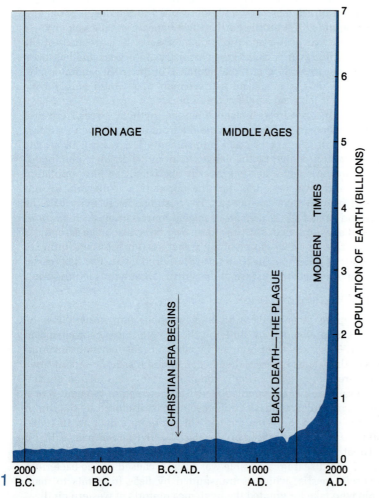

1 HUMAN POPULATION has remained virtually constant in size during most of man's existence on Earth. In 1800 A.D. world population reached one billion. By 1900 it had doubled. At present rate of growth (about 70 million people per year) it will approach seven billion by the year 2000.

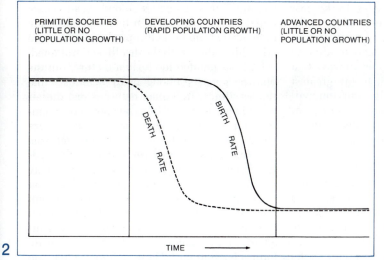

PRIMITIVE SOCIETIES (LITTLE OR NO POPULATION GROWTH)

DEVELOPING COUNTRIES (RAPID POPULATION GROWTH)

ADVANCED COUNTRIES (LITTLE OR NO POPULATION GROWTH)

DEATH RATE

BIRTH RATE

DEATH RATE

BIRTH RATE

TIME

2

DEMOGRAPHIC SHIFT is necessary to attain population stability. In primitive societies high birth rate is offset by high death rate (left) and population size remains constant. Transition to modern technology (center) slashes the death rate. Birth rate remains high, causing sharp increase in population. To restore equilibrium, birth rate must be reduced to correspond to death rate (right).

land peoples of New Guinea, which are still essentially in the Stone Age, a high death rate compensates for the naturally high birth rate. Their populations grow slowly or not at all. In the developing countries the death rate has been reduced but people still cling to the old custom of producing large families. The result is rapid population growth. Only when birth rates are brought down by contraception and deliberate population planning can the demographic shift be completed. Then low birth rates match low death rates, and the population size again stabilizes. At the present time only the most advanced industrial countries have made any progress toward completing the demographic shift. Few of these have actually reduced birth rates all the way to the point of halting population growth. To do so requires some degree of altruism and a willingness to discuss an emotional subject hedged by religious taboos. Population control will certainly be discussed with ever greater freedom and maturity in the future. It is also

likely that the subject of optimal population size, already a favorite in academic groups, will soon be debated more publicly.

The demographic shift can be terminated at any population level. Every nation must eventually decide, as a matter of public policy, what the final level should be. The great American dream of the nineteenth century, that this country contains land enough for our descendants to the thousandth and thousandth generation, has been cut cruelly short by the elementary mathematics of population growth and rising economic expectations. Will our descendants settle for one billion Americans, saving whatever might remain of the countryside by compressing the population into strip cities of megastructures extending for hundreds of miles along the coasts and rivers? Will they choose the present population size and encourage citizens to move into smaller, more livable communities? Or will they prefer to bring birth rates temporarily below death rates, in order to reduce the population size and enjoy more space than is available at the present time? One cannot predict which course will be chosen, only that the choice will have to be deliberately and carefully made in the not too distant future.

ENERGY AND RESOURCES

Overpopulation and rapid industrialization have severely depleted the natural resources of the planet. The technologically advanced nations are demanding energy for industrial and household use at an accelerating pace. In 1970 the United States generated approximately 1.5 trillion kilowatt-hours of electrical energy by tapping hydroelectric power and consuming gas, oil, coal and nuclear fuels. In the year 2000, according to conservative estimates, this nation will be generating 9 trillion kilowatt-hours per year. The needs of the developing countries will increase even faster. A variety of nonrenewable resources, from such fossil fuels as oil to copper and other metals, and even so vital an agricultural material as phosphate, are being depleted. The world supply of some of these resources, such as oil and natural gas,

The population crisis is the gravest problem that confronts mankind. Fortunately there are signs, from India to the U.S., that political leaders are willing to do something about it.

will run out within a few decades. In the near future it seems imperative for the nations of the world to set priorities on the conservation of these irreplaceable resources, and to allocate the use of the remaining supplies in ways that will provide the most benefit to the greatest number of people. Eventually mankind will have to find substitutes for these resources, or else revert, in ways painful to contemplate, to much lower levels of consumption.

MAN'S ANIMAL NATURE

Some pessimists claim that man is fundamentally too irrational to arrive at such decisions and will not find his way to a permanently secure future. They argue that human nature can't change. Or, as the seventeenth century dramatist Molière acidly expressed it for all the cynics in this world, "Man, I can assure you, is a nasty animal." This pessimistic view has gained new credence in recent years with the revelation of man's evolutionary beginnings in Africa. It is widely believed among experts on the subject that the massive enlargement of the brain during the past two million years of evolution was associated with an increasing shift to a carnivorous diet. When the ancestral man-apes took up hunting in groups, they enriched their communication system and divided labor among the group members. This theory envisions our distant ancestors as pack apes or killer apes, specialized for the pursuit and destruction of large animals. Perhaps the man-apes were also aggressively territorial. From such postulates it is easy to form a rationalization for our own bad behavior. Raymond Dart, the South African who participated in the early discoveries of the man-ape fossils, put it this way in 1953: "The blood-bespattered, slaughter-gutted archives of human history from the earliest Egyptian and Sumerian records to the most recent atrocities of the Second World War accord with early universal cannibalism, with animal and sacrificial practices or their substitutes in formalized religions and with the world-wide scalping, head-hunting, body-mutilating and necrophiliac practices of mankind in pro-

claiming this common blood lust differentiator, this mark of Cain that separates man dietetically from his anthropoidal relatives and allies him rather with the deadliest Carnivora."

Dart has painted too lurid a picture. But the opinion persists among many scientists that the beastliness in human nature is just that — beastliness held over from our man-ape ancestry, during which aggression, territorial instinct, and blind tribal loyalty became valuable adaptive traits. A few popular writers, including Robert Ardrey, Konrad Lorenz and Desmond Morris, have developed variations on the theme in the last few years. They interpret men as naked apes, to use Morris' catchy phrase, peculiar bipedal primates that are ill-adapted for the complexity of their own hastily built civilizations. These authors offer the curiously comforting proposition that our sins are only animal sins ("Original Sin" if you wish), the result of instincts that originated long ago as stereotyped, strictly inherited traits.

The killer-ape theory has not been accepted by all serious students of the subject. There are many, especially psychologists and sociologists trained in the liberal humanist tradition, who totally reject the idea that modern man is slave to the residual instincts of the man-apes. Instead, they view the shortcomings of human behavior as the result of the imperfections of society. The human mind is interpreted to be an extremely flexible, imprintable instrument, which goes wrong only when it is subjected to abnormal stress or is trained incorrectly in the first place.

The great debate about human nature reveals both the shortcomings of the social sciences, particularly human psychology and sociology, and their critical importance for the future. Perhaps you have recognized that this debate is a continuation of the old nature-versus-nurture controversy which has troubled philosophers since the seventeenth century. Man is obviously higher than a man-ape and nowhere close to being an angel. But exactly where he stands, in what dimensions of innate morality and intelligence, is an extraordinarily

Biology is now entering its golden age. Gaining daily in depth and competence, it will be used to direct the coming stabilization of the world ecosystem.

difficult matter to assess. The crucial question is: To what degree is man's nature flexible by education and the constraints of law, and conversely to what extent should his governments and institutions be shaped to conform to his instinctive nature? Biology can find no higher goal than to join the social sciences in the search for a workable answer.

Future historians may judge that science in the remaining part of the twentieth century was marked by a turning inward, an increasing appreciation of the uniqueness of the Earth as opposed to all other accessible worlds, and of living things as opposed to nonliving. Biology is now entering its golden age. Gaining daily in depth and competence, it will be used to direct the coming stabilization of the world ecosystem. Population growth must be halted at a level that permits a pleasant life for every human being. Material cycles must be carefully assayed and managed to ensure the steady supply of essential resources for countless millenia. Within a few more decades medical science will be able to control the remaining diseases and to postpone senility to a maximum age. Genetic surgery, the transplanting of healthy genes for defective mutant alleles, will probably be attempted. If it succeeds, it will open the door to the eventual full genetic reconstitution of the human species; and this in turn will create a new moral dilemma. Shall we control our own evolution? If so, who will choose the "superior" genes with which to endow future generations? Meanwhile, the technique of transplanting entire organs, such as the heart and liver (and perhaps ultimately the brain), will be steadily improved. The search will continue for a technique to induce the regeneration of organs and limbs lost by disease or accident.

It is conceivable that the technology of medicine can be advanced to the point where death by natural causes is circumvented altogether. This achievement would produce a second moral dilemma. If the death rate is drastically reduced, the birth rate must be correspondingly reduced. The result would be a second, unnatural demographic shift, resulting in a world of "ancient ones" kept alive by innumerable anatomical and biochemical crutches. Birth and childhood would become rare episodes. As unappealing as this prospect may seem to members of our own tumultuous youth-oriented culture, it is nevertheless the goal toward which man is striving — because death must always seem to be the enemy.

Primitive societies were joined intimately with nature. The hunter and gatherer had to have expert knowledge of the plants and animals on which his daily existence depended. His tribe was a modest component of the local ecosystem. With the industrial revolution and rapid urbanization, men drew temporarily away from the land, comfortable in the belief that they had "conquered nature" and entered a wholly new form of existence. The illusion took less than a century to shatter, as the problems created by overpopulation, pollution and resource shortages became critical. Future societies will return to nature, but in a vastly more enlightened way. Man will again become a component of the stable ecosystem, perhaps even a modest one. He will look upon the millions of species of plants and animals of earth, himself included, as the ultimate wonder of his existence, deserving of unending study and esthetic appreciation.

READINGS

P.R. Ehrlich and A.H. Ehrlich, *Population, Resources, Environment,* 2nd Edition, San Francisco, W.H. Freeman & Company, 1972. An authoritative account of the ramifying problems created by runaway population growth in the human species.

G. Feinberg, *The Prometheus Project,* New York, Doubleday, 1969. A clear-eyed prophesy of science in the future service of mankind, with due attention to the recent achievements and promises of biology.

P. Handler, ed., *Biology and the Future of Man,* New York, Oxford University Press, 1970. The best thoughts of 185 leading American biologists are presented in this survey sponsored by the National Academy of Sciences.

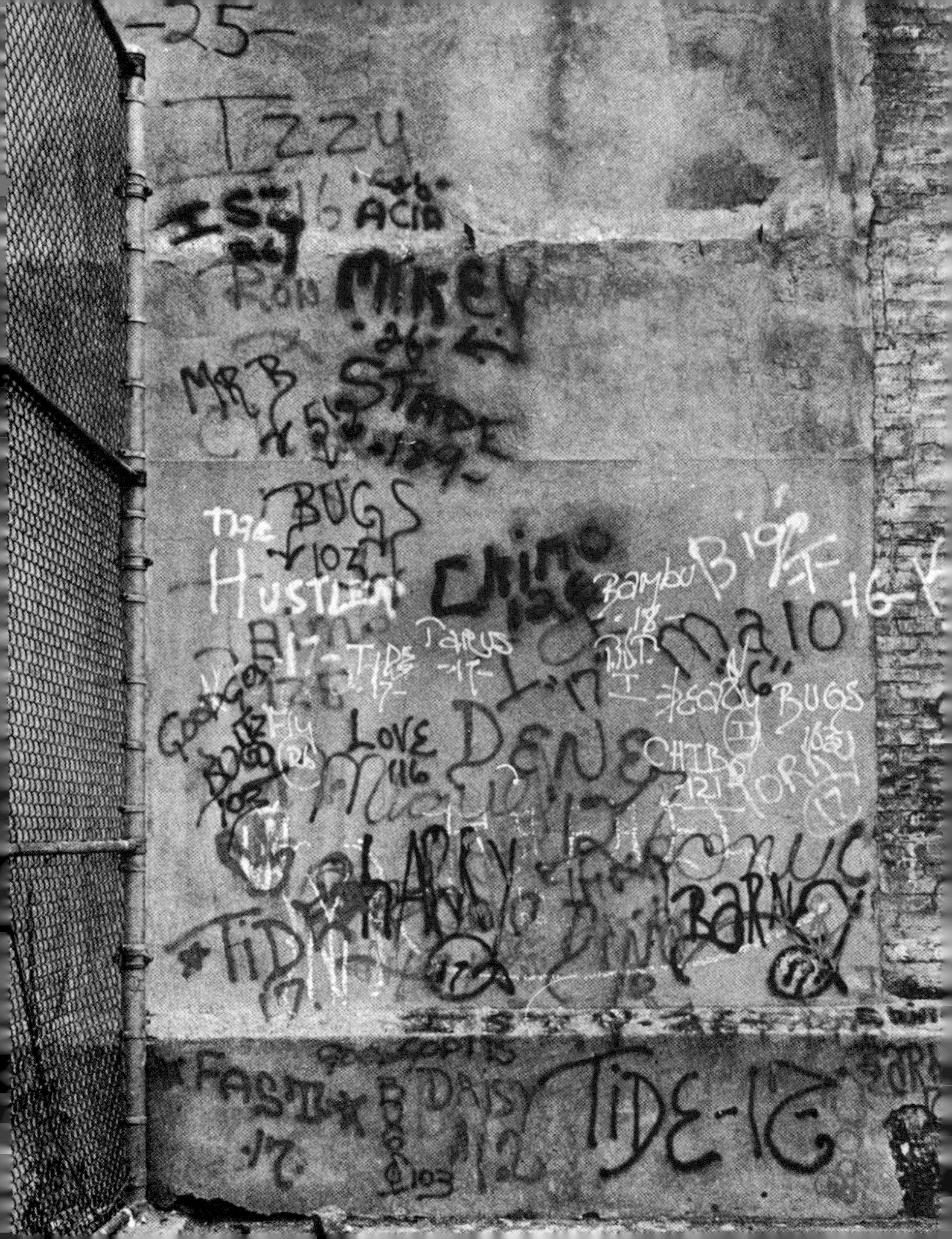

-25-
ACID
MIKEY
BUGS
LOVE
116
LARRY
DAISY

29

defeat by default

The world is being carried to the brink of ecological disaster not by a singular fault which some clever scheme can correct, but by the phalanx of powerful economic, political, and social forces that constitute the march of history. Anyone who proposes to cure the environmental crisis undertakes thereby to change the course of history.

BARRY COMMONER
THE CLOSING CIRCLE

In *The Limits to Growth*, Dennis L. Meadows and his colleagues cite a French riddle for children: A farmer's pond is gradually covered by a giant water lily that doubles in size every day. If allowed to grow unchecked it would cover the pond in 30 days, choking off all other life in the water. The farmer decides not to worry about cutting back the lily until it covers half the pond. On what day will that happen? The answer, of course, is the 29th day. He has only one day to save his pond.

Meadows uses the riddle to illustrate the suddenness with which exponential growth approaches a fixed limit. The story is also a parable of man's plight on spaceship Earth. At the present point in human history, population, pollution and the consumption of nonrenewable resources are increasing exponentially; that is, they are increasing like compound interest. World industrial output, for example, has been rising at a rate of seven percent a year, which may sound modest until one considers what it means. One hundred dollars invested at seven percent compound interest will double every ten years. Exponential growth is consuming billions of tons of resources and adding billions of tons of pollutants to the environment every year. Clearly this growth cannot long be sustained on a planet with limited resources. The choice that faces mankind is whether to try to live within the limits of the Earth's resources by accepting a self-imposed limit on growth, or to keep on growing until stopped by nature, in ways that are likely to be much harsher than those that society might choose for itself.

In theory it is possible to stabilize the human population and the economy at a level that the Earth can sustain for many millenia. But such a decision would involve radical changes in virtually every aspect of human activity. It would mean limiting the world birth rate — by coercion if necessary; diverting capital from industry to the uneconomic production of food; and probably redistributing the wealth of the rich nations. As Barry Commoner has pointed out, such a decision would be no less than a commitment to change the course of history.

The solution to the ecological crisis is not primarily technological but political and social. Most of the necessary technology already exists; what is missing is the will to apply it. Exponential growth confronts mankind with the need to make profound changes in human society in a very short time, probably within only two or three generations. And these decisions must involve a degree of global co-operation for which there is no precedent in human history.

Can human institutions react energetically enough, and swiftly enough, to avert disaster? Faced with difficult choices, human beings and human institutions usually find it easier to deny for as long as

GRAFFITI are children's reaction to the squalid anonymity of life in urban slums. Photograph on opposite page depicts names and street numbers scrawled on wall of a building in New York.

possible that a problem exists, and then to assume that if left alone, things will somehow work themselves out. Every year that the decisions are postponed, the greater the chance that the accumulated insults to the environment will become irreversible.

Recent history provides scant evidence that human society can restructure its priorities so quickly. Chapter 28 concluded with an affirmation that man will somehow find the necessary solutions simply because he MUST. No such optimism will be found here. Although we should like to believe otherwise, the authors of this chapter believe that individuals, corporations and nations will pursue their shortsighted self-interest as they have in the past, until disaster shocks them out of their complacency.

POPULATION

Chapter 28 noted that the population of the Earth will approximately double in the next three decades, approaching seven billion by the year 2000. Most of this growth will occur in the Third World nations of Asia, Africa and Latin America, which contain two-thirds of the Earth's present population. In most industrialized nations population is increasing from 0.5 to 1.0 percent a year, and shows signs of leveling off. In the Third World, population is multiplying at rates of from two to three percent a year, and in many countries the rates are still rising. Between 40 and 45 per cent of their population is less than 15 years old, which means that even if the birth rate is somehow stabilized at the replacement level, the absolute numbers of people will continue to increase for many decades.

If the developed nations can stabilize their populations at the replacement level by the year 2000, and the Third World nations by 2040 (an optimistic assumption), then about a century from now the Earth's population will be 15.5 billion. Those who believe that the Earth is already overcrowded will derive no comfort from the thought that within the lifetimes of infants now alive, there will be four people on Earth for every one alive today.

One of the great unanswered questions of demography is how to achieve population stability. The desire for children is one of the mainsprings of human behavior. Obviously the first step toward stabilizing population is to put safe, reliable, easy-to-use methods of contraception into the hands of every couple in both developed and undeveloped countries. This will enable them to have the number of children they want. The next step is somehow to persuade them to want only the number of children that Earth can support. But if voluntary methods continue to fail, what is the next step? In India, after assuming dictatorial powers, the government of Indira Gandhi has decreed that any government employee who adds a fourth child to his family after 1977 will be fired. Some workers have suggested that the only foolproof method of population control would be one that was not under the control of the couple — for example, an oral contraceptive dissolved in the public water supply, with the antidote available only upon application to the government. At birth each person might be issued a coupon entitling him or her to have one child, just as present law restricts a man to one wife.

In a democracy such coercive measures would not appear to be politically feasible in the foreseeable future. Any government that suggested them would be committing political suicide. But the alternative to these distasteful proposals is even more painful: the restoration of population equilibrium by starvation, disease, or global war. Society must make a choice; there is no way to escape that. To do nothing, or to take halfway measures, is to decide by default.

FOOD PRODUCTION

There is general agreement that perhaps one-third of the people on Earth today are undernourished. Perhaps 10 to 20 million deaths each year can be attributed to malnutrition.

Those who believe that the Earth is already overcrowded will derive no comfort from the thought that within the lifetime of infants now alive, there will be four people on Earth for every one alive today.

Data from the United Nations Food and Agriculture Organization indicate that in most of the developing countries the basic protein and caloric requirements are not being supplied. Although total world food production is rising, the gains have been wiped out by population growth.

Agronomists remain guardedly optimistic that the food supply can be doubled to feed the seven billion people who will be on Earth at the turn of the century, but they are doubtful that the Earth can sustain that population for long. Some workers argue that in terms of food supply, in the long run the Earth can sustain only 1.5 billion people, and that if one also takes into account the depletion of nonfood resources, plus the impact on the environment, the optimum long-term population of the Earth is only one billion — less than one-third the present level.

Within the past decade agronomists have introduced high-yield varieties of rice and wheat, new techniques of farming, and new ways to cope with plant diseases and insect pests. Sometimes called the Green Revolution, this "package" of seeds and techniques has enabled some farmers in countries such as Pakistan and Mexico to increase the productivity of their land, sometimes doubling local yields. The Green Revolution is not the answer to overpopulation but it will buy time, perhaps two decades, to work out a long-term solution.

The Green Revolution also has its dark side. It depends on enormous increases in the use of pesticides and nitrate fertilizers. Intensive use of pesticides is "addictive" because the pesticide selects for resistant strains of insects and destroys many of the beneficial parasites and predators that would otherwise limit the pest population. More and more spraying becomes necessary to maintain a given level of crop yields. The American agricultural system is already "hooked" on pesticides and the Green Revolution is spreading this dependence to the poorer countries.

The present world production of pesticides is more than one million tons a year. To double world food production by the end of the century perhaps six million tons a year will be necessary because of their diminishing effectiveness. Pesticides ultimately accumulate in the oceans. As much as 25 percent of the DDT produced so far may have already found its way to the sea, and has begun to show up in the cells of marine organisms. A sixfold increase in the use of pesticides threatens to disrupt the marine ecosystem, perhaps reducing or contaminating the fish harvest upon which the world now depends for 20 percent of its protein.

Some of the worst pest problems are still to come. In the developing countries vast areas are being planted in single crops of the new cereal grains, creating splendid opportunities for pest epidemics. Much has been written about new approaches to pest control, particularly about narrow-spectrum insecticides and biological control, but so far these methods have proved to be expensive and of limited effectiveness.

The new high-yield strains of wheat and rice require enormous applications of fertilizer — from ten to 27 times more than the traditional varieties. At present the global use of fertilizers totals about 62 million tons a year, almost all of it in the advanced countries. U.S. farmers use nitrates at the rate of 70 kilograms per arable hectare (about 2½ acres), compared with eight kilograms in India and five in Africa. In 30 years the world consumption of fertilizers could exceed 700 million tons a year, with serious impact on inshore waterways. Not only does the runoff of excess nutrients threaten to choke lakes and streams with blooms of foul-smelling, fish-killing algae, but nitrates also pose a serious health hazard when they seep into drinking water. Nitrates are relatively harmless to humans but they can be converted by intestinal bacteria, especially in infants, to poisonous nitrites.

As with pesticides, there is a point of diminishing return in the use of fertilizer. Illinois farmers used about 100,000 tons of nitrates to obtain an average yield of about 70 bushels of corn per acre in 1958. In 1965 they used 400,000 tons to obtain a yield of 95 bushels per acre. The extra 25 bushels required the use of four times as much fertilizer. There is mounting evidence that intensive application of nitrates eventually damages the soil, particularly the poorer soils found in many of the developing countries, reducing its response to additional fertilizer and impairing its future food-producing capacity. Moreover, the price of fertilizers and pesticides has increased sharply in the last few years. To buy them, the impoverished nations of the Third World must use their meager supply of foreign exchange. Despite the Green Revolution, overpopulation confronts the Third World with a dismal choice of risks: ignore the new agricultural technology and starve now, or apply it and starve later.

THE SOCIAL IMPACT

In little more than a decade, the Green Revolution has begun to tear at the social fabric of the undeveloped nations. The thrust of the Green Revolution is toward efficient large-scale farming, which favors large landowners who can afford the increased investment in fertilizers, pesticides and machinery. In the Third World as in the U.S., more food is being produced by fewer hands. In one village in India, for example, tenant farmers and sharecroppers are being squeezed off the land by landowners who prefer to cultivate the land with hired labor and pocket the profits. In short, the Green Revolution has aggravated the classical political tensions between the haves and the have-nots. And throughout the world unemployed rural laborers are migrating to cities.

Because the city must absorb the first shock of the population explosion, the quality of life in urban areas has deteriorated appallingly in the last decade. Growth is fastest around the most crowded metropolises of the Third World. One-third of all Argentinians now live in greater Buenos Aires, and the population of Jakarta has soared from 550,000 to 4.5 million since the end of World War II. Despite the efforts of many governments to encourage the pioneering of undeveloped regions, the migration to the cities continues. Ironically, even as living conditions become more desperate, the tide of aspiration throughout the world continues to rise. For some the city holds the promise of excitement and glamor; for most, at least the hope of a job and relief from loneliness and rural drudgery — a hope that for the unskilled worker all too often withers to despair in a decaying slum.

The fate of American cities is a portent of the fate of all cities. Analysis of the 1970 census data reveals what is happening to New York. In the previous decade almost a million middle class whites fled the city, to be replaced by about the same number of blacks, Puerto Ri-

cans and immigrants — legal and illegal — from the Caribbean and Latin America. Many of these people are from rural areas and have no marketable skills. The welfare rolls quadrupled to 1.3 million; roughly one of every six New Yorkers is now on welfare. Unemployment, drug addiction, crime and delinquency skyrocketed. The middle class fled to the suburbs, followed by many large corporations which found it increasingly difficult to attract young executives to the city. Hundreds of city blocks deteriorated into slums, particularly in Brooklyn and the Bronx, as the middle class moved out. The exodus of taxpaying citizens and businesses, and the abandonment of slum buildings by landlords (thus removing them from the tax rolls) were major factors in driving the city to the edge of bankruptcy. The story is being repeated in most of the older cities of the U.S. The American city has become, in the words of *The New Republic*, the domain of "the very rich, the very poor, the old, and the peculiar."

THE AUTOMOBILE

The sprawl of suburbs has been facilitated by the automobile. In Los Angeles, a collection of suburbs in search of a city, the tyranny of the automobile has become almost absolute. In the absence of even the rudiments of a mass transit system, the private automobile (or two or three of them) has become a necessity. Other cities in America are well along the road to Los Angelization. Between 1960 and 1970 the automobile population of the U.S. increased twice as fast as the human population, and the growth was concentrated in the suburbs. Even the Arab oil embargo of the mid-1970's, with the resulting shortage of gasoline, produced no more than a pause in the upswing of car sales. Highways, gas stations and parking lots now cover more of the surface area of the U.S. than do homes, stores and schools.

Automobiles burn more than 35 percent of the total annual energy budget of the nation, and they spew out microscopic soot and noxious fumes at mind-boggling rates. Most of the gains from the new emission-control devices

Because cities must absorb the first shock of the population explosion, the quality of life in urban areas has deteriorated appallingly in the past decade.

have been wiped out by the increase in the number of cars; each car pollutes less, but there are more cars to do the polluting. Even in death a car is a polluter. Unprofitable as scrap, it is abandoned on the street or stacked in hideous junkyards. It is probably the greatest single contributor to the trashing of America.

The automobile provides a case history of the difficulty of working out political solutions to environmental problems. British economist E. J. Mishan considers the invention of the private automobile "one of the great disasters to have befallen the human race." He argues that for a fraction of the money the nation is currently spending on the maintenance of private cars and the government services necessary to keep traffic moving, the country could provide comfortable, frequent and highly efficient public transport service (bus, train or subway) in all the major population areas; restrain and gradually reverse suburban sprawl, the futile attempts of commuters to get away from it all; restore quiet and dignity to cities, enabling people to stroll unobstructed by traffic and to enjoy once more the charm of towns and villages.

Mishan believes that the U.S. is almost hopelessly locked into its commitment to the automobile as the primary means of transit. One of every six Americans earns his living from an industry directly related to automobiles. Of the top 25 U.S. corporations (ranked according to their 1972 revenue), 17 derive a major portion of their income from products or services related to automobiles. Millions of jobs and billions of dollars add up to enormous political clout, and so nothing is done. So far no American city has dared to ban private automobiles, and the highway lobby has succeeded in blocking legislative attempts to channel highway funds into mass transit.

The rest of the world, which once mocked America's obsession with the automobile, now seems hellbent to catch up. Other nations are repeating our mistakes with the automobile, dooming themselves in turn to repeat our mistakes with highways, pollution and suburban sprawl. A political solution to the automobile problem appears to be nowhere in sight. Even an absolute dictator might hesitate at the thought of the fierce reactions that would be stirred up by an attempt to curtail the use of automobiles in America. It seems unlikely that if we lack the foresight and the will to get out of our cars and onto bicycles, buses and trains, we are not going to be able to make the much

tougher readjustments in our priorities that are necessary to restore environmental sanity.

ECONOMIC GROWTH

The industrial output of the world is increasing at an even faster rate than population, and most economists see this as a good thing. Most of the growth is occurring in advanced countries, widening the economic gap between rich and poor nations. The Japanese economy, with a gross national product (GNP) of $1,190 per capita, is growing at almost ten per cent a year; the Soviet Union, with a GNP of $1,100 per capita, is growing at 5.8 per cent a year; and the U.S., with a GNP of almost $4,000 per capita, at 3.4 per cent a year. Conversely the economy of India, with a GNP of $100 per capita, is growing at one per cent a year.

The exponential economic growth of the industrial nations makes enormous demands on the nonrenewable resources of the planet, and one can only wonder how long it will be before the rest of the world catches on to what this means. The U.S., for example, consumes 44 per cent of the total annual world output of coal, 28 per cent of its iron output, 33 per cent of its petroleum, and 63 per cent of its natural gas. If the present rates of consumption persist, the known reserves of coal will be exhausted in 300 years, oil in 70 years, most metals in 50 years, and natural gas in 30 years. The diminishing supply has already begun to push prices upward. The price of mercury has increased 500 per cent in the last 20 years and the price of lead has increased 300 per cent in the last 30.

An industrial economy not only depletes resources, but creates massive quantities of pollutants, some of them highly toxic. One of the most ominous pollution threats is radioactive waste from nuclear power plants. By the year 2000, according to current plans, roughly half of the electric power needs of the U.S. will be generated by nuclear plants. If the present safety standards are still in effect, these plants will release into the environment each year some 25 million curies of nuclear waste, mostly in the form of krypton gas and tritium (H^3) in coolant water. A curie is an enormous amount of radiation, equivalent to that emitted by one gram of radium. The "hottest" wastes from reactors, too dangerous to release, must be stored in sealed containers in abandoned mines, on the sea bottom, or in giant tank farms until their radioactivity decays, which will take centuries. Satisfactory sites for storage of radioactive waste are already in such short supply that a former AEC commissioner suggested rocketing the wastes into orbit around the sun. By the year 2000, U.S. nuclear plants will produce roughly a trillion curies of stored waste per year.

At present no one knows how much mercury, lead, radioactivity or other pollutants the world ecosystem can absorb without irreversible damage, or how one exotic pollutant may interact with others to produce nasty surprises. But without doubt there is an upper limit, and the danger of reaching that limit is increased by the long lag between the time a pollutant is dumped into the environment and the first appearance of biological damage. By then a "reservoir" of the pollutant may have accumulated in the environment, as was the case with DDT. With the continued expansion of industry and agriculture, the 15 billion people on the planet by the middle of the 21st century might well push the pollution burden to 100 times its present level. Technological developments might allow the expansion of industry with decreasing pollution, but only at high cost. The U.S. Council on Environmental Quality has estimated that even a partial cleanup of American air, water and solid-waste pollution would require an expenditure of $105 billion. A nation can defer such expenses indefinitely, provided it is willing to risk the health of its citizens and irreversible damage to the environment.

The notion of a zero-growth economy is heresy to both capitalist and socialist societies. Because most of the economic growth of the next few decades will occur in capitalist countries, the primary focus of any effort to limit resource depletion must be the corporations of the U.S., Europe and Japan.

CORPORATION VERSUS PUBLIC

It is an axiom of classical economics that the primary goal of a business venture is to maximize profit, but Harvard economist John Kenneth Galbraith argues that the primary aim of the modern corporation is growth. Skeptics might well ask themselves what corporation president would boast of a DECREASE in sales. The present trend toward the formation of giant conglomerate corporations further supports Galbraith's contention. With so powerful a commitment to growth the corporation stands as the most serious obstacle to the attainment of a steady-state economy.

In theory the role of regulating corporations falls to the government, but in the words of ecologist William W. Murdoch, "large corporations not only have the power to pollute, they have the economic and political power to prevent, delay and water down regulatory legislation. They also have the power and connections to ensure that the regulatory agencies don't regulate as they ought to."

Because of the danger that the corporate system poses to the environment, several political theorists have maintained that the corporation as we know it is obsolete. Unfortunately the performance of the socialist nations — with the exception of China — does not promise a viable alternative. The industrial-minded bureaucracy that governs the Soviet Union appears to be as unresponsive to the welfare of the biosphere as any giant corporation. Protests from Soviet biologists failed to halt the pollution of Lake Baikal, and the Caspian Sea has become a sink for industrial waste, threatening the caviar-producing sturgeon with extinction. Pollution in the socialist countries is not yet so severe as in the West, but with expanding industrialization it is rapidly catching up.

In Mishan's words, "The spreading suburban wilderness, the near traffic paralysis, the mixture of pandemonium and desolation in the cities, a sense of spiritual despair scarcely concealed by the frantic pace of life" are part of the price one pays for an expanding economy. To men born in a century shaped by the assumption that growth is good, the idea of an equilibrium population and a spaceship economy in which resources are continually

The ecological crisis will eventually lead to severe limitation of freedoms that individuals and businesses now take for granted. In the 21st century the citizens of even the richest nations will live under restrictions that would seem intolerable by 20th century standards.

recycled may seem revolutionary, and indeed it is. The idea of a zero-growth economy has only recently begun to be discussed seriously, and as one might expect, it has encountered fierce opposition from business. It seems unlikely that either the advanced industrial nations or the underdeveloped ones will voluntarily abandon their drive for growth and "progress" — the advanced countries because their political structure is dominated by business, and the poor ones because of a fear that a halt to expansion might lock them into perpetual poverty.

INDIVIDUALISM AND AUTHORITY

Political theorists have recently recognized that the ecological crisis will eventually lead to severe limitations of freedoms that individuals and businesses now take for granted. The freedom to be selfish exacts a terrible price in overpopulation, pollution, the plundering of resources, and human misery. As a political goal, the halt of economic growth in the advanced nations and population growth in the poor nations is probably unattainable under democratic rule. In a steadily worsening situation, the need for speedy and drastic solutions favors the rise of increasingly authoritarian governments. Even now one Third World government has dictated how many children a couple can have (India), and another has decreed where its citizens must live (the revolutionary regime in Cambodia deported virtually the entire population of Phnom Penh to rural areas). Throughout the world, private ownership of land — the ultimate resource — is likely to become a thing of the past, as it already has in most of the socialist nations. Other scarce resources will be rationed, particularly such "necessities" as gasoline and meat. In the 21st century the citizens of even the richest nations will live under restrictions that would seem intolerable by 20th century standards. Whether these restrictions are imposed by the economics of scarcity or by the decree of government planners is a matter of debate, but they are unlikely to be accepted voluntarily by a citizenry accustomed to thinking primarily in terms of short-sighted self-interest.

Harvard psychologist B. F. Skinner looks at the problem in a different way. In his book *Beyond Freedom and Dignity,* he argues that human survival depends on reshaping the attitudes and behavior of mankind, particularly the substitution of altruism for individualism.

Aggressive individualism, an effective form of behavior for taming a frontier or resisting the domination of tyrants, becomes dangerous to the species on a crowded and civilized planet. Man's freedom to be selfish exacts a hideous price in violence, overpopulation, and the plundering of resources.

Skinner maintains that human behavior is a product of long and subtle social conditioning. He advocates the development of a technology of behavior: the engineering of psychological tools to manipulate human behavior toward socially desirable patterns. Such techniques have already been used successfully in a limited way with mental patients and juvenile criminals. But whether behavioral engineering can control the conduct of all mankind is open to doubt. The prospect stirs uneasy thoughts of George Orwell's *1984,* and fears about how such a powerful tool might be used by political leaders such as the Chairman of the People's Republic of China, a Latin American dictator, or even the President of the United States. Can man be programmed to act with the altruism characteristic of an insect society? Should he be? And to what ends?

WAR

The most immediate threat to the future of life on Earth is annihilation by thermonuclear weapons or by chemical or biological warfare. Each of the perennial confrontations between major powers — in Berlin, Hungary, Czechoslovakia, Cuba, Laos, Vietnam and the Middle East — carries the risk that a strategic escalation of threat and counterthreat will trigger a war that would destroy hundreds of millions of people and contaminate the entire planet. Five or ten more confrontations between the great powers might leave us only a 50-50 chance of surviving until the year 2000.

In the near future, the impoverished hordes of the Third World threaten to erupt as a political force that could endanger the precarious balance between the superpowers. The ability of a few fanatics to disrupt the activities of advanced nations has been demonstrated by skyjackers, terrorists and guerillas. How much greater will be the disruptive power of tomorrow's wretched billions in Asia, Africa and Latin America? In Vietnam a tiny Asian nation stalemated the world's greatest superpower. The nations of the Third World (except China) lack hydrogen bombs and intercontinental missiles, and none seems likely to develop them. But atomic bombs are another matter.

Within a few decades many poor nations will have at least one nuclear reactor capable of manufacturing enough plutonium to assemble a bomb. A missile is not necessary to deliver such a bomb; the hold of a freighter would serve just as well. A tankerful of nerve gas or a saboteur with a few canisters of botulinum toxin might be able to decimate a city no matter how carefully it is guarded.

We might ask ourselves how long the nations of the Third World will tolerate the widening gap between their wretchedness and the prosperity of the advanced nations. Our survival depends on their co-operation, and theirs on ours. Will we be willing to lower our standard of living so that they can rise from squalor? An international redistribution of wealth would require a degree of altruism that does not seem to be characteristic of our species. It seems more probable that in the next few decades the advanced nations will, like American suburbanites, try to segregate themselves in privileged enclaves, responding only halfheartedly to the misery around them.

What then are the prospects for the future of mankind? In the long run, we do not believe that the human species will be destroyed, because it is extraordinarily resilient. But in the words of C. P. Snow, "there is likely to be a prolonged period of hardship, sporadic famine rather than widespread, commotion rather than major war. To avoid worse," he adds, "will require more foresight and will than men have shown themselves capable of up to now." In short, we believe that mankind is about to enter one of the Dark Ages of human history. Most of the ecological problems that threaten us will be resolved by the end of the 21st century, either by human intelligence or by nature's ruthless indifference. The civilization that emerges from that century will be vastly different from the one we know today.

READINGS

B. COMMONER, *The Closing Circle*, New York, Bantam Books, 1972. A lucid account of what has gone wrong and what must be done about it. Commoner points out that industrial society ignores one of the cardinal rules of ecology: everything must go somewhere.

D.H. MEADOWS, *et al.*, *The Limits to Growth*, New York, Universe Books, 1972. A team of scientists at MIT extrapolated present trends into the future, using a computer model of the world ecosystem. This easy-to-read paperback summarizes the results. The book has provoked fierce controversy. *Science* called it "a computer view of doomsday." People just don't want to believe it, which is part of the problem.

E.J. MISHAN, *The Costs of Economic Growth*, Harmondsworth (England), Penguin Books, 1969. In the tradition of Thorstein Veblen, Mishan is both an economist and a witty critic of society. He is one of the pioneer advocates of a zero-growth economy. A hardcover edition is available in the U.S. from Praeger, New York.

W.W. MURDOCH, ed., *Environment*, 2nd Edition, Sunderland, Massachusetts, Sinauer Associates, 1975. Amateurs spin a lot of wooly talk about ecology. This is a sourcebook by a score of professionals.

C.P. SNOW, *Public Affairs*, New York, Scribners, 1971. Lord Snow, novelist and scientist, was praised for his essay, *The Two Cultures and the Scientific Revolution*. This book is a collection of essays on the interface between science and politics, with emphasis on the future of man.

L.S. STAVRIANOS, *The Promise of the Coming Dark Age*, San Francisco, Freeman, 1976. A distinguished historian compares the immediate future to the Dark Age that followed the fall of the Roman Empire. He argues that it will be a dynamic period of change leading to a new golden age.

glossary

ABSOLUTE TEMPERATURE SCALE. A temperature scale in which the degree is the same size as in the centrigrade scale, and zero is the state of no molecular motion. Absolute zero is $-273°$ on the centrigrade scale.

ACCLIMATION (a cli ma' shon). A change in an organism which improves its ability to tolerate a changed "climate" (environmental situation).

ACTIN [Gr. *aktis*: a ray]. One of the two major proteins of muscle; it makes up the thin filaments.

ACTION POTENTIAL. An impulse in a nerve which is propagated nondecrementally, taking the form of a wave of depolarization or reverse polarization imposed on a polarized cell surface.

ACTIVATION ENERGY. The energy barrier which blocks the tendency for a set of chemical substances to react. A reaction is speeded up if this energy barrier is surmounted by adding heat energy or is lowered by finding a different reaction pathway with the aid of a catalyst. Designated by the symbol, E_a.

ACTIVE SITE. The region on the surface of an enzyme where the substrate binds, and where catalysis occurs. Active sites have been found which are shallow depressions, grooves (for binding chains), and deep cavities (for trapping side chains or the end of a polypeptide chain).

ACTIVE TRANSPORT. The transport of a substance across the plasma membrane against a concentration gradient, that is, from a region of low concentration to a region of high concentration. Active transport requires the expenditure of energy and is a "saturable" process (see carrier-facilitated diffusion and free diffusion).

ADAPTATION (a dap tay' shun). In evolutionary biology, a particular structure, physiological process, or behavior that makes an organism more fit to survive and reproduce. Also, the evolutionary process that leads to the formation of such a trait.

ADAPTIVE RADIATION. The division of a single species into many species specialized to diverse ways of life.

ADENOSINE TRIPHOSPHATE. (see ATP)

AFFERENTS [L. *ad*: to + *ferre*: to bear]. Neurons which carry impulses toward the central nervous system.

ALGA (al' gah). Any one of a wide diversity of plants belonging to the phyla Pyrrophyta, Chrysophyta, Phaeophyta, Rhodophyta, and Chlorophyta. Most live in the water, where they are the dominant plants; most are unicellular, but a minority are multicellular (the "seaweeds" and similar plants).

ALLELE. A particular form of a gene, distinguishable from other forms or alleles of the same gene.

ALLOSTERY (al' lo ster y) [synthesized from Gr. *allo*: different + *stereos*: structure]. Regulation of the activity of an enzyme by the binding, at a site other than the catalytic active site, of an effector molecule which does not have the same structure as any of the enzyme's substrates. If the catalyzed reaction is an early step in the ultimate synthesis of the effector molecule, then the effector molecule regulates its own synthesis in a form of feedback control.

ALTERNATION OF GENERATIONS. The succession of haploid and diploid phases in a sexually reproducing organism. In animals, the haploid phase consists only of the gametes. In plants, however, the haploid phase may be the more prominent phase (as in fungi and mosses) or may be as prominent as the diploid phase (see the cycle of *Ulva* in text). In higher plants, the diploid phase is the more prominent.

ALVEOLUS (al ve' o lus) [L. *alveus*: a cavity]. A small bag-like cavity, especially of the kind which constitute the blind sacs of the lung.

AMINO ACID (a mee' no). An organic compound of the general formula: $H_2N-CHR-COOH$, where R can be one of twenty or more different side chains. An amino acid is so named because it has both a basic amine group, $-NH_2$, and an acidic carboxyl group, $-COOH$.

AMPHIBIAN (am fib' ee-an). A member of the vertebrate class Amphibia, such as a frog, toad, or salamander.

AMOEBA (a mee' bah) [Gr. *amoibe*: change]. Any one of a large number of different kinds of unicellular animals belonging to the phylum Sarcodina, characterized among other features by its ability to change shape frequently through the protrusion and extension of soft extensions of cytoplasm called pseudopodia.

AMOEBOID (a mee' boid) [Gr. *amoibe*: change]. Referring to any cell that behaves more or less like an amoeba, constantly changing its shape by the protrusion and retraction of soft extensions of cytoplasm called pseudopodia.

ANAEROBIC (an air row' bic) [Gr. *an*: not + *aer*: air + *bios*: life]. Occurring without the use of oxygen; as "anaerobic fermentation".

ANAPHASE (anne' a phase) [Gr. *ana*: indicating upward progress]. The stage in cell division at which the separation of sister chromosomes (or, in the first meiotic division, of paired homologues) occurs. Anaphase lasts from the moment of first separation to the time at which the moving chromosomes converge at the poles of the division.

ANGIOSPERM (an' jee o spurm) [Gr. *angion*: vessel + *sperma*: seed]. One of the "higher plants"; literally one whose seed is carried in a "vessel," which is the fruit.

ANION (an' ey on) [Gr. *ana*: back + *ienai*: to go; from anode, the pole in a battery at which electrons flow back into the external circuit]. A negatively charged ion.

ANNELID (ann' el id). A member of the phylum Annelida; one of the segmented worms, such as an earthworm or a leech.

ANTHER (an' ther) [Gr. *anthos*: flower]. A pollen-bearing portion of the stamen of a flower.

ANTHERIDIUM (an' ther id' i um) [Gr. *antheros*: blooming]. The multicellular structure that produces the male gamete in plants other than seed plants.

ANTICODON. A "triplet" of three nucleotides in trans-

fer RNA that is able to pair with a complementary triplet in messenger RNA (a codon), thus aligning the transfer RNA on the proper place on the messenger.

APEX (a' peks). The tip or highest point of a structure, as the apex of a growing stem or root.

APICAL (a' pi kul). Pertaining to the apex or tip, as the apical meristem, which is the actively growing tissue at the tip of a stem or root.

AQUATIC. Pertaining to fresh water as opposed to salt water. (Contrast with marine and terrestrial.)

ARCHEGONIUM (ar' ke goe' nee um) [Gr. *archegonos*: first of a kind]. The multicellular structure that produces eggs in bryophytes and some tracheophytes.

ARCHENTERON (ark en' ter on) [Gr. *archos*: beginning + *enteron*: bowel]. The earliest primordial animal digestive tract, it first appears during gastrulation.

AREA-SPECIES CURVE. The graphical representation of the relation between the area of an island (or any other circumscribed part of the earth's surface) and the number of species that inhabit it.

ARTHROPOD (arth' row pod). A member of the phylum Arthropoda, such as a crustacean, a spider, a centipede, or an insect.

ARTIFACT [L. *ars, artis*: art + *facere*: to make]. Something made by human effort or intervention. In biology, something that was not present in the living cell, but was unintentionally produced by the experimental procedure.

ASCUS (ass' cuss) [Gr. *askos*: bladder]. In higher fungi belonging to the group ascomycetes, the club-shaped sporangium within which spores are formed by meiosis (usually of a single zygote nucleus).

ASEXUAL REPRODUCTION. A form of reproduction, such as budding or simple fission, that does not involve the fusion of gametes.

ASSORTMENT (genetic). The random separation during meiosis of nonhomologous chromosomes and of genes carried on nonhomologous chromosomes. For example, if genes A and B are borne on nonhomologous chromosomes, meiosis of diploid cells of genotype *AaBb* will produce haploid cells of the following types in equal numbers: *AB, Ab, aB, ab*.

ATOMIC NUMBER. The number of protons in the nucleus of an atom, also equal to the number of electrons around the neutral atom. Important in determining chemical properties of the atom.

ATOMIC WEIGHT. The weight of one atom of a particular substance. Approximately equal to the total number of protons and neutrons in its nucleus.

ATP (adenosine triphosphate). A compound containing adenine, ribose, and three phosphates. It is the "common currency" of energy for most cellular processes.

AURICLE (or' i cal) [L. *auris*: an ear]. The thin-walled chamber, also called "atrium", of the heart which contains blood and feeds it to the ventricle for pumping; also, the outer ear.

AUTONOMIC NERVOUS SYSTEM. The system (which in

vertebrates is made up of sympathetic and parasympathetic subsystems) which controls such involuntary "housekeeping" functions as that of the guts and glands.

AUTORADIOGRAPHY. The detection of a radioactive compound or organelle in a cell putting it in contact with a photographic emulsion and allowing the compound to "take its own picture." The emulsion is developed and the location of the radioactivity in the cell is seen by the presence of silver grains in the emulsion.

AUTOSOME. (see sex chromosome)

AUTOTROPHIC (au' tow trow' fik) [Gr. *autos*: self + *trophe*: food]. Capable of living exclusively on inorganic materials, water, and some energy source such as sunlight; for example, most plants and blue-green algae and many bacteria. (Contrast with heterotrophic.)

AUXIN (awk' sin) [Gr. *auxein*: increase]. In plants, a substance that regulates growth by affecting cell elongation or other alterations.

AXON [Gr. *axon*: an axle]. Process of the neuron which can carry action potentials; is often the longest and least branched process of the cell body; and usually carries impulses away from the cell body of the neuron.

BACTERIOPHAGE (bak teer' e o faj') [Gr. *bakterion*: little rod + *phagein*: to eat]. One of a group of viruses that infect bacteria and ultimately cause their disintegration.

BACTERIUM (plural: bacteria) [Gr. *baktron*: little rod]. A procaryote — that is, a cell with a genome consisting of a simple DNA molecule not contained in a nuclear membrane — that has a cell wall and that does not carry out photosynthesis at all, or, if it does photosynthesize, that does not produce molecular oxygen as a product of photosynthesis.

BASIDIUM (bass id' ee yum) [Gr. *basis*: base, L. diminutive]. In higher fungi belonging to the group basidiomycetes, the characteristic sporangium in which four spores are formed by meiosis, and then born briefly externally before being shed.

BATESIAN MIMICRY. Mimicry by a relatively harmless kind of organism of a more dangerous one, by which the mimic enjoys protection from predators that mistake it for the dangerous model. (Contrast with Mullerian mimicry.)

BEHAVIORAL BIOLOGY. The scientific study of all aspects of behavior, including neurophysiology, ethology, and sociobiology.

BILATERAL SYMMETRY (bye lat' e ral sym' me tree). The condition in which only the right and left sides, divided exactly down the back, are mirror images of each other. (Contrast with radial symmetry.)

BIOGEOCHEMICAL CYCLE. The route of passage of particular materials, such as water or carbon, from the physical environment into organisms and back out again, in an unending alternation.

BIOGEOGRAPHY. The scientific study of the geographic distribution of organisms. Ecological biogeography is concerned with the habitats in which organisms live, historical biogeography with the complete geographic ranges of organisms and the historical circumstances that determine the ranges.

BIOLOGICAL SPECIES CONCEPT. The conception of the basic unit of classification as a population or series of populations of organisms capable of freely interbreeding with each other and reproductively isolated from other species.

BIOLUMINESCENCE. The production of light by biochemical processes in an organism.

BIOMASS (bye' oh mass). The total weight of all the living organisms, or some designated group of living organisms, found in a given area.

BIOME (bye' ome). A major portion of the living environment of a particular region, such as a fir forest or grassland, characterized by its distinctive vegetation and maintained by local conditions of the climate.

BIOME TYPE. One of a broad category of biomes, such as all of the grasslands of the world taken together.

BIOTA (bye oh' tah). All of the organisms, including fauna, flora, and microorganisms, found in a given area (see fauna and flora).

BIOTIC (bye ah' tik). Pertaining to any aspect of life, especially to characteristics of entire populations or ecosystems.

BLASTOCOEL (blass' toe seal) [Gr. *blastos*: sprout + *koilos*: hollow]. The hollow central cavity of a blastula.

BLASTODISC (blass' toe disk) [Gr. *blastos*: sprout + disc]. A disc of cells forming on the surface of a large yolk mass, comparable to a blastula, but occurring in forms in which the massive yolk restricts cleavage to only one side of the egg.

BLASTULA (blass' chu luh) [Gr. *blastos*: sprout]. An early stage in animal embryology, usually a hollow sphere of cells surrounding a central cavity.

BUD PRIMORDIUM [L. *primordium*: the beginning]. In plants, a small mass of potentially meristematic tissue found in the angle between the leaf stalk and the shoot apex. Will give rise to a lateral branch under appropriate conditions.

BUDDING. Asexual reproduction in which a more or less complete new organism simply grows from the body of the parent organism and eventually detaches itself.

BUFFERING (buf' er ing). A process by which a system resists changes, particularly in pH, in which case added acid or base is partially converted to a less active form.

CALLUS [L. *calleo*: to be thick skinned]. In plants, wound tissue, or relatively undifferentiated proliferating cell mass, frequently maintained in tissue culture.

CALORIE. The amount of heat required to raise the temperature of one gram of water by one degree Celsius (1°C). In many physiological measurements, including those used in nutrition studies, the "calorie" refers to the kilocalorie, which is the amount of heat required to raise one kilogram of water by 1°C.

CALYX (kay' licks) [Gr. *kalyx*: cup]. All of the sepals of a flower collectively.

cAMP (CYCLIC AMP). A compound formed from ATP which mediates the effects of numerous hormones in animals. It is also needed for the transcription of catabolite-repressible cistrons or operons in bacteria.

CAPILLARIES (cap' il eryz) [L. *capillaris*: hair]. Very small tubes, especially the smallest blood-carrying vessels of animals between the termination of the arteries and the beginnings of the veins.

CARBOHYDRATES (kar bo hi' drate) [E. carbon + hydrate]. Organic compounds with the general formula: $(CH_2O)_n$. The most common examples are sugars, starches, and cellulose.

CARBOXYLIC ACID (car box sill' ik). An organic acid containing the carboxyl group: $-COOH$, which dissociates to the carboxylate ion: $-COO^-$.

CARDIAC (kar' dee ak) [Gr. *kardia*: heart]. Pertaining to the heart and its functions.

CAROTENOID (ca rot' e noid) [L. *carota*: carrot]. A class of photosynthetic accessory pigments, including xanthophylls and carotenes.

CARPEL (kar' pel) [Gr. *karpos*: fruit]. The organ of the flower that contains one or more ovules.

CARRIER-FACILITATED DIFFUSION (passive transport). Transport of a substance across the plasma membrane by carrier molecules, but without a "pump". This process is "saturable", but cannot cause the net transport of a substance from a region of low concentration to a region of high concentration (see free diffusion and active transport).

CARRYING CAPACITY. In ecology, the largest number of organisms of a particular species that can be maintained indefinitely in a given part of the environment.

CARYOKINESIS (carry oh kin ee' sis) [Gr. *karuon*: kernel + *kinein*: to move]. The division of the nucleus of a dividing cell. As opposed to cytokinesis.

CASTE. In social insects, any set of individuals of a particular anatomical type, or age group, or both, that performs specialized labor in the colony.

CATALYST (kat' a list) [from Gr. *cata-*, implying the breaking down of a compound]. A chemical substance which accelerates a reaction, without itself being consumed in the overall course of the reaction. Enzymes are biological catalysts.

CATION (kat' ey on) [Gr. *cata*: out + *ienai*: to go; from cathode, the pole in a battery from which electrons flow into the solution]. A positively charged ion. (Contrast with anion.)

CELL WALL. A relatively rigid structure that encloses cells of plants, fungi, and most bacteria. The cell wall gives these cells their shape and limits their expansion in hypotonic media.

CELLULOSE (sell' you lowss) [E. cell + ose, suffix indicating a sugar or other carbohydrate]. A straight-chain polymer of glucose molecules, used by plants as a structural supporting material.

CENTRAL NERVOUS SYSTEM. That part of the nervous system which is condensed and centrally located, e.g. the brain and spinal cord of vertebrates; the chain of cerebral, thoracic and abdominal ganglia of arthropods.

CENTRIFUGE [L. fleeing from center]. A device in which a biological preparation can be spun around a central axis at high speed, creating a centrifugal force that mimics a very strong gravitational force. Suspensions will tend to settle out under these conditions, the larger, denser particles sedimenting more rapidly.

CENTRIOLE (sent' ree ole) [L. *centrum*: center]. A cylindrical organelle of animal cells (and some plant cells) containing microtubules in a characteristic arrangement consisting of nine triplets. A centriole is at each pole of the spindle apparatus in dividing cells, but the exact role of the centriole in the elaboration of the spindle is not known (see kinetosome).

CENTROMERE (sent' row mere) [L. *centrum*: center + Gr. *meros*: part]. The point on a chromosome to which spindle fibers attach at cell division. Generally seen as a constriction. Not necessarily central in its location on the chromosome.

CEPHALOPOD (sef' a low pod). A member of the mollusk class Cephalopod, such as a squid or an octopus.

CHIASMA (kie as' muh) (plural: chiasmata) (kie as' muh tuh) [Gr. cross]. An "x"-shaped connection between paired homologous chromosomes at meiosis. A chiasma is the visible manifestation of crossing-over between homologous chromosomes.

CHITIN (kye' tin) [Gr. *chiton*: tunic]. The characteristic flexible organic component of the exoskeleton of insects and other arthropods, consisting of a complex nitrogen-containing polysaccharide.

CHLOROPLAST [Gr. *chloros*: green + *plast*: a combining form meaning "a particle"]. An organelle bounded by a double membrane and containing the enzymes of photosynthesis. Chloroplasts occur only in eucaryotes.

CHROMATID (kro' ma tid'). Each of a pair of new sister chromosomes from the time at which the molecular duplication occurs until the time at which the centromeres separate at the anaphase of cell division.

CHROMATIN [Gr. *chroma*: color]. The nucleic-acid protein complex found in eucaryotic chromosomes.

CHROMOSOME (krome' oh zome) [Gr. *chroma*: color + *soma*: body]. Complex structure found in the nucleus of a eucaryotic cell, composed of nucleic acids and proteins (primarily basic histones), and bearing part of the genetic information (genes) of the cell.

CHROMOSOME ABERRATION. Any larger change in the structure of the chromosome, including duplication of chromosomes (aneuploidy) or of complete sets of chromosomes (polyploidy), usually gross enough to be detected with the light microscope.

CILIATE (sil' ee ate). A member of the protistan phylum Ciliophora, unicellular organisms that propel themselves by cilia.

CILIUM (sil' ee um) (plural: cilia) [L. *cilium*: eyelash].

Hairlike organelles used for locomotion by many unicellular organisms and for moving water and mucus by many multicellular organisms.

CIRCADIAN RHYTHM (sir kade' ee an) [L. *circa*: approximately + Gr. *dies*: day]. A rhythm in behavior, growth, or some other activity that recurs about every 24 hours. (The prefix *circa*- refers to the lack of precision in the timing.)

CISTRON. The genetic unit of function, often considered loosely equivalent to a gene. Generally each cistron contains the genetic information for a single polypeptide chain.

CLASS. In taxonomy, the category below the phylum and above the order; a group of related, similar orders.

CODON. A "triplet" of three nucleotides in messenger RNA that directs the placement of a particular amino acid into a polypeptide chain.

COELOM (see' lome) [Gr. *koiloma*: cavity]. The body cavity of higher metazoan animals, which is lined with epithelium of mesodermal origin.

COFACTOR (ko' fak tor). An organic molecule bound to an enzyme and necessary for its catalytic activity. Flavin molecules derived from riboflavin (vitamin B_2) are cofactors for many enzymes.

COITUS (koe' i tus) [L. *coitus*: a coming together]. The act of sexual intercourse.

COLEOPTILE (koe' lee op' tile) [Gr. *koleos*: sheath + *ptilon*: feather]. A pointed sheath covering the shoot of grass seedlings; part of the cotyledon.

COLLAGEN [Gr. *kolla*: glue]. A fibrous protein found extensively in bone and connective tissue.

COMMENSALISM. The form of symbiosis in which one species benefits from the assocation, while the other is neither harmed nor benefited.

COMMUNITY. Any ecologically integrated group of species of microorganisms, plants, and animals inhabiting a given area.

COMPANION CELL. Specialized cell found adjacent to each sieve tube element in flowering plants. Probably provides the energy for driving materials through the phloem.

CONIFER (kon' e fer) [Gr. *konos*: cone + *phero*: carry]. One of the cone-bearing plants, mostly trees, such as pines and firs.

CONJUGATION (con' jew gay' shun) [L. *conjugare*: yoke together]. The close approximation of two cells during which they exchange genetic material, as in *Paramecium* and other ciliated protozoans.

CONSTITUTIVE ENZYME. An enzyme that is present in approximately constant amounts in a system, whether substrates for it are present or absent. Thus it is part of the "constitution" of the cell (see inducible enzyme).

CONSUMMATORY ACT. One of a set of very specific, stereotyped patterns of motor responses. The behavior of most animal species consist largely or wholly of such consummatory acts and is loosely referred to as "instinctive."

CONTINENTAL DRIFT. The gradual drifting apart of the world's continents that has occurred over a period of hundreds of millions of years.

CONTRACTILE VACUOLE. An organelle, often found in protozoa, which pumps excess water out of cells and keeps them from being "flooded" in hypotonic media.

COROLLA (koh rol' lah) [L. *corolla*, diminutive of *corona*: wreath, crown]. All of the petals of a flower collectively.

CORPUS [L. *corpus*: body]. The portion of the shoot apical meristem which gives rise to the bulk of the stem tissues, e.g. cortex, vascular tissue, pith.

CORTEX [L. *cortex*: bark or rind]. In plants, the tissue between the epidermis and the vascular tissue of a stem or root.

COTYLEDON (kot' ee lee' don) [Gr. *kotyledon*: a hollow space]. The leaf of the seed which unfolds as the seedling sprouts.

COVALENT BOND. A chemical bond which arises from the sharing of electrons between atoms. Usually a strong bond.

CRISTAE. Small shelf-like projections of the inner membrane of the mitochondrion. The cristae are the site of oxidative phosphorylation.

CROSS-POLLINATION. The fertilization of one plant by pollen from another plant. (Contrast with self-pollination.)

CROSSING-OVER. The mechanism by which linked markers undergo recombination. In general, the term refers to the reciprocal exchange of corresponding segments between two homologous chromatids. However, the reciprocality of crossing-over is problematical in procaryotes and viruses and, even in eucaryotes, very closely linked markers often recombine by a nonreciprocal mechanism.

CRUSTACEAN (crus tay' see-an). A member of the arthropod class Crustacea, such as a crab, shrimp, or sowbug.

CRYPTIC APPEARANCE. The resemblance of an animal to some part of its environment, which helps it to escape detection by predators.

CUTIN (cue' tin) [L. *cutis*: skin]. A mixture of long straight chain hydrocarbons and waxes secreted by the plant epidermis, providing an impermeable coating on aerial plant parts.

CYCLIC AMP. (see cAMP)

CYTOCHROMES (cy' to chromes) [Gr. *kyto*: cell + *chroma*: color or pigment]. Iron-containing red proteins, components of the electron-transfer machinery in photosynthesis and respiration.

CYTOKINESIS (site' oh kin ee' sis) [Gr. *kytos*: vessel + *kinein*: to move]. The division of the cytoplasm of a dividing cell. (Contrast with caryokinesis.)

CYTOKININ (site' oh kye' nin) [Gr. *kytos*: vessel + *kinesis*: movement]. A member of a class of plant growth hormones playing a role in cell division, retarding of senescence, bud expansion, and so on.

CYTOPLASM. The contents of a cell, excluding the nucleus.

DARWINISM. The theory of evolution by natural selection, as propounded by Charles Darwin.

DECIDUOUS (de sid' you us) [L. *decidere:* fall off]. Referring to a plant that sheds its leaves at certain seasons. (Contrast with evergreen.)

DECOMPOSER. An organism such as a bacterium, fungus, or carrion beetle, that consumes dead organic matter, recycling it through the ecosystem.

DELETION (genetic). A mutation resulting from the loss of a continuous segment of a gene or chromosome. Such mutations never revert to wild-type and act in genetic crosses as if they occupied several clearly distinct sites marked by other mutations. (See point mutation).

DENDRITE [Gr. *dendron:* a tree]. Process of the neuron which often cannot carry action potentials; is commonly much branched and relatively short compared with the axon; and commonly carries impulses to the cell body of the neuron.

DENSITY GRADIENT. A solution, usually contained in a centrifuge tube, that varies continuously from a higher density at the bottom to a lower density at the top.

DEOXYRIBONUCLEIC ACID. (see DNA)

DETERMINATION. Process whereby a group of embryonic cells, a single cell, or part of a cell becomes fixed into a predictable pathway of development.

DICOT (short for dicotyledon) [Gr. *dis:* two + *kotyledon:* a cup-shaped hollow]. Any member of the angiospem class Dicotyledonae, plants in which the embryo produces two leaves prior to germination. Leaves of most dicots have major veins arranged in a branched or reticulate pattern.

DIFFERENTIATION. Process whereby originally similar cells follow different developmental pathways. The actual expression of determination.

DIOECIOUS (die eesh' us) [Gr. two houses]. Organisms in which the two sexes are "housed" in two different individuals, so that eggs and sperms are not produced in the same individuals. Examples: humans, fruit flies, oak trees, date palms (see monoecious).

DIPLOID (dip' loid) [Gr. *diploos:* double]. Having a chromosome complement consisting of two copies (homologues) of each chromosome. A diploid individual (or cell) usually arises as a result of the fusion of two gametes, each with just one copy of each chromosome. Thus, the two homologues in each chromosome pair in a diploid cell are of separate origin, one derived from the mother and one from the father.

DISPLACEMENT ACTIVITY. The performance of a behavioral act, usually in conditions of frustration or indecision, which is not directly relevant to the situation at hand.

DNA (deoxyribonucleic acid). The fundamental hereditary material of all living organisms. Except for bacteria and bluegreen algae, stored in the cell nucleus. A nucleic acid polymer using deoxyribose rather than ribose.

DOMINANCE. In genetic terminology, the ability of one allelic form of a gene to determine the phenotype of a heterozygous individual, in which the homologous chromosome carries a different allele. For example, if A and a are two allelic forms of a gene, A is said to be dominant to a if AA diploids and Aa diploids are phenotypically identical and are distinguishable from aa diploids. The a allele is said to be recessive.

DOMINANCE HIERARCHY. The set of relationships within a group of animals, usually established and maintained by aggression, in which one individual has precedence in eating, mating, and so on; a second individual has precedence over all but the highest-ranking individual, and so on down the hierarchy.

DRIVE. A loose term used to describe the tendency of an animal to seek out an object (for example, a mate, a particle of food, or a nest site) and to perform the appropriate response toward it.

DUPLICATION (genetic). A mutation resulting from the introduction into the genome of an extra copy of a segment of a gene or chromosome.

ECDYSONE (eck die' sone) [Gr. *ek:* out of + *dyo:* to clothe]. In insects, a hormone inducing molting.

ECHINODERM (e kine' oh durm). A member of the phylum Echinodermata, such as a starfish or sea urchin.

ECOLOGY (ee kol' oh jee) [Gr. *oikos:* house + *logos:* discourse, study]. The scientific study of the interaction of organisms with their environment, including both the physical environment and the other organisms that live in it. In recent years, ecology is also often used in a loose manner to indicate the environment itself.

ECOSYSTEM (eek' o sis tum). The organisms of a particular habitat, such as a pond or forest, together with the physical environment in which they live.

ECTODERM [Gr. *ektos:* outside + *derma:* skin]. The outermost of the three embryonic tissue layers first delineated during gastrulation. Gives rise to the skin, sense organs, nervous system, etc.

EFFECTOR. Any organ or cell that moves the organism through the environment or else alters the environment to the organism's advantage. Examples include muscle, bone, and a wide variety of exocrine glands.

EFFERENTS [L. *ex:* out + *ferre:* to bear]. Neurons which carry impulses from the central nervous system.

ELECTRON (e lek' tron) [L. *electrum:* amber (static electricity), from Gr. *elektor:* bright sun (color of amber)]. One of the three most fundamental particles of matter, with mass approximately 1.850 a.m.u. and charge − 1.

EMBRYO SAC. In angiosperms, the female gametophyte. Found within the ovule, it consists of eight or fewer cells.

EMIGRATION. The deliberate and usually oriented departure of an organism from the habitat in which it has been living.

ENDOCRINE GLAND (en' do krin) [Gr. *endon*: inside + *krinein*: to separate]. Any gland, such as the adrenal or pituitary gland of vertebrates, that secretes certain substances, especially hormones, into the body through the blood or lymph.

ENDOCYTOSIS. A collective word for two very similar processes, pinocytosis and phagocytosis, involving the uptake of liquids or solids, respectively.

ENDODERM [Gr. *endon*: within + *derma*: skin]. The innermost of the three embryonic tissue layers first delineated during gastrulation. Gives rise to the digestive and respiratory tracts and structures associated with them.

ENDODERMIS [Gr. *endon*: within + *derma*: skin]. In plants, a specialized cell layer marking the inside of the cortex in roots and some stems. Frequently a barrier to free diffusion of solutes.

ENDOPLASMIC RETICULUM, ROUGH [Gr. *endon*: within + L. *plasma*: form; L. *reticulum*: little net]. A system of folded membranes with ribosomes attached, found in the cytoplasm of eucaryotic cells. The ribosomes attached to the rough endoplasmic reticulum engage primarily in the synthesis of secretory proteins.

ENDOPLASMIC RETICULUM, SMOOTH. A system of folded membranes found in the cytoplasm of eucaryotic cells. Smooth and endoplasmic reticulum is devoid of associated ribosomes. It plays a role in the packaging of secretory products and in many other cellular processes.

ENDOSPERM (en' do sperm) [Gr. *endo-*: within + *sperma*: seed]. A specialized seed tissue found only in angiosperms, which contains stored food for the developing embryo.

ENDOTHERMIC REACTION (en' do therm ik) [Gr. *endo*: in + *therme*: heat]. A chemical reaction which absorbs heat.

ENTROPY (en' tro py) [Gr. *en*: in + *tropein*: to change; "concerned with chemical changes"]. A measure of the degree of disorder in any system. A perfectly ordered system has zero entropy; increasing disorder is measured by positive entropy. Spontaneous reactions in a closed system are always accompanied by an increase in disorder and entropy. Designated by the symbol S.

ENZYME (en' zime) [Gr. *en*: in + *zyme*: yeast]. A globular protein, on the surface of which are found chemical groups arranged so as to make the enzyme a catalyst for a chemical reaction.

EPIDERMIS [Gr. *epi*: upon + *derma*: skin]. In plants and animals, the outermost cell layers. Only one cell-layer thick in plants.

EPINEPHRINE (ep i nef' rin) [Gr. *epi*: upon + *nephros*: a kidney]. The hormone secreted principally by the medulla of the adrenal gland, which is situated on the kidney; epinephrine is also called adrenaline. It is secreted as a result of fear or excitement and produces effects on the circulatory system, on glucose mobilization, etc.

EPOCH (ep' ok). A lesser division of geological time; for example, the Miocene epoch, which lasted about 12 million years.

EQUILIBRIUM, CHEMICAL. A state in which forward and reverse reactions are proceeding at counterbalancing rates, so there is no observable change in the concentrations of reactants and products.

ERA (ehr' ah). One of the major divisions of geological time; for example, the Mesozoic era, which lasted about 170 million years.

ESTRUS (es' truss) [L. *oestrus*: frenzy]. The period of heat, or maximum sexual receptivity, in the female. Ordinarily the estrus is also the time of the release of eggs in the female.

ESTROUS CYCLE. The repeated series of changes in reproductive physiology and behavior in the female, culminating in estrus.

ETHOLOGY (ee thol' o jee) [Gr. *ethos*: habit, custom + *logos*: discourse]. The study of whole patterns of animal behavior in natural environments, stressing the analysis of adaptation and evolution of the patterns.

EUCARYOTES (you car' ry ots) [Gr. *eu*: true + *karyon*: kernel or nucleus]. Organisms whose cells contain their genetic material inside a nucleus. Includes all life above the level of bacteria and bluegreen algae.

EVERGREEN. A plant that retains its leaves through all seasons. (Contrast with deciduous.)

EVOLUTION. Any gradual change. Organic evolution, often referred to as evolution for short, is any genetic change in organisms from generation to generation.

EVOLUTIONARY BIOLOGY. The collective branches of biology that deal with the evolutionary process and the characteristics of populations of organisms, as well as ecology, behavior, and systematics.

EXOBIOLOGY. The study of the probable existence of life on other planets.

EXOCRINE GLAND (ek' so krin) [Gr. *exo*: outside + *krinein*: to separate]. Any gland, such as the salivary gland, that secretes to the outside of the body or into the alimentary tract.

EXOSKELETON (eks' oh skel' e ton) [Gr. *exo*: outside + skeleton]. A hard covering on the outside of the body; the exoskeleton of insects and other arthropods has essentially the same functions as the bony internal skeleton of vertebrates.

EXOTHERMIC REACTION (eks' oh therm' ik) [Gr. *exo*: outside + *therme*: heat]. A chemical reaction that gives off heat.

EXPONENTIAL GROWTH. Growth, especially in the number of organisms of a population, which is a simple function of the size of the growing entity: the larger the entity, the faster it grows.

EXTRINSIC ISOLATING MECHANISM. Any barrier in the environment, such as a river bed, desert, or ocean, that isolates one population from another.

FAMILY. In taxonomy, the category below the order and above the genus; a group of related, similar genera.

FAUNA (faw' nah). All of the animals found in a given area. (Contrast with flora.)

FAUNAL DOMINANCE. The ability of species originating from one part of the world to spread their range to other parts of the world and displace other species already living there.

FEEDBACK CONTROL. Control of a particular step of a multi-step process, induced by the presence or absence of a product of one of the later steps. A thermostat regulating the flow of heating oil to a furnace in the home is a feedback control device.

FERMENTATION (fer men ta' shun) [L. *fermentum*: yeast]. The degradation of a molecule such as glucose to smaller molecules with the extraction of energy, without the use of oxygen (that is, anaerobically).

FLAGELLATE (flaj' e late). A member of the phylum Mastigophora, which are unicellular organisms that propel themselves by flagella.

FLAGELLUM (fla jell' um) (plural: flagella) [L. whip]. Long, whiplike appendage that propels procaryotic or eucaryotic cells.

FLORA (flore' ah). All of the plants found in a given area. (Contrast with fauna.)

FLORIGEN. A plant hormone involved in the conversion of a vegetative shoot apex to a flower.

FLOWER. The total reproductive structure of an angiosperm; its basic parts include the calyx, corolla, stamens, and carpels.

FOOD CHAIN. A portion of a food web, most commonly a simple sequence of prey species and the predators that consume them.

FOOD WEB. The complete set of food links between species in a community; a diagram indicating which are eaters and which are eaten.

FOSSIL. Any recognizable structure originating from an organism, or any impression from such a structure, that has been preserved from prehistoric times.

FRAME-SHIFT MUTATION. A mutation resulting from the addition (or deletion) of a single base-pair in the DNA sequence of a gene. As a result of this, mRNA transcribed from such a gene is translated normally until the ribosome reaches the point at which the mutation has occurred. From that point on, codons are read out of proper register and the amino acid sequence bears no resemblance to the normal sequence.

FREE DIFFUSION. Diffusion directly across the plasma membrane without the involvement of carrier molecules. Free diffusion is not "saturable", and cannot cause the net transport from a region of low concentration to a region of higher concentration (see carrier-facilitated diffusion and active transport).

FREE ENERGY. That energy which is available for doing useful work, after allowance has been made for the increase or decrease of disorder. Designated by the symbol G.

GAMETE (gam' eet) [Gr. *gamete*: wife or *gametes*: husband]. The mature sexual reproductive cell: the egg or the sperm.

GAMETOCYTE (ga meet' oh site) [Gr. *gamete*: wife, *gametes*: husband + *kytos*: cell]. The cell that gives rise to sex cells, either the eggs or the sperm. (See oocyte and spermatocyte.)

GAMETOGENESIS (ga meet' oh jen' e sis) [Gr. *gamete*: wife, *gametes*: husband + *gignomai*: be born]. The specialized series of cellular divisions that leads to the production of sex cells (gametes). (See oogenesis and spermatogenesis.)

GAMETOPHYTE (ga meet' o fyte). In plants with an alternation of generations, the haploid phase that produces the gametes. (Contrast with sporophyte.)

GASTROPOD (gas' troh pod). A member of the molluscan class Gastropoda, such as a snail or conch.

GASTROVASCULAR CAVITY. Serving for both digestion (gastro-) and circulation (vascular); in particular the central cavity of the body of jelly fish and other Coelenterata.

GASTRULA (gas' true luh) [Gr. *gaster*: stomach]. An embryo forming the characteristic three cell layers, ectoderm, mesoderm, and endoderm, which will give rise to all of the major tissue systems of the adult animal.

GENE. A unit of heredity, usually one that carries the information for a single polypeptide.

GENE FLOW. The exchange of genes between different species (an extreme case referred to as hybridization) or between different populations of the same species. A principle driving force of evolution.

GENE POOL. All of the genes in a population.

GENETIC DRIFT. Evolution (change in gene proportions) by chance processes alone.

GENOPHORE (jean' oh for) [Gr. *gen*: to produce + *phorain*: to bear]. The DNA molecule that carries the genes of a virus or procaryote.

GENOTYPE (jean' oh type) [Gr. *gen*: to produce + *tupos*: impression]. The genetic makeup of an individual, either with respect to a single trait or a larger set of traits. (Contrast with phenotype.)

GENUS (jean' us) (plural: genera) (jen' e rah) [Gr. *genos*: stock, kind]. A group of related species.

GEOTROPISM [Gr. *ge*: the earth + *trope*: a turning]. A directed plant growth response to gravity.

GIBBERELLIN. One of a class of plant growth regulators playing a role in stem elongation, seed germination, flowering of certain plants, etc. Named for the fungus *Gibberella* from which gibberellins were first isolated.

GIZZARD (giz' erd) [L. *gigeria*: cooked chicken guts]. A very muscular part of the stomach of birds that grinds up food, sometimes with the aid of fragments of stone.

GLIA (Gr. for glue). Cells found only in the nervous system, which do not conduct nerve impulses.

GLUCOSE (glue' kose) [Gr. *gleukos*: sweet wine mash for fermentation]. The most common carbohydrate sugar, with the formula $C_6H_{12}O_6$.

GLYCOGEN (gly' ko gen). A branched-chain polymer of glucose molecules, similar to starch but of lower molecular weight. Used in the bloodstream of animals as a fast-access energy storage compound.

GLYCOGENOLYSIS (gly' ko ge nol' i sis) [Gr. *lusis*: loosening]. The enzymatic breakdown of glycogen to make glucose.

GLYCOLYSIS (gly kol' li sis) [from glucose + lysis]. The enzymatic breakdown of glucose during the process of anaerobic fermentation. One of the oldest of energy-yielding mechanisms in living organisms.

GOLGI COMPLEX (goal' gee). A system of concentrically folded membranes found in the cytoplasm of eucaryotic cells. Plays a role in the production and release of secretory materials such as the digestive enzymes manufactured in the acinar cells of the pancreas. First described by Camillo Golgi (1844–1926).

GONAD (goh' nad) [Gr. *gone*: seed, that which produces seed]. An organ that produces sex cells: either an ovary (female gonad) or testis (male gonad).

GROWTH. Irreversible increase in volume (probably the most accurate definition, but at worst a dangerous oversimplification).

GYMNOSPERM (jim' no sperm) [Gr. *gymnos*: naked + *sperma*: seed]. A plant, such as a pine or other conifer, whose seeds do not develop within an ovary (hence, the seeds are "naked").

HABITUATION (ha bich' oo ay shun). The simplest form of learning, in which an animal presented with a stimulus without reward or punishment eventually ceases to respond.

HAPLOID (hap' loid) [Gr. *haploeides*: single]. Having a chromosome complement consisting of just one copy of each chromosome. This is the normal "ploidy" of gametes or of asexual spores produced by meiosis or of organisms (such as the gametophyte generation of plants) that grow from such spores without fertilization.

HARDY-WEINBERG LAW. The law that the basic processes of Mendelian heredity (meiosis and recombination) do not alter either the frequencies of genes or their diploid combinations. The Law also states how the percentages of diploid combinations can be predicted, on the basis of a simple formula, from a knowledge of the percentage of genes.

HEMOGLOBIN (hee' moh glow' bin) [Gr. *haima*: blood + L. *globus*: globe]. The colored protein of the blood which transports oxygen.

HERMAPHRODITISM (her maf' row dite' ism) [Gr. *hermaphroditos*: a person with both male and female traits]. The coexistence of both female and male sex organs in the same organism.

HERTZ (Abbreviated as Hz). Cycles per second.

HETEROCARYON (het' er oh care' ee ahn) [Gr. *heteros*: different + *karyon*: kernel]. A cell or organism carrying a mixture of genetically distinguishable nuclei. A heterocaryon is usually the result of the fusion of two cells that occurs without fusion of their nuclei.

HETEROSPORY (het' er os' poh ree) [Gr. *heteros*: different + *spora*: seed]. The condition in plants of producing two kinds of spores, the small microspores and large megaspores. (Contrast with homospory.)

HETEROTROPHIC (het' er o trow' fik) [Gr. *heteros*: different + *trophe*: food]. Requiring organic materials for food. (Contrast with autotrophic.)

HETEROZYGOUS (he' ter oh zie gus) [Gr. *heteros*: different + *zugotos*: joined]. Of a diploid organism having different alleles of a given gene on the pair of homologues carrying that gene (see homozygous).

HOMEOSTASIS (home' ee o sta' cis) [Gr. *homoios*: like, sameness + *stasis*: position]. The maintenance of a steady state, such as a constant temperature or a stable social structure, by means of physiological or behavioral feedback responses.

HOMEOTHERM (home' ee o therm) [Gr. *homos*: same + *therme*: heat]. An animal which maintains a constant body temperature by virtue of its own heating and cooling mechanisms.

HOMOLOGUE (home' o log') [Gr. *homos*: same + *logos*: word]. One of a pair, or larger set, of chromosomes having the same overall genetic composition and sequence. In diploid organisms, each chromosome inherited from one parent is matched by an identical (except for mutational changes) chromosome — its homologue — from the other parent.

HOMOLOGY (ho mol' o jee) [Gr. *homologi(a)*: agreement]. A similarity between two structures which is due to inheritance from a common ancestor. The structures are said to be homologous. (Contrast with analogy.)

HOMOZYGOUS (home' o zie' gus) [Gr. *homos*: same + *zugotos*: joined]. Of a diploid organism having identical alleles of a given gene on both homologous chromosomes. An organism may be a "homozygote" with respect to one gene and, at the same time, a "heterozygote" with respect to another (see heterozygous).

HORMONE (hore' mone) [Gr. *hormon*: excite, stimulate]. A substance, secreted by an endocrine gland into the blood or lymph, that affects the physiological activity of organs in other parts of the body.

HYDROGEN BOND. A chemical bond which arises from the attraction between the slight positive charge on a hydrogen atom, and a slight negative charge on a nearby fluorine, oxygen, or nitrogen atom. Weak bonds, abundant in proteins and other biological macromolecules.

HYDROLYZE (hi' dro lize) [Gr. *hydro*: water + *lysis*: cleavage]. To break a chemical bond, such as a polypeptide bond, with the insertion of the components of water, -H and -OH, at cleaved ends of a chain. The digestion of proteins is an hydrolysis.

HYDROPHOBIC [Gr. *hydro*: water + *phobia*: fear; "disliking water"]. Molecules and amino acid side chains which are mainly hydrocarbons — compounds of C and H with no charged groups or polar groups — have a lower energy when they are clustered together than when they are distributed through an aqueous solution. Because of their attraction for one another and their reluctance to mix with water they are called "hydrophobic." Oil is a hydrophobic substance; phenylalanine is a hydrophobic side chain in a protein.

HYPERTONIC [Gr. higher tension]. A medium with a higher concentration of osmotically active particles than is present in cells. In hypertonic media, water will flow out of cells, causing shrinkage or plasmolysis (see hypotonic and isotonic).

HYPHA (high' fuh) [Gr. *hyphe*: web]. In the fungi, any single filament. May be multinucleate (phycomycetes) or multicellular (basidiomycetes).

HYPOTONIC [Gr. lower tension]. A medium that has a lower concentration of osmotically active particles than does a cell. In hypotonic media, water will tend to flow into cells (see hypertonic and isotonic).

IMBIBITION [Lat. *imbibo*: to drink]. The binding of a solvent to another molecule. Rubber will imbibe gasoline, dry starch and protein will imbibe water. Binding forces can be very large, leading to extensive swelling of the imbibing system.

IMPRINTING. A rigid form of learning, in which an animal comes to make a particular response only to one other animal or object.

INDUCER. A small molecule which, when added to a system or growth medium, causes a large increase in the level of some enzyme. Generally it acts by binding to repressor and changing its shape so that the repressor does not bind to the operator.

INDUCIBLE ENZYME. An enzyme that is present in much larger amounts when a particular compound (an inducer) has been added to the system (see constitutive enzyme).

INHIBITOR. A substance which binds to the surface of an enzyme and interferes with its action on its substrate molecules.

INSTINCT. Behavior that is relatively highly stereotyped, more complex than the simplest reflexes, and usually directed at particular objects in the environment. Learning may or may not be involved in the development of the behavior; the important point is that the behavior develops toward a narrow, predictable end-product.

INSULIN (in' si lin) [L. *insula*: an island]. An animal hormone synthesized in islet cells of the pancreas, which promotes the conversion of glucose to the storage material, glycogen.

INTEGUMENT [L. *integumentum*: covering]. In gymnosperms and angiosperms, a layer of tissue around the ovule which will become seed coat. Gymnosperm ovules have one integument, angiosperm ovules two.

INTERPHASE. The period between successive cell divisions during which the chromosomes are diffuse and the nuclear membrane is intact. It is during this period that the cell is most active in transcribing and translating genetic information. However, because the nucleus of an interphase cell is less interesting to look at than the nucleus of a dividing cell, this phase has been misnamed "the resting stage", a term that is still sometimes used.

INTRINSIC ISOLATING MECHANISM. A genetically based trait that helps prevent members of one species from breeding with members of other species.

INVERSION (GENETIC). A rare mutational event that leads to the reversal of the order of genes within a segment of a chromosome, as if that segment had been removed from the chromosome, turned through 180°, and then reattached.

INVERTEBRATE. Any animal that is not vertebrate, that is, whose nerve cord is not enclosed in a backbone of bony segments.

IN VITRO [L. in glass]. In a test tube, rather than in a living organism (see *in vivo*).

IN VIVO [L. in the living state]. Contrast with *in vitro* [L. literally, in glass]. Many processes that occur *in vivo* can be reproduced *in vitro* with the right selection of cellular components.

ION (ey' on) [Gr. *ion*: wanderer]. An atom with electrons added or removed, giving it a negative or positive charge. (See anion and cation.)

IONIC BOND. A chemical bond which arises from the electrostatic attraction between positively and negatively charged groups or ions. Usually a strong bond.

ISOTONIC [Gr. same tension]. A medium that has the same concentration of osmotically active particles as a cell, so that there is no net inflow or outflow of water (see hypotonic and hypertonic).

ISOTOPE (ey' so tope) [Gr. *isos*: equal + *topos*: place — i.e. same place in the chemical periodic table]. Two isotopes of the same chemical element have the same number of protons in their nuclei, but differ only in the number of neutrons.

JUVENILE HORMONE. In insects, a hormone maintaining larval growth.

KERATIN (ker' a tin) [Gr. *keras*: horn]. A protein which contains sulfur, and is part of such hard tissues as horn, nail, and the outermost cells of the skin.

KEY. In taxonomy, a device for quickly identifying a specimen down to the species or a higher category to which it belongs. The key consists of a series of choices made according to whether the specimen possesses one trait as opposed to another.

KINESIS (ki nee' sis) [Gr. movement]. Orientation behavior in which the organism does not move in a particular direction with reference to a stimulus but instead simply moves at an increasing or decreasing rate until it ends up farther from the object or closer to it. (Contrast with taxis).

KINETOSOME (ki nee' to zome) [Gr. *kinein*: to move + *soma*: body]. A cylindrical, microtubular organelle found at the base cilia and flagella of eucaryotic cells. In structure, the kinetosome is twin to the centriole.

LAMARCKISM (Lah mark′ iz-um). The theory of evolution by acquired characteristics, as propounded by Jean Baptiste de Lamarck.

LARVA [L. *larva*: ghost, early stage]. An immature stage of any invertebrate animal that differs strongly in appearance from the adult.

LEAF AXIL. The upper angle between a leaf and the stem, site of axillary or lateral buds which under appropriate circumstances become activated to form lateral branches.

LEAF PRIMORDIUM [L. *primordium*: the beginning]. A small mound on the flank of a shoot apical meristem; will give rise to a leaf.

LIFE CYCLE. The entire span of the life of an organism from the moment of fertilization (or asexual generation) to the time it reproduces in turn.

LINKAGE (genetic). Association between markers on the same chromosome or genophore such that they do not show random assortment. Linked markers recombine with one another at frequencies less than 0.50; the closer the markers on the chromosome, the lower the frequency of recombination.

LIPIDS (lip′ id) [Gr. *lipos*: fat]. The substances in a cell which are easily extracted by organic solvents. Fats, oils, waxes, steroids, and other large organic molecules, including those which with proteins make up the cell membranes.

LOGISTIC GROWTH. Growth, especially in the size of an organism or in the number of organisms that comprises a population, which slows steadily as the entity approaches its maximum size. (Contrast with exponential growth.)

LUMEN (loo′ men) [L. *lumen*: light]. The cavity inside any tubular part of an organ, such as a piece of gut or a kidney tubule.

LYMPHATICS (lim fat′ ics) [L. *lympha*: water]. The system in animals of (1) spaces between the capillaries and the body cells which contain filtered blood (lymph) and (2) the vessels which collect the lymph and return it to the blood system.

LYSOSOME (lie′ so zome) [Gr. *lusis*: a loosing + *soma*: body]. A membrane-bounded inclusion found in eucaryotic cells. Lysosomes contain a mixture of enzymes that can digest most of the macromolecules found in the rest of the cell.

MACRO- (mack′ roh) [Gr. *makros*: large, long]. A prefix commonly used to denote something large; contrast with micro-.

MACROEVOLUTION (mack′ ro ev o loo′ shun). A large amount of evolutionary change, involving many elementary changes in gene proportions or chromosome structure. (Contrast with microevolution.)

MAMMAL (mam′ el) [L. *mamma*: breast, teat]. Any animal of the class Mammalia, characterized by the production of milk by the female mammary glands and having a body covered by hair.

MARINE. Pertaining to the sea. (Contrast with aquatic and terrestrial.)

MARSUPIAL (mar soo′ pee al). A mammal belonging to the mammal subclass Metatheria, most of which, like the opossums and kangaroos, have a pouch (the marsupium) that contains the milk glands and serves as a receptacle for the young.

MATERNAL INHERITANCE (or cytoplasmic inheritance). Inheritance in which the phenotype of the offspring depends on factors, such as mitochondria or chloroplasts, that are inherited from the female parent through the cytoplasm of the female gamete.

MATURATION. The automatic development of a pattern of behavior, which becomes increasingly complex or precise as the animal matures. Unlike learning, the development does not require experience to occur.

MEGA- [Gr. *megas*: large, great]. A prefix often used to denote something large. (Contrast with micro-.)

MEGAPHYLL [Gr. *megas*: large + *phyllon*: leaf]. In vascular plants, a leaf thought to be derived from a flattened branch system, its vascular tissue forming a leaf gap, where it attaches to the stem vascular tissue.

MEGASPORANGIUM. The special structure (sporangium) that produces the megaspores.

MEGASPORE [Gr. *megas*: great, very large + *spora*: seed]. In plants, a haploid spore that produces a female gametophyte. In many cases the megaspore is larger than the male-producing microspore.

MEIOSIS (my oh′ sis) [Gr. diminution]. Cell division of a diploid cell to produce four haploid daughter cells. The process consists of two successive cell divisions with only one cycle of chromosome replication.

MENSES (men′ sees) [L. *menses*, plural of month]. The "period" of a sexually mature girl or woman; the days in each reproduction cycle during which blood and mucosal tissue flow from the uterus.

MERISTEM [Gr. *meristos*: divided]. A plant tissue made up of actively dividing cells.

MESENCHYME (mes′ en kyme) [Gr. *mesos*: middle + *enchyma*: infusion]. Embryonic or unspecialized cells derived from the mesoderm.

MESODERM [Gr. *mesos*: middle + *derma*: skin]. The middle of the three embryonic tissue layers first delineated during gastrulation. Gives rise to skeleton, circulatory system, muscles, excretory system, and most of the reproductive system.

MESSENGER RNA (mRNA). A disposable copy of one of the strands of DNA, it carries information for the synthesis of one or more proteins.

METAMORPHOSIS (met′ a mor′ fo sis) [Gr. *meta*: beyond + *morphe*: form, shape]. A strong change occurring from one developmental stage to another, as for example from a tadpole to a frog or an insect larva to the adult.

METAPHASE (met′ a phase) [Gr. *meta*: between]. The stage in cell division at which the centromeres of the highly condensed chromosomes are all lying on a plane (the metaphase plane or plate) perpendicular to a line connecting the division poles.

METAZOAN [Gr. *meta*: between, among + *zoon*: animal]. Pertaining to any or all of the multicellular animals with the exception of the sponges, which are often distinguished as the "parazoan" animals.

MICRO- (mike' roh) [Gr. *mikros*: small]. A prefix often used to denote something small; often contrasted with macro- or mega-.

MICROBIOLOGY (mike' roh bye ol' o jee) [Gr. *mikros*: small + *bios*: life + *logos*: discourse]. The scientific study of microscopic organisms, particularly bacteria and other monerans, unicellular algae, and protistans.

MICROEVOLUTION (mike' roh ev o loo' shun). A small amount of evolutionary change, consisting of minor alterations in gene proportions, chromosome structure, or chromosome numbers. (A larger degree of change would be referred to as macroevolution or simply as evolution.)

MICROFILAMENT. Minute fibrous structures found in the cytoplasm of eucaryotic cells. They appear to play a role in the motion of cells, acting like small muscles.

MICROORGANISM (mike' roh or' ga niz'-um). Any microscropic organism, such as a bacterium or one-celled alga.

MICROPHYLL [Gr. *mikros*: small + *phyllon*: leaf]. A leaf possibly derived from a scale-like outgrowth of the stem. Its vascular tissue does not form a leaf gap where it attaches to the stem vascular tissue.

MICROPYLE (mike' roh pile) [Gr. *mikros*: small + *pyle*: gate]. Opening in the integument(s) of a seed plant ovule through which pollen grows to reach female gametophyte within.

MICROSPORANGIUM. The special structure (sporangium) that produces the microspores.

MICROSPORE (mike' roh spore) [Gr. *mikros*: small + *spora*: seed]. In plants, a haploid spore that produces a male gametophyte. In many cases the microspore is smaller than the female-producing megaspore.

MICROTUBULE: Minute tubular structures found in centrioles, spindle apparatus, cilia, flagella, and other places in the cytoplasm of eucaryotic cells. These tubules play a role in the motion and maintenance of shape of eucaryotic cells.

MIDDLE LAMELLA. A thin layer of pectic compounds (polymers of galacturonic acid) cementing together two adjacent plant cell walls.

MIMICRY (mim' ik ree). The resemblance of one kind of organism to another, which serves the function of discouraging potential enemies. See Batesian mimicry and Mullerian mimicry.

MITOCHONDRION (plural: mitochondria) [Gr. *mitos*: thread + *chondros*: cartilage, or grain]. An organelle that occurs in eucaryotic cells and contains the enzymes of the Krebs tricarboxylic acid cycle, the respiratory chain, and oxidative phosphorylation. A mitochondrion is bounded by a double membrane.

MITOSIS (my toe' sis) [Gr. *mitos*: thread]. Cell division in eucaryotes leading to the formation of two daughter cells each with a chromosome complement identical to that of the original cell.

MOLE. A quantity of a compound whose weight in grams is numerically equal to its molecular weight expressed in atomic mass units. Avogadro's number of molecules, 6.023×10^{23} molecules.

MOLLUSK (mol' lusk). A member of the phylum Mollusca, such as a snail, clam, or octopus.

MONERAN (moh ner' an). A member of the Kingdom Monera, which consists of bacteria and bluegreen algae.

MONOCOT (short for monocotyledon) [Gr. *monos*: one + *kotyledon*: a cup-shaped hollow]. Any member of the angiosperm class Monocotyledonae, plants in which the embryo produces but a single leaf prior to germination. Leaves of most monocots have their major veins arranged parallel to each other.

MONOECIOUS (mo neesh' us) [Gr. one house]. Organisms in which both sexes are "housed" in a single individual, which produces both eggs and sperms. Examples: corn, peas, earthworms, hydras (see dioecious).

MORPHOGENESIS (more' fo jen' e sis) [Gr. *morphe*: form + *genesis*: origin]. The development of form. Morphogenesis is the overall consequence of determination, differentiation, and growth.

MORPHOLOGY (more fol' o jee) [Gr. *morphe*: form + *logos*: discourse]. The scientific study of organic form, including both its development and function.

mRNA. (see messenger RNA)

MULLERIAN MIMICRY. The resemblance of two or more unpleasant or dangerous kinds of organisms to each other; the mimicry gives each added protection because its potential enemies find it easier to recognize as a member of a group. (Contrast with Batesian mimicry.)

MULTICELLULAR [L. *multus*: much + *cella*: chamber]. Consisting of more than one cell, as for example a multicellular organism. (Contrast with unicellular.)

MUSCLE CELL. The basic cellular unit of muscle, a highly specialized cell packed with contractile protein which functions to move other organs in the body.

MUTAGEN (mute' ah jen) [L. *mutare*: change + Gr. *gignomai*: causing]. An agent, especially a chemical, that increases the mutation rate.

MUTATION. In the broad sense, any discontinuous change in the genetic constitution of an organism. In the narrow sense, the word usually refers to a "point mutation": a change along a very narrow portion of the nucleic acid sequence.

MUTATION PRESSURE. Evolution (change in gene proportions) by different mutation rates alone.

MUTUALISM. The type of symbiosis, such as that exhibited by fungi and algae in forming lichens, in which both species benefit from the association.

MYCELIUM (my seal' ee yum) [Gr. *mykes*: mushroom]. In the fungi, a mass of hyphae.

MYOGLOBIN (my' o globe' in) [Gr. *mys*: muscle + L. *globus*: sphere]. The special hemoglobin found in muscle, possessing less molecular weight and carrying less oxygen than blood hemoglobin.

MYOSIN [Gr. *mys*: muscle]. One of the two major proteins of muscle, it makes up the thick filaments.

NATURAL SELECTION. The differential contribution of offspring to the next generation by various genetic types belonging to the same population. The mechanism of evolution proposed by Charles Darwin.

NEMATODE (nem' ah toad). A member of the phylum Nematoda; a roundworm.

NEPHRIDIUM (nef rid' ee um) [Gr. *nephros*: kidney]. An organ which is involved in excretion, and often in water control, involving a tube which opens to the exterior at one end.

NEPHRON (nef' ron) [Gr. *nephros*: kidney]. The basic component of the kidney, which is made up of numerous nephrons. Its form varies in detail, but always has at one end a device for receiving a filtrate of the blood, and then a tubule which absorbs selected parts of the filtrate back into the bloodstream.

NEURAL CREST [Gr. *neuron*: nerve]. Cells which become detached from the forming neural tube and wander great distances to fulfil a wide range of developmental roles.

NEURULA (nure' yu la) [Gr. *neuron*: nerve]. Embryonic stage during formation of the dorsal nerve cord by two ectodermal ridges.

NEUROPHYSIOLOGY (nure' oh fiz' ee ol oh jee). The scientific study of the nervous system.

NEUTRON (new' tron) [E. neutral]. One of the three most fundamental particles of matter, with mass approximately 1 a.m.u. and no electrical charge.

NUCLEAR MEMBRANE. The envelope, consisting of two layers of unit membrane, that encloses the nucleus of eucaryotic cells.

NUCLEIC ACID (new klay' ik) [E. nucleus of cell]. A long-chain alternating polymer of ribose sugar rings and phosphate groups, with organic bases (adenine, thymine or uracil, guanine, cytosine) as side chains. DNA and RNA are nucleic acids.

NUCLEOID (new' clee oyd). The region, not bounded by any membrane, that harbors the DNA genophore of a procaryotic cell.

NUCLEOLUS (new klee' oh lus) [L. little kernel]. A small, clear body, generally spherical, found within the nucleus of eucaryotic cells. It is site of synthesis of ribosomal RNA.

NUCLEOTIDE. The basic chemical unit in a nucleic acid. A nucleotide in RNA consists of one of four nitrogenous bases linked to the sugar, ribose, which in turn is linked to phosphate by ester bonds. In DNA, deoxyribose is present instead of ribose, and the base thymine is present instead of uracil.

NUCLEUS (new' klee us) [from L. diminutive of *nux*: kernel or nut]. (1) The dense central portion of an atom, made up of protons and neutrons, with a positive charge. Surrounded by a cloud of negatively charged electrons. (2) Centrally located chamber of eucaryotic cells that is bounded by a double membrane and contains the chromosomes. The information center of the cell.

ONTOGENY (on toj' e nee) [Gr. *onto*: from "to be" + *gignomai*: be born, produce]. The develop-

ment of a single organism in the course of its life history. (Contrast with phylogeny.)

OOCYTE (oh' oh site) [Gr. *oon*: egg + *kytos*: cell]. The cell that gives rise to eggs.

OOGENESIS (oh' oh jen e sis) [Gr. *oon*: egg + *gignomai*: be born]. Female gametogenesis, leading to production of the egg.

OPERATOR. The region of an operon that acts as the binding site for the repressor.

OPERON. A genetic unit of transcription, typically consisting of several structural genes or cistrons that are transcribed to give a single messenger RNA molecule; the operon contains at least two transcriptional control regions: the promoter and the operator.

ORDER. In taxonomy, the category below the class and above the family; a group of related, similar families.

ORGAN. A formed body, such as the heart, liver, brain, root, or leaf, composed of different tissues integrated to perform a distinct function for the body as a whole.

ORGANELLES (or' gan els') [L. little organ]. Organized structures that are found in or on cells. Examples: ribosomes, nuclei, mitochondria, chloroplasts, cilia, contractile vacuoles.

ORGANIC. Pertaining to any aspect of living matter: to its evolution, structure, or chemistry. The term is also applied to any chemical compound that contains carbon.

ORGANISM. Any living creature.

OSMOSIS. The tendency of water to move from a region in which it is more concentrated (e.g., pure water or a hypotonic solution) to a region in which it is less concentrated, that is, in which the concentration of dissolved molecules or ions is higher.

OVARY (oh' var ee). Any female organ, in plants or animals, that produces an egg.

OVULE (oh' vule) [L. *ovulum*: little egg. A small egg.]. In plants, an organ that contains a gametophyte and, within the gametophyte, an egg; when it matures, an ovule becomes a seed.

OVUM (oh' vum) [L. *ovum*: egg]. The egg, the female sex cell.

OXIDATION (ox i day' shun). Relative loss of electrons in a chemical reaction; either outright removal to form an ion, or the sharing of electrons with substances having a greater affinity for them such as oxygen. Most oxidations, including biological ones, are associated with the liberation of energy.

PALEOBIOLOGY. The attempted reconstruction, from fossil evidence, from knowledge of the properties of chemical substances, and from theories about primitive Earth conditions, of the early stages in the evolution of life.

PALEOBOTANY (pale' ee oh bot' ah nee) [Gr. *palaios*: ancient, old]. The scientific study of fossil plants and all aspects of extinct plant life.

PALEONTOLOGY (pale' ee on tol' oh jee) [Gr. *palaios*:

ancient, old + *logos:* discourse]. The scientific study of fossils and all aspects of extinct life.

PALISADE PARENCHYMA. In leaves, one or several layers of tightly packed columnar photosynthetic cells, frequently found just below the upper epidermis.

PARASITISM. The form of symbiosis in which one species lives at the expense of the other, but not ordinarily to the point of killing its host (such destruction would then be designated predation).

PARAZOAN [Gr. *para-:* near, close by + *zoon:* animal]. Pertaining to the sponges (phylum Porifera), which are so different in basic structure from the rest of the multicellular animals (the metazoans) that they are often distinguished by this special name.

PARENCHYMA (pair en' kyma) [Gr. *para:* beside + *enchyma:* infusion]. A plant tissue composed of relatively unspecialized cells without secondary walls.

PARTHENOGENESIS (par then' oh jen' e sis) [Gr. *parthenos:* virgin + *gignomai:* be born]. The production of an organism from an unfertilized egg.

PASSIVE TRANSPORT. (see carrier-facilitated diffusion)

PATHOGEN (path' o jen) [Gr. *pathos:* suffering + *gignomai:* causing]. An organism that causes disease.

PENIS (pee' nis) The male organ used in coitus (sexual intercourse).

PEPTIDE BOND. The connecting group in a protein chain, -CO-NH-, formed by the removal of water during the linking of amino acids, -COOH to -NH$_2$. Also called an amide bond.

PERICYCLE [Gr. *peri:* around + *kyklos:* ring or circle]. In plant roots, tissue just within the endodermis, but outside of the root vascular tissue. Meristematic activity of pericycle cells produces lateral root primordia.

PERIOD. A division of time of intermediate duration, less than an era and greater than an epoch; for example, the Triassic period, which lasted for about 50 million years.

PERMEASE. A protein in membranes that specifically transports a compound or family of compounds across the membrane.

PETAL. In an angiosperm flower, a sterile modified leaf, nonphotosynthetic, frequently brightly colored, and often serving to attract pollinating insects.

PHAGE (faj). Short for bacteriophage.

PHAGOCYTOSIS [Gr. eating cell]. The uptake of a solid particle by forming a pocket of cell membrane around the particle and pinching off the pocket to form an intracellular particle bounded by membrane (see pinocytosis and endocytosis).

PHENOTYPE (fee' no type) [Gr. *phaenein:* to show + *tupos:* impression]. The observable properties of an individual as they have developed under the combined influences of the genetic constitution of the individual and the affects of environmental factors. (Contrast with genotype.)

PHEROMONE (fer' o mone) [Gr. *phero:* carry + *hormon:* excite, arouse]. A chemical substance used in communication between organisms of the same species.

PHLEOM (flo' im) [Gr. *phloos:* bark]. In the more advanced plants, the food-conducting tissue, which consists of sieve cells or sieve tubes, fibers, and other specialized cells.

PHOSPHOLIPID. Cellular materials that contain phosphorus and are soluble in organic solvents. An example is lecithin (phosphatidyl choline). Phospholipids are important constituents of cellular membranes (see lipid).

PHOTOPERIOD (foe' tow peer' ee od). The duration of a period of light, such as the length of time in a 24-hour cycle in which daylight is present.

PHOTOSYNTHESIS (foe tow sin' the sis) [literally, "synthesis with light"]. Metabolic processes by which visible light is trapped and the energy used to synthesize energy-rich compounds such as ATP or glucose.

PHOTOTROPISM [Gr. *photos:* light + *trope:* a turning]. A directed plant growth response to light.

PHYLOGENY (fy loj' e nee) [Gr. *phyle:* tribe, race + *gignomai:* be born, produce]. The evolutionary history of a particular group of organisms; also, the diagram of the "family tree" that shows which species gave rise to others. (Contrast with ontogeny.)

PHYLUM [Gr. *phylon:* tribe, stock]. In taxonomy, a high-level category just beneath the kingdom and above the class; a group of related, similar classes.

PHYSIOLOGY (fiz' ee ol' oh jee) [Gr. *physis:* natural form + *logos:* discourse, study]. The scientific study of the functions of living organisms and the individual organs, tissues, and cells of which they are composed.

PHYTOCHROME (fy' to krome) [Gr. *phyton:* plant + *chroma:* color]. A plant pigment regulating a large number of developmental and other phenomena in plants, can exist in two different forms, one of which is active and the other of which is not. Different wavelengths of light can drive it from one form to the other.

PHYTOPLANKTON (fy' to plangk' ton) [Gr. *phyton:* plant + *planktos:* wandering]. The autotrophic portion of the plankton, consisting mostly of algae.

PINOCYTOSIS [Gr. drinking cell]. The uptake of liquids by engulfing a sample of the external medium into pocket of plasma membrane followed by pinching off the pocket to form an intracellular vesicle (see phagocytosis and endocytosis).

PISTIL [L. *pistillum:* pestle]. The female structure of an angiosperm flower, within which the ovules are borne. May consist of a single carpel, or of several carpels fused into a single structure. Usually differentiated into ovary, style, and stigma.

PITH. In plants, relatively unspecialized tissue found within a cylinder of vascular tissue.

PLACENTA (pla sen' ta). The organ, found in most mammals, that provides for the nourishment of the fetus and elimination of the fetal waste products. It is formed by the union of membranes of the mother's uterine lining with membranes from the fetus.

PLACENTAL (pla sen' tal). Pertaining to mammals of the subclass Eutheria, a group which is characterized by the presence of a placenta in the female and which contains the great majority of living species of mammals.

PLANKTON (plangk' ton) [Gr. *planktos*: wandering]. The free-floating organisms of the sea and freshwater that for the most part move passively with the water currents. Consisting mostly of microorganisms and small plants and animals.

PLANULA (plan' yew la) [Diminutive of L. *planum*: something flat]. The free-swimming, ciliated larva of the coelenterates.

PLASMA MEMBRANE (plasmalemma, or cell membrane). A lipid-containing membrane that regulates the entry and exit of molecules and ions. Every cell has a plasma membrane.

PLASMODIUM [Gr. *plasma*: mold or form]. In the non-cellular slime molds, a multinucleate mass of protoplasm, surrounded by a membrane, characteristic of the vegetative feeding stage.

POINT MUTATION. A mutation that results from a small, localized alteration in the chemical structure of a gene. Such mutations can give rise to wild-type revertants as a result of reverse mutation. In genetic crosses, a point mutation behaves as if it resided at a single point on the genetic map (see deletion).

POLAR BODY. A nonfunctional nucleus produced by meiosis, accompanied by very little cytoplasm. The meiosis which produces the mammalian egg produces in addition three polar bodies.

POLLEN [L. *pollen*: fine powder, dust]. The fertilizing element of flowering plants, containing the male gametophyte and the gamete, at the stage in which it is shed.

POLLINATION. Process of transferring pollen from the anther to the receptive surface (stigma) of the ovary in plants.

POLYCISTRONIC MESSENGER. A messenger RNA molecule containing information for the synthesis of more than one polypeptide chain. Generally such messengers are transcribed from an operon.

POLYMORPHISM (pol' lee mor' fiz-um) [Gr. *polys*: many + *morphe*: form, shape]. In genetics, the coexistence in the same population of two distinct hereditary types based on different alleles. In social organisms such as colonial coelenterates and social insects, the coexistence of two or more functionally different castes in the same colony.

POLYPLOID (pol' lee ploid). A cell or an organism in which the number of complete sets of chromosomes is greater than two.

POLYRIBOSOME. A complex consisting of a threadlike molecule of messenger RNA and several (or many) ribosomes. The ribosomes move along the mRNA, synthesizing polypeptide chains as they proceed.

POLYSOME. (see polyribosome)

POPULATION. Any group of organisms, capable of

interbreeding for the most part, and coexisting at the same time and in the same place.

POSTMATING ISOLATING MECHANISM. Any intrinsic isolating mechanism, for example hybrid sterility, that operates after mating.

PREDATOR. An organism that kills and eats other organisms. Predation is usually thought of as involving the consumption of animals by animals, but in the broad usages of ecology it can also mean the eating of plants.

PREMATING ISOLATING MECHANISM. Any intrinsic isolating mechanism that operates prior to the completion of mating.

PRIMATE (pry' mate). A member of the order Primates, such as a lemur, monkey, ape, or human.

PRIMARY CELL WALL. The first formed plant cell wall, a fabric of cellulose microfibrils embedded in a gelatinous matrix of pectic compounds and hemicelluloses. The cellulose microfibrils may have a preferred orientation, but not a very rigid one.

PRIMITIVE STREAK. A line running axially along the blastodisc, the site of inward cell migration during formation of the three ply embryo.

PRIMORDIAL GERM CELLS [L. *primus*: first]. Cells determined early in embryonic growth to become gametes (or to produce gametes on their ultimate arrival at the gonads).

PROCAMBIUM [Gr. *pro*: before + L. *cambiare*: to exchange]. In plants, the embryonic tissue which will give rise to the first xylem and phloem.

PROCARYOTES (pro car' ry otes) [L. *pro*: before + Gr. *karyon*: kernel or nucleus]. Organisms whose genetic material is distributed throughout the cell, rather than being isolated in a nucleus. Includes bacteria and bluegreen algae. Considered an earlier stage in the evolution of life than the eucaryotes.

PROMOTER. The region of an operon that acts as the initial binding site for RNA polymerase.

PROPHAGE (pro' faj'). The noninfectious phage units that are linked with the chromosomes of the host bacteria and multiply with them but which do not cause the dissolution of the cell. Prophage can turn into the virulent stage to complete the virus life cycle.

PROPHASE (pro' phase) [Gr. *pro*-: before]. The first stage of cell division, during which chromosomes condense from diffuse, thread-like material to discrete, compact bodies.

PROPLASTID [L. *pro*: before + Gr. *plastos*: molded]. A plant cell organelle which under appropriate conditions will develop into a plastid, usually the photosynthetic chloroplast. If plants are kept in the dark, proplastids may become quite large and complex.

PROTEASE (pro' te ase). A proteolytic enzyme.

PROTEIN (pro' teen, less often pro' te in) [Gr. *protos*: first]. One of the most fundamental building substances of living organisms. A long-chain polymer of amino acids with twenty different common side chains. Occurs with its polymer chain extended, in fibrous proteins, or coiled into a compact macromolecule in enzymes and other globular proteins.

PROTEOLYTIC ENZYME (pro te o lit' ik) [from protein + Gr. *lysis*: cleavage]. An enzyme whose main catalytic function is the cleavage and digestion of a protein or polypeptide chain. The digestive enzymes trypsin, pepsin, and carboxypeptidase are all proteolytic enzymes.

PROTISTAN. Pertaining to the kingdom Protista, which embraces most of what used to be included in the old phylum Protozoa (and still is by some biologists), including the flagellates, amoebas, ciliates, and a few other eucaryotic, unicellular organisms.

PROTON (pro' ton) [Gr. *protos*: first]. One of the three most fundamental particles of matter, with mass approximately 1 a.m.u. and charge +1.

PROTONEMA (pro' tow nee' mah) [Gr. *protos*: first + *nema*: thread]. The hair-like growth form which constitutes an early stage in the development of the moss gametophyte.

PROTOPLASMIC STREAMING. The motion of the contents of a eucaryotic cell. This process of "mixing" allows transport of materials, over distances that are too great to allow mixing by diffusion alone.

PROTOPLAST. A cell which would normally have a cell wall, but from which the wall has been removed by enzymatic digestion or by special growth conditions.

PROTOZOA. A group of single-celled organisms classified by some biologists as a single phylum; includes the flagellates, amoebas, and ciliates. This textbook follows most modern classifications in elevating the protozoans to a distinct kingdom (Protista) and each of their major subgroups (for example, the amoebas) to the rank of phylum.

PSEUDOPLASMODIUM [Gr. *pseudes*: false + *plasma*: mold or form]. In the cellular slime molds such as *Dictyostelium*, an aggregation of single amoeboid cells. Occurs prior to formation of fruiting structure.

PSEUDOPOD (soo' do pod) [Gr. *pseudes*: false + *podos*: foot]. A temporary, soft extension of the cell body which is used in locomotion, attachment to surfaces, or engulfiing particles.

PUPA (pew' pa) [L. *pupa*: doll, puppet]. In higher insects (the Holometabola), the inactive developmental stage that intervenes between the larva and the adult.

QUIESCENT ZONE. In root or shoot apical meristems, a central group of cells which divide only infrequently, and have relatively low levels of protein and nucleic acid synthesis.

RADIAL SYMMETRY (ray' dee al sym' me tree). The condition in which two halves of a body are mirror images of each other regardless of the angle of the cut, providing the cut is made along the center line. Thus a cylinder cut lengthwise down its center displays this form of symmetry. (Contrast with bilateral symmetry.)

RECEPTACLE [L. *receptaculum*: reservoir]. In an angiosperm flower, the end of the stem to which all of the various flower parts are attached.

RECESSIVENESS. (see dominance)

RECOMBINANT. A chromosome carrying at two or more sites markers that were originally carried on two separate chromosomes. Any one of the offspring emerging from a genetic cross that can be shown to carry such a recombinant chromosome. As opposed to parental, a chromosome (or individual member of the progeny) that emerges from a cross with markers arranged exactly as they were at the start.

REDIRECTED ACTIVITY. The direction of some behavior, such as aggression, away from the primary target and toward another, less appropriate object.

REDUCTION (re duk' shun). Gain of electrons; the reverse of oxidation. Most reductions lead to the storage of chemical energy, which can be released later by an oxidation reaction. Energy-storage compounds such as sugars are highly reduced compounds.

REDUCTIONISM (re duk' shun ism). An overly simplified attitude which holds that the properties of a complex system can be explained completely from a knowledge of the properties only of its components. Usually ascribed by psychologists to biologists, by biologists to chemists, and by chemists to physicists.

REFLEX [L. *reflexus*: reflected]. An automatic action, involving only a few neurons (in vertebrates, often in the spinal cord) in which a motor response swiftly follows a sensory stimulus.

REGULATOR GENE. A gene that contains the information for making a regulatory macromolecule, often a repressor protein.

RELEASER. A sign stimulus used in communication. Often the term is used broadly to include any sign stimulus.

REPRESSIBLE ENZYME. One whose synthesis can be decreased or prevented by the presence of a particular compound.

REPRESSOR. A protein coded by the regulator gene. The repressor can bind to a specific operator and prevent transcription of the operon.

RESPIRATION (res pi ra' shun) [literally, "breathing"]. The oxidation of the end products of glycolysis to produce much more energy than the simple glycogenic process can liberate. The oxidant in the respiration of all higher organisms is oxygen gas. Some bacteria can use nitrate or sulfate ion instead of O_2.

RHIZOIDS (ry' zoyds) [Gr. *rhiza*: root]. Hair-like extensions of cells in mosses, liverworts, and a few vascular plants that serve the same function as roots and root hairs in higher plants. The term is also applied to branched root-like extensions of some fungi and algae.

RHIZOME (ry' zome) [Gr. *rhizoma*: mass of roots]. A special underground stem (as opposed to root) that runs horizontally beneath the ground.

RIBONUCLEIC ACID. (see RNA)

RIBOSE (rye' bose). A sugar of chemical formula:

$C_5H_{10}O_5$, one of the building blocks of nucleic acids.

RIBOSOME. A small organelle that is the site of protein synthesis.

RNA (ribonucleic acid). A nucleic acid polymer using ribose. Serves as the genetic storage material in some viruses. In other organisms, used in the copying and translation of genetic information into protein structure.

ROOT PRESSURE. Movement of water and ions up the stem, driven by osmotic entry of water into the roots. Rate and pressure insufficient to account for massive water movement up stems of trees, a movement which requires transpiration.

SAP. An aqueous solution of nutrients, minerals, and other substances that passes through the xylem.

SAPROPHYTE (sa′ pro fight) [Gr. sapros: putrid + phyton: plant]. An organism (normally a bacterium or fungus) which obtains its carbon and nitrogen directly from dead organic matter.

SEGREGATION (genetic). The separation of alleles, or of homologous chromosomes, from one another during meiosis so that each of the haploid daughter cells produced by meiosis contains one or the other member of the pair found in the diploid mother cell, but never both.

SELF-POLLINATION. The fertilization of a plant by its own pollen. (Contrast with cross-pollination.)

SEMEN (see′ men) [L. seed]. The thick, whitish liquid produced by the male reproductive organ, containing the sperm.

SEMISPECIES. Populations that are reproductively isolated from one another to an intermediate degree. They are neither fully isolated, as true species, nor freely interbreeding along their boundaries, which would make them subspecies belonging to the same species.

SEPAL (see′ pul). One of the outermost structures of the flower, usually protective in function and enclosing the rest of the flower in the bud stage.

SEX CHROMOSOME. In organisms with a chromosomal mechanism of sex determination, one of the chromosomes involved in sex determination. One sex chromosome, the X-chromosome, is present in two copies in one sex and only one copy in the other sex. The autosomes, as opposed to the sex chromosomes, are present in two copies in both sexes. In many organisms, there is a second sex chromosome, the Y-chromosome, that is found in only one sex — the sex having only one copy of the X.

SEX LINKAGE. The pattern of inheritance characteristics of genes located on the sex chromosomes of organisms having a chromosomal mechanism for sex determination. The sex that is "diploid" with respect to sex chromosomes can assume three genotypes: homozygous wild-type, homozygous mutant, or heterozygous carrier. The other sex, "haploid" for sex chromosomes, is either hemizygous wild-type or hemizygous mutant.

SEXUALITY. The ability, by any of a multitude of mechanisms, to bring together in one individual genes that were originally carried by two different individuals. The capacity for genetic recombination.

SIEVE CELL. Specialized cell found in the phloem of gymnosperms and spore-bearing vascular plants, through which organic matter, principally carbohydrate, moves from leaves to the rest of the plant. Lateral walls have local areas of perforations leading to adjacent sieve cells.

SIEVE TUBE. A column of specialized cells found in the phloem, specialized to conduct organic matter, principally carbohydrate from photosynthesizing leaves to other plant parts. Found principally in flowering plants.

SIEVE TUBE ELEMENT. A single cell of a sieve tube, containing some cytoplasm but few organelles, with highly specialized perforated end walls leading to elements above and below.

SIGN STIMULUS. The single stimulus, or one out of a very few stimuli, by which an animal distinguishes key objects, such as an enemy, or a male, or a place to nest, etc.

SOCIAL INSECT. One of the kinds of insect that form colonies with reproductive castes and worker castes; in particular, the termites, ants, social bees, and social wasps.

SOCIETY. A group of individuals belonging to the same species and organized in a cooperative manner; in the broadest sense, includes parents and their offspring.

SOCIOBIOLOGY. The scientific study of animal societies and communication.

SOMITE (so′ might) [Gr. soma: body]. One of the segments into which an embryo becomes divided longitudinally, leading to the eve tual segmentation of the animal as illustrated by the spinal column, ribs, and associated muscles.

SPAWNING. The direct release of sex cells into the water.

SPECIATION (spee′ shee ay′ shun). The processes of diversification of populations and the multiplication of species.

SPECIES (spee′ shees). The basic lower unit of classification, consisting of a population or series of populations of closely related and similar organisms. The more narrowly defined "biological species" consists of individuals that are capable of interbreeding freely with each other but not with members of other species.

SPECIES EQUILIBRIUM. A condition in which the number of species going extinct in a given area per unit time equals the rate at which new species are arriving; thus the number of species in the area remains constant.

SPERM [Gr. sperma: seed]. A male reproductive cell.

SPERMATOCYTE (spur mat′ oh site) [Gr. sperma: seed + kytos: cell]. The cell that gives rise to the sperm.

SPERMATOGENESIS (spur mat′ oh jen′ e sis) [Gr. sperma: seed + gignomai: be born]. Male gametogenesis, leading to the production of sperm.

SPERMATOZOON (spur′ ma toh zoe′ an). A sperm.

SPHINCTER (sfingk′ ter) [Gr. sphigkter: shut tight].

A ring of muscle which can close an orifice, for example at the anus.

SPINDLE APPARATUS. An array of microtubular fibers stretching from the metaphase plate to the poles of a dividing cell and playing a role in the movement of chromosomes at cell division. Named for its shape.

SPONGY PARENCHYMA. In leaves, a layer of loosely packed isodiametric photosynthetic cells with extensive intercellular spaces for gas diffusion. Frequently found between the palisade parenchyma and the lower epidermis.

SPONTANEOUS REACTION. A chemical reaction which will proceed on its own, without any outside influence.

SPORANGIUM (spor an' gee um) [Gr. spora: seed + angeion: vessel or reservoir]. In plants, any specialized structure within which one or more spores is formed.

SPORE [Gr. spora: seed]. Any asexual reproductive cell, capable of developing into an adult plant, without gametic fusion. Haploid spores develop into gametophytes, diploid spores into sporophytes. In procaryotes, a resistant cell capable of surviving unfavorable periods.

SPOROPHYTE (spo' roh fyte). In plants with an alternation of generations, the diploid phase that produces the spores. (Contrast with gametophyte.)

STAMEN (stay' men). A male (pollen-producing) unit of a flower, usually composed of an anther, which bears the pollen, and a filament, which is a stalk supporting the anther.

STARCH [O.E. stearc: stiff, from Gr. stereos: solid]. A branched-chain polymer of glucose molecules, used by plants as a means of storing energy.

STATOLITHS [Gr. statos: positioned + lithos: stone]. In plants, subcellular organelles (presumably starch grains) that sense the direction of gravity, helping to orient the plant.

STEADY STATE. An apparently unchanging condition which is due to balanced synthesis and degradation of all components in the system.

STIGMA [L. stigma: mark, brand]. The part of the pistil, at the apex of the style, which is receptive to pollen, and on which pollen germinates.

STOMA (plural: stomata) [Gr. stoma: mouth]. Small openings in the plant epidermis which permit gas exchange; bounded by a pair of guard cells whose osmotic status regulates the size of the opening.

STROBILUS (strobe' a lus) [Gr. strobilos: a cone]. The cone, or characteristic multiple fruit, of the pine and other gymnosperms. Also, a cone-shaped mass of sporophylls found in club mosses and ferns. Same as strobile.

STRUCTURAL GENE. A gene that encodes the primary structure of a protein (usually an enzyme).

STYLE [Gr. stylos: pillar or column]. In flowering plants, a column of tissue extending from the tip of the ovary, and bearing the stigma or receptive surface for pollen at its apex.

SUBSPECIES (sub' spee' shees). A subdivision of a species. Usually defined more narrowly as a geographical race: a population or series of populations occupying a discrete range and differing genetically from other geographical races of the same species.

SUBSTRATE (sub' strayte). The molecule or molecules on which an enzyme exerts catalytic action.

SUCCESSION. In ecology, the gradual and predictable series of changes in species composition of organisms, from the time of colonization of an empty space to the maturing of the final, stable, climax community.

SUSPENSOR. In plants, a cell or group of cells derived from the zygote, but not actually part of the embryo proper, which in some higher plants pushes the young embryo deeper into nutritive gametophyte tissue by its growth.

SYMBIOSIS (sim' bee oh' sis) [Gr. symbiosis: to live together]. The living together of two or more species of insects in a prolonged and intimate ecological relationship. (See parasitism, commensalism, mutualism.)

SYMMETRY. In biology, the property that two halves of an object are mirror images of each other. (See bilateral symmetry and radial symmetry.)

SYMPATRIC (sim pa' trick) [Gr. syn: together + patria: homeland]. Referring to populations whose geographic ranges overlap at least in part.

SYNAPSIS (sin ap' sis). The highly specific parallel alignment (pairing) of homologous chromosomes during the first division of meiosis.

TAXIS (tak' sis) [Gr. taxis: arrange, put in order]. The movement of an organism in a particular direction with reference to a stimulus. A taxis usually involves the employment of one sense and a movement directly toward or away from the stimulus, or else the maintenance of a constant angle to it. Thus a positive phototaxis is movement toward a light source, negative geotaxis is movement upward (away from gravity), and so on.

TAXONOMY (taks on' oh me) [Gr. tasso: arrange, classify]. The science of classification of organisms.

TELOPHASE (teel' oh phase) [Gr. telos: end]. The final phase of mitosis or meiosis during which chromosomes become diffuse, the nuclear membrane reforms, and nucleoli begin to appear in the daughter nuclei.

TEPAL. In an angiosperm flower, a sterile modified leaf. This term is used to refer to such flower parts when one is unable to distinguish between petals and sepals.

TERRESTRIAL (ter res' tree-al). Pertaining to the land. (Contrast with aquatic and marine.)

TERRITORY. A fixed area from which an animal or group of animals excludes other members of the same species by aggressive behavior or display.

TESTICLE (tes' ti kul) [L. testiculus: diminutive of testis]. Same as testis.

TESTIS (tes' tis) (plural: testes) [L. testis: witness]. The male gonad, that is, the organ that produces the male sex cells.

TETRAD (te' trad) [Gr. tetras: foursome]. Paired homologous chromosomes during the first prophase and metaphase of meiosis. Each chromosome at these stages will be a double structure, consisting of two chromatids joined at a not-yet-divided centromere.

TISSUE. A group of similar cells organized into a functional unit and usually integrated with other tissues to form part of an organ such as a heart or leaf.

TORNARIA (tor nare' e ah) [L. *tornus:* lathe]. The free-swimming ciliated larva of certain echinoderms and hemichordates; its existence indicates the evolutionary relationship of these two groups.

TRACHEA (tray' key a) [Gr. *trakhoia:* a small rough artery]. A tube which carries air to the bronchi of the lungs of vertebrates, or to the cells of arthropods.

TRACHEID (tray' key id). A distinctive conducting and supporting cell found in the xylem of nearly all vascular plants, characterized by tapering ends and walls that are pitted but not perforated.

TRANSCRIPTION. The synthesis of RNA, using one strand of DNA as the template.

TRANSDUCTION. Transfer of genes from one bacterium to another with a bacterial virus acting as the carrier of the genes.

TRANSFER RNA (tRNA). A category of relatively small RNA molecules (about 75 nucleotides). Each kind of transfer RNA is able to accept a particular activated amino acid from its specific activating enzyme, after which the amino acid is added to a growing polypeptide chain.

TRANSFORMATION. Mechanism for transfer of genetic information in bacteria in which pure DNA extracted from bacteria of one genotype is taken in through the cell surface of bacteria of a different genotype and incorporated into the genophore of the recipient cell. By extension, the term has come to be applied to phenomena in other organisms in which specific genetic alterations have been produced by treatment with purified DNA from genetically marked donors.

TRANSITION MUTATION. A base substitution mutation in which a purine (A or G) replaces a purine, and a pyrimidine (T or C) replaces a pyrimidine. (Contrast with transversion mutation.)

TRANSLATION. The synthesis of a protein (polypeptide). This occurs on ribosomes, using the information encoded in messenger RNA.

TRANSPIRATION [L. *spirare:* to breathe]. The evaporation of water from plant leaves and stem, driven by heat from the sun, and providing the motive force to raise water (plus ions) from the roots.

TRANSVERSION MUTATION. A base substitution mutation in which a purine (A or G) replaces a pyrimidine (T or C) and *vice versa.* (Contrast with transition mutation.)

TRICHOCYST (trike' o sist) [Gr. *trichos:* hair + *kystis:* cell]. A threadlike organelle ejected from the surface of ciliate protozoans, used both as a weapon and an anchoring device.

TRIPLET. (see codon)

tRNA. (see transfer RNA)

TROCHOPHORE (troke' o fore) [Gr. *trochos:* wheel + *phoreus:* bearer]. The free-swimming larva of some annelids and mollusks, distinguished by a wheel-like band of cilia around the middle, and indicating an evolutionary relationship between these two groups.

TROPHALLAXIS. The exchange of liquid substances from one member of an insect society to another.

TROPHIC LEVEL. The position of a species in a food chain, indicating which species it consumes and which consume it.

TURGOR PRESSURE. The pressure which a plant cell exerts against its constraining wall, a function of the osmotic concentration of the cell, water availability, and the relative elasticity and plasticity of the wall itself.

UNICELLULAR (yoon' e cel' yew ler) [L. *unus:* one + *cella:* chamber]. Consisting of a single cell; as for example a unicellular organism. (Contrast with multicellular.)

UTERUS (yoo' ter us) [L. *uterus:* womb]. The womb, a specialized portion of the female reproductive tract in certain mammals, which receives the fertilized egg and nurtures the embryo in its early development.

VACUOLE (vac' you ole) [Fr. small vacuum]. A liquid-filled cavity in a cell, enclosed within a unit membrane. Vacuoles play a wide variety of roles in cellular metabolism, some being digestive chambers, some storage chambers, some waste bins, some cellular "bladders," and so forth.

VASCULAR (vas' kew lar). Pertaining to organs and tissues, such as blood vessels in animals and phloem and xylem in plants, that conduct fluid.

VERTEBRATE. An animal whose nerve chord is enclosed in a backbone of bony segments, called vertebrae. The principal groups of vertebrate animals are the fishes, amphibians, reptiles, birds, and mammals.

VESSEL [L. *vasculum:* a small vessel]. In botany, a tube-shaped element of the xylem which consists of hollow cells (vessel elements) placed end to end and connected by perforations. Together with the tracheid, it conducts water and minerals in the plant.

VILLUS (vil' us) (plural: villi) [L. *villus:* shaggy hair]. A hair or a hair-like projection from a membrane, for example from many gut walls.

VIRION. The virus particle, the minimum unit capable of infecting a cell. The equivalent of the cell in true organisms.

VIRUS [L. poison, slimy liquid]. Any of a group of ultramicroscopic, infectious particles that can reproduce only in living cells and do not contain enough distinct structures themselves to be ranked as living cells.

VITAMIN [L. *vita:* life]. Any one of several structurally unrelated organic molecules which an organism cannot synthesize itself, but nevertheless requires in small quantity for normal growth and metabolism.

WALL PRESSURE. The physical resistance of the plant cell wall to turgor pressure.

WILD-TYPE. Geneticists' term for standard or reference type. Deviants from this standard, even if found in the wild, are said to be mutants.

XYLEM (zy' lem) [Gr. *xylon*: wood]. In evolutionarily more advanced plants, the woody tissue that conducts water and minerals; xylem consists of tracheids, vessels, fibers, and other highly specialized cells.

YOLK. The stored food material in animal eggs, usually rich in protein and lipid.

ZOOLOGY (zoe ol' o jee) [Gr. *zoon*: animal + *logos*: discourse]. The scientific study of animals.

ZOOPLANKTON (zoe' o plangk' ton) [Gr. *zoon*: animal + *planktos*: wandering]. The animal portion of the plankton.

ZOOSPORE (zoe' o spohr) [Gr. *zoon*: animal + *spora*: seed]. In algae and fungi, any swimming spore. May be diploid or haploid.

ZYGOTE (zye' gote) [Gr. *zygotos*: yoked]. The cell created by the union of two gametes, in which the gamete nuclei are also fused. The earliest stage of the diploid generation.

credits

ABOUT THE BOOK

The text of this book is set in Melior, a typeface based on the designs of the German typographer Hermann Zapf. Although it is a modern face, Melior resembles Renaissance designs in its subdued contrast between the thick and thin elements of letters. Zapf succeeded in producing an alphabet whose pristine character can survive offset printing on mediocre papers. Chapter titles are set in Helvetica, and part titles in Helvetica Medium Outline.

The book itself was designed by Wladislaw Finne, and the drawings were made by a team of artists, including Patricia Chaudhuri, Joshua Clark, Irving Geis, Sarah Landry, Laszlo Meszoly, Lorelle Raboni, Frederic J. Schoenborn, Sandra Sevigny and Vantage Art Studios.

The book was set in type by the DEKR Corporation, and was printed and bound by the Murray Printing Company. The production of the book was managed by Joseph J. Vesely.

CHAPTER 7

Opener: D.W. Fawcett & E. Anderson 1. J.A.L. Cooke 2. L. Caro & R. Curtiss 3. Thomas Eisner 6. W.E. Boggs 7. W.H. Hodge 10. Thomson Associates by J.V. Herbert 11. Stanley & Kay Breeden 12. J.A.L. Cooke 13. Ederic Slater 14. Richard C. Kern 15. Virginia Page

CHAPTER 8

Opener: Thomas Eisner 1. J.H. Troughton & F.B. Sampson 2. J.H. Troughton & L.A. Donaldson 3. Walter H. Hodge 4. Harold J. Evans 5. Allen H. Bennett 6. Thomas Eisner 9. Charles Walcott 10. E.W. Strauss 13. Walter H. Hodge

CHAPTER 9

Opener: Julius H. Comroe 1. After G.M. Smith 3. B.E. Juniper 4. J.H. Troughton & L.A. Donaldson 5. Walter H. Hodge 6. Thomson Associates by J.V. Herbert 8. Philadelphia Zoological Society 13. (*left & center*): Thomas Eisner (*right*): D.S. Smith 14. Thomas Eisner

CHAPTER 10

Opener: P. Dayanandan 1. Thomson Associates by J.V. Herbert 2. Barbara Parnessa 3. P. Dayanandan 4. Thomson Associates by J.V. Herbert 8. Thomas Eisner 11. Thomas Eisner

CHAPTER 11

Opener: Pfizer Inc. 6. Thomas Eisner

CHAPTER 12

Opener: The Bettmann Archive 3. C.F. Cleland

CHAPTER 13

Opener: R.A. Steinbrecht 3. K.D. Roeder 8. T.J. McDonald 15. R.A. Steinbrecht

CHAPTER 14

Opener: David Phillips 2. (*right*): D.W. Fawcett 3. Gjon Mili 4. Thomas Eisner 6. (*center*): Thomas Eisner (*left & right*): Carolina Biological Supply House 8. H.E. Huxley 12. Hans R. Haefelfinger 14. Yata Haneda

CHAPTER 15

Opener: Hermann Kacher 1.Thomas Eisner 3. Frederic A. Webster 6. Peter Marler 8. Thomas Eisner 9. Thomas Eisner 10. Thomas Eisner 11. Thomas Eisner 15. Hermann Kacher

CHAPTER 16

Opener: Martin F. Glaessner 3. E.S. Barghoorn & J.W. Schopf 4. E.S. Barghoorn & S.A. Tyler 5. E.S. Barghoorn & S.A. Tyler 6. E.S. Barghoorn & S.A. Tyler 7. (*top*): E.S. Barghoorn & S.A. Tyler (*bottom*): S.M. Siegel *et al.* 8. G. Licari 9. Martin F. Glaessner 10. After S.L. Miller 11. Sidney Fox

CHAPTER 17

Opener: Joseph G. Farris, © 1976, *The New Yorker* Magazine, Inc. 6. Bernhard Kummel, *History of the Earth*, 2nd Ed. © 1970 W.H. Freeman & Co. 7. F.M. Carpenter

CHAPTER 18

Opener: Gregory Antipa 1. Norma J. Lang 2. John T. Bonner 4. After W.R. Sistrom 5. After Scagel *et al.* 6. After *The Structure of Viruses* by R.W. Horne. © January 1963 by *Scientific American, Inc.* All rights reserved. 8. (*left*): Thomas Eisner (*center*): Douglas P. Wilson (*right*): Eric V. Gravé 9. R.D. Allen 11. Eric V. Gravé 12. Eric V. Gravé 14. R. Jakus 15. Eric V. Gravé

CHAPTER 19

Opener: Harvey J. Marchant 1. Ross E. Hutchins 2. Ross E. Hutchins 4. Walter H. Hodge 7. J.H. Troughton 11. William C. Steere 12. William C. Steere 14. Walter H. Hodge 17. J.H. Troughton 18. Walter H. Hodge 21. Reed C. Rollins 22. W.E. Boggs

CHAPTER 20

2. Martin Glaessner 4. Douglas P. Wilson 5. Ralph Buchsbaum 6. Lev Fishelson 7. Douglas P. Wilson 8. L.W. Mullinger Box B: J.A.L. Cooke 13. American Museum of Natural History

CHAPTER 21

Opener: National Maritime Museum, Greenwich, England

CHAPTER 22

Opener: Robert E. Silberglied 2. From T. Dobzhansky, *Evolution, Genetics and Man*, 1955. By permission of John Wiley & Sons, Inc.

CHAPTER 23

Opener: Irven DeVore/Anthro-Photo

CHAPTER 24

Opener: Eugene L. Nakamura 9. From D. Davenport, in S.M. Henry, ed., *Symbiosis*, Vol. 1, Academic Press, 1966 10. From M. Caullery, *Parasitism and Symbiosis*, Sidgwick and Jackson, Ltd., 1952. 11. W.E. Boggs

CHAPTER 25

Opener: Lev Fishelson 3. Douglas P. Wilson 6. After Ralph Buchsbaum 13. After *Toxic Substances and Ecological Cycles* by George M. Woodwell. © March 1967 by *Scientific American, Inc.* All rights reserved.

CHAPTER 26

Opener: Carl W. Rettenmeyer 3. From E.O. Wilson, *The Insect Societies*, Belknap Press, Harvard University Press, 1971 5. From E.O. Wilson, *The Insect Societies*, Belknap Press, Harvard University Press, 1971 6. From E.O. Wilson, *The Insect Societies*, Belknap Press, Harvard University Press, 1971.

CHAPTER 27

Opener: Irven DeVore/Anthro-Photo

CHAPTER 28

Opener: From Paolo Soleri, *Arcology*, 1969. By permission of the M.I.T. Press, Cambridge, Mass. 1. After A. Desmond

CHAPTER 29

Opener: Steven Balkin

index

the authors

Edward O. Wilson is Professor of Zoology at Harvard University. Born in Birmingham, Alabama, he earned his bachelor's and master's degrees from the University of Alabama in 1949 and 1950 respectively, and his doctorate from Harvard in 1955. He joined the Harvard faculty the following year. For four years Wilson taught a general education course in elementary biology, and currently is in charge of the first half of the introductory course for majors in biology. His research interests span many of the topics in the last two sections of *Life* from the general theory of systematics, evolution and biogeography to the social behavior of insects. A member of the National Academy of Sciences and American Philosophical Society, he is the author of: *The Theory of Island Biogeography*, with Robert H. MacArthur; *A Primer of Population Biology*, with William H. Bossert; *The Insect Societies*; and *Sociobiology*, both of which were nominated for the National Book Award.

Thomas Eisner is Professor of Biology at Cornell University. Born in Berlin, Germany, he made his way to the U.S. by way of Uruguay. He did his undergraduate work at Champlain College and Harvard University, and earned his Ph.D. in biology at Harvard in 1955. After two years as a postdoctoral fellow at Harvard, Eisner moved to Cornell in 1957. He has published extensively on the behavior and chemical ecology of insects, particularly on the chemical signals employed by insects for communication with one another. He has taught the introductory biology course at Cornell, as well as courses in neurobiology and behavior, and chemical ecology. A member of the National Academy of Sciences, he has also served on the editorial board of *Science*. Happier outside the laboratory than in it, he has carried out much of his work at field stations in tropical and subtropical settings.

Winslow R. Briggs is Director of the Department of Plant Biology of the Carnegie Institution of Washington, located in Stanford, California. A native of Minnesota, he did both his undergraduate and graduate work at Harvard, receiving his Ph.D. in 1955. He moved to Stanford University the following year, and remained there until he returned to Harvard in 1967. At Harvard he taught plant physiology and introductory botany. He assumed his present position in 1973. His chief research interests are plant growth and development, and the biophysics and biochemistry of plant responses to light.

Richard E. Dickerson is Professor of Chemistry at the California Institute of Technology. He did his undergraduate work at the Carnegie Institute of Technology (now Carnegie-Mellon University), and earned a Ph.D. in physical chemistry from the University of Minnesota in 1957. Journeying to England, he worked with J. C. Kendrew on the first high-resolution X-ray analysis of the structure of a crystalline protein. In 1959 he joined the faculty of the University of Illinois, and in 1963 he moved to Caltech. His main research interests are the structure of proteins, the relationship between the structure and function of proteins, and the study of the evolution of protein molecules, particularly cytochrome c and its relatives. He is the author or co-author of four other books, including *The Structure and Action of Proteins* and *Chemistry, Matter and the Universe*, both with Irving Geis, an artist who prepared some of the drawings for this book.

Robert L. Metzenberg is Professor of Physiological Chemistry at the University of Wisconsin Medical School. He earned an A.B. degree in chemistry from Pomona College in 1951 and a Ph.D. in biochemistry from Caltech in 1956. Since then he has been more or less continuously at the University of Wisconsin, where he has taught medical students, medical technologists, graduate students, and first-year undergraduates in the Biology Core Curriculum. His current research interest is the problem of how genes regulate metabolism in eucaryotes.

Richard D. O'Brien is Director of the Division of Biological Sciences at Cornell University, and also Professor of Neurobiology. He was graduated from Reading University, England, in 1950 and earned his Ph.D. from the University of Western Ontario, Canada, in 1954. During his six years in a research institute of the Canada Department of Agriculture he was a visiting professor at the University of Wisconsin and a fellow at the Institute of Animal Physiology in Cambridge, England. He moved to Cornell in 1960, where he has served as Professor of Entomology, chairman of the biochemistry department and chairman of the Section of Neurobiology and Behavior. He assumed his present post in 1970. His main research interests are the molecular basis of chemical transmission and the action of nerve poisons.

Millard Susman is a professor in the Departments of Genetics and Medical Genetics at the University of Wisconsin at Madison; for five years he was chairman of the two departments. He received a B.A. degree in Zoology at Washington University in 1956 and a Ph.D. in genetics from Caltech in 1962. After a year's postgraduate work with the Microbial Genetics Research Unit at Hammersmith Hospital in London, he went to the University of Wisconsin, where he has been ever since. His research interest is bacteriophage genetics and development. He teaches courses in microbial genetics and was a member of the team that created and taught the Biology Core Curriculum at the University of Wisconsin.

William E. Boggs is an author and editor. A native of Pennsylvania, he earned his undergraduate degree at the University of Pittsburgh in 1954, and after two years of military service returned there for a year of graduate work. He began his editorial career with McGraw-Hill in 1957, and two years later became one of the editors of *Scientific American*. Eventually he returned to McGraw-Hill as biology editor. His articles have appeared in a number of publications, particularly *The New Republic*, where he was one of the earliest opponents of the supersonic transport.

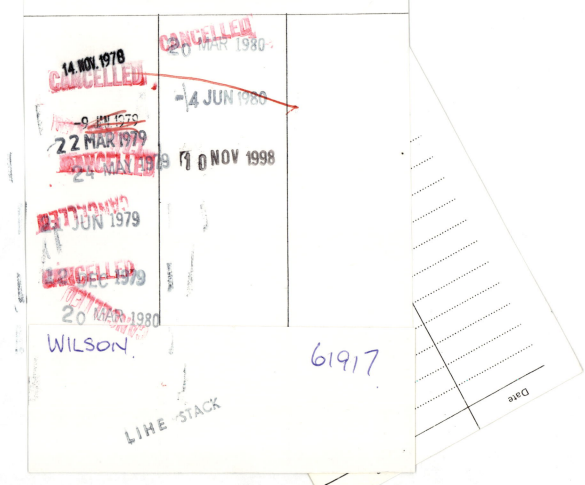